*A volume in the Hyperion reprint series*
**THE HISTORY AND POLITICS OF OIL**

**HYPERION PRESS, INC.**
*Westport, Connecticut*

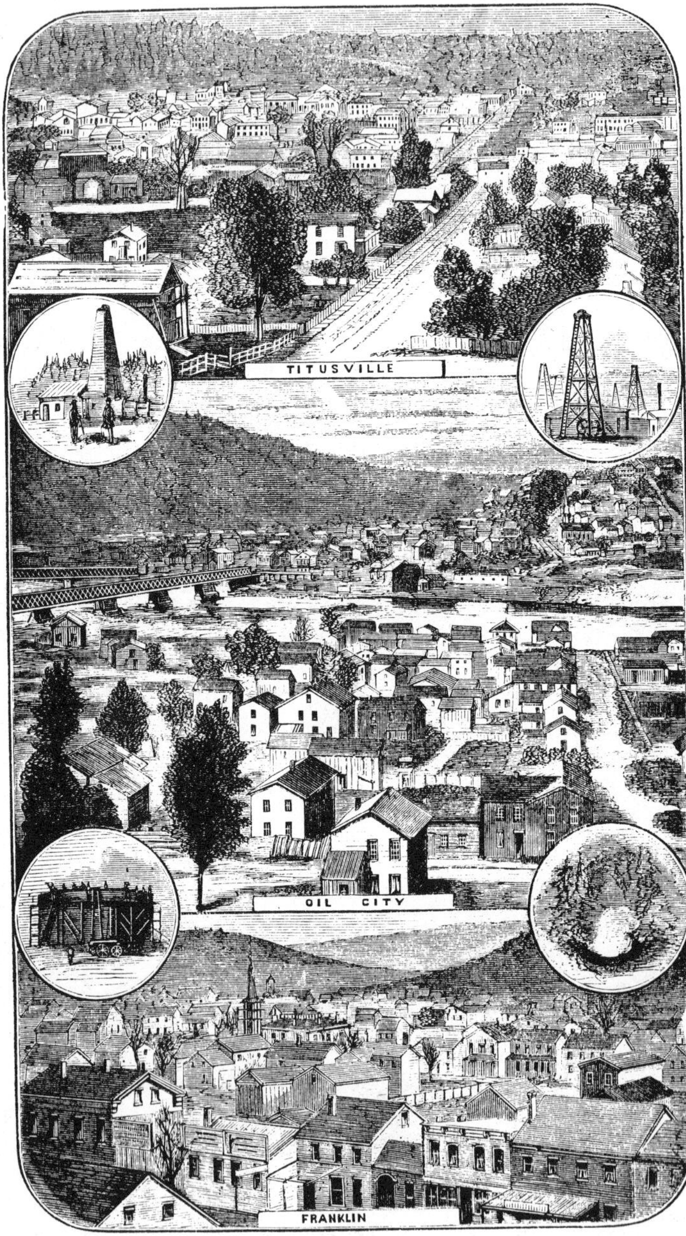
TITUSVILLE
OIL CITY
FRANKLIN

# PETROLIA:

A BRIEF HISTORY OF THE

# PENNSYLVANIA PETROLEUM REGION,

*ITS DEVELOPMENT, GROWTH, RESOURCES, ETC.,*
*FROM 1859 TO 1869.*

BY

ANDREW CONE AND WALTER R. JOHNS.

EDITED BY WALTER R. JOHNS.

NEW YORK:
D. APPLETON AND COMPANY,
90, 92 & 94 GRAND STREET.
1870.

Published in 1870 by D. Appleton and Company, New York

Hyperion reprint edition 1976
Library of Congress Catalog Number 75-6466
ISBN 0-88355-285-X
Printed in the United States of America

*Library of Congress Cataloging in Publication Data*

*Cone, Andrew.*
*Petrolia.*

*(The History and politics of oil)*
*Reprint of the 1870 ed. published by D. Appleton, New York.*
*1. Petroleum – Pennsylvania – History. I. Johns, Walter R., joint author. II. Title. III. Series.*
*TN872.P4C7 1975 338.2'7'28209748 75-6466*
*ISBN 0-88355-285-X*

# CONTENTS.


| | PAGE |
|---|---|
| CHAPTER I.—Introductory | 9 |
| CHAPTER II.—Petroleum—Its History, Ancient and Modern | 20 |
| CHAPTER III.—Early History of the Pennsylvania Petroleum Region | 39 |
| CHAPTER IV.—Dawning of the Era of Artificial Light | 46 |
| CHAPTER V.—Commencement and Growth of the Petroleum development | 69 |
| CHAPTER VI.—Geology and Physical Features of the Petroleum Region | 88 |
| CHAPTER VII.—Transportation—Early and Later Modes of | 96 |
| CHAPTER VIII.—Method of Petroleum Development—Boring Oil Wells | 116 |
| CHAPTER IX.—Obstacles met with in Boring Oil Wells, and How Overcome | 138 |
| CHAPTER X.—Pumping and Flowing | 146 |
| CHAPTER XI.—Various Phenomena of the Oil Wells | 155 |
| CHAPTER XII.—Theories of the Origin of Petroleum | 173 |
| CHAPTER XIII.—Social Characteristics of the People of the Petroleum Region | 189 |
| CHAPTER XIV.—Oil Farms of the Petroleum Region | 194 |
| CHAPTER XV.—Flood, Fire, and Financial Disaster | 528 |
| CHAPTER XVI.—General Description of Coal, Bituminous Clays, Bitumen, Petroleum—Table of Volatile Matter, Coke, Crude Oil from Coals, &c. | 542 |
| CHAPTER XVII.—The Principal Cities and Towns of the Petroleum Region | 562 |
| CHAPTER XVIII.—Method of Refining and the Refining Interest | 573 |
| CHAPTER XIX.—General Results of the Petroleum Development—Statistical, &c. | 586 |

# ILLUSTRATIONS.


No. 1.—FRONTISPIECE—
Vignettes—Drake Well—Modern Derrick—Iron Tanks—Burning Well on Allegheny River.

No. 2 to 5.—Drilling Tools and Well Machinery.

" 6.—View of Well ready to commence drilling

" 7.—Wells drilling and pumping.

" 8.—Pumping Well in operation.

" 9.—Phillips and Woodford Flowing Wells—Tarr Farm.

" 10.—Early Mode of Transportation—Mouth of Oil Creek in time of high water.

" 11.—Pond Freshet Jam at mouth of Oil Creek, May, 1864.

" 12.—View at Pleasantville in 1868.

# PETROLIA.

## CHAPTER I.

### INTRODUCTORY.

THE material wealth, and consequent prosperity, of a nation, can be estimated by the extent and richness of its mineral resources. Taking it for granted, that the assertion is correct, how almost limitless seems the mineral wealth of the vast American continent. Within its boundaries are embraced inexhaustible deposits of almost every mineral product used in arts and manufactures, or required by the commercial world. Coal, iron, gold, silver, lead, exist in vast quantity, over large portions of the country, as well as many other minerals of lesser note. The extent or area of the American coal field, is asserted to be eight times larger than all the other coal deposits of the world, extending from the Allegheny to the Rocky Mountains.

Though scarce half a century has elapsed since the commencement of the working of these mineral deposits, the rate of progress has been so great, that the amount of their production at the present time surpasses that of other countries, where the same have been actively worked for centuries.. In our own favored land, it would seem, has an ever bountiful Providence deposited the great reserve, in quantity abundant, for the wants of a population many times greater than the present world can now boast of, for countless ages yet to come.

Coal to supply heat, and iron, so indispensable to the wants of civilization, have heretofore been considered the most essential of the mineral products of the world. Co-equal with heat, in the necessities of the human race, is light. To supply this, beyond what is furnished by Nature's God, the world, until within a few short years, has relied upon illuminating substances derived from both animal and vegetable products. The commencement of the development in Pennsylvania, during the year 1859, in large quantity, and the utilization of the mineral oil, known as Petroleum, combining in a super-eminent degree heat, and for illuminating purposes, furnishing the best and cheapest artificial light yet known, and its rapid adoption by all portions of the civilized world, created a revolution in the general branches of arts, manufactures, and domestic economy, that has seldom been equalled.

The records of mining present no parallel to the rapid development of such vast resources as has characterized the American Petroleum. Scarcely ten years ago, the first artesian oil well was drilled in the Pennsylvania Petroleum region. From this small beginning has sprang into existence a mining business, second only in extent and value to coal and iron. Scarcely inaugurated ere the commencement of the civil war, that for a time threatened to destroy our Government and unity, requiring all the patriotism and the resources of the country to subdue, and for the various years subjected, perhaps, to more depressing influences than has fallen to the lot of any other mineral development, the Petroleum business has borne up against adversity in every form, and to-day constitutes the third estate among the chief mineral products of our country.

A description of the extent, value, mode, and manner of mining the mineral resources of a country, cannot fail to prove of interest, not only to its own people, but also to the enlightened portion of the civilized world. Especially is this the case of a mineral product possessing such important and essential properties, and susceptible of such general utility, as Petroleum. All information, relating to the sources from whence are derived the basis of a nation's wealth, has ever been eagerly sought for by the intelligent classes of mankind. No mineral product of our country has become so widely and favorably known in so brief a period, as the one we treat of. Such facts as could be collected of the nature, extent of resources, manner and mode of development, its growth from year to year, and all matters of practical interest connected therewith, are comprised in the different pages of this work. Though not so full in some details as could be wished, the difficulty of arriving at conclusions relative to a substance drawn from a great depth beneath the earth, only visible when coming to the surface of the same, will be apparent to the intelligent reader.

The presence of Petroleum, or rock oil, manifested itself to the earliest settlers in the several localities of the different States of America where it is now found. To the native Indians it had been known and used for so long a period that even their traditions fail to indicate its beginning, if indeed it ever had any, save with the Divine chemist himself, who has entire worlds for his laboratory. By the settlers and Indians it was used as a medicine, both for afflictions externally, and even internally, with highly beneficial effects. Beyond this no other use was discovered for the substance until its later development in the decadence of the present century. Petroleum proved a source of great annoyance to the early salt miners. Its manifestations were sometimes in the form of a highly volatile gas, that came out of the well, after the same had been drilled to a certain depth, with considerable force, and had a strong pungent odor, and in other instances in the shape or form of a dark-green oil. The substance in either case served to disgust the operators, and rendered the water of such wells unfit for manufacture. In subsequent years, the gas from these wells was used as fuel in the evaporation of salt, and answered the purpose admirably. The important part the product has performed, and service rendered to the civilized world, since its recent development, is so apparent as to need no elucidation from us. The attention of the public was especially directed to its utilization as an illuminator, in the years 1849 and 1850, by Mr. Samuel W. Kier, of Pittsburgh, Pa., who in the latter year erected a small refinery, and commenced its manufacture. The supply of the crude article was obtained from the salt wells of Mr. Kièr, and others, at Tarentum, Allegheny county, a short distance from Pittsburgh, on the Allegheny river, where it had been previously discovered. The manufacture was necessarily limited in consequence of the scant supply. There being no oil wells as yet drilled, the oil was pumped out of the salt wells at the usual depth from which salt water was obtained. This discovery of an important use for the article, caused an extensive demand, and the obtaining of the crude material in larger quantities became requisite. Pits were sunk in various places where Petroleum manifested its presence on the surface of the ground. Oil collected in these, from whence it was taken out at stated seasons.

The sinking of an artesian well, similar in most general respects to those of the salt regions, near Titusville, on the lands of the Pennsylvania Rock Oil Company, by members of the same, was the first commencement of the present Petroleum development. The striking in August, 1850, in this well, of a vein of carburetted hydrogen gas, mingled with oil, at the depth of sixty-nine feet from the surface, yielding at the rate of 400 gallons of oil per day, had the tendency to concentrate public attention to that particular locality. Hundreds of people from all parts of the country flocked into the Oil Region, then confined to the vicinity of Titusville, and commenced sinking of wells, similar to Colonel Drake's. But few of the wells drilled in that immediate vicinity were successful in obtaining oil. The field of operations was soon extended, by the striking of good paying wells at various other points along the valley of Oil Creek, and along the Allegheny River.

The first operators contented themselves with drilling to a moderate depth, where they obtained small pumping wells. The number of these increased rapidly until the latter part of 1860 and in 1861. About this time some reflective operator expressed the opinion that, as the supply of oil seemed to come from great depths below the earth's surface, deeper wells would reach the main reservoir, or source of supply, and greater quantities be obtained. The theory so opportunely stated, was soon put to a practical test. A well was drilled to the depth between 400 and 500 feet, and the "third sandrock," where the greatest supply has been found, was reached. The result of the experiment astonished the operator, and the world too, at least the business portion of it. When the operator's drill penetrated the fissure or recess in the sand-rock, where the oil and gas had collected, the heavy drilling apparatus was hurled out of the aperture, it had forced through the earth, with terrific power, far above the top of the derrick, upon the surface. The drilling apparatus was soon followed by a stream of gas and oil, the force of which was so great, as to prevent the operators, for several days, from introducing into the well the necessary iron tubing, or pipe for it to flow through. After a time this was done, and the well flowed steadily, for not only days and weeks, but even for months, at the rate of hundreds of barrels of oil per day. During 1861 and 1862, a number of these large wells were struck, some of them yielding at the rate of from 3,000 to 4,000 barrels per day. This sudden increase of the product had the effect to lower the prices of the same to a mere nominal figure. The small pumping wells, by scores, were compelled to cease operations, the prices being too low to merit their continuance. The first oil obtained from the Drake Well, sold for one dollar per gallon. During the two succeeding years, the price had fallen to fifty cents. In 1861, after the large flowing wells had been struck, and began to pour forth thousands of barrels of oil per day, the price fell to fifteen and twenty cents *per barrel* at the wells, and sales were made as low as ten cents per barrel. In 1861, and the commencement of 1862, a number of refineries were built along Oil Creek, and in various parts of the country, for the refining of Petroleum, and a large amount of oil was shipped to Europe.

Though the small wells, in great numbers, were abandoned as unprofitable, the development continued unabated, and the oil producing region was continually being enlarged by the striking or obtaining of new wells. While many of the early operators lost heavily on their first ventures, owing to the slight production of their wells and low prices, this fact did not prevent others from sinking new wells. An abiding faith in the future utilization of the product to a sufficient extent to consume the large supply, and enhance its value to a paying figure, encouraged those engaged in the development. Each successive season found new parties to take the places of those who had become discouraged or been unfortunate. By the fall of 1861, a large business had been built up. Hundreds of teams were employed in the transportation of the oil from the different wells to the principal shipping points. From these latter, the oil was conveyed, by means of flat-bottomed boats and barges, similar in shape to those of the Ohio and Mississippi, to Pittsburgh, the principal oil market. Various lines of railroad were also being projected into the Oil Region, having connection direct with the leading cities and sea-board. The low prices caused the introduction of the product, as an illuminator, to almost every portion of the civilized world, and its nature and uses were rapidly becoming universally known.

Great indeed was the tide that flowed into the Oil Region during the first years of the development of its marvellous resources. The governing principle of mankind, rapid acquisition of wealth, was here apparently presented, greater in degree of certainty than has ever before been offered to the cupidity of man. In almost unending phalanx, came the devoted pilgrims to worship at the greasy Mecca. Every State, and almost every country had its representatives in the land of oil. No sacrifice was too great to make, in order to raise the necessary capital to obtain a lease in the modern Ophir, sink a well, and secure a fortune, all within ninety days from the date of the commencement of the fascinating task. It was not a business that required years of training, and ample experience to succeed in. The chances of success were equal for all, and for once, Dame Fortune seemed impartial to her ardent worshippers. The fortunate ones stumbled into good luck, as it were, because they could not help it, while the unfortunate had an equal chance with those who outstripped them in the race. But with singular consistency Fortune has since amply revenged herself for this seeming act of forgetfulness. The riches acquired during those early years remained but a brief time with its new owners. And even the fortunate few who retain their first acquisitions, have found it necessary to use the most strenuous efforts to keep the treasure so won. Amid the wild rush for wealth, no system or concert of action was possible. The devil take the hindmost, was the ruling characteristic. The result can be easily foreseen. California first, and the gold and silver regions of our territories on the Pacific slope afterward, afford ample parallels, which, it was hoped, would act as beacons of warning to men. Thousands who made their first venture here, were ruined, and many a comfortable household reduced to penury and want. A volume could be written of the heroic struggles of the early operators, but what would it avail. Life's daily battle furnishes examples in abundance, in every pursuit, and in general respects similar. The lesson learned was a severe one, but the danger was ventured into understandingly. We have no desire to enact the part of Old Mortality to the "slain, wounded, and missing," who fell on the first skirmish line, any further than to show the extent of their mining operations. The Petroleum excitement of 1861 culminated in the Spring and Summer of 1865, the grand era of the Oil Stock Companies..

The general appearance of the Oil country in the Fall of 1861, was one of unexampled activity. The valley of Oil Creek then formed the only producing locality of any consequence. Upon this narrow valley, scarcely sixty rods in width in its broadest part, from Oil City to the Sherman well, a distance of about ten miles, were clusters and continuous lines of tall pyramidal oil derricks, engine houses, and board shanties, presenting the appearance of a town of that length. The sites of the several towns along Oil Creek had just been projected, and the building of the same commenced. For this whole distance all was bustle and activity. The large flowing wells were spouting forth, like huge whales, their greasy treasure, with a noise similar to the scape-pipe of a steamer heading against a swift current. Tall derricks arose on every hand, and amid the smoke and din were hundreds of men busily employed in the various occupations. The derricks, drilling machinery, &c., was the same in general respects, as is now employed, being copied from that used by the early salt miners, except that it was not so heavy, the wells then drilled being of a shallow depth. The weight of the drilling apparatus averaged from 300 to 600 pounds.

The first of the flowing wells was the Burning Well, on the John Buchanan farm, on Oil Creek, which commenced producing in April, 1861. The gas and oil, which was forced up a great distance in the air from the time that the drill penetrated the crevice containing it, caught fire from a neighboring engine in a short time after it was struck. In the explosion that followed, some twenty-five to thirty persons lost their lives. The Philips well, on the Tarr farm, was also one of the wonders of this period. When first struck, the producion of this famous well averaged between three and four thousand barrels of oil per day. The sight was one of of a life-time. From the mouth of the conducting pipe the oil and gas came forth with terrific power, propelled by nature's hydraulic forces. The oil, of a beautiful dark olive-green, was dashed into spray against the side of the huge receiving tank, forming a prism of colors, rainbow in hue, resplendent in the sun's rays, and beautiful beyond description. Here, indeed, was the rock pouring forth a fountain of oil, enriching the fortunate owners, and giving a stimulus to operators throughout the entire oil region. The depth of the Philips well is 480 feet, and proved one of the largest oil veins tapped in the Oil Region, having produced to date of October, 1865, upwards of 300,000 barrels of oil. Though afterwards flooded by the Woodford, the well never wholly ceased producing, and is still a paying well, producing from twenty to fifty barrels per day by pumping. The Empire well, on the M'Ilhenny farm, was another of the leviathans in operation at the time spoken of, being fully as great in production as the Philips. The Sherman, Blood, and many others, all flowing over 1,000 barrels per day, were in full operation.

By teams and by boats the oil was transported from the wells to the various shipping points. Every road was thronged with the oil wagons, and the surface of the Creek covered with oil boats. At the different shipping points a similar scene of activity was presented. By such incessant use the soft alluvial soil, composing the road-beds, became almost impassable. Through mud, nearly unfathomable, the drivers urged their jaded animals. The oil-boats were towed up the creek by horse-power, and when laden at the wells, came down by the force of the current in times of natural water, or on pond freshets. Amid this moving, living panorama of oil wells, derricks, gushing fountains of oil, busy machinery, mud, wealth, energy, and industry, so wonderful to behold, of human hopes and ultimate successes, sufficiently varied to suit all the vagaries of the ordinary imagination, was budding forth the germ of the mighty Petroleum business of to-day, ranking next to coal and iron in magnitude among the mineral resources of our country.

It required no great stretch of the imagination to foresee the future that has since been so rich in its fruition; of the advent of railroads, conveying the oil in pipes; of immense iron tanks for holding it; the employment of a vast ocean fleet, whose sails whiten every sea, and are in every clime. Of the extent, resources, and different subjects relating to Petroleum, we propose to give the reader an idea in the subsequent pages. After taking the journey, he may feel as we did, on the occasion of our first glimpse of the oil country, that, though we had seen more inviting abiding places, yet had never beheld one presenting a better chance to make one's first million in, and thus start fairly on the road to comfortable affluence.

# CHAPTER II.

## PETROLEUM—ITS HISTORY—ANCIENT AND MODERN.

VARIOUS authorities, ancient and scriptural, make mention or allude to Petroleum in some of its forms, from which it may be inferred that its uses were known to those who lived in the earliest periods of civilization. The name is said to be derived from the Greek words, *Petros* and *Elaion*, signifying, Rock Oil. In ancient days it was more commonly found in the form of asphal tum than in any other, and its first use of which we can find any record, was as mortar or cement for masonry. It has been found in a great variety of forms, and used for a number of purposes, and in all has been an article of general utility. It is defined, as now found, by modern science, as follows:

"Petroleum is a mixture of hydro-carbons of various compositions, all free from oxygen. It is an oleaginous fluid of a brownish or greenish color; has a strong and peculiar odor; feels smooth and greasy between the fingers; does not congeal at low temperatures, and burns with bright but very smoky flame. It varies greatly in density, all the way from 20° to 57° Beaume, and is divided into two classes, the heavier, or those below 35° being designated and used as lubricating oils, for oiling machinery, and the lighter, or those above 35°, as illuminating oils, from which are manufactured the refined oils. The general appearance and properties of the two classes are the same."

The following scriptural references are made to Petroleum:

Deuteronomy xxxii., 13: "And he made him to suck honey out of the rock, and oil out of the flinty rock."

Job xxxix., 6: "And the rock poured me out rivers of oil."

Micah vi., 7: "Will the Lord be pleased with thousands of rams, or ten thousand of rivers of oil."

Genesis xiv., 10: "The vale of Siddim was full of slime pits."

Mention is made of its use in the building of the ancient city of Babylon. Herodotus, the celebrated historian, says, in this connection: "Digging a fosse or ditch, the earth which was cast up they formed into bricks, and desiring large ones, they burned them in furnaces, using for lime or mortar, hot asphaltas, or bitumen." He further relates that this bitumen was brought from the river Is, a tributary of the Euphrates.

Cartwright, a traveller of the last century, speaks of this same river Is, as follows: "From the ruins of old Babylon, we came to a town called Ait (the modern Heet), near unto which town is a valley of pitch, very marvellous to behold, and things almost incredible, wherein are many springs, throwing out abundantly a kind of black substance, like unto tar or pitch, which serveth all the countries thereabouts, to make staunch their barks and boats, every one of which springs, makes a noise like a smith's forge, which never ceaseth night or day, and the noise is heard a mile off swallowing up all weighty things that come upon it."

A later traveller, Mr. Rich, says: "The principal bitumen pit at Heet has two sources, and is divided by a wall in the centre, on one side of which the bitumen bubbles up, and on the other, the oil of naphtha."

Curtius, Diodorus, Siculus, Bochart and Josephus, all speak of bitumen as forming a constituent of the mighty walls, lofty towers, and pensile gardens of Babylon, that were the wonder of the world.

Layard, in his "Nineveh and Babylon," gives the following account of a bitumen pit on fire, which will compare in general respects with many of the scenes witnessed on Oil Creek. He says: "Tongues of flame and jets of gas driven from the burning pit, shot through the murky canopy. As the fire heightened, a thousand fantastic forms of light played amid the smoke. In an hour the bitumen was exhausted for the time, the dense smoke gradually died away, and the pale light of the moon shone over the black slime-pit."

Kerr Porter, and other modern travellers to Assyria, state that in many place bituminous and naphtha wells exist, and that in the remains of the famous tower of Ackerouf, near the ruins of Bagdad, in ancient Chaldea, cotemporary with the tower of Babel, the very reeds mixed with this material, are still preserved, although the brick and stone-work has almost fallen in to dust.

After the lapse of thirty-five centuries, with all that time could accomplish in corroding and destroying the work of man, the remains of these Petroleum cemented walls and towers exist, and are looked on by moderns. Fragments of bricks, with the asphaltum still clinging to them, are still exhumed from the ruins of ancient cities.

Says a late writer: "The substance was used by the Egyptians, as early as history can furnish us the facts of the times. From the accounts at the close of the Book of Genesis, of the embalming of Jacob and Joseph, it is plain to infer that embalming was a common process then, 1700 years before Christ." Dr. Pettigrew in his history of "Egyptian Mummies," states, that many of the mummies he exhumed had the cavities of the bodies filled with asphaltum. A French writer, on the same subject, quoted by Pettigrew, says these were often immersed in liquefied pitch, a composition formed of common pitch and asphaltum. Modern research and observation would seem to confirm this assertion of the extensive use of Petroleum in the process of embalming. The color, the odor, the inflammable nature of the mummy, all indicate its presence. The cerements, and even the body itself, are used by the wandering Arab as fuel, and modern travellers in those regions have used them for the same purpose. It was also used in the manufacture of the ancient papyrus, as an agglutinant, to prevent the attacks of insects, and the corroding effects of time.

Thus it will be seen that from days so ancient that history would be unintelligible were it not for the light that has been shed from the Book of Books, a continuous history of the product is furnished down to modern times. On every continent, and on almost every island, it has been found, cropping out in some form or other, giving visible indications of its presence, and it has always occupied a prominent place in the arrangements of mankind, as an article of utility. In modern times Petroleum has been found in various parts of the world.

In Northern Italy, in the Duchies of Parma and Modena, Petroleum has been extracted from the earth since its first discovery in 1640, the method pursued being merely to sink pits in the ground, and collect the oil that exudes from the soil, in little basins on the bottom of the pits.

The Rangoon districts on the Irawaddy, in Burman, are quite wonderful for their immense production of

Petroleum. For an unknown period tne whole Burman Empire, and a considerable portion of India, have been supplied with oil from this source. The trade is carried on by large boats that come down the Irawaddy to the town of Rainanghong, a place inhabited by potters, who are constantly making the earthen jars in which the oil is kept. These are piled up in great pyramids about the town, ready for use. The wells are in beds of sandy clays, which rest on sand-stones or argillaceous slates, and are sunk to the depth of sixty feet. Under the slates is said to be coal; but this and the other strata may be of the tertiary epoch. Symes's "Embassy to Ava" volume 2, states that the number of wells in this vicinity exceeded 520, and the annual yield of Petroleum was over 400,000 hogsheads. The natives use the oil in lamps, for preserving timber against insects, and as a medicine.

*Extracts from the Journal of John Crawford, Esq., F. R. S., F. L. S., F. G. S., &c., Ambassador of the Governor General of India to the Court of Ava, in* 1826. *London: published for Henry Colburn*, 1834.

VOLUME I., page 93.—At three in the afternoon, our whole party proceeded to the celebrated Petroleum wells. Those which we visited cannot be farther than three miles from the village, for we walked to them in forty minutes. The cart-road which leads to them is tolerably good, at least for a foot-traveller. The wells altogether occupy a space of about sixteen square miles. The country here is a series of sand-hills and ravines, the latter torrents after a fall of rain, as we now experienced, and the former covered with a very thin soil or altogether bare. The trees, which were more numerous than we looked for, did not rise above twenty feet in height. The surface gave no indication, that we could detect, of the existence of the Petroleum. On the spot which we reached were eight or ten wells, and we examined one of the best. The shaft was of a square form, and its dimensions about four feet to a side. It was formed by sinking a frame of wood, composed of beams of the Mimosa catechu, which affords a durable timber. Our conductor, a son of the Myosugi of the village, informed us that the wells were commonly from 140 to 160 cubits deep, and their greatest depth in any case was 200. He informed us that the one we were examining was the private property of his father—that it was considered very productive, and that its exact depth was 140 cubits. We measured it with a good lead line, and ascertained its depth to be 210 feet; thus corresponding exactly with the report of our conductor, a matter which we did not look for, considering the extraordinary carelessness of the Burmans in all matters of this description. A pot of oil being taken up, and a good thermometer being plunged into it, indicated a temperature of 90°. That of the air, when we left the ship, an hour before, was 82° To make the experiment perfectly accurate, we ought to have brought a second thermometer along with us, but this we neglected. We looked into one or two of the wells, and could discern the bottom. The liquid seemed as if boiling, but whether from the emission of gaseous fluids, or simply from the escape of the oil itself from the ground, we had no means of determining. The formation where the wells are sunk, consisted of sand, loose sand-stone and blue clay. When the well is dug to a considerable extent, the laborers informed us that brown coal was occasionally found. Unfortunately, we could obtain no specimens of this mineral on the spot, but I afterward obtained some in the village.

The Petroleum itself, when taken out of the well, is of a thin, watery consistence, but thickens by keeping, and in the cold weather it coagulates. Its color, at all times, is a dirty green, and not much unlike that of stagnant water. It has a pungent, aromatic odor, offensive to most people. The wells are worked by the simplest contrivance imaginable. There is over each well, a cross-beam, supported by two rude stanchions. At the centre of the cross-beam, and embracing it, is a hollow revolving cylinder, with a channel to receive a drag-rope, to which is suspended a common earthen pot, that is let down into the well, and brought up full by the assistance of two persons pulling the rope down an inclined plane by the side of the well. The contents of the pot are deposited for the time in a cistern. Two persons are employed in receiving the oil, making the whole number of persons engaged on each well only four. The oil is carried to the village or ports in carts, drawn by a pair of bullocks, each cart conveying from ten to fourteen pots of ten viss each, or from 265 to 371 pounds avoirdupois of the commodity. The proprietors store the oil in their houses, and then vend it to the exporters. The price varies, according to the demand, from four ticals of flowered silver to six ticals per 1,000 viss; which is from five pence to seven pence half penny per 100. The carriage of so bulky a commodity, and the breakage to which pots are so liable, enhances the price in the most distant parts to which the article is transported, to fifty ticals per 1,000 viss. Sesamum oil will cost at the same place not less than 300 ticals for an equal weight, but it lasts longer, gives a better light, and is more agreeable than the Petroleum, which in burning, emits an immense quantity of black smoke, which soils every object near it. The cheapness, however, of this article is so great, that it must be considered as conducing much to the comfort and convenience of the Burmans. Petroleum is used by the Burmans for the purpose of burning in lamps, and smearing timber to protect it against insects, especially the white ant, which will not approach it. It is said that about two-thirds of it is used for burning, and that its consumption is universal until its price reaches that of sesamum oil, the only oil which is used in the country for burning. Its consumption, therefore, is universal, wherever there is water carriage to convey it—that is, in all the country watered by the Irawaddy, its tributary streams and its branches. It includes Bassien, but excludes Martaban, Tavoy, and Mergui, Aracan, Tongo, and all the northern and southern tributary states. The quantity exported to foreign parts is a mere trifle, not worth noticing. It is considered that a consumption of thirty viss per annum for each family of five and a half persons is a moderate average. If it were practical, therefore, to ascertain the real quantity produced at the wells, we should be possessed of the means of making a tolerable estimate of the inhabitants who make use of this commodity, constituting the larger part of the population of the kingdom. Of the actual produce of the wells, we received accounts not easily reconcilable to each other. The Burmans, perhaps, less from a disposition to impose than from an incapacity to state any facts of the nature with precision, could not be relied on, and we had no registers to consult. The daily produce of the wells was stated, according to goodness to vary from 35 to 500, the average giving about 235 viss. The number of wells was sometimes as low as fifty, and sometimes as high as 400. The average made about 200, and considering they are spread over sixteen square miles, as well as that the oil is well known to be a very general article of consumption throughout the country, I do not think the number exaggerated. This estimate will make the consumers of Petroleum for burning amount to 2,066,721. In the narrative of one of my predecessors, Captain Cox, the number of wells is given as high as 520, and the average daily produce of each well is reckoned at 300 viss, which makes the whole amount produced, 56,940,000. Calculating as before, this produce will give a population of 6,959,331. This is a much higher estimate than my rough data afford, but even this gives a very low estimate of the probable population of the empire. Calculations formed from such crude materials, and which would be justly disregarded, were means of gaining more accurate information within reach, have their value in a country in which exact details are never procurable upon any question of statistics.

VOLUME II., page 23.—Dr. Wallich and myself this morning visited the Petroleum wells, and examined several of them, we took the temperature of two of them carefully, with a good thermometer, the thermom eter being immersed in a pot of oil just drawn from one of these, which was 130 royal cubits, or 207 English feet in depth, rose to 80°. In the shade the temperature was at the same time 60°. In a pot of oil drawn from another well, in which the liquid was much less mixed with water, and which was 140 royal cubits, or 222 feet and 8 inches deep, the heat indicated by the thermometer was 90°. In going over the ground, we observed several old wells altogether abandoned. The natives informed us, that in digging new ones, they came, at a considerable depth, to coal and fossil-shells; of the latter we could, unfortunately, obtain no specimens, but of the former, which proved to be brown coal, we obtained one or two good ones at the village of Benangyun. The oil drawers stated to us that in cleaning out old wells, accidents sometimes happened from the fire damps, and they pointed out a particular well at which two men had lost their lives from this cause.

PAGE 178.—The celebrated Petroleum wells afford, as I ascertained at Ava, a revenue to the king, or his officers. The wells are private property, and belong hereditarily to about thirty-two individuals. A duty of five parts in one hundred is levied on the Petroleum as it comes from the wells, and the amount realized on it is said to be 25,000 ticals per annum. No less than 20,000 of this goes to contractors, collectors, or public officers, and the share of the State, or 5,000, was assigned during our visit as a pension of one of the Queens.

PAGE 206.—The Petroleum wells of Renangyorong have been already described in the Journal. From the more accurate information, which I obtained at Ava, it appears that the produce of these may be estimated at the highest, in round numbers, at 22,000,000 of viss, each of three sixty-five one hundredth pounds avoirdupois. This estimate is formed from the report of the Myo. Thugyi, who rents the tax on the wells, which is five in a hundred. His annual collection is 25,000 ticals, and he estimated, or conjectured, that he lost by smuggling 8,000, making the total 33,000. The value of the whole produce, therefore, is 660,000 ticals. The value of the oil on the spot, is reckoned at three ticals per 100 viss, and consequently its amount will be as above stated.

PAGE 238.—I should observe, that Petroleum is universally used, wherever the navigation of the Irawaddy and Ryendwen, with their tributary streams, will allow of its being conveyed, and that it is also carried to a considerable extent, by land carriage. It is universally consumed in Pegu, from Bassien to Martaban, and throughout the whole of Upper Ava, embracing the greatest portion of the area of the kingdom, and unquestionably all the best inhabited part of it.

*Extracts in reference to Petroleum from the Narrative of Major Michael Symes, of the English Army, who was sent by the Governor-General of India as Embassador to the Court of Ava, in* 1765. *Published by Bulmer & Co., in London, in* 1800.

PAGE 261.—After passing various sands and villages, we got to Yaynangheoum or Earth Oil (Petroleum) Creek, about two hours past noon. We were informed that the celebrated wells of Petroleum, which supply the whole empire and many parts of India with that useful product, were five miles to the east of this place. The mouth of the creek was crowded with large boats, waiting to receive a lading of oil, and immense pyramids of earthen jars were raised within and around the village; disposed in the same manner as shot and shells are piled in an arsenal. This is inhabited only by potters, who carry on an extensive manufactory, and find full employment. The smell of the oil is extremely offensive. We saw several thousand jars filled with it, ranged along the bank; some of these were continually breaking, and the contents, mingling with the sand, formed a very filthy consistence. Mr. Wood had the curiosity to walk to the wells, but, though I had felt the same desire, I thought it prudent to postpone visiting them until my return, when I was likely to have more leisure, and to be less the object of observation.

PAGE 441.—We rode until two o'clock, at which hour we reached Yaynangheoum, or Petroleum Creek, a place already noticed in our journey up the river. Dr. Buchanan partook of an early dinner with me, and when the sun had descended so low as to be no longer inconvenient, we mounted our horses to visit the celebrated wells that produce the oil, an article of universal use throughout the Empire.

PAGE 442.—The evening being far advanced, we met but few carts, those we did observe were drawn by a pair of oxen, and of a length disproportionate to the breadth, to allow space for earthen-pots that contained the oil. It was a matter of surprise to us, how they could convey such brittle ware with any degree of safety over so rugged a road. Each pot was packed in a separate basket and laid in straw, notwithstanding which precaution, the ground, all the way, was strewn with broken fragments of the vessels, and wet with oil, for no care can prevent the fracture of some in every journey. As we approached the pits, which were more distant than we had imagined, the country became less uneven, and the soil produced herbage. It was nearly dark when we reached them, and the laborers had retired from work. There seemed to be a great many pits within a small compass. Walking to the nearest, we found the aperture about four feet square, and the sides lined, as far as we could see down, with timber; the oil is drawn up in an iron pot, fastened to a rope passed over a wooden cylinder, which revolves on an axis, supported by two upright posts. When the pot is filled, two men take hold of the rope by the end, and run down a declivity, which is cut in the ground, to a distance, equivalent to the depth of the well. Thus, when they reach the end of the track, the pot is raised to its proper elevation, the contents, water and oil, together, are then discharged into a cistern, and the water is afterwards drawn through a hole in the bottom. Our guide, an active, intelligent fellow, went to a neighboring house, and procured a well rope, by means of which we were enabled to measure the depth, and ascertained it to be thirty-seven fathoms, but of the quantity of the oil at the bottom we could not judge. The owner of the rope, who followed our guide, affirmed that when a pit yielded as much as came up to the waist of a man, it was deemed tolerably productive, if it reached his neck, it was abundant; but that which rose no higher than the knee, was accounted indifferent. When a well is exhausted, they restore the spring by cutting deeper in the rock, which is extremely hard in those places where the oil is produced. The government farms out the ground which supplies this useful commodity, and it is again let to adventurers, who dig wells at their own hazard, by which they sometimes gain and often lose, as the labor and expense of digging are considerable. The oil is sold on the spot for a mere trifle—I think 200 or 300 pots for a tackal, or half a crown. The principal charge is incurred by the transportation and purchase of vessels. We had but half gratified our curiosity, when it grew dark, and our guide urged us not to remain any longer, as the road was said to be infested with tigers, that prowled about at night among the rocky, uninhabited ways through which we had to pass. We followed his advice, and returned with greater risk, as I thought, of breaking our necks from the badness of the road, than of being devoured by wild beasts. At ten o'clock we reached our boats without any misadventure."

In the vicinity of the Caspian Sea, the Bakoo springs have yielded large quantities of oil from time immemorial, and are widely celebrated throughout that region The oil, under the name of naphtha, is very generally burned for its light.

As far back into the ages of the past as we have any record, asphaltum has been found on the shores of the Dead Sea. This sea, as is well known, is of supposed volcanic origin; and is the probable site of the cities of Sodom and Gomorrah. Its surface is 1,300 feet below the surface of the ocean, and it has been fathomed to the depth of 2,000 feet. In several places, no bottom has been found, and owing to internal convulsions, the depth changes from time to time. The water is very dense, holding in solution twenty-five per cent. of solid matter, of which seven per cent. is salt. The bituminous substance is thrown up from below, and towards the centre of the sea, it is found in a liquid state, like Petroleum; but is probably solidified by evaporation, as it appears on the shore in hard, compact masses.

Upon the West India Islands, or, rather, several of them, it exists in various forms, known as Chapapote, Barbadoes tar, &c. Its existence is also apparent in various portions of South America. The most remarkable natural fountain of Petroleum known, is on the island of Trinidad, in the West Indies. It is called by the inhabitants Tar Lake. Bitumen, in a hot state, is continually boiling up, and it has formed a lake several miles in circumference. In the centre, or at the mouth of the fountain, the oil is hot and liquid, but as it recedes in every direction, it gradually cools and thickens, until, on the shores, it becomes solid. Humboldt also reports the spontaneous product of Petroleum in the West India Islands to be large, and as it ran to waste, it covered a large surface of the sea with its unctious tide. This report was made in 1799.

Having thus traced Petroleum in several of its connections, we will now turn to others no less curious. China, the land of everything old and new, next claims attention. That famous traveller, the Abbe Huc, gives a literally glowing description of the natural gas found issuing from the surface of the soil in various portions of the Celestial Empire. Every rich Chinese, in the regions where the product is discovered, if he knows nothing else, knows enough to have his own fire well, as he graphically calls it. It is often terrible work, as he has generally to perforate the earth to a depth of 1,500 to 2,000 feet. But John Chinaman is a patient fellow, and seldom fails. He has a long tube, about six inches in diameter to drill or bore with, and by the employment of some very odd machinery, bores on till he reaches the volcanic cavities below. Then rushes forth the inflammable gas, in a huge jet, that leaps like a black arrow, hundreds of feet into the air, and is finally mastered at no little expense, or difficulty, and danger. The least touch of flame, and an awful catastrophe is the result.

A noted Catholic Missionary, Monseigneur Imbert, who was in 1833, assigned to the province of Slo-Tchouch, celebrated for its fire wells, and who is quoted by the great Humboldt, in his "Fragments of Geology," describes a catastrophe of this kind that occurred at the Tsee-Lieon-Tsing (or well flowing of itself), in the great oil and gas regions. It caught fire in a twinkling, a frightful explosion was heard, followed by a shock of earthquake.

The narrator says: "The flame, which was about twenty feet high, flitted about without burning anything. Four men volunteered to risk their lives in endeavoring to arrest it. They cast a large stone on the mouth of the well, but it was instantly hurled far into the air. Three of the men were burned, the fourth one only escaping by almost a miracle. Neither water nor earth would extinguish the flames, until at length, after two weeks of incessant toil, a sufficient quantity of water was conveyed to the adjacent heights, where it was collected in a little lake, and suddenly let loose on the well in one volume, with success."

When the mouths of these wells have been properly secured, the gas can be conducted to any quarter through tubes of bamboo, and, once lighted, it burns with a steady bluish and dark reddish flame, about four inches in height and one in diameter. It so continues, day and night, until artificially extinguished. The light is heavy, but sufficient for ordinary purposes. Monseigneur Imbert also describes the uses to which this gas is put in the province of Kite-Sing-Tan, about 200 leagues from Canton. There the wells are very numerous, and the gas is used to light the great salt mines, which also abound in that district, and for fuel. One foot below the surface of the ground, on the four sides or faces of the well, are inserted four enormous bamboo tubes, which carry the gas beneath the boilers. Each boiler has its bamboo tube or fire-conductor, at the extremity of which is an earthen-ware tip several inches in length; and an orifice in the centre of about one inch in diameter. This tip keeps the bamboo from burning. Other tubes, leading to the outside, illuminate the great halls and chambers of the workshops. It is impossible to utilize all the flame, and the excess is conducted to the outer air through immense chimneys, above which leap their long tongues of flame. The surface of the ground above the tubes, is made very warm, and even burns the feet. In the very depth of January, the workmen are all half-naked, wearing a short pair of drawers. The flame of this gas produces no smoke; but a smell of bitumen can be perceived for miles around. The flame is of a dark red, like that of coal, and does not cling to the end of the tube, as that of an ordinary lamp, but hovers at the height of about two inches above it. In the winter time, the poor, in order to warm themselves, dig round cavities in the sand, a foot deep. A dozen of them crouch about it, then, with a handful of lighted straw, set fire to the cavity, and thus warm themselves as long as they wish. When they see fit, they fill up the hollow with sand, and so extinguish the flames.

As early as 1618, Jean Tardin, a physician, of Tournon, in France, gave to the world a curious, and now very rare work, entitled, "Natural History of the Burning Fountain, near Grenoble, with an Inquiry into the Causes and Principles, and an ample Treatise on Subterranean Fires." Tardin went to work, enlightened by his own investigations, and after trying various substances, placed coal in a close vessel, over a slow fire, and so obtained the vaporous exhalations, which is our ordinary gas.

Tardin's experiments were followed by those of Thomas Shirley, who, in 1667, tested the contents of the Burning Well, so called, near Wigan, Lancashire, England, which he found burned like oil. He was succeeded by the Rev. John Clayton, who wrote a singular letter to Sir John Boyle, "Concerning the spirit of coals," and therein states, that he had seen a ditch, where "the water burned like brandy." The country people boiled eggs and even meat in it. This letter was written previous to 1691, and is quoted in the "Philosophical Transactions of 1739," on pages 59 to 61 of the forty-first volume.

In Oriental Georgia, near to Bakoun, in the place known as Atesch Gah, or the Dwelling of Fire, are found wells, and pits of gas, and naptha, used by the weavers of that country for heat and light.

In Hungary, in the "circle" of the Marmarosch, a

gallery in the salt mine of Szalino, gave vent to a stream of bituminous gas, now used to illumine the mine, and in Belgium similar accidents have occurred at Liege, at Warmer, and near to Charleroy. Van Nooth and Van Campen add to the narratives already given from China, by the description they offer of the fire wells in the city of Xen Si.

It also appears, to advert to a still more recent date, that the inhabitants along the Little Muskingum River, in Ohio, came very near discovering the importance of Petroleum in 1819. Dr. S. P. Hildreth, of Marietta, in an account of the region, written in that year, and published in the "American Journal of Sciences," (1826), speaking of the borings for salt water, says: "They have sunk two wells, which are now more than 400 feet in depth. One of them affords a very strong and pure water, but not in great quantity. The other discharges such vast quantities of Petroleum, or, as it is vulgarly called, 'Seneca Oil,' and besides, is subject to such tremendous explosions of gas, as to force out all the water, and afforded nothing but gas for several days, so that they could make but little or no salt. Nevertheless, the Petroleum affords considerable profit, and is beginning to be in demand for lamps, for workshops and manufactories. It affords a brisk, clear light when burned this way, and will be a valuable article for lighting the street lamps in the future cities of Ohio."

From the traces we thus find of bitumen, naptha, and inflammable carbonic gas, in the most diverse and widely separated quarters of the globe, we are irresistibly led to infer, that the interior of our planet is one huge laboratory, in which the various carboniferous deposits are being separated into oil and the gases that precede or accompany it, and that the lapse of time required by a far vaster population than the earth can now boast, to consume the internal accumulations already existing, would be infinitely more than enough to have them replaced by fresh filterings, from the surface, of carbonic matter, the result of both animal and vegetable decay.

The next conclusion is, that, as the general supply must be practically inexhaustible, and as on this continent have been found the freshest, purest, and most copious deposits of this wonderful fluid, the United States of America alone have it as a source of wealth for ages to come. In it, with it, and about it, are found, either separately or united, all the other bituminous products, and while it can be used for fuel, it can be converted into gas, thus forming a sort of magic circle, giving light and heat all around the periphery of its fantastic transformations.

It is something to know that a cargo of Petroleum may navigate a river, cross a lake or ocean, in a vessel propelled by steam it has generated, acting upon an engine it lubricates, and directed by an engineer who may grease his hair, anoint his body, perfume his clothing, enrich his food, rub his bruises, freshen his liver, and waterproof his boots, with the same article.

# CHAPTER III.

## EARLY HISTORY OF THE PENNSYLVANIA PETROLEUM REGION.

FROM the remains of circular, square, and oval pits, sunk in the earth, to a depth of fifteen and twenty feet, cribbed or walled with timber, still visible in many places in the valley of Oil Creek, hundreds in number, and covering an aggregate surface of hundreds of acres, are indicated unmistakably the operations of a race possessing in some degree the elements of modern civilization. From the number of these pits, and their systematic arrangement, Petroleum was doubtless obtained in considerable quantities. These pits have been excavated with care, and with reference to one design. Of the date of their excavation, we can only vaguely conjecture. Trees, of not less age than from two to three centuries, are found growing out of these excavations, and there may possibly have been a previous growth. The Petroleum has protected the timber against the ravages of time, and their forms still remain, an unsolved problem to modern comprehension. By whom, and in what age of the world's history, these extensive works were executed, is a theme upon which the imagination can find abundant food to speculate.

The monuments of a race, superior in all respects to those who inhabited this continent when discovered, can be found from the Alleghanies to the Pacific. And stranger still, these vestiges are characterized by progressive degrees of civilization. The ancient mounds of the Ohio, and the remains of ancient cities on the Gila River, amply attest the rate of progress made.

To us, these ancient vestiges would seem to indicate the stages of the migration of successive races from a northern direction, west as far as the Rocky Mountains, which may have formed to them an obstacle too difficult to overcome, and from thence in a southerly course, until the more tropical countries were reached. In "Prescott's Conquest of Mexico," volume I., we find, in relation to the primitive races of Mexico, the following:

"Of these races, the most conspicuous were the Toltecs, advancing from a northerly direction, but from what region it is uncertain. They entered the territory of Anahuac, probably before the close of the seventh century. Of course, little can be gleaned, with certainty, respecting a people, whose written records have perished, and who are known to us only through the traditionary legends of the nations that succeeded them. By the general agreement of these, however, the Toltecs were well instructed in agriculture, and many of the most useful mechanic arts; were nice workers of metals; invented the complex arrangement of time adopted by the Aztecs; and, in short, were the true fountains of the civilization which distinguished this part of the continent in later times."

According to the same eminent authority, "the Toltecs, who had extended their sway over the remotest borders of Anahuac, having been greatly reduced by famine, pestilence, and unsuccessful wars, disappeared from the land as silently and mysteriously as they had entered it. * * * Their shadowy history reminds us of those primitive races, who preceded the ancient Egyptians in the march of civilization; fragments of whose monuments, as they are seen at this day, incorporated with the buildings of the Egyptians themselves, give to these latter the appearance of almost modern construction. * * * After the lapse of another hundred years, a numerous and rude tribe, called the Chicimecs, entered the deserted country from the *regions of the far Northwest.* They were speedily followed by others of higher civilization, perhaps of the same family with the Toltecs, whose language they appear to have spoken. The most noted of these were the Aztecs, or Mexicans, and the Acolhuans. The latter, better known in later times by the name of Tezcucans, from their capital, Tezcuco, were peculiarly fitted by their comparatively mild religion and manners, for receiving the tincture of civilization from the few Toltecs that still remained in the country. This, in their turn, they communicated to the barbarous Chicimecs, a large portion of whom became amalgamated with the new settlers as one nation."

From Clavigero, an authority quoted in the same work, the following dates are assigned of the arrivals and departures of the races spoken of. He was a most inquisitive and laborious historian of earlier times, and his testimony, elicited after intelligent research, is worthy of respect, at least. He gives the dates as follows:

| | A. D. |
|---|---|
| The Toltecs arrived in Anahuac | 648 |
| They abandoned the country | 1015 |
| The Chicimecs arrived | 1170 |
| The Acolhuans arrived | 1200 |
| The Mexicans or Aztecs reached Tula | 1196 |
| They founded Mexico | 1325 |

We make the foregoing extracts, not for the purpose of proving anything by the strong points of comparison presented, but merely as a curious coincidence.

The country around the Great Lakes may have possibly been the region from which the primitive races spoken of, originally came, as well as the source of all the races that had preceded them in their march of emigration, who once peopled the valley of the Mississippi, and even beyond. If the analogy could be established, the building of the monuments left by the mysterious race of "mound-builders," as they are called, and the reasons and explanations therefor, would be clear. With so important a link missing, we are not inclined to make such faint resemblance a basis.

We have beheld, so far as pits sunk in the earth are concerned, ample evidences of operations of a mysterious race, whose occupation of this country ante-dated the one found by our own, so far, that even tradition has failed to give any sign of them. A race possessing intelligence to sink and afterward crib the walls of these primitive oil wells, had certainly arrived at a sufficient state of civilization to utilize it. These uses may have been somewhat similar to those of the present. The artesian wells are unknown, so far as we can learn, in the famous oil region of the Burman Empire, at the present day. Yet for centuries, a large supply of Petroleum has been obtained from their wells, similar in general respects to the pits left in our own locality. In the age in which these pits were sunk, it may have been that the country, from the Alleghanies to the Rocky Mountains, teemed with millions of inhabitants. Thousands of towns and cities dotted the valley and plain, requiring a commerce co-equal to our own. A Petroleum development, similar to our own, may have taken place, and the same eventful scenes have been re-enacted as within the last few years, "Stock Companies," not excepted. Pits, or wells, were sunk by hundreds, "dry holes," occurred then as now, and "obstacles" of a similar nature pestered the oil miner, as in the present day; landowners received a "royalty," lessees grew thin and the landowners fat; dealers sold "short" and "long"; "bulls," and "bears," had their brief seasons of triumph or of tribulation, while the latest market intelligence from the far off commercial centres, brought by "pony expresses," or "shanks'" tried and trusty steed, was scanned by excited operators. A revenue may have been realized by the ruling government from its taxation; large towns or cities built then, as now, by the trade, and operators who had "struck oil," invested their suddenly acquired wealth in "faro" or "five-twenties." All this may have been, and more too. With traces so plain as the existence of these ancient oil pits, the hope is certainly not futile, that some authentic record of their history will yet be discovered and the wonderful achievements of the race that constructed them, made known.

The present product, known as Petroleum, has been found, in not only the Pennsylvania Oil Region, but also in portions of Virginia, Kentucky, and Ohio, by the early settlers. It was found in springs of water, coming to the surface of the same in the form of globules, and its presence was not unfrequently manifested on the surface of streams, and occasionally, in fissures of the rocks. It was met with, in the form of gas and oil in the boring of salt wells, and annoyed the early salt miners greatly, rendering the water from the well useless for salt, by the impregnation of its offensive odor. It was known to the Indian tribes, who used it both as a medicine and for toilet purposes, for dressing their wounds, and for mixing their paint. Red paint rock, resembling ochre, is still found in localities on the Alleghany River, which, no doubt, was used by the red men. Its use as a medicine was made known to the first settlers by the Indians, and adopted by them to a great extent. It was used for sprains, cuts, bruises, wounds of all kinds, and sometimes taken internally for various bodily diseases. It acts as a cathartic, and is sufficiently disgusting to the taste, to suit the most ardent lover of nostrums.

Ever since the settlement of this portion of Pennsylvania, hints of its existence have been manifest on the surface of springs, streams, and from the rocks themselves. Though the quantity gathered was small, the supply was generally equal to the demand, as a few barrels would glut the market. The manner of gathering the oil has been described as follows:

"A point was selected where the oil appeared to bubble up to the surface of the water most freely, when a pit was excavated to the depth of two or three feet. Sometimes this pit was rudely walled up; sometimes not. Sometimes it was near the edge of the water on the bank of the stream, and sometimes in the bed of the stream itself. In these pits the oil and water would collect together, until a stratum of the former would collect upon the surface of the latter, when a coarse blanket or a piece of flannel was thrown in. This blanket or flannel soon became saturated with oil, but rejected the water. The blanket was then taken out, wrung into a tub, and the operation repeated as often as desirable."

The first shipments of Petroleum to Pittsburgh, are related by EATON as follows:

"Mr. Cary, one of the first settlers of Oil Creek, possessing, perhaps, a little more enterprise than his neighbors, would collect or purchase a cargo of oil, and proceed to Pittsburgh, and exchange it for commodities needed in his family. This cargo consisted of two five gallon kegs, that were slung, one on each side of a horse, and thus conveyed by land, a distance of seventy or eighty miles. It was a small beginning. * * * At a period later, General Hays, who settled in Franklin in the year 1803, relates, that at one time he purchased the entire product of the region, amounting to sixteen barrels, which he sold in Pittsburgh for about one dollar per gallon." This is the first "corner" in the Petroleum business of which we have any record. It proved, doubtless, more successful than many of those of recent date of far greater magnitude.

The same writer says, "that a well was bored for the purpose of procuring salt water for the manufacture of that article, at Horse Creek, five miles above Oil City, on the Allegheny River, some forty years since. At the depth of seventy or eighty feet, a strong vein of oil was struck. Disgusted, the miners abandoned the well. Oil was also discovered in Franklin, in a well sunk for household uses, at the depth of thirty feet. An oil well was afterwards drilled through the bottom of this, but no considerable amount of oil was obtained. Its presence was also manifest to the workmen engaged in making the excavations for the lock of the canal at Franklin.

Such, in brief, is the material portion of the early history of the Pennsylvania Petroleum product, so far as known.

# CHAPTER IV.

## DAWNING OF THE ERA OF ARTIFICIAL LIGHT.

From the able report, made by the United States Commissioners, in 1866, we glean the following interesting facts:

"Rev. John Clayton, at the close of the sixteenth century, discovered coal gas, and its utility for illuminating purposes, but no application of the discovery was made until the year 1792, when Mr. Murdock, of Cornwall, in England, commenced a series of experiments, the result of which was so encouraging, that Dr. Henry and others became interested in their further prosecution. Gas was at length introduced into some of the manufacturing establishments. In the years 1803 and 1804, the Lyceum Theatre, in London, was lighted with gas, and by the year 1816, it had become quite common in England and France, both of which countries claimed the honor of its discovery. In a few years more its use had extended to all parts of the civilized world.

"This discovery, and the common process of distillation, used for production of gas, and various experiments with different coals, peats, and oils, made in connection therewith, and with different lamps and burners for the more economical use of gas, naturally resulted in the manufacture of hydro-carbon oils from coals and shales, in the art of purifying and refining the oils as now practised, and in the invention of the Kerosene or Petroleum lamp, which has removed the objection to the use of these oils for illuminating purposes.

"These inventions, with the application of the artesian well, by which the existence of extensive subterranean oil deposits have been demonstrated, and the present enormous production brought about, have added greatly to the wealth of the world, and indirectly to the advancement of civilization, by reducing the cost of artificial light. When we reflect that artificial light adds, perhaps, on an average, one-eighth to each day for all the inhabitants of the earth, and when we consider the inestimable value of the time thus gained, not only for the prosecution of industrial pursuits, but for social enjoyment and the cultivation of the mind, we can appreciate the immense utility of these inventions and discoveries, by which is being brought into general use a better light, unlimited in its supply, and at a greatly reduced cost.

"De Saussure, of Switzerland, the Chervan Brothers, and Selligue, of France, are the most conspicuous of those who invented, improved, and utilized the processes for producing illuminating oils from coals, peat, shales, and schists. They deserve the credit of having created on the continent of Europe that branch of manufacture, which had become quite large, more than fifteen years ago, and is now of very great importance.

"Mr. James Young, of Bathgate, in Scotland, took out a patent in 1850, now expired, but is continued by letters-patent obtained in the United States, and in force here until 1871. They are understood to be for the destructive distillation of coals, shales, and asphalta from the lowest temperature of decomposition up to a dull red heat, for the production of paraffine oil, or oils containing paraffine.

"The first lamps, so indispensable for burning of Petroleum, it is stated, were brought from Vienna, in Austria. They were preceded by American inventions, on the same principle, but less perfect.

"Soon after Mr. Young had perfected his invention the manufacture of artificial oils from different minerals, but principally Cannel coal, by process of destructive distillation, was commenced in the United States. This oil was refined and deodorized, and proved to be a valuable illuminator. Cannel coal is found in large quantity in Beaver and other counties in Western Pennsylvania, in Kentucky, and other States. It is rich in oil, producing forty gallons to the ton. The process was carried on by placing the coal in large iron retorts, inclined at a slight angle to the horizon, and applying heat. About the year 1858, the business began to assume a large and growing importance, and large amounts of Cannel coal lands were bought up by capitalists. Inquiry and investigations, as well as utilization had previously ensued. The natural oil that had for long ages been percolating through fissures of the rocks, and making its presence known in the waters of springs and streams, began to attract the attention of scientific and practical men. Analyzation and experiments indicated the same illuminating qualities, only in a higher degree, than that of any of the previous substances."

Several parties claim the honor of originating the Venango Petroleum development, or, in other words, the drilling of the first well, and the pumping of the hidden treasure, from the lower depths of earth. We have no data whereby the illustrious discoverer can be known, but the development of the product, and its application to uses of general utility, was reserved for a late date. Disclaiming all intention of interference with the claims of others, we give the following facts as a matter of history. We claim that the honor of originating the present Petroleum development, of Venango County, to which, too, the world owes its present source of light, clearly belongs to Mr. George H. Bissell, formerly of the firm of Eveleth & Bissell, now a resident of New York City.

In the year 1853, Mr. Bissell, who was then a resident of New Orleans, acting under the advice of his physician, took up his residence in the North, his health having become impaired by a long residence in a tropical climate. In the Summer of that year, he paid a visit to Hanover, N. H., the site of Dartmouth College, where Mr. Bissell graduated in 1845. One evening, Prof. Dixie Crosby, of Dartmouth College, exhibited to him a bottle of crude Petroleum, which he inferred, was gathered on the lands of his nephew, Dr. F. B. Brewer, on Oil Creek, near Titusville, Pa. Being well acquainted with Mr. Brewer, and becoming interested in the product, Mr. Bissell wrote to him, and obtained all the information in regard to the locality, and the product itself, of which he was possessed. Some time after, he sent a young man to Titusville, to obtain further information, and as his report was favorable, he determined to examine into the matter more fully. About this time, he met Mr. Eveleth, whom he had formerly known in New Orleans, and stated to him what he had heard. Mr. Eveleth signified a desire to join Mr. Bissell in examining into the matter.

Messrs. Eveleth & Bissell went on to Titusville, in 1854, and while there purchased of Brewer, Watson & Co., the territory where the principal oil wells were found, and obtained from Brewer, Watson & Co., a lease for ninety-nine years, for oil purposes, of all their boring land in Venango County, free of royalty, paying them therefor, the sum of $5,000,—this being the first purchase of land, for oil purposes, made in Venango County. They then hired a man named Angier, to trench the lands and pump the surface oil and water into vats. The pumping was effected by an apparatus attached to the working gear of a sawmill. The first three barrels raised, were taken to New Haven, Conn., and Prof. B. Silliman, Jr., employed to analyze the same. In the fall of 1855, Messrs. Eveleth & Bissell published Professor Silliman's very complete report. The total expense of the same being borne by these two gentlemen. The report attracted attention in New Haven, and finally, some capitalists there, purchased of Eveleth & Bissell, one-third in the property, Eveleth & Bissell retaining two-thirds of the stock, and the whole was placed in a company, called the Pennsylvania Rock Oil Company, of which Professor Silliman was elected president. The work of trenching went on with in different success, until 1857, when some members of the company agreed to sink an artesian well, and pay the company twelve cents a gallon, for forty-five years, on all oil raised. They employed a gentleman, named Drake, then a conductor on the New Haven railway, to superintend the work. Mr. Drake owned one forty-eighth part of the stock of the company. After a delay of over a year, and on the 28th day of August, 1859, the first vein of oil was struck, at a depth of sixty-nine and one-half feet, which produced for a time 400 gallons per day.

From this source, the entire development of Petroleum dates, and takes its origin. Having thus in brief given its history, we take pleasure in appending the elaborate and highly interesting report of Professor Silliman:

## REPORT.

MESSRS. EVELETH & BISSELL:

GENTLEMEN: I herewith offer you the results of my somewhat extended researches upon the Rock Oil, or Petroleum, from Venango County, Pa., which you have requested me to examine with reference to its value for economical purposes.

Numerous localities, well known in different parts of the world, furnish an oily fluid, exuding from the surface of the earth, sometimes alone in "tar springs," as they are called in the Western United States, while frequently it is found floating upon the surface of water in a thin film, with rainbow colors, or in dark globules, that may, by mechanical means, be separated from the fluid on which it swims.

In some places, wells are sunk for the purpose of accumulating the product in a situation convenient for collection by pumping the water out. The oil exudes, on the shores of lakes and lagoons, or rises from springs beneath the beds of rivers. Such are the springs of the Baku, in Persia, and the wells of Amiano, in the duchy of Parma, in Italy. The usual geological position of the rocks furnishing this natural product, is in the coal measures, but it is by no means confined to this group of rocks, since it has been found in deposits much more recent, and also in those that are older; but in whatever deposits it may occur, it is uniformly regarded as a product of vegetable decomposition. Whether this decomposition has been effected by fermentation only, or by the aid of an elevated temperature, and distilled by heated vapor, is perhaps hardly settled.

It is interesting, however, to remember in this connection, that the distillation, at an elevated temperature, of certain black bituminous shales in England and France, has furnished large quantities of an oil having many points of resemblance with naptha, the name given to this colorless oil, which is the usual product of distilling Petroleum. The very high boiling point of most of the products of the distillation of the

Rock Oil from Venango County, would seem to ind·cate that it was a pyrogenic (fire produced) product.

Bitumen, asphaltum, mineral pitch, chapapote, &c., are names variously given to the more or less hard, black, resinous substance, which is produced usually from the exposure of Petroleum. The most remarkable examples of the occurrence of these substances so intimately connected with the history of Rock Oil, are the Lake Asphaltites, or the Dead Sea, so memorable in history, the well-known bitumen lake in Trinidad, and the deposits of mineral pitch or chapapote in Cuba. In one of the provinces of India, vast quantities of Petroleum are annually produced, the chief consumption being local, for fuel and light, but a portion is also exported to Europe for the production of naptha. In the United States, many points on the Ohio and its tributaries, are noted as producing this oil; nearly all of them within the coal measures. A detailed history of these various localities, can be found recorded in books of science, and their repetition here would be out of place

### GENERAL CHARACTERS OF THE CRUDE PRODUCT.

The crude oil, as is gathered on your lands, has a dark-brown color, which, by reflected light, is greenish or bluish. It is thick even in warm weather, about as thick as thin molasses. In very cold weather it is somewhat more stiff, but can always be poured from a bottle even at 15° below zero. Its odor is strong and peculiar, and recalls to those who are familiar with it, the smell of bitumen and naptha. Exposed for a long time to the air, it does not thicken or form a skin on its surface, and in no sense, can it be called a drying oil. The density of the crude oil is 882, water being 1,000. It boils only at a very high temperature, and yet it begins to give off a vapor at a temperature not greatly above that of boiling water. It takes fire with some difficulty, and burns with an abundant smoky flame. It stains paper with the appearance of ordinary fat oils, and feels smooth and greasy between the fingers. It is frequently used in its crude state to lubricate coarse machinery. In chemical characters, it is entirely unlike the fat oils. Most of these characters are common to Petroleum from various places. In one important respect, however, the product of your lands differ from that obtained in other situations, that is, it does not, by continued exposure to the air, become hard and resinous like mineral pitch or bitumen. I have been informed by those who have visited the locality, that on the surface of the earth about the springs, which furnish your oil, there is no crust or deposit of this sort, such as I have seen in other situations where Petroleum or mineral tar is flowing. This difference will be seen to be of considerable importance, as it is understood and represented that this product exists in great abundance upon your property, that it can be gathered wherever a well is sunk in the soil, over a great number of acres, and that it is unfailing in its yield from year to year. The question naturally arises, of what value is it in the arts, and for what uses can it be employed? To enable you to answer these inquiries, has been the object of my researches.

### EXAMINATION OF THE OIL.

To determine what products might be obtained in the oil, a portion of it was submitted to fractional distillation.* The temperature of the fluid was constantly

* Fractional distillation is a process intended to separate various products in mixture, and having unlike boiling points, by keeping the

regulated by a thermometer, the heat being applied first by water bath, and then by a bath of linseed oil. This experiment was founded upon the belief that the crude product contained several distinct oils, having different boiling points. The quantity of material used in this experiment, was 304 grammes. The thermometer indicated the degree of the Centigrade scale, but for convenience, the corresponding degrees of Fahrenheit's scale are added. The water bath failed to distill any portion of the oil at 100°C. (=212° Fah.), only a small quantity of acid water came over. An oil bath was then substituted, and the temperature was regularly raised by slow degrees until distillation commenced. From that point the heat was successively raised by stages of ten degrees, allowing full time at each stage for complete distillation of all that would rise at that temperature before advancing to the next stage. The results of this tedious process are given in the annexed table—304 grammes of crude oil, submitted to fractional distillation, gave:

1st product at 100°C.=212°Fah. (acid water) 5 Grammes.

2d. product at 140°C. to 150=283° to 302 Fah. 26 Grammes.

3d product at 150°C. to 160°C.=302° to 320 Fah. 29 Grammes.

4th product at 160°C. to 170°C.=320° to 338 Fah. 38 Grammes.

5th product at 170°C. to 180°C.=338° to 356° Fah. 17 Grammes.

6th product at 180°C. to 200°C.=356° to 392° Fah. 16 Grammes.

7th product at 200°C. to 270°C.=392° to 428° Fah. 17 Grammes.

mixture contained in an alembic at regulated successive stages of temperature as long as there is any distillate at a given point, and then raising the heat to another degree, &c.

8th product at 220°C. to 270°C.=128° to 518° Fah. 12 Grammes.

Whole quantity distilled by this method, 160 Grammes, leaving a residue in the retort of 144, out of the orginal quantity 304.

Product No. 1, as above remarked, was almost entirely water, with a few drops of colorless oil, having an odor similar to the orginal fluid, but less intense.

Product No. 2 was an oil perfectly colorless, very thin and limpid, and having an exceedingly persistent odor, similar to the crude oil, but less intense.

Product No. 3 was tinged slightly yellow, perfectly transparent, and apparently as limpid as the 2d product, with the same odor.

Product No 4 was more decidedly yellowish than the last, but was in no other respect distinguishable from it.

Product No. 5 was more highly colored, thicker in consistence, and had a decided empyreumatic odor.

Product No. 6. This and the two subsequent products were each more highly colored, and denser than the preceding. The last product had the color and consistency of honey, and the odor was less penetrating than that of the preceding oils. The mass of crude product remaining in the retort, equal to 47.4 per cent., was a dark, thick, resinous-looking varnish, which was so stiff, when cold, that it could be inverted without spilling. This showed no disposition to harden or skin over by exposure to the air. The distillation was arrested at this point in glass, by our having reached the limit of temperature for a bath of linseed oil. The density of the several products of this distillation, shows a progressive increase, thus:

No. 2............................density, .732
No. 3.............................. " .752
No. 4.............................. " .766

| | | |
|---|---|---|
| No. 5............................... | density, | .776 |
| No. 6............................... | " | .800 |
| No. 7............................... | " | .848 |
| No. 8............................... | " | .854 |

To form an idea of the comparative density of these several products, it may be well to state, that sulphuric ether, which is one of the highest fluids known, has a density of .736, and alcohol, when absolutely pure, .810.

The boiling points of these several fluids present some anomalies, but are usually progressive, thus, No. 2 gave signs of boiling at 115°C. (=239 Fah.), and boiled vigorously, and remained constant at 225°C. to 228°C. (=437° to 420 Fah.). No 3 began to boil at 120°, (=248° Fah.), rose to 270° (=518° Fah.), where it remained constant. No. 5 began to vaporise at 140° (=284° Fah.), rose to 290° (=554° Fah.), where it remained constant. On a second heating, the temperature continued to rise, and passed 350° (=581 Fah.). No. 5 gave appearance of boiling at 160°(=320 Fah.), boiling more vigorously as the heat was raised, and was still rising at 308°(=581° Fah.). No 6 commenced boiling at 135° (=275° Fah.), boiled violently at 160° (=320 Fah.), and continued rising above the range of the mercurial thermometer. No. 7 commenced ebullition at the same temperature as No. 6, and rose to 305° (=581° Fah.), where the ebullition was not very active. Much time was consumed in obtaining these results. We infer from them that the Rock Oil is a mixture of numerous compounds, all having essentially the same chemical constitution, but differing in density and boiling points, and capable of separation from each other, by a well-regulated heat.

The uncertainty of the boiling points indicates that the products obtained at the temperatures named above, were still mixtures of others, and the question forces itself upon us, whether these several oils are to be regarded as *educts*, (*i. e.*, bodies previously existing, and simply separated in the process of distillation), or whether they are not rather produced by the heat and chemical change in the process of distillation. The continued application of an elevated temperature alone is sufficient to effect changes in the constitution of many organic products, evolving new bodies not before existing in the original substance.

### PROPERTIES OF THE DISTILLED OILS.

Exposed to the severest cold of the past winter, all the oils obtained in this distillation remained fluid. Only the last two or three appeared at all stiffened by a cold of 15° below zero, while the first three or four products of distillation retained a perfect degree of fluidity. Exposed to air, as I have said, they suffer no change. The chemical examination of these oils, showed that they were all composed of hydrogen and carbon, and probably have these elements in the same numerical relation. When first distilled they all had an acid reaction, due to the presence of a small quantity of sulphuric acid, derived from the crude oil. This was entirely removed by a weak alkaline water, and even by boiling on pure water. Clean copper remained untarnished by the oil which had thus been prepared, showing its fitness for lubrication, so far as absence of corrosive quality is concerned. The oils contain no oxygen, as is clearly shown by the fact that clean potassium remains bright in them. Strong sulphuric acid decomposes and destroys the oil entirely. Nitric acid changes it to a yellow, oily fluid, similar to the changes produced by nitric acid on other oils. Hydrochloric, acetic acids, do not affect it. Litharge and other metallic oxides do not change it, nor convert it in any degree to a drying oil. Potassium remains in it unaffected, even at a high temperature. Hydrates of potash, soda, and lime, are also without action upon it. Chloride of calcium and many other salts manifest an equal indifference to it. Distilled with bleaching powders (chlorid of lime), and water, in the manner of producing chloroform, the oil is changed into a product having an odor and taste resembling chloroform. Exposed for many days in an open vessel, at a regulated heat below 212°, the oil gradually rises in vapors, as may be seen by its staining the paper used to cover the vessel from dust, and also by its sensible diminution. Six or eight fluid ounces, exposed in this manner, in a metallic vessel for six weeks or more, the heat never exceeding 200°, gradually and slowly diminished, grew yellow, and finally left a small residue of dark-brown lustrous-looking resin, or pitchy substance, which in the cold was hard and brittle. The samples of oil employed were very nearly colorless. This is remarkable when we remember that the temperature of the distillation was above 500° Fah. The oil is nearly insoluble in pure alcohol, not more than 4 or 5 per centum being dissolved by this agent. In ether the oil dissolves completely, and on gentle heating is left unchanged by the evaporization of the ether. India rubber is dissolved by the distilled oil to a pasty mass, forming a thick, black fluid, which, after a short time, deposits the india rubber. It dissolved a little amber, but only sufficient to color the oil red. It also dissolves a small portion of copal in its natural state; but after roasting, the copal dissolves in it as it does in other oils.

### USE FOR GAS MAKING.

The crude oil was tried as a means of illumination.

For this purpose a weighed quantity was decomposed, by passing it through a wrought-iron retort filled with carbon, and ignited to full redness. The products of this decomposition were received in a suitable apparatus. It produced nearly pure carburetted hydrogen gas, the most highly illuminating of all the carbon gases. In fact the oil may be regarded as chemically identical with illuminating gas in a liquid form. The gas produced equalled ten cubic feet to the pound of oil. It burned with an intense flame, smoking in the ordinary gas jet, but furnishing the most perfect flame with the argand burner.

These experiments were not prosecuted further, because it was assumed that other products, now known and in use, for gas making, might be employed at less expense for this purpose, than your oil. Nevertheless, this branch of inquiry may be worthy of further attention.

### DISTILLATION AT A HIGHER TEMPERATURE.

The results of the distillation at a regulated temperature in glass, led us to believe, that in a metallic vessel capable of enduring a high degree of heat, we might obtain a much larger proportion of valuable products. A copper still, holding five or six gallons, was therefore provided, and furnished with an opening, through which a thermometer could be introduced into the interior of the vessel. Fourteen imperial quarts (or, by weight, 560 ounces), of the crude product were placed in this vessel, and the heat raised rapidly to 280°C. (=536° Fah.), somewhat higher than the last temperature reached in the first distillation. At this high temperature, the distillation was somewhat rapid, and the product was easily condensed without a worm. The product of the first stage was 130 ounces (or over

28 per cent.), of a very light-colored thin oil, having a density of .792. This product was also acid, and as before, the acid was easily removed by boiling with fresh water. The temperature was now raised to somewhat above 300°C. (=572° Fah.), and 123 ounces more distilled, of a more viscid and yellowish oil, having a density of .895. This accounts for over 43 per cent. of the whole quantity taken. The temperature being raised now above the boiling point of mercury, was continued at that until 170 ounces, or over 31 per cent., of a dark-brown oil had been distilled, having a strong empyreumatic odor. Upon standing still for some time, a dark blackish sediment was seen to settle from this portion, and on boiling it with water, the unpleasant odor was in a great degree removed, and the fluid became more light-colored and perfectly bright. (It was on a sample of this that the photometric experiments were made.) The next portion, distilled at about 700° Fah., gave but about 17 ounces, and this product was lighter in color and more fluid than the last. It now became necessary to employ dry hickory wood as a fuel, to obtain flame, and a sufficient heat to drive over any further portions of the residue remaining in the alembic.

It will be seen that we have already accounted for over 75 per cent. of the whole quantity taken. There was a loss on the whole process of about 10 per cent. made up, in part, of a coaly residue that remained in the alembic, and partly of the unavoidable loss resulting from the necessity of removing the oil twice from the alembic, during the process of distillation, in order to change the arrangements of the thermometer, and provide means of measuring a heat higher than that originally contemplated.

About 15 per cent. of a very thick, dark oil, completed this experiment. This last product, which came off slowly, at about 750° Fah., is thicker and darker than the original oil, and when cold is filled with a dense mass of pearly crystals. These are paraffine, a peculiar product of the destructive distillation of many bodies in the organic kingdom. This substance may be separated, and obtained as a white body, resembling fine spermaceti, and from it beautiful candles have been made. The oil in which the crystals float, is of a very dark color, and by reflected light, is blackish green, like the original crude product. Although it distills at so high a temperature, it boils at a point not very different from the denser product of the first distillation. The paraffine, with which this portion of the oil abounds, does not exist ready formed in the original crude product, but it is a result of the high temperature employed in the process of distillation, by which the elements are newly arranged.

I am not prepared to say, without further investigation, that it would be desirable for the company to manufacture this product in a pure state, fit for producing candles (a somewhat elaborate chemical process), but I may add, that should it be desirable to do so, the quantity of this substance produced may probably be very largely increased, by means which it is now unnecessary to mention.

Paraffine derives its name from the unalterable nature of the substance, under the most powerful chemical agents. It is white, in brilliant scales of a greasy lustre; it melts at about 116°, and boils at over 700° Fah.; it dissolves in boiling alcohol and ether, and burns in the air with a brilliant flame. Associated with paraffine are portions of a very volatile oil, eupion, which boils at a lower temperature, and by its presence renders the boiling point of the mixture difficult to determine. I consider this point worthy of further examination than I have been able at present to give it, *i. e.*, whether the last third, and possibly the last half, of the Petroleum may not be advantageously so treated as to produce from it the largest amount of paraffine which it is able to produce.

The result of this graduated distillation, at a high temperature, is that we have obtained over 90 per cent. of the whole crude product in a series of oils, having valuable properties, although not all equally fitted for illumination and lubrication.

A second distillation of a portion of the product which came over in the latter stages of the process (a portion distilled at about 650° Fah., and having a high color), gave us a thin oil of density about .750, of light yellow color and faint odor.

It is safe to add that, by the original distillation, about 50 per cent. of the crude oil is obtained in a state fit for use as an illuminator without further preparation than simply clarification by boiling a short time with fair water.

### DISTILLATION BY HIGH STEAM.

Bearing in mind that by aid of high steam, at an elevated temperature, many distillations in the arts are effected, which cannot be so well accomplished by dry heat, I thought to apply this method in case of the present research. Instances of this mode of distillation are in the new process for stearine candles, and in the preparation of rosin oil. I, accordingly, arranged my retort in such a manner that I could admit a jet of high steam into the boiler, and almost at the bottom of the contained Petroleum. I was, however, unable to command a jet of steam above 275° to 290° Fah., and, although, this produced abundant distillation, it did not effect a separation of the several products, and the fluid distilled had much the same appearance as the Petroleum itself, thick and turbid. As this trial was made late in the investigation, I have been unable to give it a satisfactory issue, chiefly for want of steam of a proper temperature. But I suggest, for the consideration of the Company, the propriety of availing themselves of the experience already existing on this subject, and particularly those who are concerned in the distillation of rosin oil—a product having many analogies with Petroleum in respect to its manufacture.

### USE OF THE NAPTHA FOR ILLUMINATION.

Many fruitless experiments have been made in the course of this investigation which it is needless to recount. I will, therefore, only state those results which are of value.

1. I have found that the only lamp in which this oil can be successfully burned is the camphene lamp, or one having a button to form the flame, and an external cone to direct the current of air, as is now usual in all lamps designed to burn either camphene, rosin oil, sylvic oil, or any other similar product.

2. As the distilled products of Petroleum are nearly or quite insoluble in alcohol, burning fluid (*i. e.*, a solution of the oil in alcohol), cannot be manufactured from it.

3. As a consequence, the oil cannot be burned in a hand lamp, since, with an unprotected wick, it smokes badly. Neither can it be burned in a Carcel's Mechanical Lamp, because a portion of the oil being more volatile than the rest, rises in vapor on the elevated wick required in that lamp, and so causes it to smoke.

I have found all the products of distillation from the copper still, capable of burning well in the camphene lamp, except the last third or fourth part (*i. e.*, that portion which came off at 700° Fah., and rising, and which was thick with the crystals of paraffine.) Freed from acidity by boiling on water, the oils of this distillation burned for 12 hours without injuriously coating the wick, and without smoke. The wick may be elevated considerably above the level required for camphene, without any danger of smoking, and the oil shows no signs of crusting the wick tapes with a coating of resin, such as happens in the case with camphene, and occasions so much inconvenience. The light from the rectified naptha, is pure and white, without odor. The rate of consumption is less than half that of the camphene, or rosin oil. The imperial pint, of 20 fluid ounces, was the one employed—a gallon contains 160 such ounces. A camphene lamp, with a wick one inch thick, consumed of rectified naptha in one hour, 1¾ ounces of fluid. A Carcel's Mechanical Lamp, of ⅞ inch wick, consumed of best sperm oil, per hour, 2 ounces. A "Diamond Light" Lamp, with "Sylvic Oil," and a wick 1½ inch diameter, consumed, per hour, 4 ounces.

I have submitted the lamp burning Petroleum to the inspection of the most experienced lampists who were accessible to me, and their testimony was, that the lamp burning this fluid gave as much light as any which they had seen, that the oil spent more economically, and the uniformity of the light was greater than in camphene, burning for 12 hours without a sensible diminution and without smoke. I was, however, anxious to test the amount of light given more accurately than could be done by a comparison of opinion. With your approbation I proceeded, therefore, to have constructed a *photometer*, or apparatus for the measurement of light, upon an improved plan. Messrs. Gunow, scientific artists of this city undertook to construct this apparatus, and have done so to my entire satisfaction. This apparatus I shall describe elsewhere—its results only are interesting here. By its means I have brought the Petroleum light into rigid comparison with the most important means of artificial illumination. Let us briefly recapitulate the results of these.

### PHOTOMETRIC EXPERIMENTS.

The *unit* adopted for comparison of intensities of illumination is Judd's Patent Sixes Sperm Candle.

The sperm oil used was from Edward Mott Robin son, of New Bedford—the best winter sperm remaining fluid at 32° Fah. The Colza Oil and Carcel's lamps were furnished by Dardonville, lampist, Broadway, New York.

The gas used was that of the New Haven Gas Light Company, made from best Newcastle coal, and of fair average quality.

The distance between the standard candle and the illuminator sought to be determined was constantly 150 inches—the photometer traversed the graduated bar in such a manner as to read, at any point where equality of illumination was produced, the ratio between the two lights. I quote only single examples of the average results, and with as little details as possible, but I should state that the operation of the photometer was so satisfactory that we obtained constantly the same figures when operating the same way, evening after evening, and the sensitiveness of the instrument was such that a difference of one-half inch in its position was immediately detected in the comparative illumination of the two equal discs of light in the dark chamber. This is, I believe, a degree of accuracy not before obtained by a photometer.

Table of illuminating power of various artificial lights compared with Judd's Patent Candles as a unit.

Gas burning in Scotch fish-tail tips, 4 feet to the hour.......................................1 : 5.4
Gas burning in Scotch fish-tail tips, 6 feet to the hour.......................................1 : 7.55
Gas burning in Cornelius fish-tail tips, 6 feet to the hour.......................................1 : 6.2
Gas burning in English Argand burner, 10 feet to the hour.......................................1 : 1.6
Rock Oil, burning in 1 inch wick, Camphene Lamp, consuming 1¾ ounces fluid the hour...1 : 8.1
Carcel's Mechanical Lamp, burning best sperm oil, 2 ounces fluid to the hour, wick ⅞ of an inch.......................................1 : 7.5
Carcel's Mechanical Lamp, burning Colza Oil....1 : 7.5
Camphene Lamp, same size as Rock Oil above, burning best camphene, 4 ounces fluid per hour.......................................1 : 1.1
"Diamond Light" by "Sylvic Oil," in ½ inch wick, 4 ounces per hour....................1 : 8.1

From this table it will be seen that the Rock Oil Lamp was somewhat superior in illuminating power to Carcel's Lamp of the same size, burning the most costly of all oils. It was also equal to the "Diamond Light" from a lamp of one-half greater power, and consequently is superior to it in the same ratio in lamps of equal power.

The camphene lamp appears to be about one-fifth superior to it, but, on the other hand, the Rock Oil surpasses the camphene by more than one-half in economy of consumption, (*i. e.*, it does not consume one-half so much fluid by measure), and it burns more constantly. Compared with the Sylvic Oil and the sperm, the Rock Oil gave on the ground glass diaphragm the whitest disc of illumination, while in turn the camphene was whiter than the Rock Oil light. By the use of screens, of different colored glass, all inequalities of color were compensated in the use of the photometer, so that the intensity of light could be more accurately compared. Compared with gas, the Rock Oil gave more light than any burner used, except the costly Argand, consuming ten feet of gas per hour. To compare the cost of these several fluids with each other, we must know the price of the several articles, and this varies very much in different places. Thus, gas in New Haven costs $4 per 1,000 feet, and in New York $3 50 per 1,000; in Philadelphlia, $2 per 1,000, and in Boston about the same amount.

Such sperm oil as was used costs $2 50 per gallon, the Colza about $2, the Sylvic Oil, 50 cents, and the camphene, 68 cents—no price has been fixed upon for the rectified Rock Oil.

I cannot refrain from expressing my satisfaction at the results of these photometric experiments, since they have given the oil of your Company a much higher value as an illuminator than I had dared to hope.

### USE OF THE ROCK OIL AS A LUBRICATOR FOR MACHINERY.

A portion of the rectified oil was sent to Boston to be tested upon a trial apparatus there, but I regret to say, the results have not been communicated to me yet. As this oil does not gum or become acid, or rancid, by exposure, it possesses in that, as well as in its wonderful resistance to extreme cold, most important qualities for a lubricator.

### CONCLUSION.

In conclusion, gentlemen, it appears to me that there is much ground for encouragement in the belief that your Company have in their possession a raw material from which, by simple and not expensive process, they may manufacture very valuable products.

It is worthy of note, that my experiments prove that nearly the whole of the raw product may be manufactured without waste, and this solely by a well directed process, which is in practice, one of the most simple of all chemical processes.

There are suggestions of a practical nature, as to the economy of the manufacture, when you are ready to begin operations, which I shall be happy to make, should the Company require it—meanwhile I remain, gentlemen,

Your obedient servant,

B. Silliman, Jr.,

*Professor of Chemistry in Yale College.*

New Haven, April 16, 1855.

# CHAPTER V.

## COMMENCEMENT AND GROWTH OF THE PETROLEUM DEVELOPMENT.

WITH the drawing of the drill from Drake's Well came the dawning of the light, whose source had been hidden so long in its rock-bound caverns. The mystery was now to be solved that had in previous years proved incomprehensible. The problem of nature providing every article of utility to man, in practically inexhaustible quantities, was interpreted in unmistakable terms, by the tapping of its rich fountains by the drill of the persevering oil miner. The fact, too, was demonstrated, beyond a peradventure, that large reservoirs of oil that had been giving hints of its existence in the springs and upon the streams of Venango County, from the memory of its oldest settlers, and was known to their predecessors, the Indians, so long, that the beginning was older than their traditions, existed in the rocks, many feet below the surface of the earth, and could be obtained in what was then thought to be very large quantities, by boring artesian wells of sufficient depth. A wonderful excitement was created all over the land by the striking of this well, pumping oil from subterranean deposits, that was in a few short years to become an article so important and indispensable to the wants and uses of mankind. Thousands, from all sections of the country, came to see the wonderful phenomena that had been reserved for this, the most enlightened age of the world.

Hundreds secured leases in the different localities, and very soon scores of new oil wells were commenced. Derricks sprang up, as it were almost by magic, in the valleys and ravines. The treasure that had so long remained dormant, and existed, as was supposed by practical men, only in the minds of the visionary, had budded forth into a reality. At the time of the striking of Drake's Well, the oil so obtained was selling at about one dollar per gallon. By the investment of a few thousand dollars, the operator had before him the chance of securing an ample fortune, not by the slow accumulations of years of toil, but in the course of a few months, and with a comparatively trifling outlay. Never before, in all the history of modern civilization, had a more tempting bait been offered to the cupidity of man.

The machinery used by the first operators was generally of the most primitive character, and the progress made in sinking a well was necessarily slow, and the labor tedious. Yet the wells were generally of moderate depth, operators being content with the supply of oil obtained from the first and second sand-rocks, the average depth being from 200 to 300 feet. In a short time a number of wells, yielding a daily average of from ten to fifty barrels each, were struck at various points along the valley of Oil Creek, and on the Allegheny River, all of which were pumping wells. The fact of obtaining oil in such quantities in localities so remote from each other, furnished ample evidence of the general distribution of its subterranean reservoirs. Developments were successfully made at Tidioute, at Franklin, and various points along the Allegheny. Speculators thronged all the producing localities, and the excitement that soon prevailed was unequalled by any previous mineral discovery, not excepting that of California. During these first years, men of abundant wealth were slow to make any investment in a development that came into existence so suddenly. The first operators with few exceptions, were men of moderate means, relying more upon their industry and skill than upon a lavish expenditure.

Owing to the impetus given by developments, lands around, and even remote from, producing localities, began to appreciate in value. Many of the farmers of Venango County, who had been content in previous years to wrest a bare subsistence from its rugged hills and valleys, were now bewildered at the golden prospects that loomed up before them. Intense as the desire may have been with many, to retain the old homestead, which for generations had sheltered their forefathers, and endeared by all the ties of local attachment, the temptation for acquisition of wealth was too great for the large majority. Their farms were purchased by men eager only to obtain the liquid treasures supposed to lay beneath their surface. The derrick and engine-house frequently occupied the site of the old homestead; orchards were levelled for fire-wood, or to make room, and the whole face of the country was in a few years so entirely changed as to be scarcely recognizable by its former possessors.

But few of the landowners at first developed the same at their individual expense. They leased to operators wishing to put down wells, the size of the lease ranging from one to ten acres, according to the means of the latter, for a term of years, varying from twenty to ninety-nine, for a certain proportion of all the oil obtained. The portion obtained by the landowner, or land interest, as it is called, was from an eighth to one-quarter. The lessee was bound by the terms of the lease, to commence operations in a specified time, and pursue the same with due diligence, to success or final abandonment. Abandoning the lease worked forfeiture.

When an individual operator secured a lease or leases, and lacked the means to develop the same, he formed a company or association of individuals. These companies consisted frequently of fifty and sometimes more. In such cases the assessment was ten dollars per share for each member at a time. The cost of sinking a well in the early days was about $1,000, the machinery for boring and pumping costing about $1,000 more. These companies were not chartered, nor had they any authority by law for holding or chartering real estate, any farther than a simple partnership for the transaction of business. Each individual member of a company was, of course, held responsible for the debts of the whole concern, should the creditor demand it at his hands. In case of failure, those members possessed of means, were selected by the creditors to pay all the debts. The regularly chartered stock companies were of subsequent date, and will be spoken of hereafter.

The number of wells continued to multiply in many portions of Venango County, but more especially along the valley of Oil Creek. The total daily product of all the wells in June, 1860, was estimated at 200 barrels. From that time the production increased rapidly. By January, 1861, the daily production reached 700 barrels, and in the Spring of 1861, 1,200 to 1,500 barrels per day.

About this time, as before stated, it occurred to some genius, that the pumping well was a slow and expensive mode of procuring oil from beneath the ground. As the supply seemed to come from great depths below, by reaching the source or fountain, a greater supply could be obtained without increasing the working expenses. To do this, the theorist argued that it was necessary to drill or sink the well to a greater depth. As steam-power had been generally introduced, this could be done without material difficulty. The experiment that gave promise of such important results was not very expensive. A well was, accordingly, commenced, and, regardless of the good "show" of oil in the first and second sand-rock, the operator persevered, until the third sand-rock, which, to present date, has been ascertained to contain the largest amount of oil, was reached at the depth of between 400 and 500 feet. At this depth a cavity, containing gas and oil, was tapped by the miner's drill. This came to the surface with such force as to throw the drilling tools out of the well, far above the top of the derrick. A column of gas and oil followed. The well was tubed, and a continuous stream flowed forth, at the rate of from 1,000 to 4,000 barrels per day, and continued to yield at this rate for months. Other parties drilled similar wells in different localities, with similar results. The Burnt Well, on the J. Buchanan Farm; the Phillips Well, on the Tarr Farm; the Empire Well, on the M'Ilhenny Farm; the Sherman, just above, and several others, ranging in daily yield from 1,000 to 4,000 barrels per day, followed in rapid succession. The production was increased in a very short time from 1,200 to between 8,000 to 10,000 barrels per day.

The effect of this large production upon the business was, for the time, disastrous. The prices declined to a mere nominal rate, oil selling as low as ten cents per barrel at the wells, being cheaper than the water peddled in our streets. Thousands of barrels were wasted, being allowed to run into Oil Creek, and no material demand existed for the product, even at the low rates. An incident will illustrate the slight value set upon it by the producers. An acquaintance of the writer sold a boat-load of oil at ten cents per barrel. The tanks being adjacent to the Creek, the boat was run up alongside, and loaded, in bulk; that is, the oil ran into the open boat, direct from the tanks. The boat had been previously measured. After loading, the buyer complained of the measure, claiming that he lacked *ten* barrels of the necessary amount. The producer, a great wag, by the way, gave a signal to his workmen, who let a full stream run into the boat, and sunk it. The small pumping wells were forced to cease operations, and scores became disheartened and abandoned their wells.

The production during the Winter and Spring of 1861 and 1862, has been variously estimated at from 12,000 to 20,000 barrels per day. Fifteen thousand barrels would be, we judge, a fair average, for the daily production of that period. A careful estimate of the entire oil field, made in May, 1862, and published in the *Oil City Register* of that date, gives as the daily production at that time, 5,717 barrels; the number of wells then flowing, 76; the number of wells that had formerly flowed and pumped, 62; sunk, and in process of being drilled, 358; total number of wells, 495; amount of oil on hand, 92,450 barrels; amount of oil produced previous to the date mentioned, 1,000,000 barrels; cost of sinking wells, $498,000; cost of machinery, buildings, tanks, &c., $500,000.

The production of 1863 was scarcely half that of the beginning of 1862, and that of 1864, still less. In May, 1865, the production had declined to less than 4,000 barrels per day; the valley of Oil Creek being the only producing locality at that time. Soon after this the marvellous developments upon Pithole Creek began to take place.

In the Winter of 1864 and 1865, the Fraser Well, at Pithole, the first in that locality, was struck, and flowed, as estimated, January 7th, 1865, 650 barrels per day. The striking of this well was the cause of the Pithole excitement, which will ever hold a prominent place in the history of Petrolia. Operators thronged to the locality. The Holman and adjoining farms were soon surveyed into suitable lots, and scores of wells were commenced. The results were all that could be desired. A production of from 6,000 to 7,000 barrels per day was soon attained. The United States Petroleum Company, of New York, had secured a lease of this and several of the adjoining farms on Pithole Creek, for twenty years, sometime before this.

A city sprang up on the bluff adjoining the producing wells. The town plat of Pithole City was commenced May 24th. November 25th, six months afterward, between 400 and 500 houses had been erected. Among these were some fifty hotels. The population was in a short time after this, estimated at 8,000 It had miles of streets, lined with buildings, including banking-offices, school-houses, churches, an opera house, and other appendages of a first-class modern city. Several of its hotels were palatial in size, and truly gorgeous in their equipment. The cost of the Chase House, complete, was over $80,000; the cost of the Morey and Bonta Houses, or hotels, was equally as large.

Twenty-four thousand dollars bonus was given for the privilege of drilling a well on a half acre lease adjacent to the United States Well, the royalty being one-half the oil. Strange to relate, the purchaser made a handsome profit on his investment by selling it again.

After being resold a number of times, the lease was, we think, abandoned, nothing but the erection of a derrick having been done in the shape of development. Sanguine operators located it as the great oil centre, and thought they had here found the main source of Petroleum supply. Fortunes were made and lost rapidly in the wild speculation that followed.

"Solitary horsemen," and horsemen in groups, were met on all the roads leading to the new Mecca. Pilgrims came from every land to worship at its greasy shrine. Wagons, loaded with every conceivable article, household furniture, drygoods and groceries, engines and well-machinery, blocked the roads. Pedestrians plodded steadily through mud and mire. Bearded horsemen, water-proof externally, so far as india-rubber garments and high-topped boots could make them, floundered through the almost unfathomable depths of Holmden Street, then the main thoroughfare of the magic city of Petroleum. A great scarcity of females at first prevailed. The residents soon brought on their families, and began to live more in accordance with the ideas of modern civilization. "Interests for Sale," stared one in the face from almost every house-front, and crowds of eager-looking men, of every degree and profession, thronged the hotels, saloons, and sidewalks. Like Tadmor, and other ancient cities of note, after a brief season of unequalled splendor, as far as hopes for its future was concerned, Pithole began to show signs of decay. The production of oil decreased to a mere nominal figure; fire swept away whole streets of the town. A general collapse took place, and at date of present writing the main portion of the remaining buildings, including the Chase House, have been removed to the new oil field of Pleasantville, to build up a similar city with like results.

This and the discovery of new oil fields at Bennehoff Run, and the Stevenson Farm increased the daily production to an average from 10,000 to 12,000 barrels, which has been the average yield for the last three years.

The excitement caused by the first wells struck, was but slightly diminished by the decline in prices, and other unfavorable causes. Each succeeding year has brought a fresh crop of operators, eager to invest their skill and capital in the venture. An ever bounteous Providence has likewise each year supplied a new oil field, giving better promise than any that had preceded.

Commencing at Titusville in 1859, the tide of development swept over the valley of Oil Creek, and along the Allegheny River, above and below Oil City, for a considerable distance. Cherry Run, in 1864, furnished the first subsequent excitement. Then came Pithole Creek. Bennehoff and Pioneer Run, the Woods and Stevenson farms, on Oil Creek, near Petroleum Centre, came, in like succession, in 1865 and 1866. Tidioute, or rather Dennis Run and Triumph Hill, was a promising candidate for public favor in 1867, and in the latter part of the same year, Shamburgh, on Upper Cherry Run, made its brilliant debut. For 1868, the Pleasantville oil field furnished the chief excitement. And so the development progressed. Each year furnished a new extent of territory for operation and investment, and we suppose will continue so, until all the available territory is drilled over. When the wave of reaction rolls back, more system and economy of management will ensue.

The exodus of skilled operators from the older oil fields to the new ones, caused a perceptible decrease in the production of the latter. The production of the new fields, consequently, does not more than maintain the daily average. Hence, with the new fields of Shamburgh and Pleasantville, each of which reported a daily production of over 2,000 barrels, we find the daily average for 1868 to foot up but little more than 12,000 barrels. This we give from the most careful estimates. In order to keep up this production, each month of 1868 has shown an average of about 300 new wells in process of drilling. We cannot, of course, in the present stage of the development limit the amount of production that may be obtained by more extensive development. This will be regulated, to a material extent, by the ruling prices, and the increase of consumption which regulates the whole. The prices have, of course, exercised a ruling influence upon the extent of development of each year. A statement of these and their effects, will give the reader a clearer conception of the business in all its different bearings.

The first oil obtained from the well of the Pennsylvania Rock Oil Company, in 1859, sold at fifty cents per gallon. By July, 1860, the price at the wells had declined to seven cents per gallon. In October, it was ten cents per gallon. It advanced steadily until January 1st, 1861, when it reached twenty-five cents per gallon. It continued at about that figure until toward the end of February. On the 1st of March, it was fifteen cents; on the 18th of March, ten cents per gallon. The decline continued with the increasing supply, until just before the discovery of the large flowing wells. In the Summer, the price had fallen to five cents per gallon or two dollars per barrel. The sudden and immense increase of the production almost entirely destroyed its value. Thousands of barrels were allowed to run to waste, and the sales made in August and September were as low as fifty, twenty-five, and thirty cents per barrel, some sales, during July, being made at ten cents per barrel. Sales were made at forty cents per barrel, in October, November, and December, and some sales at thirty-five cents per barrel. In January, 1862, the price had advanced to five cents per gallon.

Nearly all these purchases proved unprofitable. The great flood of oil reduced the price at the seaboard to nine cents per gallon in May, 1862, causing a loss to the seller of the first cost of the oil, and from one to two dollars per barrel in addition.

The want of demand in the home-market, caused the enterprising merchants who had ventured in the trade, to seek in Europe for new fields for its consumption. The first shipments abroad were made in October, 1861. The exports for that year were 1,112,476 gallons, or 27,812 barrels. When this flood reached the European ports in the Summer of 1862, the same destruction of values took place that had occurred in this country. Parties who had bought in New York, at nominal prices, suffered heavy losses. But the article had been forced upon public attention; and, although for a time, the markets were overstocked, the way was opened for future and increasing demand.

The suspension of specie payments in the Spring of 1862, and the subsequent rapid advance in the price of gold, greatly contributed to reanimate the oil business, speculation revived, and, in October, the price was carried as high in New York as fifty cents per gallon. By December it had receded to twenty-five cents per gallon. During the year 1863, it ranged between eighteen and twenty-five cents. In 1864 the advance in price was rapid, from twenty-nine and a quarter cents in January, to fifty-six cents per gallon in July. It continued high with some fluctuations until January, 1865, when crude was selling at forty-nine to fifty cents. The advance in the price of gold and exchange, which began in the Spring of 1862, and continued until it reached its maximum in the Summer and Fall of 1864, soon carried up prices to a point at which the oil would pay all expenses of transportation, and give the producer a profit of from three to seven dollars, and at one time even ten dollars, per barrel at the wells. This soon stimulated development. Wells that a few months before were unprofitable, owing to the preceding low prices, became of immense value.

Speculation in oil lands, and the organization of stock companies, under the laws of the different States, followed to an enormous extent. Lands were bought at high prices, and resold again at advanced rates, or were revalued by the holders at an immense advance, and used by them in the formation of stock companies, by means of which the stock was sold wherever the spirit of speculation had been sufficiently excited by flaming prospectuses, newspaper reports of sudden fortunes made by the operators, or other means used for the purpose. Many of these companies were fraudulent, but the majority, we are willing to believe, were honestly organized, and conducted with integrity. The capital stock of these companies, over 1,000 in number, as stated in their prospectuses, was as near as we can approximate, about $600,000,000.

The amount of capital thus withdrawn from other pursuits, or from the savings of the community, and applied to the purchase and development of oil land, cannot be accurately estimated, but is supposed to have exceeded $100,000,000.

The speculation was at its greatest height when our generals, using the immense resources placed by the country at their disposal, brought the rebellion to a sudden close, and restored to something like its true value, the national currency which, under the apprehension of the failure of the Union, had depreciated as low as forty cents on the dollar. Gold, which compared with the currency, had been worth over 250 per cent., and even as high as 285, declined to 130. The returns from the shipments of Petroleum and the currency price at home, were reduced to a great extent, and as the cost of transportation and other expenses continued with but little change, the effect was again disastrous upon the business. The great flood of March, 1865, destroyed a large amount of oil and other property, numbers of wells, and reduced the production. The war tax, of one dollar per barrel, which went into effect in April, 1865, still further depressed the business, and disheartened those engaged in it. The exposure of the fraudulent nature and worthlessness of many of the oil companies added to the depression.

In the meantime, prices began to advance abroad, owing to a large increase of the consumption, a generally increased demand, and diminished supply. A large number of wells were sunk by the companies organized the Fall and Winter previous. Although most of these proved unprofitable, owing to various causes, chief among which was the lack of experience on the part of those entrusted with their management, or insufficiency of means, some few were highly profitable.

The total production of Petroleum for the year 1865, was estimated at 2,830,000 barrels crude, which, added to the stock on hand, January 1st, 1865, was equivalent to 2,312,000 barrels refined, of which the home consumption was 839,000 barrels, the exports, 677,000 barrels, and the leakage, 154,840 barrels, leaving a stock on hand, January 1st, 1866, of about 640,000 barrels refined. The average price per barrel of crude on Oil Creek, are given as follows:

| CURRENCY. | | GOLD EQUIVALENT. |
|---|---|---|
| 1862 | $1.15 | $1.02 |
| 1863 | 3.25 | 2.24 |
| 1864 | 8.13 | 4.00 |
| 1865 | 6.71 | 4.27 |

The following general remarks are given respecting the course of the trade during 1865, from a carefully prepared report, made at its close:

"The production was exceedingly light until summer, in consequence of accidental causes. These being removed, a steady increase was gained later, which carried the daily yield to a full average of any of the previous years. Nevertheless, the supply in our markets was inadequate throughout; during the first half of the year, on account of the light production; later, because there were no facilities to forward the quantities gained from the new oil fields, except at an expense too high to be generally incurred. The demand for home consumption was steady and increasing; for export, very light during the first five months (the shipments for 1864 having been excessive), but later, large and too heavy for the limited supply on our markets. The value of the article was steady during the first part of the year, later, continually advancing, and during the last months it was sustained at a height which had in previous years been but occasionally reached by speculative movements. The average value of the article during 1865 was about twenty per cent. higher than in 1864, although our paper currency appreciated on an average about twenty-nine per cent. as against 1864.

"After the production has, during 1865, been above twenty per cent. larger than in 1864, while the demand for the article has increased about twenty-five per cent., we have now ample stocks left in the country, and if the yield of the wells continues large, the supply will, therefore, probably be for the current year, plentiful if not excessive, and a reduction of our present exceptional prices necessary. Under these circumstances it is more to be regretted that the cost of producing and forwarding the oil is now very high, far above the rate at which we will in future be able to export our over supply."

The above sets forth some of the principal reasons for the falling off in prices in 1866. The daily production for the Fall and Winter of 1865 and 1866 was estimated at about 12,000 barrels. By May, 1866, this production had fallen off about one-third. This falling off was mainly attributable to the rapid decrease of the production of the Pithole oil field.

Prices ruled steady for 1866, ranging, for the average, from four to seven dollars per barrel. The development was considerable, owing to the efforts made by the numerous oil companies to develop their lands.

The impetus given the trade during the previous years of 1864 and 1865, carried it through a greater portion of 1866, notwithstanding the many premonitions of a general collapse. Early in May, 1866, the Government tax on crude was repealed, but the abatement of the tax failed to induce higher prices. The bursting of the oil bubble was imminent. One company after another failed to meet the expectations of sanguine stockholders. The prices, however, were remunerative.

Considerable anxiety was manifested by holders of all kinds of property to realize on the same. As a consequence, a rapid reduction of the values of the two preceding years took place. A flood in the Spring, a number of very destructive conflagrations in various parts of the Oil Region, amounting in all to a loss equivalent to from $4,000,000 to $5,000,000, expedited the general collapse. The oil companies fell, in the Fall and Winter of 1866 and 1867, one after another, like a row of bricks. Thousands who had purchased the stocks were overwhelmed in the ruin that followed. Its reaction upon the Oil Region itself, was terrible; all classes were affected by it.

The construction of the Oil City and Pithole Railroad, from Oil City to Pithole, a distance of about fifteen miles, along the Allegheny River, the Reno and Oil Creek Railroad, from Reno, midway between Oil City and Franklin, to the upper end of Cherry Run, crossing the valley of Oil Creek at Rouseville, and the extension of the Meadville Branch of the Atlantic and Great Western Railway from Franklin to Oil City, were among the notable events of 1866. The completion of these roads, and the extending of the Oil Creek Railroad, with its important connections at Corry, down the valley of Oil Creek to Shaffer, and the laying of a number of oil pipes for transportation of the oil from the wells to the various shipping points, changed the entire mode of transporation. Thousands of teamsters and boatmen were thrown out of employment. These classes constituted a large element of the consuming part of the general population, and all branches of domestic trade dependent thereon, were rendered unprofitable and speedily abated. The transportation of oil from the wells to the various shipping points was dependent chiefly upon teams in former years, while the Allegheny River furnished the main outlet for the product to Pittsburgh, the principal oil market. The average load of oil to each team was from five to seven barrels, the weight of a barrel of oil being about 360 pounds. The cost of transportation by this mode was from one dollar and fifty cents to three dollars and fifty cents per barrel, according to distance. In some cases higher rates were paid. The introduction of more reliable, cheaper, and rapid transportation sensibly affected prices, but this was more than compensated by placing the general trade upon a more permanent basis.

The year 1867 was one of general depression and low prices. Thousands of acres of the lands of defunct oil companies, and scores of engines, as well as a vast amount of machinery, were sold by the sheriff of Venango County for debt and taxes. A general thinning out of superintendents, operators, and business men, from all the principal localities took place. Pithole with its numerous hotels, vast expectations and large productions, which had shown visible signs of decay in 1866, vanished into thin air. Other magnificent embryo cities followed, or, rather, had preceded it. Oil was sold at the wells as low as one dollar and twenty-five cents per barrel, and the highest price attained during the year, was four dollars and eighty-five cents per barrel. The price of refined was reduced in like ratio. The low prices induced the building of iron tanks at Oil City and other shipping points, of a capacity from 6,000 to 15,000 barrels, for the storing of the oil. The amount of iron tankage thus constructed reached over 1,000,000 barrels in 1868, fully half of which is at Oil City.

The building of the Farmers' Railroad from Oil City to Petroleum Centre, a distance of seven miles, along the valley of Oil Creek, connecting at Petroleum Centre with the Oil Creek Railroad, early in the season, afforded cheap transportation to Oil City. At this point connections were made with the Atlantic and Great Western Railway, and with the Warren and Franklin, which company had obtained possession of the Oil City and Pithole Railroad, forming connection at Irvincton, their terminus, with the Philadelphia and

Erie Railroad, and affording abundant transportation to all the principal points, East and West.

The Allegheny Valley Railroad, which had been commenced at the beginning of 1867, was finally completed to Oil City during the Winter of 1867 and 1868. This road, extending from Pittsburgh to Oil City, following the course of the Allegheny River for the entire distance, completed the railway circle, by furnishing connection with the only required outlet. Late in the season, the Warren and Franklin Railroad obtained control, by lease and purchase, of the Farmers' Railroad, and consolidated it with their own, and commenced to run shortly afterward over the Oil Creek Railroad to Corry, connecting with the Philadelphia and Erie Railroad at that place, thus forming a complete circle of railway through the most productive portions of the Oil Region.

The sacrifices made in the sale of property by the sheriff during 1867, and the Winter of 1868, was very great. Engines, the first cost of which was from $2,500 to $3,000, sold as low as twenty-five and fifty dollars each. Other machinery and even the lands, were sold at similar rates, in porportion to their first cost.

The new oil field at Dennis Run, adjacent to Tidioute, which at its maximum reached a production of from 2,000 to 3,000 barrels per day, increased, or, rather kept up the aggregate production to a fair figure, while the low prices stimulated consumption and export. The cheap rates at which engines and other machinery for oil wells, as well as skilled labor could be obtained, invited a large amount of capital, and preparations were made for extensive development in 1868. The prices ruling during 1867, owing to the greatly reduced rates of transportation, enabled oil producers to weather the storm, while those engaged in business, established on a legitimate basis, by the adoption of a system of rigid economy, managed to get through the season without serious loss. The reaction, and the consequent reduction of former values to their real worth, was a serious matter to many, but upon the ruins of the exploded oil bubble the business was being built up on a permanent basis.

At the commencement of 1868 prices ruled quiet at $2.70@$2.80 per barrel, closing for the month firm at $2 @$2.15. The market appreciated steadily during the subsequent months until June. At the commencement of that month there was a material advance, and prices reached $4.50, and at the close of the upward movement, $4.90@$5. The highest figures attained during the month on the Creek was $4.90@$5.25, and at Oil City, $5.00@$5.25. During the balance of the year, prices alternated beween $3.75 and $4.90 per barrel, and occasionally reached $5@$5.10. Owing to the improved means of transportation, and cheaper rates of machinery, labor, and the uniform successes of the general development, the general business for the year 1868, the first upon the new basis, may be stated as a highly prosperous one. All classes engaged in the trade reaped an ample share of the rich harvest. The foreign demand for export and the home consumption required about the entire product, the stock on hand January 1st, 1869, being scarcely 500,000 barrels.

The best commencement the business has yet had is shown at the commencement of 1869. Prices advanced to $5@$6 per barrel at the commencement, and sales during subsequent months were at a higher figure. All branches of business have been re-invigorated; and the season of 1869 promises to be one of unexampled prosperity, and the development will exceed that of any of the preceding years.

# CHAPTER VI.

## GEOLOGICAL AND PHYSICAL FEATURES OF THE OIL REGION.

The geology of the oil country has not been as satisfactorily explained, as could be desired, owing to the fact of no complete survey having ever been made, except the State Survey made by Professor Rogers, previous to the Petroleum development. Subsequent eminent geologists have expressed various opinions, and made partial surveys, resulting in fragmentary information. This has been published in such a form, that we are unable to give the credit that is due, for the reason that several uncompleted investigations, or, rather, investigations confined to certain localities, have been made. Professor Ridgway made a partial survey of the valley of Oil Creek several years since. Professor Rogers' report is, after all, the most reliable authority on the subject, and his survey was made previous to the oil discovery. Therefore we give from the best sources we can get the following abstract report, published a few years since, as the best yet made:

"The geological strata of Oil Creek and vicinity has been described by eminent geologists to consist of conglomerates, slates, and shales.

"Conglomerate rock is made up of pebbles, mixed with more or less sand, and all cemented into a close, hard rock. These pebbles, vary in size and quality in various localities, being usually of quartz, though sometimes of sandstone; and they are found from the size of a pea to that of a goose-egg, and occasionally even, though not in our portion of the State, with a diameter measuring four to five inches. They have evidently been formed into the shape in which we now find them, by the action of the water tumbling, pushing, and rolling them together, and sweeping them along, by which the sharp angles they must have had when first torn from their native bed, have been broken and worn away by attrition, until they present the well-known smooth and rounded form, which so distinctly characterizes them.

"The conglomerates of this vicinity belong to what is called the vespertine formation in the Pennsylvania Survey. It is found *in situ*, or in its native bed, only upon the tops of the highest hills, but pieces of it, which have been broken off by their own weight, after the softer rock beneath had been decomposed and washed away, are found scattered over the hillsides, sometimes in immense masses or blocks, which are so enduring as to defy the action of the elements, and bear record in their ruins of the former conditions and changes which their more yielding neighbors, the sandstones and slates could not survive.

"As found here it is not coarse, the pebbles being rarely larger than hickory nuts; and they become smaller as we trace this formation westward, while the opposite will hold good if we go eastward. The accompanying vespertine sandstones and slates also become finer in their textures, and the whole formation becomes thinner as it spreads westward from 2,600 feet on the Susquehanna River to not over 100 to 150 feet on Oil Creek.

"From this thinning down of the mass towards the west, and a corresponding change in the texture, from coarse to fine, we are led to believe that the materials from which the rocks of this formation are composed, were derived from a continent lying on the east or northeast of the Appalachian range, previous to their upheaval; and that these materials, after being brought down to the sea through the channels of rivers flowing west or southwest, were distributed to their present location by the powerful ocean currents that were subject, doubtless, to laws similar to those which govern our present great rivers of the sea.

"For a familiar illustration take a long mill-pond or a lake with a creek flowing into it at one extremity and out at the opposite—the creek will bring down, especially at the time of a flood, large quantities of loose stones, pebbles, sand, black mud, or vegetable mould, and blue mud or clay; and it will dispose them over the bottom of the pond or lake in the order in which we have named them; that is, at the upper end of the lake, at the mouth of the creek, will be found the large stones, then the smaller ones or pebbles, then, as the current became less, the black mud was deposited, and finally, the blue clay, which the water held longest, and carried farthest; and the beds will be found to become thinner as they become finer in texture, thus corresponding to the conglomerates, sandstones, slates, and shales of the New York State and Pennsylvania formations of the secondary series."

"It will be readily inferred from the above, that a sandstone is only a very fine conglomerate; also that black carbonaceous slates may be attributed to vegetable origin; and that argillaceous shales or the soapstones of the Oil Regions, are derived from clayey formations.

"The vergent series of rocks, so-called by Professor Rogers, is immediately below the vespertine, and it corresponds to the Chemung and Portage groups of the New York State geologists. This formation consists of sandstones, slates and shales, interspersed with the sandstones in their layers, varying from five to fifty feet in thickness, while the slates and shales are found in immense deposits, sometimes 800 or 1,000 feet in thickness. To this series, doubtless, belong the sandstones, slates, and shales which appear in the bluffs along the valley of Oil Creek, throughout its entire length; also the first, second, and third sandrocks of the wells, with their intervening slates and shales, as far as the drill has yet penetrated, and how much deeper it extends is unknown. It is not improbable the fourth sandrock of Pithole corresponds to the third of Oil Creek, and, in fact, the later developments, where the third sandrock has been found at the depth of 750 to 900 feet, in localities adjacent, proves this, as well as that the first at Pithole is identical with the one found above the bottoms along the bluffs of our valleys; although it is by no means impossible that the continuity does not exist, for the reason that there may have been causes operating at the time when these rocks were deposited, which produced local changes and variations of greater or less importance. A third sandrock is found on Church Run, while no trace of one is found on the flats around Titusville, even at the depth of 1,200 feet; and the third sand, or great oil bearing rock of the valley of Oil Creek, disappears at the upper end of the Foster farm, and we have not learned that any has been found in any portion of the valley above."

The rocks above described are stated to belong to the paleozoic rocks, because containing the most ancient remains of ancient animal and vegetable life yet discovered, stretching all the way between the gneissic formations beneath and the lowest of the coal deposits above. Sometimes they are denominated "fossiliferous," "sedimentary," or "secondary" rocks.

"In Pennsylvania," says Professor Rogers, "this class has been deposited during all the four earliest periods of the great European divisions, namely, the Cambrian, the Silurian, the Devonian, and the Carboniferous. No traces of the fifth, or Permian group, have as yet been discovered in North America. * * * The prolonged succession of sedimentary action ceased with the close of the Cambrian system, being terminated by the upheaval of the ocean, in whose broad bed and around whose margin these deposits had collected."

The same eminent authority states that the vergent series abounds in the remains of marine vegetation, and also, that the aggregate thickness of all the rocks belonging to the classes above described, measured at their greatest depths, is not less than *thirty-five thousand feet!*

Those of our readers who are fearful that the several sandrocks of to-day may eventually become exhausted of their deposit, can surely take courage from this last statement, for it is possible that oil-bearing sandrocks may exist at intervals for this entire distance. Improved machinery and a cable of length and weight will be required, in such wells. What language would be adequate to describe the feeling of reverential awe that would fill the frame of the oil miner of some future period, as he commenced to "sand-pump" from the last one of this series at the above-mentioned depth!

There is nothing very remarkable about the physical features of the Oil Region. The most productive portion has been aptly described as in the shape of an irregular quadrangle, each of its sides being from twenty to thirty miles in length, the axial line nearly corresponding with the course of Oil Creek. Subsequent developments have changed this to nearly an oblong square. The general face of the country is rugged and hilly. As far as present development extends, operations have been confined to the valleys of Oil Creek and the Allegheny River, and a portion of its northeastern slope. The Allegheny is the principal stream, and flows nearly through the centre of Venango County, but from the peculiar structure of the land, it runs towards every point of the compass in its course. According to EATON, Franklin is elevated about 750 feet above Pittsburgh, so that there is a fall in the Allegheny of five and a third feet to the mile. From Franklin to Meadville, about thirty miles by the course of French Creek, there is an ascent of 130 feet, or four and a third feet, on an average, to the mile.

Oil Creek, from Oil City to Titusville, a distance of about seventeen miles, is estimated to have a fall somewhat greater than French Creek.

The main producing portion is an elevated table land, sloping toward the lakes and Allegheny River, with the summit at or about Pleasantville—the hills become higher in descending the valley of Oil Creek, from 150 feet in height just below Titusville to 450 to 500 feet at Oil City. The approaches to the Allegheny River from Meadville to Franklin, consist of a grade, or downward slope. Starting from a nearly level country, the same phenomena are perceptible as along the valley of Oil Creek, heights gradually increasing in altitude until the objective point, the Allegheny is reached.

The principal streams are the Allegheny River; French Creek, which enters the Allegheny River at Franklin; Oil Creek entering the Allegheny at Oil City, seven miles above Franklin. Besides these are numerous other small streams. The tributaries of French Creek are Patchell's Run, Sugar Creek, Mill

Creek, and Deer Creek. The tributaries of Oil Creek are Cherry Run, Cherry Tree Run, and Cornplanter Run. Pithole Creek is sixteen miles above Franklin; Hemlock, twenty-one miles; Horse Creek, eleven miles; Tionesta, thirty miles.. Above these are East, West, and Little Hickory. Two Mile Run, is two miles above Franklin. Below Franklin, are East Sandy, Big Sandy, and Scrubgrass.

Abundance of good water prevails. Springs gush forth from the hillsides, and even from the tops of the highest hills. The proportion of good farming land is small, the general surface of the country being unfavorable for extensive agricultural operations. Venango County was taken from Allegheny and Lycoming Counties, by an Act of Assembly, passed March 12, 1800, and organized for judicial purposes by Act of April 1, 1805. In 1839 its proportions where somewhat lessened by the organization of Clarion County from a portion of its eastern territory, and still later by the addition of a portion of its northern territory to the new county of Forest. The county now forms an irregular figure with many angles; and contains as estimated about 800 square miles. The population in 1800 was 1,130; in 1810, 3,060; in 1820, 4,915; in 1830, 9,470; in 1840, 17,900; in 1850, 18,310; in 1860, 25,044. The present population, transient and resident, we should estimate at from 75,000 to 100,000.

The climate is in the highest degree healthy, although the seasons are somewhat irregular. The general humidity of the atmosphere from the heavy Spring and Fall rains, is not conducive, so far as our observation has extended, to any material extent, to rheumatic and pulmonary complaints. Bilious and typhoid fevers prevail during some seasons, though generally in a mild form, and never as yet to any great extent. To the city resident, worn out with toil, the pure mountain breezes of the Oil Region are bracing and invigorating. Many who came here in impaired health have, after a short residence, attained a degree of robust vigor they have scarcely ever possessed before. This fact will be apparent to the stranger who will observe the appearance of the active residents of Petrolia. The health of those engaged around the oil wells, who are constantly inhaling the pungent odor of Petroleum, fresh from its hidden depths, has long since passed into a proverb. The degrees of cold and heat are similar, in most general respects, to that of elevated table lands in all portions of the world.

# CHAPTER VII.

## TRANSPORTATION—EARLY AND LATER MODES OF.

THE principal business of this portion of Pennsylvania, during the earlier years, was the preparing and bringing to market the lumber obtained from the Upper Allegheny and its tributaries, which were bordered by heavy pine forests.

The lumber prepared during the season was conveyed to Pittsburgh, and other points along the Ohio and Mississippi, by making the logs or sawed material into large rafts, and floating them to their destination whenever the water in the river was high enough for successful navigation. The heavy Fall and Spring rains rendered these seasons the most auspicious. The headwaters of Oil Creek were one of the chief lumbering localities, and a large number of saw-mills were erected thereon. In order to get the lumber to the Allegheny from its tributaries, when the latter did not afford a sufficient depth of water for the purpose, the usual practice, as among lumbermen generally, was to dam up the water at the head of the stream sufficiently to furnish an artificial rise in the stream, when it was let loose, for the entire length of the same, the lumber and logs being floated down on the artificial floods thus created. In order to do this, coöperation on the part of mill-owners and all parties interested, was necessary

The *modus operandi* was as follows:

A general understanding was had among the lumbermen and mill-owners, and a day fixed on. At the specified time, the sluice gates of all the upper and lower dams were raised, and the water allowed to accumulate at the main dam, a short distance below Titusville, the water-way, or sluices of all these being so constructed as to allow the passage of lumber, logs, &c. The lower or main dam was so arranged, by means of strong uprights, called "brackets," or "splash-boards," that on a given signal they could be cut loose, and the entire accumulation of water allowed to flow into the narrow stream of the Creek, furnishing an average depth of about three feet in the shallowest places, and having a duration of several hours, generally three to four. The arrangements having been completed, on the specified day the rafts were all collected in the various basins above the main dam. The signal was given, the brackets or splash-boards of the main dam cut away, and after waiting for a few minutes, until the water had attained full headway, in order to prevent the rafts from running ahead of the "rise," in which case they grounded, or became fast, upon the first shallow place, furnishing a fair target for all who followed, to crash against or run over—the lines were cast loose, and guided by long oars, with sweeps fastened on each end, and in charge of competent pilots and ample crew, the rafts started on their journey, swiftly borne along on the swirling waters.

Oil Creek has an average width of scarcely sixty yards, and it required no small degree of nerve and skill in the management of these unwieldy crafts, to avoid collision and great loss of property, as well as imminent danger to life and limb. The sound of the rushing waters, bearing upon their surface the numerous rafts of boards and logs, the long sweep-blades flashing in the sunlight as the stalwart raftsmen pulled starboard and larboard, or right and left, as the case required, around the bends of the Creek, the occasional wreck of some devoted craft, demolished in a moment by those behind acting as huge battering rams which no human structure could successfully resist, not unfrequently followed by a general "jam," the shouts of direction and excitement resounding among the rugged hills that lined the banks of the stream for its entire distance, furnished a scene that was grand in its picturesque novelty, and of excitement to the participants in the highest degree exhilarating.

The journey to the mouth of the Creek being successfully made, the sections that had been thus floated, were fastened together in large rafts, and in charge of a competent river pilot and crew, embarked for the ports along the Ohio and Mississippi, that furnished the chief markets. The raftsmen not required to form the regular crew, returned to the forests again, and there remained until a succeeding occasion required their services.

Such, in brief, was the pond-freshet of early days; and in all general respects, those used by oil operators in later years to convey their oil from the wells to Oil City, the principal shipping point, were similar, except that the pond-freshet was furnished exclusively for oil boats, and these were some distance below the lower or main dam.

The mode adopted by early operators, and in use to present date, of storing or saving the oil as it came from the wells, was by means of wooden vats or tanks, cylindrical in shape, made of jointed pine planks, and hooped with iron, ranging in capacity from 100 to 1,200 barrels each. In the days of the leviathan spouters, these, in localities, were so numerous as to cover acres. The transportation furnished at the present date, saves to the operator the expense then incurred, one or two tanks, of moderate size, being all that is required, each farm having one or more large iron tanks for permanent or transient storage.

The only mode of land-carriage at the date of which we are writing, consisted of hauling the oil from the wells to Oil City and the other shipping points, by means of two-horse wagons. The gross weight of a barrel of oil averages about 360 pounds. From five to seven barrels of oil constituted a load for a wagon. The main and by-roads, when much used, especially in wet weather, soon attained a depth of mud that, in any other locality but this, would have been deemed impassable, being composed for the most part of the debris of the sand-stone that becoming loosened from the hill-sides, rolled down upon the level flats at their base. This soon became decomposed and incorporated readily with the alluvial soil of the latter, affording but slight foundation for the heavily-laden wagons that thronged them. The oil dripping from the barrels, mixed thoroughly with the dirt of the road-bed, the travel of horses and wagons kept it from drying, thus insuring, in perpetuity, the most disagreeable, pasty mass that ever man or beast forced a way through.

Oil Creek mud attained a fame in the earlier and subsequent years, that will ever be fresh in the memory of those who saw and were compelled to wade through it Teamsters and horsemen swore both loud and deep at it. Newspaper correspondents exhausted all their adjectives, epithets, and expletives in essaying to give a faint description of its demerits. Weary pedestrian pilgrims, (the original "carpet-baggers,") like Bunyan's Christian, were inclined to part with their knapsacks after a brief experience; ministers of the Gospel and devoted laymen, earnestly desired sustaining grace while urging their weary beasts over and through it Mud, deep, and indescribably disgusting, covered all the main and by-roads in wet weather, while the streets of the towns composing the chief shipping points, had the appearance of liquid lakes or lanes of mud. When it is taken into consideration, that from 1,000 to 2,000 laden oil teams traversed the main street of these daily, a sufficient reason for its existence is manifest.

The moving of a few thousand barrels of oil from the wells to the different shipping points, never over ten to twelve miles distant, and generally from three to five, was a herculean task. Broken wagons, oil barrels, filled and empty, lined the roads during a rainy season. Tankage at first was expensive and the prices of oil low. The general price of oil barrels was from three to four dollars, even when oil sold at the wells at ten and twenty cents per barrel, the contents scarcely averaging one-twelfth of the value of the package containing it. At the commencement of the oil development, the oil wagon furnished the only mode of transportation from the wells to the shipping points. Though slow and annoying to dealer and shipper, it was the best that offered, and continued to monopolize the carrying trade for some time. But finally a material change was brought about. In what age of the world's history has the genius of man failed to be equal to the emergency! Some one, unknown to the present historians, and, consequently, lost to undying fame, to whose memory the oil trade should erect a suitable monument, solved the existing difficulty, and provided the means of practically carrying out his conceived ideas. The plan was, in effect, the employment of flat-bottomed boats, to hold the oil in bulk or barrels, and the employment of the pond-freshets to float the said boats, when laden with oil, from the wells to that great-

EARLY MODE OF TRANSPORTATION – MOUTH OF OIL CREEK IN TIME OF HIGH WATER

ly desired haven, Oil City, at the mouth of Oil Creek. The idea was speedily adopted. Flat-bottomed boats were procured from the Upper Allegheny, and from every other point where they were built, or were brought up the river from Pittsburgh. These consisted of boats of every size and kind, from large barges, and metal boats, used to convey pig-iron from the furnaces to the manufactories, of a capacity of from 1,000 to 1,200 barrels, to the diminutive "guiper," as it was called, a small flat boat, holding from twenty-five to fifty barrels. When prices ruled low, oil was shipped in bulk, that is, run into boats and barges. When prices ruled high, shippers did not care to risk so much, and used barrels. At first the oil was run into open boats. This plan was soon found to be impracticable, from the fact that any slight agitation or rocking motion of the water, set the large body of loose oil in motion, and often capsized the boat, spilling out the greasy cargo upon the surface of the water of the Creek or the Allegheny, along which it floated to the Ohio, or even beyond, to the intense disgust of the worthy burghers of all river cities that obtained their water supply from said streams.

To obviate all such accidents, compartments, water-tight, were built across the oil barges, each compartment of a capacity of from 80 to 100 barrels, the same securely decked over, and the obstacle surmounted. The late Mr. Richard Glyde, of Pittsburgh, built the first boat of the kind, applied for, and secured a patent for the application of this principle to oil boats. But he found it impossible to enforce his claims, and soon gave up the idea of collecting the moderate per centage claimed upon all oil so run, as a royalty for the use of his patent.

Arrangements were made with the mill-owners at

the head-waters of the Creek, for the use of their surplus water at stated intervals. A large fleet of boats, suitable for both river and creek, were soon collected, and ample transportation furnished for all the wells adja cent to the Creek. The boats were towed up the Creek by horses from the mouth of Oil Creek, at Oil City, to various points on the farms along its banks. When laden with oil from the wells, they were floated to Oil City by the pond-freshets, in the same manner as the lumber-rafts before described, with all the attending incidents and accidents, but far greater loss, owing chiefly to the nature of the cargo, and the frail material used in the construction of the boats.

The mill-owners furnishing the water used for pond-freshets, were paid by the shippers for the same. The prices for the several years averaged from $100 to $250 each, and in exceptional instances $300 to $400 were paid for a single pond-freshet. This amount was collected from the shippers by a superintendent, employed for the purpose, and who had general charge of all the arrangements; a tax of a few cents per barrel upon the shipments being levied for the purpose, the amount varying according to the price paid for the freshet, the sum paid the mill-owner and the superintendent's salary, being the only expenses incurred.

A large number of boats were employed in this branch of the business. The amount of oil brought out upon one of these pond-freshets averaged from 15,000 to 20,000 barrels; the highest amount ever brought out of the Creek upon a single freshet, not exceeding 40,000. The number of freshets had depended upon the season, price, and the stage of water in the river. When the latter was in navigable order to Pittsburgh, the number was greater. From two to three per week was the average then; in other seasons one each week, or less, as the occasion required. The class of boats employed in the transportation of oil from the wells on the Creek, were smaller than those used on the river, the cargo of the Creek boats being transferred at the oil wharves at Oil City.

In the best days of boating, the number of these on Creek and river, was estimated at 1,000. Besides these were an average of from twenty to thirty steamers, passenger, and tow-boats. All the passenger boats carried oil as freight, while the tow-boats towed down the full boats and brought back those that had discharged their cargo of oil, these being generally laden on the return-trip with empty barrels, coal, and acids, for refiners' use. The average price of oil boats and barges ranged from seventy-five cents to one dollar per barrel of their tonnage capacity. The value of the steamers employed in the trade averaged from $25,000 to $30,000 each. The entire oil fleet furnished employment to about 4,000 men. The number of teams employed, at the height of the business, in 1864 and 1865, was not less than 4,000, employing more than double that number of horses. The number of horses employed in towing, both upon the river and Creek, was also large. No one would like to believe in the doctrine of the transmigration of souls, after witnessing the hardships of the tow-horses. Compelled day after day, to toil against the rapid current of Oil Creek, attached to heavy boats and barges, in all seasons when that stream was not closed by ice, their lot was, indeed, a hard one. In a very short time the hair was removed from their legs, by the action of the greasy water, which also caused a species of eruption on their limbs, very painful to the poor beasts. A pair of horses lasted but a short time, and their carcasses were frequently seen floating on the stream. On land, the destruction of wagons, teamster's morality, and horse-flesh, was equally great. But as the price of a team was frequently earned in a week, the loss of these animals was not deemed a serious matter. The teamsters brought their horses and wagons from this and the adjoining counties, and even from adjoining States.

Rich, indeed, was the golden harvest reaped by teamsters and boatmen in those days. They made their own terms, regardless of the ruling prices of the product, never failing to take advantage of the necessities of the shipper. As a consequence, the latter generally had a light margin left, after all expenses were paid. But a far less amount of capital was required to transact business upon than now, the producer, as a general thing, being willing to sell on reasonable time, giving the buyer a chance to realize on the product, and pay for it out of the proceeds. Confidence in each other was the ruling trait of those engaged in the business. The disastrous reaction following close upon the wild speculation of 1864 and 1865, that swept over the oil country in 1866 and 1867, destroyed, to a material extent, men's confidence in each other's solvency, and the financial failures of those years caused the adoption of a strictly cash system in the buying and selling of oil, as well as in all other business transactions.

The pond-freshet of the oil boats was a novel and interesting scene. The number of boats generally averaged about 200, and the distance from Oil City to the Noble Well, the highest point on Oil Creek to which they were towed, about eleven miles. The larger class of boats were generally from 80 to 120 feet in length, by 15 to 20 in breadth; the smaller ones from 20 to 50 feet in length, and of corresponding width. All these boats were of light draught, generally from fifteen to twenty inches, and never exceeding three feet. As before stated, public notice was given of each pond-freshet, and the boats towed up the Creek, loaded and made ready. The first notice had after this, was the rushing flood. Then it was that all the nerve of the boatmen and pilots was called into play. When a fair start was had, but little difficulty occurred. But a great anxiety prevailed among the boatmen to be among the first to start, and thus avoid the danger of collisions and "jams." A laden boat, starting with the rise in a stream, will always run ahead of the water, and ground fast on the first shallow place. This not unfrequently happened during the pond-freshets of Oil Creek, and was the general cause of the numerous accidents that occurred. The boat thus aground was generally swung around by the force of the stream broadside to the current, where it soon filled with water and sunk. As the boats coming after had but little steerage way, even a small boat thus aground presented a serious obstacle, unless it should be totally demolished by the first collision. If the boat were of large size, a French Creeker, for instance, built with heavy oaken gunwales, that projected like a prow from each of its ends, an effectual barrier was formed, the size of these boats, much used on the Creek on account of their strength, being such as to occupy a greater portion of the channel. Against this obstacle the advancing boats dashed with great force, the weaker ones becoming splintered from the concussion, and the stronger ones being wedged fast, in the order which they came, and thus formed what is known among raftsmen and boatmen as a "jam."

The smaller boats were run over and sunk by the larger ones; the weaker, crushed by the stronger boats, were splintered, the surface of the Creek deluged with their contents, and in a few moments a large number of boats of all sizes were entangled in an almost inex tricable mass. The confusion, dire and indescribable, that reigned for a time, was only equalled by that of a battle. Finally, a passage would be effected by some of the leading boats that were intact, and those that were not disabled followed, perhaps to meet with similar disasters during the remainder of their brief voyage. In cases where the fall of the water, caused by the freshet, was rapid, the boats in the "jam" remained, and the oil had to be taken out of them, that is, all that had not been carried off by the current. Several narrow places, where islands occurred in the Creek, the old Forge Dam, at the Clapp farm, the pier at the M'Clintockville Bridge, the pier of the bridge at Oil City, and the sand-bar at the mouth of the Creek, were localities where these "jams" generally occurred. The centre piers of the bridges, and the bar formed at the mouth of Oil Creek, being considered the most formidable. The boatmen frequently miscalculated their distance, missing the current which formed the channel between the pier and abutments, were forced broadside upon the pier of the bridge, where they remained, similar to a pair of old-fashioned saddle-bags, being soon broken in twain by the force of the current or advancing boats. When a boat thus fastened, was laden with bulk oil, or oil in bulk, the cargo, in a very few moments, was emptied into the stream. A "jam' frequently ensued from this cause.

The freshets were seldom had without loss, in cases amounting to thousands of barrels. The freshet of May, 1864, resulted in a loss of from 20,000 to 30,000 barrels. Upon several occasions, the ladened boats at the upper end of the Creek broke loose at the first rush of the pond-freshet torrent, sweeping the banks clear of all those fastened below, breaking their strong cables

POND FRESHET JAM AT MOUTH OF OIL CREEK, MAY, 1864.

like pack-threads, making a general wreck along the Creek for its entire length. When the wrecked boat contained oil in bulk, the cargo proved a total loss to the shipper, and if the boatman insured its delivery, he lost his trip; if in barrels, the greater portion was generally recovered, the boatman, in all cases, having to bear the loss or damage to his boat. When the water had receded after the freshet, the boat if not seriously damaged, was raised and repaired for the next freshet. When this was not possible, a new one was purchased. In the latter years, shippers employed their own boats. Pond-freshet days were regarded by the residents of the Creek as a general holiday and a frolic. All who could spare the time came down on the boats. These, together with the boatmen, would frequently number from 1,000 to 1,200. The influx of so large a crowd in the course of a few hours, gave to Oil City a busy appearance. The wharves were crowded with boats; the bar-rooms, saloons, and streets thronged with the hurrying multitude, and thousands of barrels of oil were bought and sold in a few hours. After a brief sojourn, the producers and boatmen returned to their homes and avocations, and the town was again comparatively quiet.

Many of the terms in use among boatmen were highly expressive. Prominent among these was the term "snubbing." When necessary to check the headway of a boat, one of the crew jumps ashore with the line, which he makes fast by passing the rope round a tree or stump, and pays out the line gradually, the peculiar hitch allowing the line to revolve or slip, at the will of the boatman holding the end of it. The headway of the boat was thus checked gradually. The operation is one requiring both skill and nerve, especially when it was desired to check the headway of a heavily laden boat or a raft. It was done usually in order to effect a suitable landing, or to lower a craft from one locality to another a short distance below. The crew of a Creek boat was from three to five, according to the size, and on river boats a larger number were required. The average rates of freight from the wells to Oil City, was from fifteen to seventy-five cents, and occasionally one dollar per barrel, according to distance. By flat-boat, from Oil City to Pittsburgh, the average during the year was from forty cents to one dollar per barrel. In times of extreme low water, the rates were higher. By steamboat, fifty to seventy-five cents per barrel. The general duration of the pond-freshets was from three to four hours.

The wagon-train of the Oil Region merits something more than a passing notice. During all seasons the roads were thronged with them. Summer's scorching rays and winter's fierce blasts, caused but slight diminution in their numbers. Though the prices charged by them were generally higher than the market rates of oil would justify, their labors were severe, and we do not think, as a class, the teamsters were overpaid. In the height of the Pithole excitement, the enterprising citizens of Titusville constructed a plank road from that place to the producing wells of Pithole by way of Pleasantville. General Avery, of military fame, organized a wagon-train that was employed between the two places, and realized handsomely from the venture. The greater number of these teams were drawn from agricultural pursuits in this and the adjoining counties, greatly interfering with farming operations. Neither wagons nor boats are now used to any material extent. The first scream of the locomotive whistle, and the introduction of iron pipes to convey the oil from the wells to the shipping points, rendered their further use unprofitable.

It was not long before the immense freight traffic of Oil Creek and the Allegheny River, attracted the attention of the managements of the various railroads centreing toward the Oil Region. The Atlantic and Great Western Railway first built a branch road from their main stem, at Meadville, to Franklin, a distance of about thirty miles. Their main line made its Eastern connection with the Erie Railroad at Salamanca, with the Philadelphia and Erie Railroad at Corry, and its Western connections were made at Cleveland, O. The Oil Creek Railroad projected about the same time, connecting with the Atlantic and Great Western and the Philadelphia and Erie Railroad at Corry, was built from the latter place to Titusville. The Oil Creek Road gradually extended its line down the valley of Oil Creek, reaching Shaffer in 1865, and its final terminus, at Petroleum Centre, in 1867. In the Fall and Winter of 1865 and 1866, the Atlantic and Great Western commenced the extension of the Meadville and Franklin branch of their road from Franklin to Oil City. During the same Fall and Winter, the Warren and Franklin Railroad Company built their road along the valley of the Allegheny to Oleopolis, at the mouth of Pithole Creek. The Oil City and Pithole Railroad Company also built their road, from Oil City to Pithole, connecting with the Warren and Franklin Railroad at Oleopolis, and extending up Pithole Creek to the producing wells. About the same time, the Reno and Oil Creek Railroad, was built by the Reno Oil and Land Company. This road crossed the valley of Oil Creek at or near Rouseville, and from thence up Cherry Run to Plumer. The grade was completed to Pithole, but the failure of the company prevented its completion beyond the point named. All these roads were in operation at the opening of the Spring of 1866, the Atlantic and Great Western having been completed to Oil City in March of that year. The interests of the general business were materially advanced by the com pletion of these roads. Early in 1867, the Farmers' Railroad, extending from Oil City to Petroleum Centre, a distance of about seven miles along the valley of Oil Creek, was built by some New York capitalists. This completed the circle, and the Pennsylvania Oil Region had now a direct connection by rail, with all the principal railroads East and West, North and South, and with the commercial cities of the country. In January, 1868, the Allegheny Valley Railroad, originally projected from Pittsburgh to Franklin, and which had the year previous been finished to the mouth of Mahoning on the Allegheny, was completed to Oil City. The Warren and Franklin Railroad Company, which had sometime previous obtained possession of the Oil City and Pithole Railroad, purchased the Farmers' Railroad, and took possession of the same, January 1st, 1868, and arrangements were made with the Oil Creek Railroad, for the use of their track to Corry, and thence, by Philadelphia and Erie Road, to Irvineton, the terminus of the Warren and Franklin Railroad. The Warren and Franklin, Farmers' and Oil Creek Roads, have lately been consolidated under the name of the Oil Creek and Allegheny River Railroad. A third rail allowing the passage of broad gauge cars, has lately been laid on the line of the Farmers' and Oil Creek Roads, thus giving a decided advantage to the Atlantic and Great Western and the Erie Roads. The railroad of the Cranberry Coal Company was also built in 1867, from their mines to Oil City, a distance of four and a half miles. The Jamestown and Franklin Railroad is now being extended to Oil City. This will be an important road for the interests, as it comes through the immense coal beds of Mercer County, adjoining Venango, and also furnishes a short route to Cleveland as another outlet for the product.

The advent of railroads in such rapid succession, soon revolutionized the mode of doing business. A pair of wooden tanks, of about forty barrels' capacity each, mounted on a platform car, and securely fastened, furnished the means of transportation for oil from the wells to Oil City and the other shipping points, as the same could be taken direct to the distant cities of New York and Philadelphia, whence the export trade is chiefly supplied, without change of cars.

The general depression of all kinds of business, and consequent low prices of the Petroleum product, in 1867, was the chief cause of the commencement of the storage business in the Oil Region. Experience of former years had satisfactorily demonstrated that tanks of boiler iron, securely rivetted, similar in shape to the wooden ones then in use, but of larger capacity, afforded the best means of preserving oil securely from fire, leakage, and rapid evaporation. Early in 1867, the building of iron tanks was commenced. They are of cylindrical shape, composed of boiler iron, in sheets, well rivetted, varying in capacity from 6,000 to 20,000 barrels each. Top and bottom are of iron or wood, generally the latter. When the top is of wood, it is set in just below the edge of the tank, and from four to six inches of water kept constantly upon it. Stop-cocks at the bottom allow the egress of the oil or water. The oil is pumped in and out of them by powerful steam pumps. Switches are built to those of the tanks not directly on the line of the railroad, or connection made with iron pipes. The oil from the tank cars is run direct into the receiving tanks and pumped from these into the iron tanks, and is loaded into tank cars by the same means, a platform, supporting the supply pipes, being erected at a proper height alongside or near the iron tank. Pipes have also been laid across the Allegheny, connecting with iron tanks on the Venango City side, or direct with the cars of the Allegheny Valley Railroad.

The amount of iron tankage in the Oil Region at the present date is a few hundred barrels in excess of a million. Its distribution will be found in our statistics. The qualities of the iron tanks for resistance to fire, have been often and amply tested. A tank of this kind, belonging to Fisher Brothers, of Oil City, was subjected to the most intense heat from a burning tank of oil a few feet distant from it, built of wood and sunk in the ground, for eleven hours, neither contents nor tank sustaining serious injury, the iron tank being nearly full of oil. The tank itself was frequently red hot on the side next to the fire. So intense was the heat, that the wooden supports of the roof of the tank were charred as if by action of fire. One or two tanks of this kind have been destroyed by fire, but they were nearly empty. Those that were full, and had their tops securely fastened, have so far, come out of the fiery ordeal unscathed. The iron tanks are also free from leakage, and allow but a slight degree of evaporation, as compared with the old-fashioned wooden ones.

Owners of oil farms have erected one or more of these tanks on most of the farms, in which are stored their own, and the oil of the lessees, until it is sold. Iron tanks, for both crude and refined, are used on the different railroads. Shipments are made in these from the wells or refineries of the Oil Region, direct to the eastern cities. By means of these and other facilities, the transportation business, so far as oil is concerned, has been reduced to a science, as it were, rendering loss almost impossible, and insuring economy and the greatest possible despatch.

No material product of the country has a more comprehensive and perfect system of transportation by railroad, oil pipes, and by water, than has Petroleum. This system, too, is becoming better each year, keeping pace with the necessities of the trade. The railroads traverse the entire Oil Region, forming as it were a circle, having four outlets: At Oil City, by the Allegheny Valley and Atlantic and Great Western; at Corry with the same and Philadelphia and Erie; at Meadville, by the Atlantic and Great Western; and at Irvineton, by the Philadelphia and Erie. The Allegheny River, likewise, furnishes ample means of transportation to Pittsburgh during seasons of high water, at which time it is a formidable railroad rival. In former years, before the railroad lines were completed through the Oil Region, oil was shipped exclusively in barrels. As a consequence, fully one-third of the capital of the shipper was thus invested. The present mode of shipment, by means of tank cars, dispenses with this expense, and also saves in bulk, the tank car having an average capacity of 80 barrels.

Oil pipes for the transportation of oil from the wells in localities where railroads have not yet been constructed, are now laid from all such localities to the shipping points on the railroads. These consist merely of a two-inch iron pipe laid in a trench a few feet in in depth, in the ground. Where the gravitation is insufficient, powerful pumps are placed at proper distances, which forces the oil through. No difficulty has yet been found in running the oil at all seasons. In the coldest weather, crude oil, like alcohol, does not freeze, at least we have never known of such an occurrence. Elevation or distance, seem alike powerless to prevent its successful working. At the time of the adoption of this system of oil transportation, it was proposed to lay pipes from the wells to the principal cities, and thus have a complete monopoly of the carrying trade. Still later, the project has been seriously considered by capitalists, of conveying the surplus gas of the wells to New York, Philadelphia, and other cities, to be used in place of the illuminator now furnished from coal gas. Petroleum gas will furnish a better and far cheaper light than is now furnished, and the supply is inexhaustible. The project of conveying it from the wells to any given distance, is feasible, in fact it can be easily done. It could also be used as fuel for manufactories, and for domestic purposes, in cities or densely populated districts.

The Oil Pipe Company ships the oil from gauge-tanks located on the farms, contracting to deliver at their tanks, the amounts as shown by the gauge of the same at their receiving tanks upon the railroads, subject to the shipper's order. At many places small rotary pumps are attached to the engine, and the oil pumped, as it comes from the well, to the gauge-tanks. The average rate charged by the Pipe Company is twenty to thirty cents per barrel. The system works admirably. Some complain that the Pipe Company manages to secure a monopoly in the selling market, and generally has the fixing of the market rates. But such will ever be the case where capital controls, and results in no material disadvantage to the general business.

In the foregoing pages we have endeavored to show, from the commencement, the means of transportation used to convey the Petroleum product from the wells, the place of production, to the principal markets of this continent. The rapid progress made in this essential branch of the business, is one of its most remarkable features. The magnitude of the shipment can be seen by reference to the statistical portion of this work. Never have the necessities of a trade, as regards means of transportation, been more promptly met. The aggregate capital stock of the different railroads built for its almost exclusive transport, amounts to from $25,000,000 to $30,000,000, and the investment has thus far proved one of the best of the kind in the country, netting to the different lines enormous profits. The capacity of the different railroads is about equal to the production, and can be increased to any required extent. The business now is in the ninth year of its age, and six railroads, having a length in the aggregate of over 200 miles, have been already built for the transportation of the product to the principal markets. And so far from exhibiting any signs of culmination, it does not seem to us, at least, to have fairly commenced to grow. Upon every railroad in this country, and in fact those of the greater portion of the civilized world, is the traveller greeted with the pungent odor of Petroleum, as oil trains are whizzed past. The sails of the Petroleum fleet spread themselves over every sea, and never, in so short space of time, has a product, mineral or otherwise, attained such a universal and well-deserved notoriety.

# CHAPTER VIII.

## METHOD OF PETROLEUM DEVELOPMENT—BORING OIL WELLS.

As a guide to man, Nature has in all ages shown the existence of her mineral resources by certain visible indications, or unmistakable signs upon the earth's surface. Thus indicated, mankind have not as yet failed in appliances and skill to develop the same. Coal and iron cropped out of the hills; gold, the blessing and curse of mankind, a blessing by its use, and a curse by its abuse, made its appearance in the sands of streams, and other mineral products in like manner. While, as before described, Petroleum has manifested its existence from time immemorial, by its exudations from springs, streams, the rocks, and even from the earth itself. By a special dispensation of Divine Providence, its general development, or rather discovery, in any appreciable quantity, was deferred to about the middle of the present century. At no other age of the world could Petroleum have been so universally utilized and appreciated.

The artesian system of boring deep wells, had been perfected and practiced long before the discovery of Petroleum or its uses were thought of. Thus at the commencement the oil miner had all the necessary appliances to operate with. The drilling tools and other mining apparatus used by salt miners, in a more modified and simple form, furnished all that was requisite for testing and obtaining the rich deposits of oil that had for so many ages, lain beneath the surface of the valleys and the rugged hills of Venango County. The success of Drake's well, in 1859, proved the existence of the treasure so assiduously sought for, in no stinted supply, and furnished exact information as to the means to be used for obtaining it. Of course, it was unreasonable to suppose that a deposit of this nature was confined to the limit of a few acres or square miles. Wells, similar to Drake's, sunk at various points along the valley of Oil Creek, proved amply, by striking good veins of oil, how unfounded any assumption of the kind would have been. As was natural, the first oil drillers placed great faith in "surface indications," as the outward appearance of oil bubbles in springs, streams, or from the rocks were called. The appearance of a globule of oil on the surface of a stream, seemed naturally enough to indicate the existence of a reservoir of oil in the immediate vicinity, while the flats, forming the valleys, offered the best facilities for mining operations.

Acting upon these hints, the early operators erected their derricks as near the edge of the streams as the nature of the banks of the same would allow, and even built piers or cribs of stone, upon which the derrick and engine-house was erected, in the stream itself. Only a short period of time was required to prove the fallibility of the first surface indications, so many wells located with direct reference to the same having proved failures. The valleys, great and small, formed by water courses, were from the beginning considered the best for oil production. The existence of oil upon the surface of streams, was the main reason for the adoption of this theory. In brief, any apparent surface indication, immediately around a good producing well, was eagerly scanned, and its counterpart, in any general respect, assiduously sought for in that and the surrounding localities.

A good memory often enabled its possessor to attain a high reputation as standard authority, from whom all necessary information required for the selection of a proper site to sink a well with a prospect of successful results, could be obtained. The ancient oil pits were drilled through, but in few cases was the miner rewarded for his trouble. Oil springs, whose existence was co-equal with the first occupation of the soil by the races that preceded the white man, exhibited a similar obstinacy in refusing to shed a single ray of light upon the subject sought for so earnestly. The wells drilled in the streams, likewise, failed to make a favorable response. All the indications so apparently pointing the first operators to successful results, did not prove satisfactory in all general respects. Still they were sufficient to prove a general guide to operators, scant indeed, but sufficient to stimulate those engaged in the business to vigorous effort.

The science of geology was invoked, and has shed more light upon the subject, by defining the position and nature of the different strata beneath the earth's surface, than anything else. Yet even this science, before whose mighty shadow we bow with reverential awe, has not proved a practical guide to the operator in a great majority of instances. The sinking of a shaft several feet in diameter, with drifts into adjacent territory, in different directions, open for the inspection of the human vision, would doubtless be of great utility, in fact, indispensable to successful results. The drilling or boring, as it is generally termed, a four and a half inch hole into the ground for the depth of from 500 to 800 feet, is a different affair. The surface once penetrated by the drill, the balance of the operation is as slightly visible, beyond the exhibit made by debris brought up in the sand-pump, as an opaque body could well be. Though the formations perforated by the drill at the present day, guide with tolerable certainty to the desired depth where the oil is found, it is nothing more than the results obtained in the drilling of wells in other localities, in pre vious years, and is liable to vary, even at a distance of a few feet or rods, much less a distance of several miles. The knowledge of oil mining, subject as the operator is to far less opportunities for the exact defin ing of the depth of the rocks containing the oil deposits, than has been exhibited in previous mineral discovery, cannot, with any considerable degree of accuracy, be acquired only by long years of practical application and experience.

It will be thus seen, that the early operators, with no other light than had been afforded by the drilling of salt wells, had no ordinary trial before them, in commencing and sinking an oil well, and were ready to make use of any theory, that might be advanced, provided it gave evidence of practicability. They well knew that only by experiments, many of which would prove futile, could desired results be finally obtained.

As those who professed to have no faith in signs or indications, and scouted at theory, drilling their wells wherever a suitable location offered, were on the average as successful as those who followed the former to the letter, faith became shaken in oil springs and other so-called "surface indications," and more practical theories were sought for. And strange to relate, whenever a good well, located with special reference to signs, indications, or preconceived theory, was struck, this same class of skeptics, who "did not believe in such things," were among the first to procure leases in close proximity as possible to it, erecting their derricks, so far as practicable, in a line with that of the fortunate well, upon the side which the oil vein was supposed to be.

Hydro-geology, so long known and practised by experts to designate the exact location of water veins below the earth's surface was brought into requisition, and had a host of advocates in the earlier and subsequent years. Numerous wells located by its professors, proved successful, enabling those who made it a regular business, to realize handsomely. Hydro-geology, as it is termed, is the use of the hazel or peach-tree rod to point out the depth of the water veins in the earth, for the purpose of indicating the proper place for digging water wells. It has been practised for as long a period as can be remembered, and of late has been elevated, in France, to the dignity of a science. The manner or mode of operating is as follows: The natural fork of a hazel or peach-tree twig is taken, having its limbs of equal size and length, and the leaves stripped off to the main stem. The ends of the limbs are grasped firmly, one in each hand, with the back of the hands downwards, being at the same time extended from the body. The assertion of its believers is that the existence of oil beneath the surface where the stick or rod is thus held, will be indicated by the revolving of the forked end toward the earth, having the two extremities for an axis. The explanation given by its experts is, that there is a kind of magnetism by which the rod is disturbed and set in motion, and the possession of the same is limited to certain persons. We give the explanation for what it is worth, and are free to confess our inability to see any philosophical reasoning in it. The experiment cost but a trifle, and operators, consequently, were willing to test its efficacy.

Spiritualism, likewise, had its advocates. Several good wells were obtained that had been located, as asserted, by parties who professed to be guided by revelations made by spirits or spiritualistic manifestations. The latest instance of the location of a well by this means, was the Harmonial well, at Pleasantville, one of the first to obtain oil. In this case, the owner of the well states, that some time previous to its location, while on his way from Pithole to Titusville, when at the edge of the town of Pleasantville, his horse stopped, and he was taken from his buggy by some invisible power, through or over a rail fence, into an adjoining field, and set down with no gentle force on the precise spot where the Harmonial well is located, and by some process, only known to spiritualists, caused to understand that he was the fortunate instrument selected to commence the development of the Pleasantville oil field, that has proved the most generally successful one yet operated upon. Shortly after this muscular and spiritual manifestation, the gentleman in question, a highly intelligent man by the way, commenced operations for sinking a well in the precise location thus indicated, and was rewarded by obtaining an hundred barrel well.

As usual, the skeptics were among the first to secure leases around the new well, and every one drilled or sunk, proved successful. Per contra, better wells, in other portions of the same locality, have been obtained, whose owners and projectors had been subjected to no such muscular "manifestations." But so vast and wonderful has been the Petroleum development, that it has afforded an excellent field for the exercise of the exploded superstitions of less enlightened times. It is due to state, that intelligent operators, and these constitute the mass, and have by far been the most successful, have never resorted to devices so unworthy of our civilization, and so contrary to the dictates of common sense.

The progress made in oil mining so far, is truly wonderful, taking all the attendant circumstances into consideration. The large majority, four-fifths at least, of those who commenced the Petroleum development, had no previous experience in mining of any kind. They were composed of all classes of mechanics, business men, farmers, laborers, &c. The sinking of an artesian well, such as we have in the Oil Region, was a work that had to be learned. No exact counterpart was furnished in any of the previous mining experiences of the country. Hence the knowledge so necessary could only be acquired by the expenditure of money, skill, and untiring labor. The prospect of such large gains from a comparatively slight investment, urged them on. Surely no better "School of Mining," was ever known, or had apter pupils.

Mining engineers, who had graduated in the best schools of the old world and the new, came in the subsequent years of the development. But their plans were of too costly and elaborate a nature to admit of general adaptation. None of the companies, much less individual miners, had the capital to indulge in such machinery as is in general use in all other branches of mining. They have contented themselves with the centre-bit, in use for at least a century preceding, and such other appliances as incurred the lightest expense, and combined strength and simplicity.

Numerous devices and inventions have been tested intended for drilling oil wells, without sufficient success to cause even their partial adaptation. The period has not yet come for the general adoption of costly and elaborate mining machinery. When the present reservoirs of supply are worked to depletion, or flooded by water from the numerous abandoned wells, such appliances will doubtless be introduced. The necessity will then have arrived for the adoption of different means than are now in use to obtain the subterranean deposits. The machinery to effect this result will be ready at the proper time, and its conception in some cute Yankee brain has doubtless even now taken place.

Knowledge sufficient has been gained from the development of each preceding year to furnish general principles for the guidance of the oil miner, enabling him to reach the oil deposits with greater accuracy, each succeeding one, until at the present date, save in undeveloped territory, a dry well is the exception and not the rule, as at the commencement.

The original depth of wells in the early years, seldom exceeded 300 to 500 or 550 feet, and some of the largest in amount of production, have been of less. From 700 to 800 and 900 feet, is now the common depth to which wells are drilled on the elevated lands. With this large increase of depth comes a corresponding increase of obstacles to the miner, yet persistent labor and skill overcome these. Heavier drilling machinery is of course required. In the first years of development, the percentage of successful producing wells, scarcely averaged one in twenty. At present five out of every eight is not, we consider, an over estimate. The most attractive site for the oil operator of to-day, is in close proximity to a good producing well, or in a line as near as possible with the same.

The all prevailing oil theory of 1868, the first by the way, that has given general practical favorable results, was that of the "oil belt." It has been proved by successful experiments, during the past season, that the oil-bearing strata (the third sandrock), of the Oil Region, has a direction, or general bearing, from northeast to southwest. The idea originated by observing the general direction presented by the successful wells in the different producing localities. By taking the general direction or bearing of these, the general direction of the so-called "oil belt" can be traced to any desired distance.

Since the adoption of this theory, the success in obtaining paying wells has been uniform, and is fast becoming reduced to a comparative certainty. At Pleasantville and immediate vicinity some 300 wells have been sunk, with direct reference to the new theory, and very few total failures have as yet been reported. This belt, which is supposed to contain the oil-bearing strata, so far as developed at Pleasantville, has a width of from one to two miles, and this width is being gradually increased. In other localities the width seems to vary, occasionally to a few hundred yards. From the hill on which the Wood farm, adjoining Petroleum Centre, is located, this "oil belt" can be seen clearly defined by a line of producing wells, singly and in groups, all having the same general bearing. On the northeast are Bennehoff Hill, Pioneer, and still further, Shamburgh, while beyond the latter can be seen the smoke from the busy oil field surrounding Pleasantville. In a general southwest direction, the line crosses the Story, Tarr, and finally appears on the high bluff of the Blood farm. The recent developments on the bluff of the John Buchanan farm, lower down the Creek, gives the same results, in a similar general direction. Lateral branches, as at Cherry Run, and other localities, seem to occur. Oil operators or miners, now develop in these given directions, with the success we have above stated. A single test well is generally sufficient to show the location of the "belt," but occa-

sionally several are required before the exact position is found. The extreme northeast portion of the "oil belt," so far as developed, is the oil field of Tideoute and vicinity, and the southwestern at Scrubgrass and Parker's Landing, on the lower Allegheny.

It is by such discoveries as these, all in accordance with nature's physical laws, that oil miners have been, from year to year, guided by unerring instinct as it were, to such successful results, as the present large daily production evidences. And the time will come when the theories of the past, wild as they seemed, will be appreciated as the first glimpses obtained of the way to the rich storehouse of mineral wealth, that is now gushing forth from Mother Earth in such bounteous and seemingly inexhaustible stream.

Having thus given in detail, the different theories held and acted upon by the oil miners during the successive years of Petroleum development, we will endeavor to give the reader an idea of the mode and manner in which an oil well is bored or sunk. The term drilled, would probably be more correct, but the other is the generally accepted term, and is more favorably known. In giving a description of this kind, we are aware that it will be a repetition, in sum and substance at least, written on the same subject by previous writers. As the present work would be incomplete without it, its insertion becomes a necessity. Important improvements are constantly being made in minor details.

We will suppose that the site for the well has been selected, and a suitable place made ready for the foundations of the derrick, engine-house, &c. The first step generally taken after securing and preparing the location is to procure an engine, suitable in amount of horse-power, for the effectual prosecution of the work, and transport it to the location aforesaid. Next in order is the erection of an engine-house, derrick, sampson-post, working-beam, jack-frame, and band-wheels, and placing them in proper position. The engines in use in the Oil Region are of every conceivable style, portable, stationary, upright, and oscillating. Portable and stationary are the kinds now in general use. These are from eight to thirty horse-power, principally the former. In former years it was the custom to operate several wells with one of these large engines, sometimes two; never over three. It is now considered more advantageous to have an engine to each well. The motion required to pump one well is generally different from what is required for another. Arranging motion for one well where two are operated with one engine, disarranges it for the other, consuming more time and fuel, than would pay for the extra one. Engine-houses are built of rough boards, and are of sufficient capacity to protect the engine from the weather, contain a forge for dressing tools, bunks for workmen, a small amount of fuel, necessary tools, ropes, &c.

The derrick is a tall frame work, in the shape of a pyramid. They were formerly built of rough poles, or hewn timber, the bottom being from ten to twelve feet square, the poles, four in number, being erected one at each corner, thirty feet in height, converging toward each other, forming a square at the top, of two and a half feet, with girths and braces, at suitable distances to make the structure sufficiently substantial for the purposes designated. Derricks are now built of sawed lumber or boards, two inches thick and from six to eight wide, the two edges being spiked together, forming a half square, on each corner of the foundation, which is from fourteen to sixteen feet square, and in some localities more. The derrick is now put up in sections, being braced transversely as it goes up, in order to secure the strength necessary, until it reaches the proper height, which for deep wells is about fifty-six feet; for shallow ones, less height and lighter derrick is required, and at the top it forms a square of from two to three feet.

On the top of the derrick is put a strong frame work for the reception of a pulley, over which the drill rope passes. The floor of the derrick is made strong by cross sleepers, covered with plank or boards. A roof for the protection of the workmen, is laid with boards across the girths, some ten to twelve feet above the floor. In cold weather the sides are boarded up.

The bull-wheel, as it is called, is a shaft of timber, six to eight feet long, fastened like the shaft of a common windlass, and six to eight inches in diameter, the ends of the shaft banded with iron, and a journal of inch-iron driven into each end for it to revolve upon. Mortices are made through this shaft, eight or ten inches from each end for the arms of the wheel. The wheels are usually made from six to eight inches thick on the face, with strips of plank sunk into and spiked on the outer surface, for the double purpose of receiving the rope-belt, and connecting it with the band-wheel for drawing up tools, tubing, &c., out of the well, and for the workmen to take hold of with their hands, when working it without the help of the engine. The bull-wheel is placed on the side of the derrick next to or opposite the band-wheel and engine, as the workmen may desire. The drill-rope is coiled on this shaft, between the wheels, one end passing from it over the pulley on the top of the derrick, and attached to the tools.

The sampson-post is of hewed timber, twelve to fifteen inches square, and usually twelve feet in height, erected on heavy timbers, framed, crossing each other, bedded firmly in the ground with a mortice to receive the tenon on bottom of post; also a brace on each side, reaching nearly to the top of the post. On the top of this are the irons fitted to receive the working-beam, which is balanced on the top of the sampson-post, admitting of the rocking motion required in drilling and pumping. The working beam is a stick of timber, from twenty to twenty-six feet long, eight to ten inches square at each end, eight by fourteen to sixteen inches in the middle with iron attachment in the centre, fitting to a similar one on the sampson-post. To the end over the well is an iron joint for attaching the temper-screw when drilling, and sucker-rods when pumping. On the other end of the working beam is an iron joint for attaching the pitman-bar, which connects the same with the crank, or band-wheel shaft. The band-wheel, shaft, crank, spider for wheel, pitman, and working-beam, as well as all the other parts of the machinery are shown in the illustrations in this work.

The band-wheel is usually about six feet in diameter with a six inch face, built in various styles, according to the fancy of the builder, and is placed upon a strong frame, built for its reception, called the jack-frame. The jack-frame is secured in position by two heavy timbers, bedded into the ground with gains sunk into them to receive the sills of the jack-frame, to which they are keyed fast. The engine is usually placed from eight to twelve feet distant from the band-wheel, and connected by rubber or other belting. The belting in general use is six inches in width.

When all this is completed, water supply for the engine procured, &c., we are ready to commence operations. The first thing in order is to drive the iron conductor or driving-pipe, as it is generally termed, to the bed-rock, which varies, as will be seen by reference to the body of our book, from six to seventy-five feet,

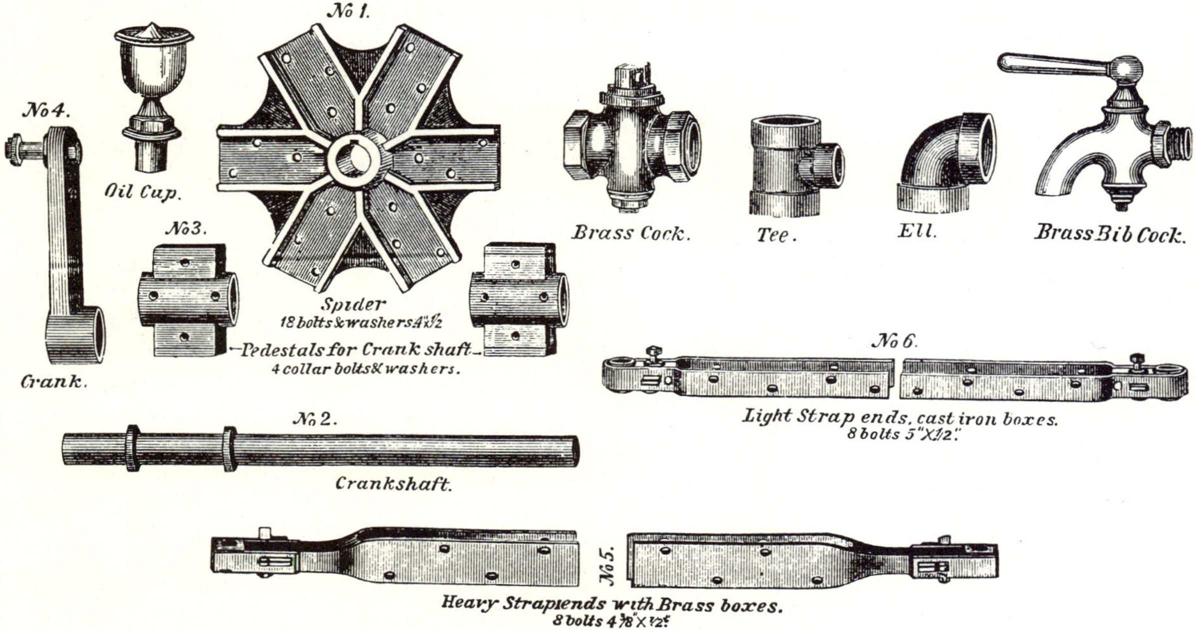
No 4.
Crank.
Oil Cup.
No 1.
Spider
18 bolts & washers 4"x1/2
No 3.
←Pedestals for Crank shaft→
4 collar bolts & washers.
Brass Cock.
Tee.
Ell.
Brass Bib Cock.
No 6.
Light Strap ends, cast iron boxes.
8 bolts 5"X1/2".
No 2.
Crankshaft.
No 5.
Heavy Strapends with Brass boxes.
8 bolts 4 3/8"X1/2".

generally from twenty to fifty feet. This pipe acts as a conductor, and prevents earth or stones from falling into the pit or hole, while the drilling is going on. The driving-pipe in general use is of cast iron, six to eight inches in diameter, having walls of about one inch in thickness, and is in joints nine or ten feet long. The driving of this pipe is a work of difficulty, requiring the utmost skill, since the pipe must be forced down through all obstructions to a great depth, while it must be perfectly vertical. The slightest deflection from a straight perpendicular line, ruins the well, as the pipe acts as the conductor for the drilling tools. The process of driving is simple but effective. Two slide-ways, made of plank, are erected in the centre of the derrick to the height of twenty or more feet, twelve to fourteen inches apart, with edges in toward each other, and the whole made secure and plumb. Two wooden clamps or followers are made to fit around the pipe, and slide up and down on the edges of the ways. The pipe is erected on end between the ways, and held perpendicular by these clamps, and a driving-cap of iron, fitted to the top. A battering-ram is then suspended between the ways, so arranged as to drop perpendicularly upon the end of the pipe. The battering-ram is of timber, six to eight feet long, and twelve to fourteen inches square, banded with iron at the lower or battering-end, with a hook in the upper end to receive a rope. When the whole is in position, a rope is attached to the hook in the upper end, passed over the pulley of the derrick, down to and around the shaft of the bull-wheel. Everything is now in readiness to drive the pipe. The belt being adjusted connecting the engine and band-wheel, and the rope connecting the band-wheel and bull-wheel, called the bull-wheel rope, the machinery is put in motion by the engineer, one man standing behind the bull-wheel shaft, grasping the rope attached to the ram and coiled around the bullwheel shaft, holds it fast, and takes it up in his hands, thus raising the ram to its required elevation when it is let fall upon the pipe, and by repeated blows it is driven to the requisite depth. When one joint of pipe is driven, another is placed upon it and the two ends secured by a strong iron band, and the process continued as before. The pipe has to be cleaned out frequently, both by drilling and sand-pumping. Where obstacles, such as boulders, are met with, the centre-bit is put in requisition, and a hole, two-thirds the diameter of the pipe, is drilled through the same. The pipe is then driven down, the edges of the obstacle being broken by the force applied, the fragments falling into the vacuum created by the passage of the bit. When this cannot be done the whole machinery and derrick is moved sufficiently to admit of the driving a new set of pipe, or abandoned. It sometimes happens that the pipe is broken, or diverted from its vertical course by some obstacle. The whole string of pipe driven has to be drawn up again and the work commenced anew. If this is not possible, a new location is sought.

After the pipe is driven, the work of drilling is com menced. The drilling-rope, which is generally one and a quarter inch hawser-laid cable, of the required length, from 500 to 1,000 feet, is coiled round the shaft of the bull-wheel, the outer end passing over the pulley on the top of the derrick down to the tools, and attached to them by a rope-socket. The tools consist of the centre-bit, auger-stem or drill-bar, jars, sinker-bar, and rope-socket. When connected, these are from thirty to forty feet in length, and sometimes more, weighing from 800 to 1,600 pounds, according to depth required to reach the third sandrock. The process of drilling, until the whole length of the tools are on, and suspended by the cable, is slow. When the depth required to suspend the tools is obtained below the surface, the attachment between the working-beam and drilling-cable is made by means of a temper-screw suspended from the end of the working-beam, and attached to the rope by a clamp. The temper-screw is from two to three feet in length, made with a coarse thread, and works in a thin iron frame, with a wheel at the lower end of the screw for the driller to let out the same as is required. As the drill sinks down into the rock, the screw is let down by a slight turn of the wheel by the driller, some allowing a full revolution every few blows of the bit, others once only in a few minutes, depending upon the hardness of the rock being drilled through.

What is termed the "jars" by oil miners, attached to the auger-stem, play a highly important part in the work of drilling. They are two long links or loops of iron or steel, sliding in each other. Drillers always have about from four to six inches play to the jars, which they call the "jar," and by this they can tell when to let down the temper-screw. With the downward motion the upper jar slides several inches into the lower one; on the upward motion this is brought up, bringing the end of the jars together with a blow like that of a heavy hammer on an anvil, making a perceptible jar. Experienced drillers can as soon as they take hold of the rope tell how much "jar" they have on.

In drilling, the tools are alternately lifted and dropped by the action of the working-beam in its rocking motion. One man is required constantly in the derrick to turn the tools, as they rise and fall, to prevent them from becoming wedged fast, and to let out the temper-screw as required. This is one of the most important duties of the work, requiring constant attention to keep the hole round and smooth. The centre-bit is run down the full length of the temper-screw. The centre-bit is about three and a half feet in length, with a shaft two and a half inches in diameter, and a cutting edge of steel, three and a half to four inches in width, with a thread on the upper end by which it is screwed on the end of the auger-stem. The reamer is about two and a half feet in length, having a blunt instead of cutting edge, with a shank, two and a half inches in diameter, terminating in a blunt extremity, three and a half to four and a half inches in width, by two inches in thickness, faced with steel. The weight of heavy centre-bits and reamers, average from fifty to seventy-five pounds each.

The centre-bit is followed by the reamer, to enlarge the hole, and make it smooth and round. The sediment, or battered rock, is taken out after each centre-bit, and again after every reamer, by means of a sand-pump let down in the well for the purpose. The sand-pump now in use is a cylinder of wrought iron, six to eight feet in length, with a valve at the bottom and a bail at the top, to which a half-inch rope is attached, passing over a pulley suspended in the derrick some twenty feet above the floor, and back to the sand-pump reel, attached to the jack-frame, and coiled upon the reel-shaft. This shaft is propelled by means of a friction pulley, controlled by the driller in the derrick, by rope attached. The sand-pump is usually about three inches in diameter. Some drillers use two, one after the centre-bit, and a larger one after the reamer, the two being preferable. When the sand-pump is lowered to a requisite depth, it is filled by a churning process of the rope in the hands of the driller, and is then drawn up and emptied. This operation is repeated each time the

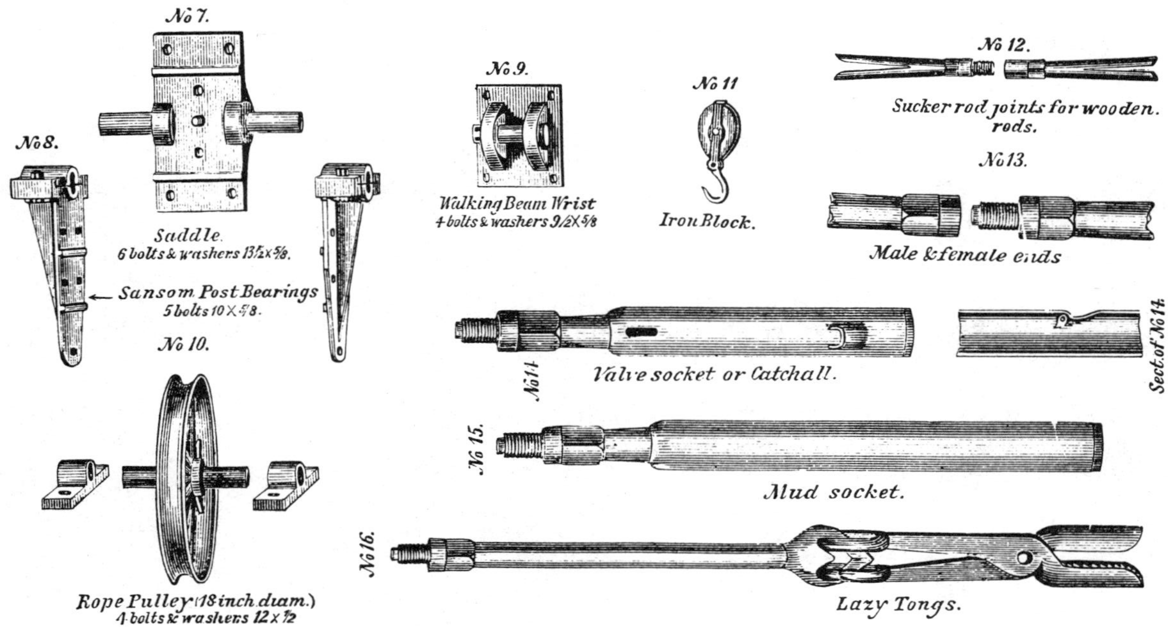

No 7.
No 8.
Saddle.
6 bolts & washers 13½ x 5/8.
Sansom Post Bearings
5 bolts 10 X 5/8.
No 10.
Rope Pulley (18 inch diam.)
4 bolts & washers 12 x ½
No 9.
Walking Beam Wrist
4 bolts & washers 9½ X 5/8
No 11
Iron Block.
No 12.
Sucker rod joints for wooden. rods.
No 13.
Male & female ends
No 14
Valve socket or Catchall.
Sect. of No 14.
No 15.
Mud socket.
No 16.
Lazy Tongs.

tools are drawn up out of the well, the pump being let down and drawn up a sufficient number of times to remove all the drillings. The fall of the tools is from two to three feet. This labor goes on, first tools and then sand-pump, until the well is drilled to the required depth. Abundance of water is found in the wells, both for rope and tools, from the commencement. It flows in from the surface veins, and from the larger ones below.

After the well is drilled, the process of tubing to prepare for testing and pumping, is next in order This is done generally with two-inch wrought iron pipe, previously tested by hydraulic power. This tubing is in joints, from twelve to fifteen feet in length, screwed together by means of a thread on each of its ends, with a close fitting thimble.

The working or pump barrel, is usually from five to six feet in length, made of brass, with a bore of from one and three quarters to one and seven-eighth inches, always smaller than the bore of the tubing. In the lower end of this is placed the lower valve, or standing-box, as it is termed. The working barrel is then screwed to the first joint of pipe by means of a sleeve or thimble, such as attaches the sections of tubing together.

The swivel is now screwed into the other end of the joint of tubing, the tackle-blocks being suspended in the derrick. The hook of the lower block is attached to the swivel, and by means of the rope, the whole is suspended over the well and let down into it until the derrick floor is reached. The clamps are placed across the mouth of the driving pipe under the thimble at the end of the joint, fitting closely, and secured by a ring at the end of the handles. The swivel is then taken out by the pipe-tongs, another joint or section of tubing attached and lowered as at first. This process is repeated until the amount required is in the well, being held securely by the clamps, as before described. The tubing is put down various depths, usually to or near the bottom of the well. All practical drillers keep a record of the rock passed through, sand, pebble, or slate, which is done by preserving a portion of each sand-pumping, its thickness and relative hardness, which is shown by the drillings. This record is essentially necessary, in order to determine the proper place for fastening the seed-bag. The seed-bag is placed upon the tubing so as to come at the depth necessary to cut off the surface water, as well as the heavier veins below from the oil veins further down. The depth generally required, being at the first sandrock, averaging from 150 to 350 feet. This seed-bag is of stout leather, made to about fit the bore of the well, and is from four to six feet in length. The tubing is passed through the seed-bag, the lower end of which is fast at the proper place on the tubing by lashing it with a stout cord. It is then filled with flax-seed, pressed in, to a trifle less than the exact bore of the well, lashed fast at the upper end, and lowered into the well to the desired depth with the tubing. The flax-seed swells in a few hours, closing the hole effectually, when properly arranged, so that water cannot pass down, nor the gas and oil up beyond it. This important appliance is used for the double purpose of shutting off the surface water from the oil veins below, when the well is pumping or flowing, and to force the current of oil and gas up through the tubing to the surface.

In pumping or testing a well, what is called sucker-rods, are used. These are of wood, about an inch in diameter, and twenty feet in length. Ash and hickory are the kind in general use. The sucker or working-

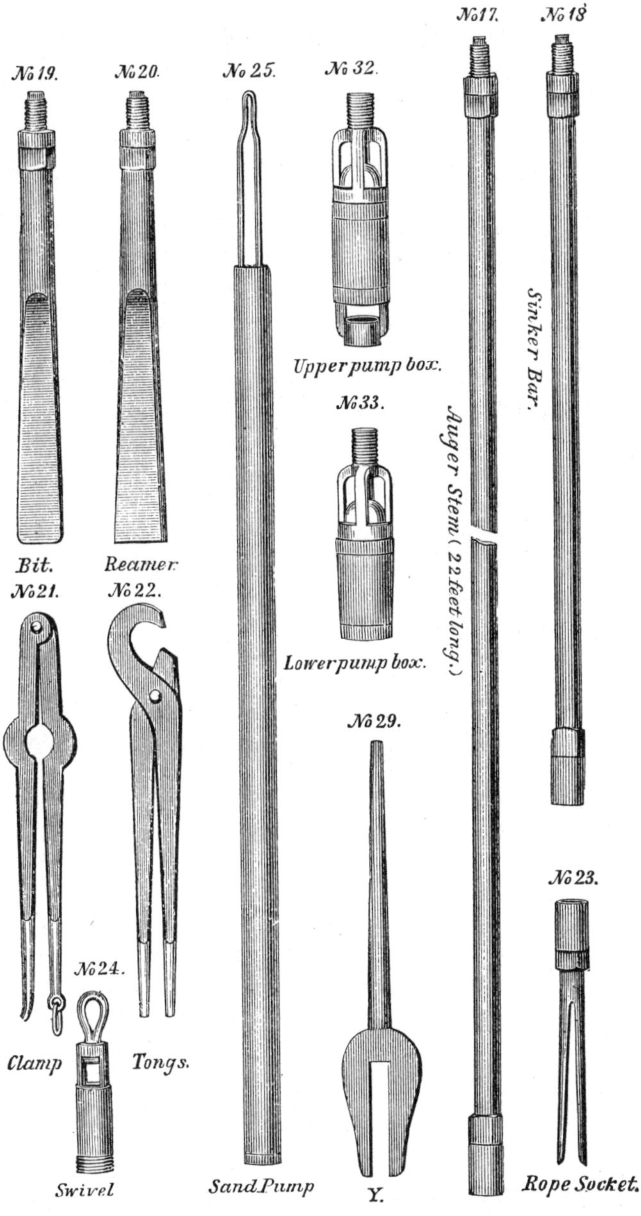

№ 19.
№ 20.
№ 25.
№ 32.
№ 17.
№ 18.
Upper pump box.
№ 33.
Lower pump box.
Sinker Bar.
Auger Stem (22 feet long.)
Bit.
Reamer.
№ 21.
№ 22.
№ 29.
№ 23.
№ 24.
Clamp
Tongs.
Swivel
Sand Pump
Y.
Rope Socket.

valve is attached to the end of one of these, each end of the rod being fitted with a screw-thread and thimble alternating, the same fitting on in the form of a socket, and lowered into the tubing. The rods are lowered one after another until the valve goes into the working barrel. The attachment is then made to the working-beam by means of a rod passing through a stuffing-box, fitted on the end of the tubing, above the driving-pipe. When these arrangements have been completed, the pump is ready to operate, which is done by the machinery in the same manner as in drilling. The sucker at the end of the rod operating upon the standing-valve in the working-barrel. In order that the uninitiated reader may clearly understand this detailed description, we refer him to the engravings, which are sufficiently explanatory.

The process of "casing" the wells is now in general use. This is done by putting three and one-quarter inch artesian tubing in the well to the first sandrock, and placing the seed-bag at or near the same. The two-inch tubing is then put on the inside of the casing without a seed-bag, and can be taken up without danger of flooding out the well with the surface water, as in the old way, by turning the seed-bag every time the tubing required moving.

In early days the conductor, was made of heavy plank, or a log of wood, bored out like a pump-log, the excavation having been made previously to the rock, or as far as the water would permit, by digging a well-hole. The iron driving-pipe now used is not only more economical, but insures greater accuracy in drilling. The tools and general machinery now in use, are of a heavier description than those formerly used, in consequence of drilling deeper wells, and insuring greater rapidity. The drilling of an oil well was formerly a labor of months. Now a well can be drilled and tested in from twenty to thirty days.

At the commencement of the development, operators were content to drill to the depth of a few hundred feet. They employed various kinds of power. The most primitive and novel, was that of the "spring-pole," first employed by the early salt-miners in sinking their artesian wells, which is described as follows: "The spring-pole consisted of a green sapling, some forty feet in length and ten inches in diameter, with the butt end made fast in the ground, or attached to an upright pole. A second post, ten or fifteen feet from the butt, acted as a fulcrum, while the pole passed over the well, and about ten feet above it. The boring implements were attached to this pole, and the 'power' adjusted near its smaller extremity. This was applied by the strength of two men throwing their weight upon the pole. Sometimes a small wooden staging, four feet square, was hinged by one of its sides to the derrick, and the other side suspended to the pole. In this case the two men stood on the staging, and brought down the pole by throwing their weight on the side attached to it, and permitted it to rise by throwing their weight on the side next to the derrick. In either case the spring of the pole brought up the implements, while the downward motion of the pole permitted the stroke. The general term for this method of drilling a well, was termed 'jigging it down,' from its resemblance to the dance so styled."

Another process, by which human muscle was brought into requisition, is spoken of. This was commonly called "kicking down a well." "A short, elastic pole, ash or hickory, ten to fifteen feet in length, was arranged over the well, working over a fulcrum, to the end of which was attached stirrups, in which

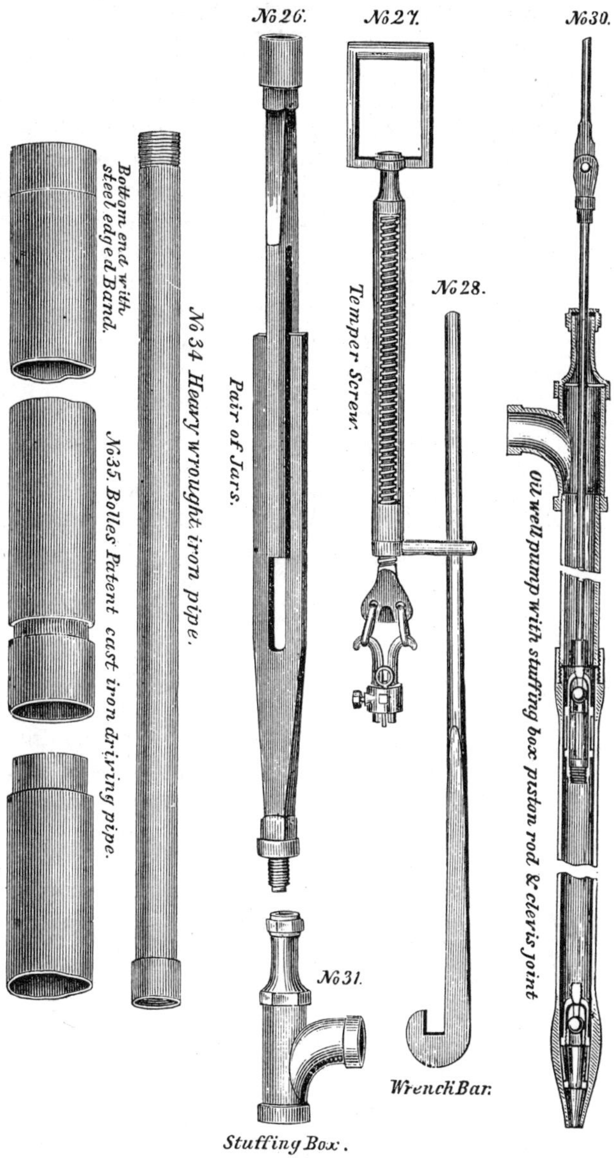

No 26.
No 27.
No 30.
Bottom end with steel edged Band.
No 34 Heavy wrought iron pipe.
No 35. Bolles Patent cast iron driving pipe.
Pair of Jars.
Temper Screw.
No 28.
Oil well pump with stuffing box piston rod & clevis joint
No 31.
Wrench Bar.
Stuffing Box.

two or three men each placed a foot, and by a kind of kicking process brought down the pole, and produced the motion necessary to work the bit. By this process the strokes were rapid. Horse-power was used, of different patterns, suitable for one horse, and sometimes for two and three. They resembled in their general features the horse-power of a threshing-machine, the horses walking around the centre, and over a tumbling-shaft, that gave the necessary perpendicular motion."

"Man-power," answered very well for the depth of a hundred feet, though extremely laborious and slow. The horse-power was a decided improvement. Water-power was used to still greater advantage in many instances, and deep wells have been thus drilled, with but slight cost to the operator. All these expedients answered the purpose so long as oil mining was comparatively an experiment, and the operators financially unable to procure more costly and efficient machinery. The crude inventions of the spring-pole and the "kicking stirrup," furnished ideas for practical minds, that has finally brought the working machinery used at the present day, to a high degree of perfection.

The early contrivances gave way before the advent of the steam engine, and have forever vanished. We have great respect for them, as they constituted the germ which has since budded forth, flooded the civilized world with an illuminator second only to the glorious orb of day itself, shedding its rays into dark places never before rendered visible by any of its successors, and one, too, that is simply as inexhaustible as time itself. It is a civilizer, too, and will eventually accomplish greater changes in manufactures and the mechanical arts, than has any previous mineral discovery, while its almost universal utility is a marvel.

# CHAPTER IX.

## OBSTACLES MET WITH IN SINKING OIL WELLS AND HOW OVERCOME.

HAVING in the preceding chapter made all the necessary arrangements, we will now commence operations, and accompany the driller on his downward journey of from 600 to 900 feet, through earth, slate, shale, and rock. The engine is fired up, and the attachments made, giving the crank motion to the working-beam, which in turn moves the cable and the drilling apparatus. The driller takes his seat on a high stool, above the chosen spot, adjusts the centre-bit or drill with great care, and down it goes through the conductor-pipe, striking thirty to forty blows per minute. Between the strokes the tools require to be moved around, to make the aperture uniform, and to prevent them from wedging fast. With this is also continued a slight downward motion every few strokes, by a turn of the temper-screw. The first operation tells the entire story. Day after day, night after night, the drill is kept moving up and down, cutting from one to six inches and even twelve inches of rock and shale per hour, according to hardness. At intervals the centre-bit is drawn up, badly worn and battered, and a reamer let down to enlarge the hole, and put on the finishing touches, by making it smooth and round, and these are followed by the sand pump.

The first few hundred feet are generally gone through without difficulty, provided all the arrange-

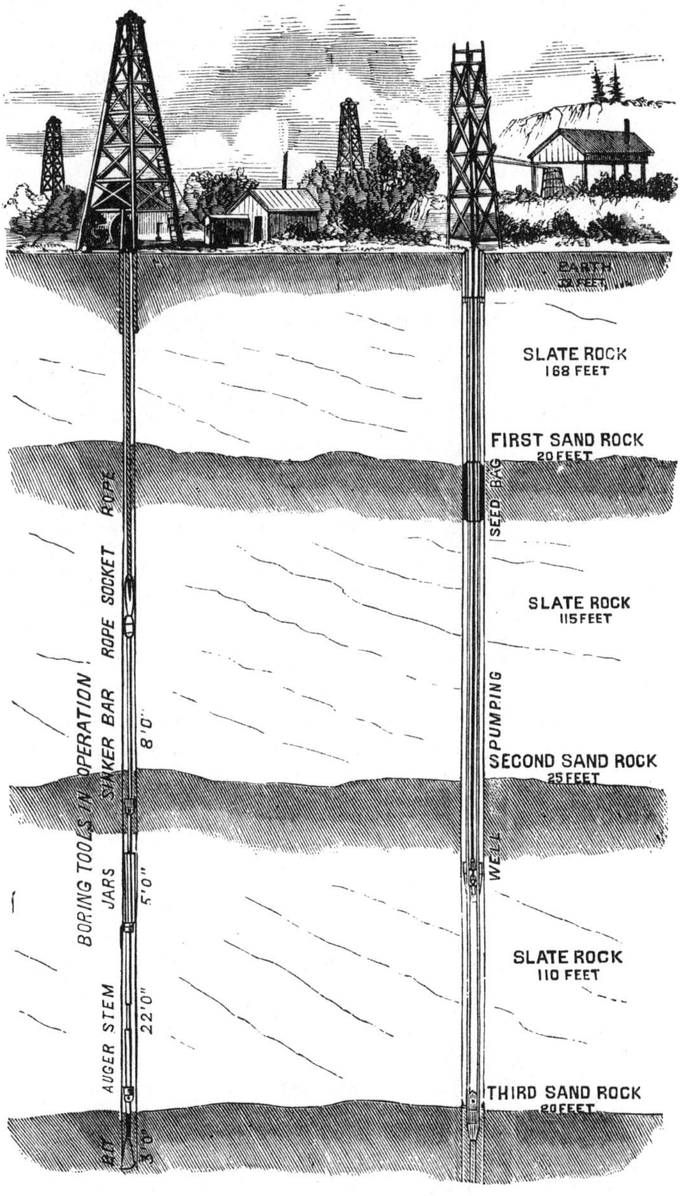


WELLS DRILLING AND PUMPING.

ments have been made with care, at the beginning, and the drillers are skilful. Occasional accidents to machinery will happen, and the best skilled men are but human. Difficulties occur further down, that test to its utmost endurance the most persistent energy. Sometimes they are attributable to a want of caution on the part of the driller, from imperfection in the material of, or improper dressing, or tempering the drill, but more often from circumstances unforeseen and unavoidable. In its passage the drill not unfrequently dislodges gravel or fragments of hard rock, that have a tendency, and often do, wedge it fast in the hole, from which it is only dislodged by the most persistent "jarring." The reamer is also subject to the same mishap, or a sand-pump breaks loose from its rope, and has to be fished up. When the bit or reamer becomes so firmly imbedded as to render its removal impossible by jarring or breaking it in pieces, the well is abandoned. Sometimes a bit or reamer breaks, leaving a piece of hard steel fastened securely in the rock several hundred feet below the surface. Where the fragment is small, it is pounded into the sides of the well, and causes no further annoyance. When it is larger the difficulty is greater, and not unfrequently insurmountable. The bit or reamer sometimes becomes detached from the auger-stem, by the loosening of the screw from its socket. This difficulty is often greatly heightened from the fact that the workman may not be aware of its displacement, and for an hour or two be pounding on the top of it with the heavy auger-stem. Various plans are resorted to in order to extract the fastened tool, and a large number of implements have been devised for "fishing up" the same. Many persons have become so expert and successful, as to adopt this as a regular calling. The first instrument used is an iron with a thin cutting edge, straight, circular or semi-circular, acting as a spear, or to cut loose the accumulations around the top, and along the sides of the refractory bit or reamer, so as to admit a spring socket, that is lowered by means of the auger-stem over the top of it, and lays hold upon the protuberance just below the thread. If the socket can be made fast, the power of the bull-wheel and engine is brought into requisition, and in a great number of cases it is brought to the surface. In the jarring and other operations rendered necessary in cases of this kind, the entire set of tools, forty to sixty feet in length, may become fastened; and cases are of frequent occurrence where two and even three sets of tools have become fastened in a well, as they were successively let down to extricate the first ones. It often happened in previous years, as well as now, that after weeks and months of incessant labor, and a considerable expenditure, the well had to be abandoned after the first few hundred feet had been drilled. The difficulty described is liable to occur at any stage of the work, and its frequency increases with the depth.

In addition to the difficulties mentioned, there is yet another, far more dreaded by the driller. This is what is called a "mud vein." It is a thin stratum of mud or clay, from one to several inches in thickness, generally met with at the depth of from 400 to 900 feet. Mud veins abound in most of the producing localities, and not a few operators regard them as invariably indicating an abundant supply of oil. This mud or clay is of a most tenacious character, is highly annoying to the operator when drilling, and in many cases disastrous. Though not deemed of much importance as an obstacle, in the beginning of the development, the mud-vein exhibits new features in different localities. The mud suddenly flows into a well while the process of drilling is going on, settling around the drill, bedding it as firmly almost as the rock itself. Its presence is often indicated to the driller by the sudden downward pressure on his rope. When drilling on or below it, the workman, when about to withdraw his drill, will have assistance at the bull-wheel, and the instant the working-beam ceases its motion, a few turns will be taken on the wheel, so as to raise the bit above the mud, as it sets almost as quickly as plaster of Paris. Sometimes this mud will flow into the hole for a depth of twenty or more feet, burying, as it were, the entire drilling tools and attachments. This renders the jars useless. By attaching a cutting instrument to sucker-rods, the rope above the sinker-bar is cut, and then a spear-pointed instrument substituted, with which, by means of a light set of tools, the substance around the tools is forced from them, an extra pair of jars lowered, and efforts made to jar the tools loose. The spear is sometimes shaped like a common wedge, faced with steel at the cutting edge, made thin. A half circular instrument, made in similar manner, is also used. The mud-socket, circular-shaped, with thin edge, terminating on the inside with an abrupt shoulder, made of steel, is sometimes forced over the head of obstinate tools, and by holding fast to the shoulder made below the thread of these, enables the workmen to extract them. A large number of appliances have been invented for the dislodgment of fastened tools. These are numerous and complicated, so much so, that a full description would be anything but interesting to the general reader. The main thing sought to have is an instrument that in the first place will remove the material around the top of the fastened implements, to be followed by others acting on the principle of a clamp, sufficiently powerful to allow the jarring of the tools loose, or the drawing of them up.

In later years the occurrence of accidents of this kind has been greatly lessened, from a more intimate knowledge of the obstacles by the driller, and corresponding care in taking all precautions to avoid them. At present a most effective instrument for the dislodgment of tools is in use. This consists of a number of heavy iron rods or bars, similar to an auger-stem, weighing from ten to eleven tons. It can be made of any desired length or weight. It is lowered over the head of the tools, and these screwed fast into a suitable socket made fast to the end of the rods, and is worked from the top. When a set of tools are fast, each separate piece is unscrewed, the apparatus acting as a *left-handed* screw. Each piece, as loosened, is brought to the surface. This is stated to be the most efficient device yet invented, and is in use in all the most extensive developing localities. By applying the full force of the engine, these two and a half inch iron rods are frequently twisted like an augur. They are lowered and raised from the top by jack-screws.

The work of drilling a well is one requiring the exercise of the greatest care and skill on the part of the workman. His operations are mainly below the surface, where the eye cannot perceive the nature of the obstacles and impediments. He has but the narrow bore of the well in which to operate, and cannot at a glance take in the whole state of the rocks through which he penetrates. Patience and skill triumph over every obstacle, and the work that was undertaken in former years, with well-founded forebodings of great labor and financial expenditure, has now become an ordinary venture, with an almost absolute certainty of success. The work is by no means so monotonous as it appears to a superficial looker-on. A register or record of the well is kept, of the various rocks drilled through, each sand-pumping carefully scrutinized, and every unusual appearance of sand, or new feature discovered, carefully noted. Veins of fresh water are met with to the depth of about 100 to 150 feet, seldom below that depth. After this come the oil veins, which are found for the balance of the depth of the well. Veins of gas and oil are struck at different depths, generally in the first and second sandrocks. The appearance of these is manifested in the pumpings brought up by the sand-pump, the gas being in the shape of globules, and the oil, of a dark green color, floating on the surface of the water in the pump, or depositing, after the manner of sediment, when the sand-pump is emptied. Frequently these appearances are exceedingly flattering, the pump coming up nearly filled with oil, seeming to insure a paying well. But these "shows" are deceptive, and never to be implicitly relied on, and the operator continues to drill until the desired "third sandrock," where the largest and most permanent supply of oil has been found, is reached. The drilling being finished, the tools are withdrawn, and preparations made for tubing and testing the well; the tubing, &c., having been already brought on the ground, it being considered of the utmost importance to tube and test the well as soon as practicable after the work of drilling has been completed.

In earlier years, when flowing wells were struck, the tools were thrown out of the well to a great height upon the penetration of the gas and oil crevice below, and a volume of gas and oil, the full size of the aperture, forced up to a height of fifty or more feet, creating a shock, with its first escape, similar to an earthquake, bursting forth with a roar like the 'scape-pipe of a steamer, covering whole acres with its greasy flood. This frequently continued for several days, and sometimes for weeks; the force of the gas being so great as to prevent the workmen from tubing the well. In such cases trenches were frequently made for the oil to run into, and the oil dipped out of these into barrels, boats, or other suitable vessels. In many cases the gas and oil give indication by bubbling up through and out at the end of the conducting-pipe. While many others, among which have been some of the best paying wells ever drilled, exhibited no sign until after they were tested. After the tubing is completed, in the manner before described, the pump is put down; a suitable tank or tanks for receiving the oil having been previously built, within a convenient distance of the well. The size of these tanks, which are made of pine planks, cylindrical in shape, and stoutly hooped with iron, varies, according to the necessity of the case, from 150 to 1,200 barrels. Connection is made with these by means of a pipe attached to the mouth of the stuffing-box.

The well is generally full of water below the seed-bag, from the surface water and veins above or the salt water below. The pump is set to work to pump this off, as also to clear out the well. This operation is called "testing." When the water is thoroughly exhausted, oil generally makes its appearance, in large or small quantities, according as fortune has favored the operator. The depth to which a well is drilled, is generally regulated by the depth of the producing wells in the immediate vicinity, and sometimes by the "show," as it is called, of the oil in the well. It is usual to sink the well several feet below the oil vein, in order to prepare a suitable receptacle for sand, gravel, and particles of earthy matter that fall from the sides of the well, thus preventing the closing up or clogging the oil vein from such causes. The trite old saw, that "all signs fail in dry weather," is amply verified by the frequent experience of the operator at this stage of the proceeding. The presence of oil and gas in fair quantity in the well, does not necessarily insure a remunerative result. The veins may be small and soon exhausted. The lack of any such indication is not proof that success will not ensue. The oil may be held back by the dense column of water upon it, and when it is freed from this, it may respond liberally to the strokes of the pump, or even flow out abundantly without its aid. The testing process is continued until the water is exhausted, and the oil makes its appearance. Where wells are cased, as before described, and the seed-bag perfectly tight, it may be a work of only a few hours or a few days. Where casing is not practised, the duration of the testing generally, though not invariably, occupies weeks, and even months, of incessant labor, and after all, perhaps, results in abandonment. But few operators continue to test beyond a month, unless the "show" of oil is very good. A few inches of rock, or other trifling obstacle, may intervene between the operator and princely affluence. If the oil gushes forth copiously, after due testing, the operator so fortunate becomes possessed of competence at least, and not unfrequently ample fortune. In case of failure the reverse, financial loss, or even ruin, is equally probable.

# CHAPTER X.

## PUMPING AND FLOWING.

After the well is tested fully, occupying days, weeks, or months, as the case may be, subject to frequent delays, caused by re-adjusting the tubing and the seed-bag, occurring in some localities oftener than in others, the bursting of an occasional section of tubing, breaking of sucker-rods, defective working-barrels, and other machinery, and a stream of pure green oil pours into the tank from the conducting pipe, the operator breathes easier. He is tolerably sure, for a few weeks or months, at least, of receiving an income from the yield of the well, that will recompense him for his outlay, and even do better. The anxiety experienced when the first strokes of the pump were made, and the consequent vexations, have vanished; and for the time he indulges in the hope that the foundation of the competence or fortune he has so long sought for, is made, and as for the structure itself, it will come afterward. Some wells commence to pump at the rate of four to five barrels per day, and gradually increase until a maximum production, say fifty barrels, is reached. Others commence at 50 to 100 barrels per day, and begin to decrease in a very short time. The decrease, after a short time, a few months, or more frequently a few weeks, after the production commences, is gradual, and sometimes rapid. But few wells yield their maximum production over two to three months, though

WELL IN OPERATION—PUMPING.

exceptional cases occur. This decrease is doubtless owing to a number of causes that are apparent, and many more that are not. Sometimes it is caused by a seeming exhaustion of the fountain the drill has penetrated, the clogging up of the oil veins by paraffine, so called, and the closing up of the tubing from the same cause, and last, but most frequent, by being flooded by the water from other wells. In case the well is affected by paraffine, it is usual to take up the tubing, which, where casing has been used, is a work of a few hours, and clean it out by the introduction of a jet of steam. A few barrels of benzine poured into the well, effectually disposes of the paraffine that may have colleted round, or clogged up the oil vein. The tubing is replaced, and the work of production goes on again. The paraffine, which is one of the constituents of the product, is of a dirty brown color as it appears in the sand-pump and on the tubing, of the consistency of lard. That of the wells is materially different from the manufactured article from coal, and is doubtless caused by the evaporation and subsequent rapid condensation of the oil coming in contact with the surface atmosphere. It forms upon the outside of pipes on the surface from a similar cause.

When all other expedients fail in resuscitating a well, the tubing is removed, and a torpedo exploded in it. This is a thin cylinder or shell of iron, four to six feet long, and two to three inches in diameter, charged with powder, gun-cotton, nitro-glycerine, or a compound of all of them, generally equal in explosive force, according to size and quality, to from 100 to 1,000 pounds of powder. This is lowered into the well by a wire to a depth necessary to bring it about opposite to the oil vein, or crevice. It is then exploded by means of a percussion cap, placed on the top of the cylinder, a heavy iron weight being sent down on the wire for the purpose. Confined between the walls of rocks, the force of the explosion is necessarily great, the reaction forcing the heavy body of water above out at the surface of the well in considerable quantity. The explosion fractures the rock and clears away any obstructions from the oil vein. The results of the torpedo have been wonderfully successful. Wells only yielding five barrels per day, have been increased by its application to seventy-five and one hundred barrels.

Flowing wells are equally as liable to obstructions of the nature above described, as pumping ones. It is the general impression among skilled operators, that wells should be pumped continuously, from the time of their testing, only stopping for repairs in case of accidents. Flowing wells are, of course, exempt from this operation at first, but are pumped when their production decreases to a material extent. Pumping wells, in amount of daily yield, range from 2 barrels to 250 barrels each. Those which commence with a small yield, and gradually increase, are deemed to be more lasting by some operators. While the average maximum yield of an oil well scarcely averages over sixty days, there are a number of wells, in different localities of the Oil Region, that have yielded almost continuously, for five, and six, and even nine years, the yield being of course proportionably small, as compared with that of the same when first struck. Pumping wells from two to three years old, are quite common. A large majority of moderate pumping wells have a duration of one to two years, depending upon locality, management, and surroundings.

Many of the pumping wells commence, as stated, with a small production, and gradually increase until a maximum is reached. A well belonging to the Lackawanna Oil Company, on Shaffer Run, near Oil City, drilled some three years since, after three months' continuous pumping yielded five barrels per day. This well gradually increased its daily yield until it reached eighty barrels per day, and continued at this rate for a long period. In many of the pumping wells the large amount of gas is a serious impediment. This is sometimes so great as to stop the engine, snapping the tough ash sucker-rods like pipe-stems, involving labor and consequent delay. Gravel and sand play havoc with the valve of the standing-box, at the lower end of the tubing, wearing away the valve, which is generally a ball-valve, thus destroying the suction necessary to impel the oil to the surface. The sucker, at the end of the pump, or sucker-rods, is packed with heavy sole-leather, compressed into rings, that fit closely around the same. These wear away very fast by the action of the water and the oil, the latter being highly destructive to leather, causing it to decay, and have to be removed often. Thus it will be seen that the business of pumping a well is one of continual anxiety to all alike, operator and employee. The larger pumping wells throw a continuous stream of oil, accompanied generally with a moderate amount of gas. The small ones, like other diminutives in the world, are more fussy, and make a considerable splutter of gas and water, with a far less yield of oil. In the smaller class it is not unusual for them to flow at the rate of several hundred barrels per day, for a few moments, churning out the oil at a rapid rate, and then subsiding as suddenly. The intervals are generally regular, the spasms, as it were, occurring in some wells every five minutes, continuing to produce for about the same length of time. The intervals in others are of longer duration, sometimes for hours, when the flow is only for a few moments. The motion required to pump wells varies, according to the best experience had, in almost every well. It is generally desirable to obtain a quick, steady motion on the working-beam, that works the pump, in order to procure the greatest possible amount of suction. The large amount of gas in some wells prevents this, the escaping gas in the tubing acting as an air-cushion against the upper and lower ends of the sucker, rendering the strain too great for the sucker-rods to bear. From breakage and other reasons, all more or less unavoidable, the average number of working-days of a pumping-well, in each week, is about four and a half. In some localities, where the most systematic management prevails, the average is greater.

The wells yielding five barrels per day and under, are pumped by what is called "heads." The well is allowed to remain idle for a greater portion of the twenty-four hours, in order that it may accumulate what is called "a head," by miners, that is, sufficient oil to wholly or partially fill the tubing for a short period, when pumped. This is pumped into the receiving tank, and is generally the work of a few hours. The operator can thus perform all his own labor, if he is a tolerable engineer. In case he is not, one man is all that is required, if he is expert, to pump off half a dozen "heads," in as many different wells, in the course of the day. The expense of production is greatly lessened by this mode, and many of these small operators are accumulating wealth more surely, if slowly, than their more pretentious neighbors, with larger wells, and consequently greater expenses. A number of these small wells also furnish a sufficient amount of gas, which is used for fuel, to run them.

Until within a few years past, bituminous coal was

exclusively used for fuel. Before the advent of railroads, the average price of this at the wells was about fifty cents per bushel, and in seasons of bad roads, more. The quality was poor, and the supply unreliable. When prices ruled low, oil was sometimes used for fuel, and found more economical than coal. The oil was conveyed to the engine by means of a pipe, having the end nearest the fire perforated with small holes; the oil running on an iron pan attached to the bottom of the fire-box under the boiler. The adoption of gas as fuel, was not general until within the last few years; the want of suitable appliances being the principal reason for this. At first the gas from the well was collected by means of a small barrel or tank placed in or near the derrick, which the conducting pipes passed through to the engine-house. From their liability to explode, this plan soon became unpopular. The apparatus now used is so arranged that the gas is taken directly from the well, an extra pipe being attached to the tubing at a slight depth, into which the gas is forced by action of the pump. In addition to the stop-cock, in use formerly, as now, to regulate the supply, a check-valve is also in use, which allows the gas to flow toward the fire, but shuts off any backward flow of the same, thus effectually preventing any accident from fire. The fire-pan is of iron, and the end of the supply pipe, usually T-shaped, perforated with holes to allow the escape of the gas upon the fire, is generally used, though the present plans are being constantly improved on. All the developed localities furnish a large amount of gas, for a time quite sufficient for all the wants. In new oil fields it is used whenever an adequate supply can be obtained As now arranged it is safe, and under the most perfect control of the engineer, who can regulate it at will.

The heat is intense, somewhat more injurious to the bottom of boilers than coal, but this consideration is of minor importance, when its economy, cleanliness, and other advantages are considered.

Petroleum gas burns with a clear, steady light, not unlike that of coal. It is used for illuminating purposes at night, and for the warming of engine-houses in cold weather. Where the gas is furnished by the well of the operator, the only cost incurred for burning is the pipe and apparatus, and it can be used not only as fuel for pumping his well, but also for the drilling of new ones in the immediate vicinity. When the gas jets are lighted at night, their bright reflection against the sky can be seen for many miles. From Oil City and other central points, after nightfall, every point of the horizon is aglow with this bright light from nature's inexhaustible lamp.

We have thus far given the reverse side of the mining branch of the oil business. Of course there is a bright side. A large percentage of the accidents which we have described, are due to causes that are constantly being remedied. Lack of experience upon the part of the workmen, and defective machinery are among the chief causes. Second-hand engines, to say the least, are doubtful economy, unless the same are thoroughly overhauled and repaired by a competent machinist before commencing operations. Tubing of the same kind, is far worse. Many valuable wells, or at least that gave bright promise of proving so, have been ruined from the use of second-hand tubing, while the delay and loss occasioned, have been very great. Operators have at last found that the best is the cheapest; buy new material, and have all their machinery in the most complete order before the work of drilling is commenced. This forethought brings a corresponding reward, but few accidents being met with when this plan is adopted.

The striking of the flowing wells, as has been stated elsewhere, was a new and startling feature in the business. One after another of these leviathans was struck, yielding from 1,000 to 3,000 barrels per day each. The oil flowed spontaneously as it were from the earth, as freely as water and nearly as limpid in appearance. The operators of small pumping wells were greatly puzzled at this marvellous phenomenon. If it were possible to continue the new mode of supply, it was argued that the source would soon become exhausted. If not, the product itself, from the great quantity forced on a limited market, would render prices unremunerative. The large wells continuing to flow week after week and month after month, in but slightly diminished supply, restored the confidence of operators to a great degree. Many of them abandoned their small pumping wells, and went earnestly to work drilling new ones to the third sandrock, where all the flowing wells had found their supply. The success was great. On a space of about twelve acres, on the Blood farm, there were, in 1861, no less than thirteen flowing wells, ranging in capacity from 150 to 1,000 barrels each.

The rapid accumulation of oil lowered prices to such an extent, that scores of owners of small wells became discouraged, and abandoned them, going back to their homes, after investing most of their means in an unprofitable venture. Cool-headed men reasoned, like the philosopher of the Noachian period, that it would not be much of a shower after all, securing in the mean time everything that promised to be of value in the future. Others there were who, like the man upon the banks of the river, concluded to wait until it should run by.

The supply kept up persistently, and in the end proved a permanent boon. The cheap rate at which the article could be furnished, induced its introduction to every portion of the civilized world. Once introduced, and its uses properly understood, it became indispensable to the wants of the people. The consumption increased at a rapid rate also, so that ten and twelve dollars per barrel were obtained in less than two years from the time of striking the large wells. The foreign exports of 1862 were 10,887,701 gallons. That of 1868 will foot up nearly 100,000,000 gallons, and this demand is increasing at the rate of nearly fifty per cent. per annum. The appreciation in value of our national currency, and the lessening of the cost of production and transportation, have reduced the price to a maximum of from four to five dollars per barrel. At this figure, even a very small well is a good paying investment to the operator, that is, if he is content with a net profit of from ten to sixty dollars per day.

## CHAPTER XI.

### VARIOUS PHENOMENA OF OIL WELLS, &C.

VARIOUS theories have been given as to the causes combining to form a flowing well. For the most part these are vague and unsatisfactory. In theorizing we are apt to overlook the physical laws of nature. These are ever simple, and admit of practical solution when once understood. The flowing well is doubtless caused by a peculiarity in the form of the crevice containing the gas, oil, and salt water, or from an elevation of the sandrock above the uniform depth. Into this elevated portion of the crevice the gas ascends from its lighter specific gravity. When the crevice is penetrated by the centre-bit, the gas forces the contents of the same through the aperture to the surface until its force is exhausted. The contents of the crevice lying at a lower elevation can only be had by use of the pump. Any agitation of the oil tends to bring it to the surface. That caused by the gas greatly exceeds the agitation occasioned by the pump. That salt water has something to do with the formation of Petroleum, combined it may be with other substances, we have but little doubt. Whether it acts merely as a decomposing agent, or supplies from its own composition the gas or oil, or both, in any respect, is a question we leave to those better versed in the field of science. The fact that oil and salt water have been invariably found at the same depth, and come to the surface so intimately commingled, indicates an affinity, at least, of the substances. The rock itself is generally of close texture, and averages in thickness from twenty to sixty feet, in the different localities. In all places where this hard, thick rock has been found, the wells seem to have a more intimate connection, as though the supply of an entire locality was drawn from a reservoir having more or less continuity. Hence, all the flowing wells had their production interfered with, and in most cases stopped, from the sinking of other wells in their immediate vicinity. The oil field of Pleasantville, the development of which was commenced the past year, exhibited a new feature. The sandrock, the third we presume, where the supply of oil has been found in that locality, is of a coarse, porous nature, and only averages about twenty feet in thickness. The wells are all pumping, with but a moderate supply of gas, plenty of salt water, and after nine months of constant developing, but few wells were flooded with water from any of the neighboring ones. But many of the same wells have so greatly decreased in production since as to be worthless. In this, the latest oil field, operators, profiting from past experience, have adopted the system of "casing" in all the wells of that locality. This has doubtless a great deal to do with the prevention of flooding. Still, we are of opinion, effectual as the system has seemed to prove in the instance named, circumstances, more or less favorable, have had considerable influence in its exemption from this prominent evil to oil miners in other localities. The oil itself obtained at Pleasantville, though possessing the illuminating and other qualities, so far as tested, in all general respects with that of Oil Creek, is of different color, being nearly black. The sandrock has no doubt something to do with this, as well as the exemption from flooding, or connection with the water-veins.

Some of our practical operators hold to the theory that the flooding of the wells, by which they are rendered worthless for production, in any locality, is only a matter of time, from six months to a year being the average. Triumph Hill, at Tidioute, and the Bennehoff and Stevenson farms, as well as other localities, have verified fully the predictions thus made at the commencement of their development. The argument advanced in support of this theory is, that the oil on the surface, or above the body of salt water, becomes exhausted, to a great extent at least, leaving but little save salt water to pump. Some wells exhausting their supply before the others, causes a general flooding with water of all the others in the vicinity. Triumph Hill became thus flooded in the course of a few days. In other localities the flooding of wells has been more gradual.

The operators so fortunate as to strike large flowing wells theorized but little. The drillers' records were anything but satisfactory, being only a record, in fact, of the depth at which the different sandrocks were found beneath the surface. Being but in few instances practical miners, their records possess little value, and shed but a dim ray of light upon the vexed question. Subsequent developments were conducted with more care, but even at this date, too great a degree of carelessness is exhibited by operators in this important particular. A brief account of some of the large flowing wells, the history of their development, and incidents connected with the same, may not prove uninteresting to the reader.

"The Funk well, on the M'Ilhenny farm, commenced to flow in June, 1861, at the rate of 250 barrels per day, and yielded a constant supply for fifteen months. The Philips well, on the Tarr farm, commenced to flow in the Fall of the same year, at the rate of from 3,000 to 4,000 barrels per day, and was followed by the Empire well, on the M'Ilhenny farm, yielding from the start about the same. A flowing well had been struck previous to any of these, on the Watson Flats, near Titusville. Upon the penetration of the drill through the crevice of the rock containing the reservoir, the oil and gas came out of the well with great force, throwing the drilling tools to a considerable distance, and igniting instantaneously upon reaching the surface, injuring the workmen severely. The usual explosion and shock occurred. So great was the force of the explosion, that a man walking along the road at some distance from the well, was thrown into a mill-pond near by. The derrick and engine-house were completely wrecked."

The Burning well, on the John Buchanan farm, said to have been the first of the large flowing wells, was struck in April, 1861. Its tragic history constitutes the most notable event in the early development. The following account, given by an eye-witness, is the most reliable description we have been enabled to procure of the sad affair. The narrator, who, it appears, had an interest in the well, states:

"We had drilled to the depth of over 300 feet, and were expecting to strike oil at any moment. We went up to the shanty where we boarded, that is Mr. R. and ourself, to supper, and were on our way back to the well, which was just below, in the hollow, when we saw the men hurrahing, and presently a jet of gas, water, and oil rushed up, fairly lifting the tools out of the well. It roared and hissed like letting off the steam from a boiler. The stream seemed to me to reach to a greater height than the derrick, which was forty feet high. The people of the neighborhood came down with their shovels, and dug a circular trench around the well, throwing up a bank to catch the oil, as we had not expected such a flood, and had no tanks ready. The gas mingled with the air, and for a distance about the well, the air was almost yellow with the gas and spray of the oil from the fountain.

"Mr. Rouse and myself looked on awhile, and then started to go to the engine-house of the next well to have the fires put out. Before we had reached it, however, the gas took fire like a flash of lightning. Mr. R——, who was passing a small tank of oil, was covered with it as it took fire, and I lost sight of him for a moment. My hair and face were burned, but I was not much hurt. The sight of the burning well was horrible. A great fountain of fire, it wavered to and fro with the wind, and threw off blazing jets of oil. The poor people who were dipping oil around the well, wilted down before it like leaves when a forest is on fire. Some tried to crawl away, but the liquid flame ran along the ground, and caught them. Several hundred barrels of oil belonging to an adjoining well caught fire. Vast clouds of smoke floated off over the hills, and when the night set in, the hills and clouds were red with the light of the conflagration. Mr. R—— died soon after."

From the best accounts the gas ignited from the engine pumping the Wadsworth well, some eight rods distant. The explosion was from the tank of the same well. All the houses on the farm were shaken as if by an earthquake. All this occurred within thirty minutes after striking the vein of oil in the well; the oil and gas pouring forth from the well with great fury, and threatening to fill the whole valley with a sea of fire. As estimated, from ninety to one hundred persons were standing around the well, many of whom were enveloped in the flames, and saturated with the burning fluid. Of thirty-eight persons burned, eighteen died of their injuries. Among the latter was Mr. H. R. Rouse, one of the proprietors of the well and the farm, an energetic and enterprising business man, highly esteemed. Many of the victims could not be removed, and were well nigh consumed. Other accounts of the fire place the number of lives lost by this accident at thirty.

The fire continued to burn for several days, and was finally extinguished by digging up earth, carrying it upon blankets, and smothering the flames. After the fire was extinguished, some sixty feet of chamber, or tubing, was put in the well, when it flowed about 20,000 barrels and ceased. After this, tubing was extended down to the second sandrock, a depth of about 300 feet from the surface, and a pump inserted, but no oil was found. Numerous trials were made afterwards, but without avail.

The striking of the Philips and Empire wells was followed by others in irregular succession in the subsequent years, among the principal of which was the Van Slyke well, on the Widow M'Clintock farm, yielding 1,500 barrels per day; the Brawley well, on the Buchanan farm, 1,000 barrels; Blood well, Blood farm, 1,000 barrels; Noble well, 2,500 barrels; Caldwell well, 1,000 barrels; Sherman well, 1,000 to 1,200 barrels; Reid well, on Cherry Run, 600 to 800; Maple Shade, Hyde & Egbert farm, 1,000 to 1,500 barrels; Coquette, on same farm, 600 to 800; the United States, on Pithole Creek, 600 to 800, and many others of less note.

The brief history of the embarrassing circumstances attendant upon the drilling of some of the wells may prove of interest in this connection. Captain Funk owned some few acres of land on Oil Creek. When the development began he concluded to have a well put down. He gave a lease to a man named M'Ilhenny, who undertook the task of drilling the well by the tedious process of the spring-pole. The labor became wearisome after the depth of 80 to 100 feet had been reached. A horse-power was substituted, and the work carried on to successful completion.

"The Sherman well was commenced by a gentleman of that name. Mr. Sherman came to the Oil Region at an early day with limited means, but had control of a small amount of capital belonging to his wife. He obtained a lease on the Foster farm, on Oil Creek, and commenced to put down a well by means of the spring-pole. He had a fine show of oil, but long before reaching any considerable vein, his funds gave out. Working by this mode was no longer possible on account of the depth of the well, and there were no means at hand for purchasing a horse and horse-power. After working and waiting for something favorable to turn up, an interest in the well was finally exchanged for a horse, and the work proceeded. But after a time the labor became too severe for the poor horse, and the well was once more at a dead stand. Another interest was sold to two men who owned a small engine, and the work once more went forward. But coal was expensive, and none of the partners being able to purchase, another sixteenth was offered for sale to raise the necessary funds to purchase fuel. After a time this interest was sold for eighty dollars and a shot-gun. The money, and perhaps the shot-gun, were about expended, and the spirits of the partners at a low ebb, when the drill penetrated a crevice that yielded at once from 1,000 to 1,200 barrels per day, and their fortunes were made. The Sherman continued to yield for two or three years. finally coming down to a pumping well."

The following account is given of the Smith farm, on Cherry Run, the property of the Cherry Run Petroleum Company:

"A man named Smith owned a farm of fifty acres on Cherry Run. Unable to make anything out of it, he endeavored to sell the farm, and try his luck at oil mining. He owed some money, and his creditors were pressing him. In a desperate mood, he applied to J. W. Sherman, who had then recently struck oil, and was reputed to be getting rich, offering to sell the farm for $250. Sherman declined, and advised him to keep the farm, in case something advantageous should turn up. But Smith was determined to sell, and eventually found a purchaser at $500. The new owner sold it for $2,400. It was afterward purchased by the Cherry Run Petroleum Company for $6,500. The company at once proceeded to develop it, and it proved to be one among the best oil-producing farms in the Oil Region, ample fortunes having been realized by its fortunate owners."

Scores of similar experiences could be related, but the preceding instances, in general terms, cover the rest. The Noble well is said to have yielded to its owners $3,000,000. The Hyde & Egbert farm, on Oil Creek, rendered its owners millionaires in a very short time. The owners of the Noble well gave $142,000 for the Caldwell well, adjacent to their own, the purchase being made as a precautionary measure merely, to prevent any interference of the wells. It is stated on good authority, that during the oil excitement of 1864 and 1865, oil stock companies, representing a capital stock of $65,000,000, were based upon interests held on the Hyde & Egbert farm, the extent of the farm being scarcely forty acres, embracing about a dozen producing wells. The yield of several of these, however, was enormous.

The phenomena exhibited by the large wells were, of course, novel, but in general respects, considerable uniformity was observed in their flow. Some of them poured out a steady stream, with scarcely a perceptible agitation, save the ceaseless splash of the oil in the tank. Others flowed by breaths as it were, regularly as a pulse beats. Others, again, were spasmodic, throwing out large quantities of oil at regular intervals, this being forced against the sides of the tank with great force, roaring and hissing like the steam from a 'scape-pipe, churning the oil in the tank into a foam. Many of the larger pumping wells of the present day exhibit similar features, but in a less degree.

The Yankee well, on the Smith farm, Cherry Run, was probably one of the most eccentric of the flowing wells. As described: "After a rest of twenty minutes, discharging neither oil nor gas, gentle puffs of gas, accompanied by small spurts of oil, were visible at the end of the conducting-pipe, both increasing in volume for two or three minutes, until they became violent and frequent. The noise could be heard for the distance of a few hundred yards. The noise and discharge would then subside gradually, and in five or six minutes cease altogether. The production of this well was about fifty barrels per day, producing at this rate for a long time. From the general working of the well, it was conjectured by its owner that the oil came from a great distance." The gas wells of the Oil Region have been quite numerous, having been struck in almost every locality. Some of these produced a medium supply of oil, seldom exceeding fifty barrels per day, and others nothing but gas. The volume of gas in all of them was very great; and some few are still in operation, having been struck recently, or rather during the present year. Several of them have produced no oil, but continued to pour forth for months a steady stream of gas. It is supposed in these cases that the drill penetrated a crevice where the gas is located, being in such instances some distance from the oil, the passage being admitted to the one substance and prevented in the case of the other. Such wells are generally regarded as a nuisance, but few of them proving permanently productive.

The method of tubing the flowing wells was similar in all respects to the description heretofore given, excepting that the pumping apparatus was dispensed with, and a conducting tube, called a "goose-neck," which it resembled in shape, placed on the mouth of the tubing at the top of the well, conducted the oil to the wooden receiving tanks. At large flowing wells these were quite numerous, 1,200 barrels being about the largest size. These were connected by means of two-inch iron pipes running from one tank to another, with suitable apparatus to draw off the water that settles to the bottom, or the oil, when barrelled. These were placed as near the banks of the Creek as circumstances would allow, in order to facilitate transportation, the oil being frequently run from the tanks into the bulk boats. When a large flowing well was struck, several days generally occurred before it could be tubed and arrangements made for saving the oil. Where boats could be obtained and the well was adjacent to the Creek, the oil was run into them. In the absence of such facilities, the oil was allowed to cover the adjacent ground, or run into the Creek, many persons realizing handsomely by digging trenches into which the oil was collected, and then dipping it from these into barrels or other suitable vessels. Even after the well was tubed the waste was great, fully one-seventh in the early years.

The presence of salt water in wells drilled below the second sandrock, and even the first, is invariable. This varies in density, in some wells being equal to that obtained in the best salt-producing localities. The operator expects to find oil wherever salt water is obtained. No satisfactory explanation has been given of this seemingly intimate relationship between salt brine and Petroleum. An eminent writer gives the following as his solution. It is:

"That the strata in which they are found, at one period in the world's history, formed parts of the ocean's bed. This may have consisted of limestones, sandstones, or conglomerates, all saturated with brine, and reposing on what had been beds of clay, containing carbonaceous ingredients; while the process of baking this clay into shales, filled it with cracks and seams, that have since become so many veins filled with salt water or Petroleum expressed from rocks subsequently formed, the whole being upheaved to their present elevation."

As a general rule, the oil and salt water are in such close proximity, that it cannot be definitely ascertained which is struck first. Operators have clearly ascertained that the oil exists in crevices or fractures of the rock, which is indicated by the rapid sinking of the bit when drilling, and that the largest quantities have been found in the stratum generally known as the "third" sandrock. This rock varies in thickness in the various localities, as will be seen by reference to the description of the farms in another portion of this work. What connection there is, or what important function this sandrock performs, beyond an oil receptacle, is unknown. A shaft sunk to a proper depth, revealing this stratum to the inspection of practical men, would doubtless solve the mystery. That a plan of this kind will be finally carried into successful execution, is not at all doubtful. Though such an enterprise is perfectly practicable, it would be costly, but the results obtained could hardly fail to be highly remunerative. Two attempts of the kind were made in 1864 and 1865, one on the Allegheny River, at Tidioute, and the other on the Hayes farm, adjoining the Hyde & Egbert, on Oil Creek. The general plan of these was to sink a shaft to the depth of the first sandrock, and then drift in any required direction. It was supposed that enough oil would be obtained at this stratum to pay the entire ex pense, while that obtained from the strata below would be clear profit. The same process, as regards drifting, to be pursued at the subsequent strata where oil is found. The air in the shaft was to be kept pure by means of fresh air from the surface being pumped in by means of powerful engines. The dimensions of the shaft at the Hayes farm were nine by seventeen feet. Seventy feet was the maximum depth attained, when the work was abandoned for want of means. Enough has been done, however, to show the entire practicability of such a scheme.

The greatest obstacle to the oil operator is the large quantity of water met with. This water comes from the surface and from the water veins beneath, generally met with at no greater depth than 150 feet, and is in such quantity as to often literally drown him out. Hundreds of wells have been rendered worthless by "flooding," as it is called. Wells cased in the manner before described, are not so seriously affected by water, for the reason that the seed-bag, which cuts off the water-veins below, remains stationary in all wells that are cased. This, in wells not cased, is moved each time a removal of the tubing is necessitated, allowing the water to descend, flooding and rendering worthless all the wells surrounding that are located on the same crevice, for considerable distances. It is an undeniable fact that the flooding of each other by wells on the same crevice, or having any connection, is only a matter of time or circumstances. The method of casing the wells with artesian tubing was especially designed to prevent this result. So far, the system has proved admirable, and no doubt such means will be devised as to prevent its recurrence entirely.

The sinking of one well in close proximity to another, before the system of casing was adopted, generally flooded, or drew from the well its oil supply, by diverting the current of oil and gas in its own direction. As the leases consist of from one to half an acre each, the operator in previous as in present years has no protection against this serious evil. When a good well was struck, parties commenced immediately to sink wells in as close proximity as possible to it, for the express purpose of flooding or obtaining its supply of oil. By this means they hoped to compel the owners of the producing well to pay them certain sums in the form of a compromise. The system of casing effectually prevents flooding by water, and greatly obviates the danger of losing their supply of oil by having the current diverted in other directions. In former years it was a custom often practised, for both the contending parties to draw their tubing at the same time, rendering their wells worthless for production. Finally a compromise would be effected. Though generally restricted by the terms of the lease from such improper action, a way was found to evade the conditions imposed.

From the fact that crevices appear more continuous in some places than others, interference, likewise, takes place to a similar extent, wherever such features exist. The entire flat portion of the valley of Oil Creek, the

Stevenson farm, on the hill back of Petroleum Centre, Pithole, Triumph Hill, at Tidioute, Bennehoff Run, in fact the greater portion of the territory developed in former and subsequent years, is all affected to a greater or less degree by this cause, and rendered more or less unproductive.

The landowners could doubtless remedy this evil, if inclined to do so. A large number, unfortunately, are not. Being generally non-residents, they seem to lease their lands with a view of obtaining all the available oil from them in as short a period as possible. Many of them argue that only a maximum amount of oil can be obtained from a certain extent of territory, and it is just as well to get it all in two years, by sinking a large number of wells, as to occupy twenty years, by sinking a smaller number, to accomplish the same results. This selfish reasoning, besides being false in point of fact, is ruinous upon the future of the entire business. Others wish to have their lands developed, in the hope of obtaining large profits and the first cost of investment from the royalty, and after this to obtain possession of the wells successively abandoned, to work for their individual benefit. In honorable contrast to these, are many others of the class who are disposed to grant leases of several acres in extent, so that lessees can be afforded some measure of protection, binding the lessees in question to drill a certain number of wells per annum.

The most remarkable case of flooding wells occurred on the Tarr farm several years since. The Woodford was sunk but a short distance from the celebrated Philips well. Upon reaching the third sandrock it was found that the Woodford flooded the Philips. The Woodford was tubed, but could obtain nothing but oil so mixed with water as to render it of slight commercial value, while the Philips pumped pure oil. But whenever the owners of the Woodford removed their tubing, the oil of the Philips well was rendered worthless. After a conflict that lasted many months, and resulted in effectually flooding the entire flat portion of the Blood and Tarr farms, rendering fully sixty wells unproductive, a compromise was effected. The Woodford produces but little or no oil at present, while the Philips pumps about fifty barrels per day, having produced almost continuously for seven years.

The system of casing now practised is to tube to the first sandrock, with four and a half to five inch artesian tubing, the bore of the well being drilled slightly smaller from that depth, leaving a shoulder for the bottom of the tubing to rest upon. A tight seed-bag is put on this at the proper depth, and in cases the sediment settling at the shoulder answers as well. This plan has so far answered the purpose for which it was devised; and when it fails, a better one will be put in operation. The tapping of the oil veins where the wells are drawing their supply from the same crevice will always ensue to a greater or less degree. The flooding of oil territory generally proceeds from the neglect to properly shut off the water from wells that are abandoned. Where all the wells on a farm are kept constantly pumping, flooding does not occur to any material extent. Though a crevice, or crevices, or, as it is now termed, a "belt," may extend in one general course, it does not necessarily follow that there is a continuous connection of the oil and water veins. If such were the case, all the wells in a locality would be simultaneously affected by like results. Experience goes to show unmistakably, that each well has certain peculiar features, different from any of those surrounding it, and these are distinctly apparent. The differences consist in the difference of the flow and in the motion necessary to pump them.

A well yielding a production of from 100 to 150 barrels per day, is often obtained in the midst of wells that are only producing a tithe of that amount, not seeming in the slightest degree to diminish the supply of the previously producing wells. A general "law of lawlessness" prevails in these strange vagaries or freaks of the wells, which admits as yet of no satisfactory explanation. The theories upon each of the different features are numerous. At the present rate of progress, but a few short years will elapse before a correct knowledge of all these seemingly impenetrable mysteries of nature will be obtained.

The theory was advanced in early and subsequent years, that the wells yielded less production in winter than in summer. We have seen no practical proof of the assertion, beyond the natural one, that cold weather interferes to a greater extent with operations carried on at the earth's surface. The state of the atmosphere can surely have but slight effect on the oil at the depth of several hundred feet from the surface.

Another remarkable feature has been mentioned by WRIGHT, "that in some wells, where it has been tested by the thermometer, the temperature of the oil as it comes to the surface, is but slightly above the freezing point." We have ourselves seen pellets of ice discharged from the conducting-pipe of a well in mid-summer, but supposed it proceeded from either the rapid condensation or evaporation of the gas as it came in contact with the atmosphere, or both. If this fact was clearly established, it would no doubt be vastly comforting to many hardened sinners in this portion of the vineyard, and cause a perceptible falling off in the contributions for home and foreign missions. The same authority says: "As the decreasing temperature is measured in ascending from the sea-level, so its increase is measured downward, as in caverns and mines. The rate of increase in descending is estimated at one degree Fahrenheit for each sixty feet. At this rate, the oil, water, and gas, brought from a depth of 600 feet, would be ten degrees warmer than on the surface. This phenomena is in direct opposition to the igneous theory advanced by eminent geologists, who have determined the exact distance from the surface that our globe is in a liquid state. The experiments made were crude, and, perhaps, not to be fully relied on. The decreasing temperature of the oil as it arrives at the surface may be owing to causes of a different nature."

It would be futile to attempt a philosophical explanation of the various phenomena we have enumerated, at this stage of the Petroleum development, subject as it is to so many contradictions or new phases each day, over so great an extent of territory, each locality differing in minor details at least, and principally from their action being evolved at so great a distance from the earth's surface. We are content to leave their solution to the results that will be obtained by the miner's drill, that being the safest premises for both author and reader.

It is a debatable question whether the oil supply becomes exhausted in any locality. The wells may become flooded, and rendered unproductive for a time at least. But we are of opinion that whenever the heavy body of water is removed, the wells will again prove productive, probably not to as great an extent as before. The employment of pumps sufficiently powerful to lift this mass of water from the oil is all, in our opinion, that will be required. To do this will require a concert of action upon the part of the landed and working interests, and the expenditure of a large amount of labor and money. Nor will it be done so long as new oil fields present themselves for development in such rapid succession as in the last two years.

The question, and oft-recurring one at this date, is whether there is any reproduction of Petroleum going on at the present time. Some there are who believe it to be a deposit formed ages since, the requisite amount manufactured, and enclosed in the rocky vaults where it is now found. While we are free to admit the antiquity of the *causes* from which its formation was first evolved, we cannot believe that its manufacture, so to speak, has ever ceased. The same causes, it is fair to presume, still exist, and are as actively at work as in any other age of the world, and the same elements are equally as abundant.

# CHAPTER XII.

## THEORIES OF THE ORIGIN OF PETROLEUM.

THE origin and the causes which lead to or unite in the formation of Petroleum, and its deposition in the strata in which it is found, have proved an inexhaustible theme for writers, and elicited no small degree of interest in scientific minds. Upon a subject of which so little can be known, theorizing, if futile so far as immediate practical results are concerned, is highly useful in inducing inquiry and investigation that will eventually result in its successful solution. We give the most reasonable theories we can obtain, not so much for the light they shed upon the subject, as to show the industry and ingenuity displayed by their different authors. As has been truthfully remarked, if the Petroleum of the different wells was of the same quality, and the strata from which they are derived of the same character, a great obstacle in the way of reasoning out a theory would be removed. But the Petroleums not only of the United States, but even of the different localities of the Oil Region, differ very materially. In theorizing upon this subject, one eminent authority says:

"The theory that the Petroleum of Canada, which occurs in the older Silurian rocks, is derived from the decomposition of vast numbers of marine animals, is not an unreasonable one. In distillation the Canada Petroleums yield acroleine, an oil which is obtained from animal oil and fats. The vapor of acroleine is very pungent, and attacks the mucous membrane of the throat and lungs, causing great irritation. Fish oil yields it by distillation. It is not found in the Petroleums of Pennsylvania."

Another learned writer states:

"The transformation of wooden fibre into oil is a chemical change, taking place always out of contact with atmospheric air, and usually under water, but by no means connected with any particular geological period, as, for example, the coal epoch, with which many intelligent people associate it."

Another authority on the subject, Dr. Gesner, says:

"During the passage of vegetable substances into coal, there is an escape of vast quantities of carbon combined with hydrogen. It is only necessary that the gases of these elements should be condensed to produce hydro-carbon oils. The operation is a decomposing and combining one, and the new combinations formed during the transmutation of wood into coal, have a close analogy to those produced during the distillation of wood without the admission of air. The gases generated in strata of coal and coal strata, are always under great pressure, which tends to their condensation, and the consequent formation of oil.

"That coal has been derived from vegetables is undoubted. Peat and wood are found to pass by insensible shades into lignite, lignite into compact, bituminous coal, and the end of the transformation appears in the anthracite, from which nearly all the hydrogen has been expelled, and carbon remains.

"From the expulsion of oxygen, carbon and hydrogen, from wood, and the variety it presents until it forms true coal, heat has not been absolutely necessary, although it has doubtless exercised a powerful influence in connection with those chemical changes ever going forward in the earth.

"The condensation of hydrogen and carbon producing oil, and the fact of strata of coal and shale before they reach the maximum of carbonization giving out these elements in great quantities under pressure, and the tendency of these gases and oils to diffuse themselves, are fair reasons for finding oil in formations bearing no traces of vegetation."

Another writer supposes: "That the Petroleum of Pennsylvania arises from the distillation by subterranean heat of the hydro-carbon agents resident in the carbonaceous strata underlying the Oil Region."

Another writer says: "That the great beds of anthracite coal of Pennsylvania, on the southerly slope of the Alleghenies, are merely the residuary coke, as it were, of a distillating process, which has converted their bituminous matter into oil, and distributed it by some convulsion of the earth through the formation beyond the mountain range."

The theory of the scientific men of the present day in explanation of the source of salt in the sea is this: "The carbonic acid of the air is constantly acting upon the rocks and on the surface of the earth, thus turning them to clay, and forming carbonates with the soda, potash, lime, and magnesia set free. These are carried down as carbonates to the sea, where the carbonate of soda decomposes the chloride of calcium of its waters, and forms common salt and carbonate of lime.

This series of actions is the source of the salt of the sea, of clays, and of limestone. Organic living beings do not generate the carbonate of lime, but appropriate it, when formed for them by chemical reactions; and thus great portions of our limestone rocks are made up of fossil remains. In a volume of limestone there is separated and condensed from the air a large amount of carbonic acid gas; the early atmosphere was, therefore, very dense and unfit for the sustenance of the higher forms of life, until by far the greater portion of this gas had been removed by the formation of the carbonate of lime and vegetable matter now constituting coal and Petroleum."

The following exhaustive treatise on the subject is taken from the report of the United States Commissioner of Agriculture, and published several years since:

It is probable that all instances of solid bitumen found on or beneath the surface of the earth have resulted from the hardening of drops or reservoirs of liquid coal oil. The lumps and crystals of graphite found in the oldest rocks, like the lumps of amber found in the newest, were doubtless oily substances involved by sand and mud. Flakes of anthracite are found in the centre of rock crystal. Gelatinous animals and mucous plants abounded in these ancient seas, and ought to have provided, by their death, plenty of animal and vegetable hydro-carbon for the mineral. The old red sandstones, like modern formations, present us, for our cabinets, innumerable flattened fish, converted into bitumen; some in so perfect a state that every scale can be counted, and every sculptured line upon them submitted separately to the microscope; others an undistinguishable mass or daub of tar. Some rocks have been so thoroughly charged with animal dead matter that they emit a fœtid odor whenever struck, and are technically known as stink stones. The bituminous limestones and shales of many different geological ages are so many reservoirs of animal and vegetable oil, produced by the death and decomposition of successive floral and faunal creations, perhaps, principally coralline. The fossiliferous black shales of the central belt of the State of New York underlie Lake Erie, cross Ohio and Kentucky into Tennessee, and return through Indiana and form the beds of lakes Michigan and Huron. In middle Kentucky, the faces of the rocks are smeared and streaked with oil, fried out of them by the sun, so that the surfaces are blackened as if by tar.

Up to the horizon of these black slates, ascending in the columns of deposits, gelatinous sea organisms, both animal and vegetable, seem to have constituted the principal, if not the sole apparatus for generating Petroleum. But Dawson has lately discovered in the sandstones over them a true angiospermous exogenous tree, not much, if any, lower in the scale of development than those of which our forests are composed. Coniferous trees began also to abound, and coal beds to be deposited in groups. Thence the higher we ascend towards and through the second, and the third or great coal measures, the more abundant became the vestiges of fresh water and land vegetation, until, in the tree stumps of the coal beds of Nova Scotia we find small land animals. The mosses and ferns, the rushes and reeds, minute and gigantic, of which the coal beds came, suggest the vegetable origin of oil. For it is near or between the three systems of coal measures proper that the amazing discoveries of subterranean reservoirs of oil had taken place. It is impossible to suppress the suspicion that Petroleum is a product of the slow decomposition of vegetable tissue.

But the oil wells are not sunk in coal measures, but through them at the edge of the great coal area. The oil is never found in coal beds; nor have the subterranean reservoirs of oil apparently any connection with coal beds, nor even with coal slates, or bituminous shales or pyroschists, as they are called. Black slate, cannel, fat coal, like lignite, peat and living wood, will yield the oil and gases by distillation, but the geological distinction must be carefully preserved between the free Petroleum of the rocks and wells, and the distilled Petroleum of the old oil works.

The connection of the oil region with the coal basins of western Pennsylvania and Virginia, and eastern Ohio, and Kentucky, is, in good measure, a geographical deception. (In the report of a geological reconnoissance of Indiana, 1859, 1860, under D. D. Owen, State geologist, and published in 1862, Professor Lesquereux expressed the opinion that the mineral oil of the borders of the coal field comes from the lowest great bed of the coal measures, I. B., page 285. The opinion of such an authority is to be carefully considered.) The Oil Creek rocks dipping southward, pass 500 or 600 feet below the coal measures. The nearest coal bed to the northern springs occurs on the highest hill tops, many miles away. The hills in the vicinity of some of the wells, are capped by the conglomerate base of the coal measures, at least 100 feet thick. The shales and sandstones of the valley belong to formations X., IX., and VIII. descending, called by the New York geologists the Catskill, Chemung and Portage groups, extending over all the southern groups of western New York. The southern dip carries down these oil bearing rocks, and the wells must deepen in the same direction. Mr. Ridgeway reports (July 10, 1862), the lowest oil-bearing sandrock, capping the hills near Waterford, on Le Bœuff Creek, and the same sandstones appear on Big French Creek, full of plant remains.

The following wells show the dip in a well-marked manner: The Philips well on Oil Creek, is 480 feet deep; the Brawley well, at the mouth of Cherry Run, 503 feet; the Cornwall well, 530 feet; the Avery well, over 700 feet; and at Titusville he estimates the proper depth at 1,000 or 1,200 feet.

In the Mahoning coal oil region in western Pennsylvania and eastern Ohio, near the line, the three oil-bearing sandrock strata are beneath the lowest coal bed. The "Continental" boring at Edenburg in Lawrence County, penetrated in descending order, the following formations before it struck the oil: First, the superficial drift, 80 feet thick. Second, sandstones and shales, 200 feet thick, the bottom layers of which consisted of fetid black shales, from which coal gas flew off with violence. Third, the first white sandstone, 50 feet thick, arranged in three strata, a softer middle between harder upper and lower formations, the whole of which was said to be thin, going east and holding abundance of gas in its crevices. Fourth, shales and slates, 45 feet thick, charged with oil and gas in its crevices. Fifth, the second white sandstone, 75 feet thick—softer, coarser, and tougher, or more difficult to bore through than the first, and full of gas; after passing through which they struck the great oil stratum, 448 feet from the surface. Crawford's boring, not far off, went down 580 feet through another shaley formation, and struck oil, supposed to come up through a crevice from the third white sandrock.

That there is an intimate connection between the character of these sand formations and the character of the oil which issues from them is indubitable. The rule among the miners is, as stated by Mr. Clark in the "Proceedings of American Philosophical Society," (June 1862, p. 57), that the harder the rock may be to drill, the lighter in color, purer in quality, and the smaller in quantity, will be the oil obtained therefrom;

and the softer the rock, the darker and more abundant the oil.

The chemist of the Canada survey, Mr. Hunt, insists strenuously "upon the distinction between lignitic and bituminous rocks, inasmuch as some have been disposed," he says, "to regard the former as the source of the bitumen found in nature, which they conceive to have originated from a slow distillation. The result of a careful examination of the question has, however, led us to the conclusion that the formation of the one excludes more or less completely that of the other, and that bitumen has been generated under conditions different from those which have transformed organic matters into coal and lignite; and probably, in deep water deposit, from which atmospheric oxygen was excluded."

Mr. Hunt instances in support of this view, that the highly inflammable pyroschists or black slates of the Utica and Hamilton groups contain no soluble bitumen, and that the Trenton and Corniferous limestones at the base of the Silurian system are impregnated with Petroleum, and gave rise to Petroleum springs, although no fossil land plant has been found in them. The fact that a considerable portion of the tissues of the lower marine animals is destitute of nitrogen, and very similar in chemical composition to the woody fibre of plants, forms another link in the chain of reasoning on this distinction between bituminous and lignitic rocks. The black slates, and even the coal beds are, in fact, layers of mud, charged slightly or to excess with lignitic matter, peat, or humus, part of which has assumed the form of glace coal and part the form of mineral charcoal, but almost none of which is soluble in benzole or sulphuret of carbon; whereas these liquids easily dissolve out the ready-formed bitumen from the rocks which may contain them. But whenever a coal bed became a repository of dead fish, like the eight-foot coal at the mouth of the Yellow Creek, at the bend of the Ohio, or as in the case of the two-foot stratum of phosphatic iron-ore deposited between the two benches of the Deep River coal bed, at Egypt, in North Carolina—how different an aspect the mineral then wears, glossy with soluble bitumen!

Mr. Hunt argues with much force that the mere fact that intermediate strata, porous enough to absorb all the floating bitumen in their vicinity, are nevertheless, destitute of any, is enough to prove that the accumulations of oil now furnishing the world with light never came from the sub-volcanic distillations of the beds of coal in their neighborhood, but that the mineral has been generated by the transformation of organic matter in the strata where it is. Mr. Wall has shown that the asphalt of Trinidad and Venezuela (belonging, however, to a much later—upper miocene or lower pliocene—tertiary age), occurs in limestones, sandstones, and shales, associated with beds of lignite or fossil wood, and is confined to particular strata which were originally shales containing vegetable remains which have undergone "a special mineralization, producing a bituminous matter instead of coal or lignite, and not attributable to heat, nor of the nature of a distillation, but due to chemical reaction at the ordinary temperature and under the normal conditions of climate." He describes, also, wood partially converted into bitumen, when removed by solution, woody fibre remains.

The theory of the genesis of coal oil is, however, far from being cleared up by such facts. It is true that the oil is not found in immediate contact with coal beds made of land or fresh water plants, but on the other hand, coal oil regions are geographically connected with coal bed regions, whether of devonian, carboniferous, oolitic, or tertiary age. Coal beds are said to underlie the Rangoon oil wells. Tertiary lignites abound in Trinidad, Venezuela, Lombardy, and middle Asia. The lower devonian horizon of the Canada black slate oil region yields coal beds in Pennsylvania. The structural difficulties attending the solution of the problem remain.

Fissures are filled with oil, and gas, and salt water, and different wells strike them at different depths. The oil bearing sandrocks seem charged from top to bottom with gas and blow off from every fissure as it is passed through by the auger. Whence comes this gas, if not by subterranean distillation? It is impossible to postulate the gas first, and oil afterwards; for that order would require the generation of pressure sufficient afterwards, and the oil would be in the condition of a mechanically explosive fluid. The gas must be a subsequent expansion of the oil, as it is in the case of coal mine fire damp. Whence, then, comes the oil, and why has it collected in reservoirs? How are such reservoirs preserved, and what is their extent? It is easy, after these questions have been answered, to describe the mechanical propulsion of the oil to the surface, partly by gravity and partly by the pressure of the gas it has itself generated, through natural fissures producing natural oil springs, or through artificial augur holes. The intermittent action of most of the flowing and spouting wells is like that of the Iceland geysers, where steam is the motive power. The oil men of the Mahoning valley say that more gas is blown off in winter than in summer.

In the almost unchanged horizontal posture of the western coal measures no considerable fracturing took place. Faults of all kinds are uncommon and very small when they exist at all. The rise of stratification from the Allegheny River towards Lake Erie is a fraction of one degree. The original contents of the rocks have, therefore, been preserved. Not so with the anthracite basins on the southeastern side of the coal area. Crushed and overturned, contorted and fractured in every part, this part of the earth's crust has been dried and hardened, and exposed to chemical action from the superincumbent drainage waters, until its various formations (the coal beds included in the number), have been metamorphosed and partially re-crystallized. The oils which they contained have been lost by dissolution and evaporation. The bituminous coals have become anthracites, and the last oil spring on the head waters of the Lehigh, the Schuylkill, the Juniata, the Potomac, or the New River, ceased to flow many millions of years ago. In the West, on the contrary, in equally ancient, nay, in identically the same rocks, the Petroleum still remains, having had no outlet, always hermetically sealed and under pressure. It remained partly condensed in coal beds and black shales, partly distributed through the sandrocks and limestones, and partly filling up the joints which the shrinking of ages has produced. Possibly a small portion of it may be held in caverns through the more soluble limestone strata. Especially important are the water bearing horizon.

The vertical cleavage planes and few downthrow fissures which exist play but a subordinate rôle to these. Rain waters percolate from every hill surface and downwards, leeching every permeable stratum that will give up its oily contents, the out-crops of every coal bed issue innumerable springs of painted water. At the base of every great sandrock, and on the top of the clayey deposits next below it, collect the mixed proceeds of the drainage in a standing sheet of oily brine. Capillary attraction and hydrostatic pressure perpetually reinforce the reservoir. The weight of rock on the top and the pressure of disengaged oil-gas sends its filaments forward and upward by every secret crack to the surface again, holding it in every part ready for an explosive rush into the air when an artificial outlet is provided. If there be no fissure in the locality, the oil wells descend to the sheet of water at about the same depth. Where fissures intercept them they are of various depths and fortune, for a well may pass a fissure where its walls are polished and tight together. A well may also pass the water sheet where some change in the porosity of the rocks above and below has taken place to oppose a like obstruction. In some parts of the western coal fields, the dip is as high as five degrees, and the basins from five to ten miles wide. Sharp flexures make local dips of thirty degrees or more, and a central subanticlinal is sure to subdivide the basin. In the secondary basins thus formed, the wells are more perfectly artesian as to the salt water; but it is upon the subdividing anticlinals that the gas and oil collect. In such regions it is asserted that all the flowing and many of the spouting wells are ranged along the summits of such anticlinals. In the case of some of the old gas-blowing salt wells, their actions demonstrate that they have been bored past one gas bearing stratum to another deeper salt water stratum, for when the water is allowed to rise in the auger hole, by stopping the pumps awhile, then the gas and oil no longer come up, the brine stopping their issue. In the case of neighboring wells of different depths striking a slanting fissure, the one which strikes it highest up will deliver gas; another, striking it lower down, will deliver oil; a third, striking it still lower down, will deliver nothing but salt water.

The compressibility of coal oil gas is one of its most dangerous qualities, increasing indefinitely the dangers of those explosions which annually cost so many valuable lives. Confined in the walls of the gangways and rooms, it issues from innumerable cells or pockets, the larger of which are called "blowers;" sometimes with the noise of heavy rain; sometimes with small reports. It collects among the timbers of the roof, in the upper galleries of the mine, in deserted portions of the colliery, and especially in those accumulations of refuse coal and slate called "gob," or "goaf," with which the miners pillar up the superincumbent rocks. These acres of worked-out and filled-up galleries bcome vast reservoirs of fire damp. The gas collects especially over the anticlinal rolls. From these great powder magazines, solicited by the least diminution of barometric pressure in the atmosphere, the gas rushes out to fill the working rooms. Long experience has shown that a falling barometer and explosions in coal mines always go together. But the mischief is accumulative. The vacuum produced by the first explosion is a new vocation to the world of back gas to leave its hiding-places, come forward afresh, and produce another, and again another, until the proportion of air to gas becomes too small to make an explosive mixture; so that, like the stroke of lightning, the coal mine explosion is not a unit, but a series, cause and effect reciprocally acting to produce the last result.

Among the most curious exhibitions of superior lightness of Petroleum to other minerals with which it is found, and of the nice train of reasoning dependent thereon, is the observation of Mr. Vanuxem that the film of black bitumen found in the cavities of the calciferous sandrock of New York, with crystals of bitter spar and quartz, occur on the upper side of the crystals, on the mother liquor of which they once floated as pellicles of oil; and, as the crystals hardened and grew, it moulded the oxydated oil to a sheet of bitumen, brittle, very pulverulent, of a shiny black, yielding little ash, and 11½ per cent. of principally water. The same mammillary surface, arguing original fluidity, characterizes the specimens obtained by the Canadian mineralogist from the Quebec group filling cavities in its limestones, sandstones, and even in the accompanying trap dykes; readily crumbling to a black powder, and, when highly heated giving off an abundance of strong-smelling, inflammable gas, condensing to a tarry oil, and leaving eighty per cent. of a black residue, which, when heated slowly, burns away, leaving only a trace of ash. The same kind of mineral found at the Acton copper mine is harder, less friable, and more like anthracite. The Petroleum which fills cavities in the Montmorency rocks is still unhardened. It flows in drops from a fossil coral of the Birdseye limestone there; and at Parkenham's it fills the cast moulds of large orthoceratites in the Trenton limestone to such an extent that a pint has been poured out of one. It is, perhaps, from these lower Silurian fossil coralline limestones that the oil makes its way to the surface through the overlying Loraine shales to form the Guilderland oil spring near Albany, according to Beck, through the Utica slate on the Great Mountain island, and through the red Medina shales at Albion mills, near Hamilton, according to Mr. Murray.

The next great limestone in the ascending series is the Niagara, and Eaton early made known the oozing of Petroleum from its fossil caste. Hall describes it in Monroe County as a granular crystalline dolomite, including small laminæ of bitumen, which gave it a resinous lustre. Bitumen, sometimes flows like tar from the lime kiln. The corniferous limestone, next above the Niagara, has the cells of its fossil corals filled with Petroleum, the remains of the gelatinous coral animal which inhabited them. Mr. Murray drew attention to this fact in 1844, and cited the gravelly bay quarries in Wainfleet, western Canada, as examples.

The oil springs of Enniskillen, as well as the lake of solid bitumen in the same township, half an acre in extent and two feet thick, no doubt have their deep-seated sources not in the black shales of the region, but in the corniferous limestone underneath. These black shales belong to the base of the Portage and Chemung group. The wells sunk in them soon strike argillaceous shales and limstones of the Hamilton group, and go through them toward the corniferous limestone, specimens of which yielded to Hunt's analysis from 7.4 to 12.8 per cent. of bitumen, fusible and readily solible in benzole.

In the blackish Marcellus shales, at the base of the Hamilton group, are found septaria or modular concretions containing Petroleum. The same phenomenon recurs at the top of the Hamilton sandstones (formation viii.), are often bituminous to the smell, and contain Petroleum in cavities or hardened into solid seams. A calcareous sandrock in Chautauque County contains more than two per cent. of bituminous matter. These are the rocks around the famous oil springs of the Seneca Indians. It is only necessary to ascend the series of these devonian sandstones to their upper part among the rocks of the Catskill group, or just beneath them, to find oneself in the oil regions of northern Pennsylvania and Ohio, described by Dr. Newberry and others.

There only remains to be noticed that anomalous deposit of the Albert coal in New Brunswick, made famous by long litigation and the discussion of geologists, described by Professor Dawson in his Acadian Geology, and called, by Dr. Wetherill, of Philadelphia, Melanasphalt.

Its position has been misinterpreted by several observers, who have reported it a volcanic injection of bitumen into a fissure of the earth, many feet in width, by the force of which large pieces of the wall rock have been torn off and carried forward in the mass. It seems, however, pretty well made out, that it was originally a horizontal bed or lake of Petroleum, hardened and covered up by sand and clay deposits of carboniferous age and afterwards upturned, bent over and fractured so as to assume its present posture. It is not properly a coal bed, therefore, but a mass of hardened coal oil,which can be, and, in fact, has been, mined like a coal bed, and the product used wholly for making gas.

# CHAPTER XIII.

## SOCIAL CHARACTERISTICS OF THE PEOPLE OF THE PETROLEUM REGION.

The characteristics as shown by those engaged in the Petroleum business of this region, has afforded a fruitful theme to scribblers and journalists in years past. From our own observations, extending from the first general commencement of the development to the present date, we find that in but few features do the representative class of the people of the Oil Region differ from those of the human race generally, or at least the enlightened portion of it. The business itself is one that requires, for its successful prosecution, the exercise of all the energy and skill of average mankind. The progress of the development, as we have shown, denotes a degree of industry as commendable as it is unparalleled. The persistent energy which in scarcely ten years, has wrested from the rocks several hundred feet below the earth's surface, a mineral product, that, besides proving an inestimable blessing to the human race, has added nearly $100,000,000 per annum to the wealth of our great country, giving employment to thousands of men, requiring railroads and fleets of ships for its transit, and many millions of dollars to transact its already vast business; converted a comparative wilderness into a teeming hive of unceasing industry, built towns, and cities, bridged rivers, and finally, homes, surrounded by all that good taste could require, needs no panegyric in these pages. It is apparent everywhere, and its influence is felt in all the great commercial centres of the world. It stands forth in bold relief, as one of the most remarkable achievements of industry of the century we live in.

The labor required to produce such wonderful results, has been unceasing, requiring skill and capital in no stinted measure. We are content, therefore, to sum up briefly the ruling characteristics of our people as those of industry, skill, perseverance, and the exercise of the best business and mechanical knowledge that our country, from the Atlantic to the Pacific, could supply, or money obtain.

In point of general morality, the average, so far as our observation has extended, is rather above that of surrounding communities, and will compare favorably with that of any other locality in the land. The commercial system is perfect, and the business men of no place have a higher regard for their commercial honor, and the failures to meet these have been so few as to form exceptions only. The cash system prevails in all branches of business throughout the Oil Region; cash upon delivery being the universal motto.

Those engaged in the business here, strange as it may appear, retain to a certain extent, the habits and tastes acquired in the places they came from, or during their previous years. These are exercised so far as the ability or inclination of the possessor will allow. Men of temperate, steady business habits, find here ample scope for the profitable employment of such qualities. Intemperance receives no encouragement, and a reform or departure from the locality is necessitated from the fact that to procure even the necessaries of life, constant industry is insuperable. Sobriety is indispensable for the successful prosecution of any branch of the business. Every hamlet, village, and town has its churches, in which the congregations of the different denominations worship. Well-organized Sunday-schools abound in each community, and the arrangements for the education of youth are ample. The different congregations are represented by a membership drawn from all classes of society. The preachers or pastors of the different congregations, so far as we know, are liberally sustained, and the church buildings paid for. The eleventh commandment, "Attend to your own business," is observed generally, all branches of the trade, requiring about all the time that can be spared from the needed hours of rest, for its transaction.

Strict integrity, respect for the laws, industry, due observance of the proprieties of life, enterprising, hopeful, ready to commence again with renewed energy after any business disaster that may befall them, possessing an abiding faith that all will be right again in a short time, severe as the conflict may prove, are the chief characteristics of the leading class. With such men all human achievement is possible, and no difficulty too great to overcome. They have already achieved much, and the beginning of the development of the vast resources that surround them on all sides, has not been fairly inaugurated. There are no drones in the hive; at least none who have a local habitation.

Recreation is had at appropriate seasons, and the capacity for healthful enjoyment fully satisfied. The Sabbath is strictly observed, save by the pumping wells. The owners of these claim that, financially speaking, their working is a necessity, on account of losing a day in the week when the well is allowed to remain idle on the Sabbath. The practice is fast being discontinued, and those who observe the Divine commandment in this particular seem to prosper fully as well as their more industrious neighbors. The large majority of those engaged in business have removed their families from their previous places of residence to the Oil Region, and the beneficial results are everywhere apparent. In all the principal towns, society is refined, and even elegant. Courses of lectures are delivered in all the principal towns during the season, debating societies organized, and in Oil City there is a public library, possessing a well-selected assortment of books. Sociability is a prevailing trait. In brief, we have here in the Oil Region an Utopia, almost, if one is disposed to seek only for that which is good. Doubtless the missionary mind, yearning to benefit the human race, even at the expense of martyrdom, might find much to reform. Vice and immorality no doubt prevail, as is common everywhere; it does not, however, parade itself in such public view, but has to be sought for. To the modern "moral statistician," we shall leave this latter branch of the subject, confident of our inability, owing to lack of familiarity and inclination, to do the subject the justice it merits.

It would be strange, indeed, if among so many who have had fortune thus lavishly thrust upon them, there should not be some inclined to foolish ostentation and display. Even these have been few, as the social structure is such, that displays of the kind attract but little or no attention from the busy multitude. Besides, it is highly necessary to treat your poorer neighbors with respect at least, for they are liable at any time to "strike oil," and soar as high in the social scale as yourself. That is, so far as money will entitle them; and surely its power is potent.

Liberality toward all deserving objects has ever been the characteristic of our people. During the war, the township of Cornplanter, in which the borough of Oil City is embraced, contributed over $100,000 to the

Philadelphia and Pittsburgh Sanitary Fairs, and three companies of soldiers were furnished from the same locality. Appeals for religious objects, and for charity, have ever been met liberally. When disaster has befallen one community, the others have always promptly tendered material aid, and this feeling is extended to individual cases where the party is worthy.

But few of the original owners of the oil farms remain. These, for the most part, were descendants of the first settlers. Content to wrest a bare subsistence from the rugged soil, they were greatly astonished at the discovery of the immense treasures that had so long remained dormant beneath their feet. While a number sought to retain the paternal acres, fraught with endearing remembrances, the offer of large sums proved too great a temptation. The money was paid, the owner taking possession, and with many misgivings, they sought new locations, bought farms, and are continuing life in the same channel they began it.

The foregoing is, in brief, the general and social characteristics of the representative people of the Pennsylvania Oil Region, as we have found them. Having no desire to be partial, we are willing to admit that many are to be found in each of our localities who would fall below the standard. We speak understandingly of the class we have moved and mingled among. The unworthy class of community we have no desire to write of, content that their own sins should find them out, and due punishment be awarded by others than ourself.

# CHAPTER XIV.

## THE OIL FARMS OF THE PENNSYLVANIA PETROLEUM REGION—TOPOGRAPHICAL AND STATISTICAL.

A BRIEF description of the location, and development had upon each farm embraced in the present area of the oil producing portion of Venango and the adjoining counties, together with such information regarding them, both of a general and local character, as could be obtained, is considered by us as the most important portion of this work. Such, indeed, was the main object for which it was first designed.

The magnitude of the task will be apparent to the reader. The plan is as comprehensive as we could make it, and though not as full in some localities as we could wish, is sufficiently so for all general reference.

By the subsequent pages, the amount and nature of the development upon each farm where any development has been inaugurated, from the time of the striking of Drake's well, in August, 1859, to the date of January 1, 1869, is fully shown. The descriptions that follow contain many items of information and phenomena that could not well be noticed in any other connection, without rendering the same fragmentary.

It has been impossible to obtain full information regarding all the farms noticed by us; so many of the same belonging to oil companies having been deserted at an early date by their owners, or changed hands in later years to satisfy hungry claimants. On a large number of others no active operations have prevailed for several years, leaving no sign to show that any development had ever been prosecuted. The oil-bearing localities are annually shifting, embracing each season lands hitherto deemed worthless for oil production. Hence the importance of preserving a record of the previous operations in all the different localities, as a guide for the operator, as well as to show the vast labor and money expended in the Petroleum development. As a matter of history, too, we regard the record we have made as invaluable.

In this connection we deem it necessary to state, that until within the last three years, no inconsiderable degree of efficiency has been attained in the mode of drilling and managing oil wells. Like all others, it was a business that had to be learned, as no other branch of mining afforded any accurate light to guide the oil operator. Money and skill without stint has been freely given to attain even the present standard of excellence. New improvements are being constantly introduced, and the culmination of perfection is yet far distant.

The numerous failures to meet with success in earlier years, is greatly attributable to the want of skill on the part of the operators. A comparison of their mode of management with that of the present day, clearly establishes this fact, and if any further proof is required, it can be found in the subsequent development of good paying wells in the same localities. The general average of producing wells, at the height of the development of 1865 and 1866, at a liberal estimate, was scarcely twenty per cent. At the present date we estimate the failures, in the general aggregate, at less than ten per cent. By this rate of progress, the operators, it will be seen, have been diligent.

The average extent of the leases granted to oil operators, has been from a half to an acre each. Half an acre furnishes ample room for the derrick, engine house, tank, &c. Acre leases are generally given out at present. Many practical operators prefer to take leases of several acres, and are thus better protected from interference from other and adjoining wells, it being generally conceded that wells drilled in close proximity soon interfere with each other. Allowing half an acre to each well that has been drilled, the reader can form an accurate idea of the limited amount or proportion of the Petroleum Region that has been really developed, and the proportion that is still susceptible of the same, from a perusal of the description of the farms that follow. When the act is taken into consideration that many of the wells, after becoming to all appearance exhausted, have produced abundantly after being abandoned for years, and the finding of green oil in a sandrock below where the black was obtained, and the not improbable theory that sandrocks equally as abundant in Petroleum, will eventually be found below the present oil-bearing strata, gives a reasonable promise of an inexhaustible supply for as long a period as it may be needed.

The depth at which the different sandrocks have been obtained in each separate locality, their thickness, depth of the wells, general and local features of each farm described, is a matter of the highest importance, not only to the operator, but to the general and scientific reader as well.

With the above explanation we shall proceed to the consideration of the subject indicated:

*Haliday Farm.*—Is located at the confluence of Oil Creek with the Allegheny River, lying on the west side of the Creek and north side of the river, in Cornplanter township. A large portion of the farm is hilly, sparsely covered with timber, and difficult of access. It is bounded on the east by Oil Creek, south by Main street, Oil City (the same running parallel with the Allegheny River), west and north by tract formerly owned by J. Nevins. Haliday Run, a small stream passes through the back portion, and along the northern side of the front, running mainly in a due south course, and empties into the Allegheny River, a short distance below the mouth of Oil Creek. The river at this point (Oil City), runs nearly due west. The road from Oil City to Dempseytown passes up Haliday Run and through this farm in a northerly direction. Original owner, F. Haliday. Present owners, F. Haliday and Midas Petroleum Company, of Pittsburgh and New York; National Oil Company of Venango County; Girard Oil Company, Philadelphia Oil Company, Empire Oil Company, and others. Farm contains 500 acres.

The Midas Company have twenty-one acres next to Oil Creek and Main street, Oil City, embracing a portion of the borough; about 200 rods on the Creek and 30 rods on Main street. The Linden well, on Main street, is on this property. Commenced to produce, by flowing, in 1861, about 30 barrels per day; is 560 feet in depth; flowed three and a half years, and is still being pumped. Amount of oil from this well 75,000 barrels. The other wells are all on the Creek side and on the Run. One well on Run belongs to the Oil Creek Petroleum Company. Those on the Creek, eight in number, belong to the companies named. In

1865 and 1866 there were on this property eight stationary and portable engines, with ordinary machinery. Depth of first sandrock, 127 feet; second, 325 feet; third, 480 feet. Thickness of sandrock: first sand, 21 feet; second, 25 feet; third, 30 feet. No active developments at present time. Several adjoining farms have been recently purchased with a view to active operations. Cost of wells average, $4,000.

*Nevins Farm.*—Located on north side of Allegheny River and west side of Haliday Run, Cornplanter township. Bounded on the north by Foster and others, on the east by Haliday, south by river and Moran, and west by Moran and Piaget, and contains about 500 acres. Formerly owned by J. Nevins, and afterwards by Plumer and Drum, of Franklin. Sold by them to the Michigan Rock Oil Company, in 1860. Present owners, T. B. Porteous, P. H. Tiernan, Geo. Cornwall, H. M'Clintock, A. D. Barbour, Wm. M. Abrams, R. Chisholm. A portion of the purchase of the Michigan Rock Oil Company forms the greater part of the western portion of the borough of Oil City. The present owners erected a number of buildings on their property, and also made the Oil City Driving Park; driving track one mile in length, on the top of the hill, with a road, ascending by easy grades, leading to the same; the park being enclosed by a substantial fence. In the flush days of 1865 a race meeting was had, and the meetings for such purposes have since been continued. The wells are on Haliday Run, six producing and four non-producing, all pumping. Average production in 1865 and 1866, 35 to 40 barrels per day. Depth of the wells, 518 to 689 feet. Average cost, $5,000. Wells all on flat and hillside, and owned by lessees. Six portable and stationary engines on the property. Depth of first sandrock, 123 feet—20 feet thick; second sand, 380 feet—25 feet thick; third sand, 495 feet—30 feet thick. Depth of driving-pipe, 12 feet. Hills high and abrupt, fronting on river and Haliday Run; on Charley Run on west side. Level on top of hill. Two refineries, Bagg, Tillson & Co., capacity, 300 barrels crude per week; Duncan, Dunlap & Co., capacity same.

Both refineries have since been discontinued, and the production of the wells on side next to Haliday Run is nominal at present. During present year three good wells have been obtained on the portion of this property lying on Charley Run, two of which claimed a production at the commencement of 100 barrels per day each. Present daily production of the three wells from 100 to 125 barrels per day. These wells flowed at their commencement, with a large amount of gas. The Blakely well, the first struck, being the heaviest. The gas from this well came forth for weeks with a roar like a small volcano, and saturated the adjacent territory with oily spray. A number of leases have been made by the owners of the land, and the wells are being drilled rapidly. It has been decided that a greater portion of this farm is in the celebrated "oil belt." At the present rate of improvement on the development, this territory promises to be as productive as any in the Oil Region, while its shipping facilities are unequalled. An oil-pipe has already been laid from these wells to the lower part of Oil City.

*Bastian Farm.*—On Allegheny River, south side, opposite mouth of Oil Creek, in Cranberry township, contains eighty-two acres of land. Lies on a gentle slope, back from the river, with a northern descent toward river. Is beautifully located for building purposes. A portion of Venango City is built upon this farm. Formerly owned by Henry Bastian. Purchased by Wm. L. Lay, in 1862. Present owner, Laytonia Town & Oil Company. Purchased by them in 1864. Bounded on north by river, east by Downing, south by Downing, and west by Lee. No streams passing through. River fronting north runs due west. Allegheny Valley Railroad runs along river front. Laytonia and Salina Turnpike runs through property in southerly course. Three producing and three non-producing wells on property. Amount of product very small. Wells on river bank and flat. Owned by company. Three portable engines. Depth of first sandrock, 230 feet—30 feet thick; second sand, 365 feet—35 feet thick; third sand, 485 feet—30 feet thick. Depth of driving-pipe, 50 feet.

Scarcely a vestige of these wells remain at the present date. Mr. Geo. V. Forman has commenced to drill two wells on the bluff, at the southern side of the town. The river bridge fronts on this property, and the Cranberry Coal Company's Railroad passes through it.

*Lee Farm.*—On Allegheny River, south side, opposite Oil City, lying on a gentle slope back from the river, like Bastian farm, with a northern slope toward river. A portion of Venango City is on this farm, and Allegheny Valley Railroad passes along the river front. Contains 180 acres. Original owner, Jas. H. Lee. Present owners, heirs of Lee, Wm. Reed, and others. Mr. Reed purchased a portion of the farm in 1864. Bounded on the north by river, east by Bastian, west and south by Moran. No stream passing through

River fronting it on the north, and runs west. State road runs south from this property. Three producing and eight non-producing wells, and one in progress of drilling. Producing wells average three barrels each. Average depth of wells, 550 feet. Commenced producing in 1863 and 1865. Wells all on river bank and flat above, and belong to C. Haines, Wm. Reed, J. Dickey, J. Riddle & Miller, and others, lessees. Five portable and stationary engines on the property. Depth of first sandrock, 230 feet—30 feet thick; second sand, 365 feet—35 feet thick; third sand, 485 feet —30 feet thick.

The majority, if not all the wells spoken of above, are producing little or nothing now, and no active development is in progress.

*Moran Farm.*—Located on both sides of Allegheny River, containing 360 acres on the north side and 160 on the south side. North side lies mostly on high hills, with a narrow bench along the river, and a narrow flat on Charley Run, which passes through the northern portion, running south into the river. The Atlantic and Great Western Railway and Oil City and Franklin Turnpike pass along this narrow table. The south side of river land lies on an easy slope back from river, similar to Lee and Bastian farms, affording a fine scope of territory for the extension of Venango City. Bounded on the north by Nevins tract, east by Nevins and Lee, south by Farran and others, west by Farran, river, and Piaget. River passes through property, running west and southwest. Railroad and turnpike taking same course. Allegheny Valley Railroad runs through property on south side following river bank.

The property was bought from Wm. Bastian by Mr Angier, from whom Mr. Moran purchased some twelve years since. Present owners, Moran heirs, and Orange Oil Company, who purchased fifty acres on south side of river, adjoining Lee farm, in 1865. On north side of river are two producing wells, and five non-producing. Average daily product, ten barrels per day. The producing wells are owned by J. K. Morange, one of which produced for four years. Both these wells on river bank. Those on Charley Run have not as yet proved paying wells. One at the mouth of the Run, belonging to J. B. Stockton & Co., has produced some oil, but is doing nothing at present. The wells are from 550 to 570 feet deep. Amount of production not ascertained. Two engines on this part of the property, with ordinary machinery. Depth to first sandrock, 202 feet—29 feet thick; second sand, 352 feet—25 feet thick; third sand, 469 feet—40 feet thick. Depth of driving-pipe, 7 feet. Sandrocks deeper as ascending Charley Run. On south side of river, Cranberry township, are two producing wells, one non-producing, and one in progress. Average daily production, 12 barrels. Average depth of wells on this side, 520 feet. Average cost of wells, $4,000. Commenced producing in the Spring of 1864. Amount of production not ascertained. Wells are on river bank and flat above. Three engines, stationary and portable, on the property. Depth of first sandrock, 225 feet—35 feet thick; second sand, 360 feet—30 feet thick; third sand, 475 feet—30 feet thick. Depth of driving-pipe, 50 feet. On this tract is some good tillable land. Moran's Eddy is in the river at this point.

But little or nothing doing on this farm at present time. The recent developments on the upper portion of Charley Run may possibly appreciate its value materially.

*Piaget Farm.*—On north and west side of Allegheny River, Cornplanter township, contains 147 acres. Bounded on north by R. M'Fatè, east by Moran, south by River, west by Howe and Hickman. Original owner, L. H. Piaget. Present owner, Oil City Petroleum Company. Purchased in 1864, for the sum of $105,000. Six and a half miles from Franklin, and one half mile west of Oil City. Allegheny River runs southwest here, and fronts the farm. The Atlantic and Great Western Railroad and Franklin and Oil City Turnpike pass through the property along the river bank. Hills high and unproductive, covered with timber and rocks. Hickory Island lies in the river in front of this property. Five producing and three non-producing wells, and one in progress. The five producing wells average four barrels each, an average total of twenty barrels per day. Have produced over 500 barrels. Average depth of wells, 500 feet. Average cost, $6,000. Commenced producing in the Spring of 1865. Are on river bank. Owned by New York & Newark Petroleum Company and J. W. Fielder & Co. Six portable engines and ordinary machinery on the property. Depth of first sandrock, 215 feet—25 feet thick; second sand, 350 feet—50 feet thick; third sand, 475 feet—25 feet thick. Depth of driving-pipe, 20 feet. The Cornplanter, the leading hotel of early times, favorably known to our pioneers for its good cheer, is on this property, and was the home of Mr. Piaget.

Nothing doing now in the shape of development, and the production merely nominal.

*Farran Farm.*—On southeast side of Allegheny River, in Cranberry township. Contains 175 acres.

Bounded on the north by Moran, east by Moran, Hayes, and others, south by Hayes, and west by river. Original owner, J. Farran. Present owners, Buchanan Farm Oil Company and Passaic Oil Company of Newark, New Jersey. Passaic Oil Company have fifty acres, and Buchanan Farm Oil Company the balance. Is one mile southwest from Venango City, on the road leading along the bank of river, which fronts it on the west. Seven producing wells, five on Passaic and two on Buchanan, and nine in progress—five on Buchanan and four on Passaic. None flowing. The seven wells are producing by pumping from fifteen to thirty-five barrels each. Daily product, 120 barrels. Have produced 2,000 barrels. Commenced producing July and September, 1856. Average depth of wells, 525 feet. Average cost, $4,000; all on the flat or table lands, owned by the companies and lessees. Fifteen engines, portable and stationary, on the property. Depth of first sandrock, 225 feet—30 feet thick; second sand, 360 feet—35 feet thick; third sand, 475 feet—30 feet thick. Depth of driving-pipe, 45 feet. Lies on a gently elevated table land, sloping back from river. All the wells tested so far have been paying ones. Well located for operating upon.

Active operations ceased some time since upon this farm. But few of the old wells are producing, and but few new ones have been drilled. It is looked upon as good territory.

*Hickory Island.*—Is in Allegheny River, below Moran's Eddy, and opposite Farran farm. Contains seven acres. Boundary, river. Original owner, J. Farran. Present owner, Hickory Island Oil Company. They purchased it in 1864. The flood of March, 1865, washed away portions of the island, so that the original seven acres are not at present visible. The works of the company were also washed away, requiring replacing with new. The island is reached by skiffs from any direction, and by teams in times of low water. One producing and three non-producing wells on the is and. The producing well flows three barrels per day; is an old well. Amount of production not ascertained. Average depth of wells, 545 feet. Average cost of wells, $6,000. Owned by company, Willings & Powell, and Cherry Run & Pithole Petroleum Company. Have two engines on property. Depth of first sandrock, 240 feet—30 feet thick; second sand, 360 feet—30 feet thick; third sand, 515 feet—15 feet thick. Depth of driving-pipe, 40 feet.

All operations have now ceased. But a small portion of the orginal dimensions of island left.

*Hayes Farm.*—On east side of Allegheny River, in Cranberry township, opposite Reno. Bounded on north by Farran, east ———, south by Huff, and west by river. Original owners, J. P. & Allen Hayes. Present owners, J. P. Hayes, Buchanan Oil Company, heirs of Allen Hayes, and Artesian Oil Company. Two miles west and south from Oil City, with a road along the river bank leading to it, and roads to State road. Six producing wells and seven non-producing wells on the property. Of producing wells one is flowing and five pumping. Average daily production, 60 barrels. Whole amount of production not ascertained. Average depth of wells, 525 feet. Average cost, $5,000. Commenced producing in 1861, and on to 1865. Wells all on flat and table land. Owned by Thompson & Turner, Adams, Taylor, Patch, and others. Six engines, portable and stationary, with ordinary machinery on the property. Depth of first sandrock, 240 feet—25 feet thick; second sand, 360 feet—25 feet thick; third sand, 485 feet—25 feet thick. Depth of driving-pipe, 30 feet. This farm lies on a flat, gently elevating as it recedes from the river, with a ravine making down through it. Apparently good oil territory. Very little doing at the present date, and production merely nominal.

*Wilson, Shaffer & Ormsby Farms.*—On west and north side of river, in Sugar Creek township. The river here changes its course from west of south to due west and north of west. These tracts embrace 1,206 acres. Original owners, Jas. Wilson, J. Shaffer, and ——— Ormsby. Present owners, Reno Oil and Land Company. Purchased in the Fall of 1865. Bounded on the north by Hickman, Rossman, and M'Fate, east by Piaget and river, south by river, and west by Brannon. Three miles west of Oil City and four miles east from Franklin. Shaffer Run passes through the property on the western side, in a southwest course. Atlantic and Great Western Railroad and Franklin and Oil City Turnpike pass through on the river side. Reno and Pithole Railroad branches off at this point, passing through the property, following on Shaffer Run, as it winds its way over the hills and valleys of Cornplanter Run, Oil Creek, Cherry Run, on to Pithole. This property has a large river front, slopes back gradually from the river, with a southern ascent, and is the location of the once famous town of Reno. Ten producing wells on the property, most of them small. Average daily production, 50 barrels. Average depth of wells, 600 feet. Average cost, $6,000. None flowing, all pumping. Amount of production not ascertained. Commenced producing in 1861 and on to 1865. Wells all on flat. Ten engines, portable and stationary, on the property. Wells owned by company in fee, working interests by lessees—Say & Co., Gardner & Co., Pacific Oil Company, and others. Depth of first sandrock, 230 feet—20 feet thick; second sand, 400 feet—25 feet thick; third sand, 550 feet—10 feet thick. Depth of driving-pipe, 12 feet. Two refineries on the property—one Kinkaid & Lockwood's, capacity 350 barrels crude per week; the other about same. Howe's Eddy, in river, is at this point. A small island also in river. The company contemplate sinking fifty new wells the coming season, (1866), some of which are already under contract.

The collapse of the Reno scheme in 1866 caused active operations to cease, and the same have not yet been resumed. Mr. Culver still resides on the property, and has, we learn, obtained control of, and intends to develop it. The railroad is now torn up. Production at present slight. Reno is but a small village, with post-office, stores, and one refinery, and a hotel. The Atlantic and Great Western Railroad have a good brick depot here.

*Huff Farm.*—On south side of Allegheny River, in Cranberry township, opposite Reno. Contains 140 acres. Original owner, J. A. S. Huff. Present owners, F. Prentice and Manross Oil Company. Leased to Ballard & Filley. Bounded on the north by river and Hayes, on east by ——, on south by Huff & Milton, west by Milton & Plowman. Four and a half miles from Franklin, two and a half miles from Oil City. Huff's Run passes through the property, running north. River bends from southwest to west. Roads leads along river to Venango City in an eastern direction. Six producing wells on the property, one flowing, and five pumping, five more in progress. Average daily production ten barrels. Have produced over 1,000 barrels. Average depth of wells, 550 feet. Average cost, $5,000. Some of the wells are old ones. Commenced producing in 1860, 1861, and 1865. All on the flat. Owned by Susquehanna Oil Company and others, lessees. Ten engines, portable and stationary, on the property. Depth of first sandrock, 233 feet—32 feet thick; second sand, 370 feet—32 feet thick; third sand, 495 feet—30 feet thick. Depth of driving-pipe, 19 feet. Here the flat on this side of the river ends against an abrupt hill, which comes to the edge of the water. On Magee farm, up Huff Run, are two wells, both non-producing. This looks like good oil land, but needs energetic development.

The production of the farm at present date is merely nominal, and no active development is in progress. The Allegheny Valley Railroad passes through the farm.

*Plowman Farm.*—On south side of Allegheny. Contains ——— acres. Original owner, J. Plowman. Present owner, Middletown Lubricating Oil Company. Time of purchase not ascertained. Bounded on the north and east by river, south by Cox, Hensely, Milton, Hay, and others, west by river and Neely. Here river bears northwest and then southwest. No road along river. Hills high and coming down abrupt to river edge, covered with timber. Three non-producing wells on the property, and two engines. Depth of wells, sandrocks, &c., not ascertained. No one on the ground to make inquiries of. Nothing doing. A narrow ravine comes down through, making to the river.

*Neely Farm.*—On south side of Allegheny River, in Cranberry township. Contains —— acres. Original owner, P. Neely. Present owners, Pennsylvania Petroleum Company, and Great Northern Oil Company.

Purchased in 1864. Great Northern Oil Company have fifty acres. Bounded on the north by Plowman and river, east by Plowman, south by Hay and others, west by river. Two miles from Franklin, and five miles from Venango City and Oil City. The Allegheny River fronts it on the northwest; river running southwest. Seneca Run comes down from the hills through the farm in a northwest course into river. Three wells have been sunk on the farm, all non-producing. No one on the ground. Depth of wells, sandrocks, &c., not ascertained. Lies on high hills, difficult of access. Is opposite Shirk & Fuller farms.

*Milton Farm.*—On south side of Allegheny River, Cornplanter township. Contains —— acres. Original owner, J. Milton. Present owner, Highgate Petroleum Company. Is triangular in shape. Bounded on north and west by river, east and south by Bowser. One mile from Franklin and six from Venango City. No road along river. Course of river south of west. Two wells, both non-producing. No one on farm to get items of. Lies opposite Hiland's farm.

*Bowser Farm.*—Located on south side of Allegheny River, Cornplanter township. Contains —— acres. Original owner, P. Bowser. Present owner not ascertained. Bounded on the north by river, Neely, and Milton, east by Hensely, south by Milton, west by Milton and river. One mile from the Franklin bridge. River bears south of west. No streams passing through. A small flat or table land on river, but little good territory for operating on. Have been four wells sunk, all non-productive. Operators evidently became discouraged, as no one was on the property to obtain information from.

*Hoge Island.*—Is in the Allegheny River, between Hilands, Plumer, and others on north side, and Neely, Bowser, and others on south side of river, extending down nearly to mouth of French Creek. Contains forty-five acres. Original owner, Thos. Hoge. Present owner, Hoge Island Lubricating Oil Company. It is six miles from Oil City, and adjoins the borough of Franklin. Course of river, south. Communication by skiffs and boats, and by teams in low water. The property is valued by the company at $200,000. It may possibly be worth this amount when they get enough producing wells. Have two producing and two non-producing wells, and are producing eight barrels of lubricating oil per day. Aggregate amount produced not ascertained. Average depth of wells, 350 feet. Average cost, $3,000. Commenced producing in 1862 and 1865. The island is but slightly above the level of the river, and in times of high water is sometimes under the surface of same. Wells owned by company. Have two engines, portable and stationary. Depth of first sandrock, 240 feet—32 feet thick; second sand, 620 feet—40 feet thick; third sand not found. Depth of driving-pipe, 37 feet. The flood of March, 1865, did much damage to the company's property by sweeping over the island.

*Martin & Epley Tract.*—Located on the north side of the Allegheny River and French Creek, in Sugar Creek township. Contains twenty-two acres, divided up and sold out in small lots to various parties. It is known as out-tract No. 8, in the borough of Franklin. Original owners, Martin & Epley. Present owners, Lycoming & Clinton Counties Oil Company, Fay, Great Northern Lubricating Oil Company, Hues, and others. Is in the borough of Franklin, and seven miles from Oil City. Atlantic and Great Western Railroad and Franklin and Oil City Turnpike pass through on river bank and table lands. There are eight producing and eight non-producing wells on the property. Average daily yield, twenty-three barrels of lubricating oil. The producing wells average from three to five barrels each. All pumping. Average depth of wells, 500 feet; one 542 feet. Average cost, $5,000 to $7,000. Commenced producing in 1861, 1862, and 1865. Wells all on flat, and owned by companies and parties as above described. Have nine engines on the property, with ordinary machinery. The Great Northern Company run three wells with the same engine that drives their lubricating oil works. Depth of first sandrock, 280 feet—40 feet thick; second sand, 510 feet—30 feet thick; third sand not found. Depth of driving-pipe, 40 feet. The Great Northern Lubricating Oil Works are located here, on the former site of Mr. Thos. Hoge's residence, and have a capacity of 200 barrels per day of lubricating oil, known as the Hendrix Lubricator. These tracts are in the borough of Franklin, and the boundaries will be by other tracts in the same. French Creek at this point runs southeast, and forms a confluence with the Allegheny River. A substantial dam is built across French Creek at this point which supplies the Venango Flour Mills with necessary motive power.

*Chambers Farm.*—Located on south and east side of Allegheny River, Cranberry township, opposite Franklin. Suspension Bridge crossing at this point. Hills high and abrupt. This tract contains —— acres. Bounded on the north by Bowser and river, east by J. Chambers, south by Fuller and west by river. River bears east of south. Franklin and Clarion Turnpike passes along river bank east of south from suspension bridge across the river. No profitable developments on the farm. But little territory for operating on, except on top of the high bluff.

*Borough of Franklin.*—(See *Towns* of *Oil Region.*) Here and at various points along French Creek, Oil Creek, and the Allegheny River are the remains of large pits, generally of a square or oval shape, made by a people who inhabited North America at a period so remote that tradition fails to shed any light upon them. It was in one of these localities that the first well was sunk in Venango County, by Colonel Drake, under the direction of a company of gentlemen of New Haven, Connecticut.

There are about fifty derricks standing in the borough limits. Many are old ones, simply monuments of the past, indicating where the explorer has searched for and in many instances obtained oil. Others show where oil is now produced in quantities varying from one to four barrels per day. A number of new ones mark the location of new wells in progress. The wells vary in depth from 120 to 700 feet. Oil is found in most of the wells between the first, or bedrock, and the first sandrock. Oil is found principally in the first sandrock, which varies in depth, in different localities, from 265 to 465 feet. The second sandrock is found on the Brown tract, in Nicholson Run Oil Company's well, at a depth of 715 feet from the surface. This is near the river, at the lower part of borough. In the Oleo Oil Company's well, back near the hill, it is found at a depth of 502 feet, showing a heavy slope in the rock toward the river. Depth of driving-pipe on the river, 30 feet; on French Creek, 35 feet. Thickness of first sandrock on the river averages 20 feet; on French Creek, 30 feet, the Mattawan well, on Creek, just drilled, taken as a guide. Most of the wells in progress are old ones, being reamed out and drilled deeper. Some old wells are still producing, as they have been for several years. Among those familiar in the past, are the Mammoth well, producing at present time four barrels of lubricating oil per day, selling at $25 per barrel. It has produced since 1860. The Broomstick well is another which has produced largely, but is idle at present.

Among the most noted is the Evans' well, the first one drilled in the borough of Franklin, and the second in the Oil Region. This well is only 120 feet deep. Flowed at first seventy-five barrels per day, but soon run down to forty barrels per day, continuing to yield at that rate for a considerable length of time. Is not in operation steadily at the present time. Produces four barrels per day of lubricating oil, when pumped. Mr. Evans still owns the well in fee, has leased to N. B. Mosely & Co., of Philadelphia. The well commenced producing in the winter of 1859 and 1860. There are three refineries within the borough limits, one belonging to S. F. Dale, known as the Dale Oil Works, having at present a capacity of refining 300 barrels of crude oil per week. They have three acres of land, are located near the Atlantic and Great Western Railroad depot, on French Creek. One on the Blakely farm, Allegheny River, in lower part of borough, has a capacity of 160 barrels crude per week. The other is an old refinery on the opposite side of the river, is not in operation. Belongs to Norfolk Oil Company. Capacity, 160 barrels crude per week. The works of the Great Northern Lubricating Oil Company, are in the borough limits, and have been previously noticed.

But few, if any, of these wells are operated at present.

*Booth Farm.*—On east side of Allegheny River, in Cranberry township, being a portion of the Fuller farm and adjoining the Chambers tract. Contains sixty acres. Original owner, —— Fuller. Present owners, —— Simonds, —— Jacobs, J. Wilkes Booth's (of in famous notoriety) heirs. Purchased in 1864. Bounded on the north by Chambers, east by Bissell, south by Fuller, west by river. Is half a mile south of Franklin bridge, Allegheny River fronting it on the west. No streams passing through. Franklin and Clarion turnpike passes through on the bluff, running north to bridge, There are one producing and two non-producing wells. The producing well gives a daily yield, by pumping, of twelve barrels per day. Average depth of wells, 500 feet; one 800 feet in depth. Average cost of wells, $3,500. Commenced producing in 1864. Wells located on river side in the bluff. Owned by parties above-mentioned. Have two portable engines. Depth of first sandrock, 330 feet—20 feet thick; second sand, 440 feet —20 feet thick. No third sandrock found. Depth of driving-pipe, 40 feet.

Nothing doing in this locality at present time.

*Fuller Farm.*—Located on the east side of the Allegheny River, Cranberry township, containing acres. Original owner, —— Fuller. Present owner, Fuller Farm Oil Company. Purchased in 1865. Bounded on the north by Booth, east by Bissell, south and west by river. Is one and a half miles from Franklin. Allegheny here has a southeast course. Franklin and Clarion Turnpike passes through it. No producing wells at present. Seven non-producing, two of which produced two barrels per day formerly. Average depth of wells, 550 feet. Average cost, $3,500. Wells all on the flat. Owned by Fuller Farm

Oil Company, Radnor Oil Company, and Boston Oil Company. Have three engines, portable and stationary. Depth of first sandrock, 60 feet—4 feet thick; second sand, 470 feet—9 feet thick. Depth of driving-pipe, 25 feet. This property is well located for oil operations and easy of access.

*Blakely Farm.*—Located on west side of Allegheny River in Sandy Creek township, containing 230 acres. Original owner, —— Blakeley. Present owners, N. B. Mosely & Co., of Philadelphia, and Murphy & Irvine. Purchased in 1864. Boundaries not ascertained in full. One half mile below, and adjoining the borough of Franklin. Two small streams pass through and empty into the river. A broad flat; good territory for oil operations, both on river and streams. Has the appearance of good property. Hills much broken. No producing wells. Eight non-producing wells. One pumping with a blower but doing little good. Two of the wells have formerly pumped some oil. One was drilled by water-power, the machinery of which is still standing Wells all on flats. Owned by companies and individuals. One portable engine, water-power, and ordinary machinery on the property. Depth of first sandrock, 75 feet—8 feet thick; second sand, 205 feet —25 feet thick; third sand not found. Depth of driving-pipe, 39 feet. One refinery, as mentioned in the description of Franklin.

*Hastings Farm.*—On west side of Allegheny River, in Sandy Creek township. Contains —— acres. Boundaries not ascertained in full. Original owner, —— Hastings. Present owner, Pacific Oil Company, of New York. Purchased in 1864. One and a half miles below Franklin. Allegheny River running east of south, is the only stream. The road to Franklin running north of west. Two producing wells, and one in progress. Pumping but very little. Are not paying wells. Average depth of wells, 500 to 600 feet. Cost, $4,000. Wells on the flat and hillside. Owned by company. Three portable engines. Depth of first sandrock, 320 feet—25 feet thick; second sand, 442 feet —20 feet thick; third sand, 600 feet. Depth of driving-pipe, 28 feet. What oil there is, is found in the second sandrock.

*Bissell & Stewart Farms.*—East side of Allegheny River and on Lower Two Mile Run, Cranberry township. Contains 181 acres. Original owners, Geo. H. Bissell & Vance Stewart. Present owner, Cameron Petroleum Company, of Philadelphia. Purchased in the Spring of 1865. Bounded on the north by Chambers, east by Brandon, south by river, and west by Fuller Farm Oil Company. Two miles from Franklin. Lower Two Mile Run passes through the property in a westerly direction. Franklin and Clarion Turnpike passes through it, running northwest to Franklin. No producing wells. Two non-producing, and ten in progress. Average depth of wells, 500 feet. Average cost, $4,000. All on the flat and on the Run. Owned by the lessees. Ten engines, portable and stationary, on the property. Depth of first sandrock, 300 feet—40 feet thick; second sand, 445 feet—25 feet thick. No third sand found. Depth of driving-pipe, 30 feet on the flat, and 5 to 10 feet on the Run. Has the appearance of good oil territory. There is an old grist mill near the mouth of the Run. The Franklin and Clarion Turnpike strikes the river bank at this point.

*Irvine Farm.*—Located on east side of Allegheny

River, in Cranberry township. Contains five acres, the company having 101 acres on the river below. Original owner, —— Irvine. Present owner, Pennsylvania Oil and Coal Company, of Baltimore. Purchased in in Fall of 1864. Full boundaries not ascertained. Three miles from Franklin. Allegheny River the only stream. Bearing of same southeast. Road to Franklin running northwest. One producing and two non-producing wells, one of which pumping ten barrels per day. Average depth of wells, 480 feet. Average cost, $3,000. Commenced producing October, 1865. Wells are on the hillside. Owned by the company. Two stationary engines are on the property. Depth of first sandrock, 300 feet—40 feet thick; second sand, 435 feet—15 feet thick; no third sandrock found. Depth of driving-pipe, 25 feet. The company have coal on their land on the river below.

*Cochran Farm.*—On Allegheny River, east side, in Cranberry township. Contains 100 acres. Original owner, —— Cochran. Present owner same. Leased to various parties and companies. Bounded on north by Brandon, east by Hoover, south by river, west by river and Bissell. Two and a half miles from Franklin. No streams passing through. Road to Franklin crosses it in a northeast direction. This territory is well located for operating upon. There are fourteen producing and five non-producing wells. All pumping wells. Produce from three to thirty barrels each. Average daily production, ninety barrels per day. Aggregate amount of production not ascertained. Average depth of wells, 450 feet. Average cost, $4,500. Commenced producing in 1861, 1862, and 1865. Wells all on flat and hillside. Owned by individuals and companies. Fourteen engines, portable and stationary, with ordinary machinery, on the property. Depth of first sandrock, 285 feet—50 feet thick; second sand, 440 feet—20 feet thick; third sand, 575 feet—5 feet thick. Depth of driving-pipe, by river, 45 feet; back of hill, 50 feet. Energy in operating seems to prevail here, and the reward of labor greater than at many other points.

*Hoover Farm.*—On east side of Allegheny River, in Cranberry township. Original owner, —— Hoover. Present owner, River Oil Company, of Philadelphia. Contains 100 acres. Boundaries not ascertained in full; fronting the river on southwest side. River bearing southeast. Franklin road passes through. No streams passing through, but a ravine makes a break through the farm. Is two miles from Franklin. Hills rise gradually back from river, presenting quite a large scope of territory for operations. Developments are rather limited. There is one producing and four non-producing, and one well in progress. The producing well is pumping four barrels per day. Amount produced, seventy-five barrels. Average depth of wells, 500 feet, one 700 feet. Average cost, $3,500. Commenced producing September, 1865. Wells all on the flat and ravine. Owned by company. Have two portable engines. Depth of first sandrock, 330 feet—20 feet thick; second sand, 440 feet—20 feet thick; third sand not found. Depth of driving-pipe, 40 feet on river, and 20 feet on ravine.

*Harmon Farm.*—On west side of Allegheny River in Sandy Creek township. Contains —— acres. Original owner, —— Harmon. Present owner, Hoover Oil Company, of Philadelphia, and Eureka Oil Company. Purchased in 1864. Boundaries not fully ascertained. Is two and a half miles from Franklin. No streams passing through. River bearing southeast. No road on river side except a foot-path. There are six wells pumping, producing a little oil, not in very extensive quantities, and seven wells in progress. Average depth of wells, 460 to 600 feet. Average cost, $5,000. Commenced producing in 1865. Wells all in bluff on the river bank. Owned by individual companies. Have thirteen portable engines and ordinary machinery. Sandrocks, or drillers, vary as to depth materially, ranging all the way from 72 to 275 feet deep for first sandrock—45 feet thick; second sand, 306 to 425 feet—18 feet thick. No third sandrock found. Depth of driving-pipe, 24 feet. Those interested here displayed a wonderful amount of energy in prospecting for oil, but the wells have returned a poor reward for the capital and labor invested. Hills high, coming down abrupt to the river, compelling operators to dig into the hills to set their engines and commence operations.

*Martin Farm.*—On west side of Allegheny River, in Sandy Creek township. Containing 160 acres. Original owner, Amos Martin. Present owners, Tulpehocken Oil Company, of Philadelphia; Honeycomb Oil, Suffolk & Venango Oil Company. Is three miles from Franklin. Hills high and abrupt, with but narrow table land to operate upon. Wells logated in bluff and on river bank. Owned by companies and lessees. There are six producing and four non-producing wells on the farm. The producing wells are producing from seven to thirty barrels each daily. Average daily production, forty-four barrels; 4,000 barrels shipped in 1863 and 1864. Recent shipments not ascertained. Average depth of wells, 450 feet. Average cost of same, $4,000. Commenced producing in 1863 and 1865. Have six engines, stationary and portable, with ordinary machinery. Depth of first sandrock, 180 feet —40 feet thick; second sand, 420 feet—40 feet thick; third sand supposed to be 640 feet deep, not known. Depth of driving-pipe, 33 feet. On the Honeycomb tract the rock comes to the surface, requiring no driving of pipe.

*Nicklin Farm.*—On Allegheny River, east side, in Cranberry township. Contains ——— acres. Original owners, Roberts & Boyd. Present owners, Mammoth Oil Company. Purchased in 1859. Boundaries not ascertained. Is three and a half miles from Franklin. River bearing in a southeast course at this point. A rough road leading along the river bank to Franklin, in a northwest direction. There are six wells, producing from one to seven barrels each daily, making a total of fifteen barrels, all pumping. Average depth of wells, 500 feet. Average cost, $4,000. Commenced producing in 1861 and 1864. All on hillside, and owned by the company. Have six stationary engines. Depth of first sandrock, 300 feet—35 feet thick; second sand, 450 feet—35 feet thick; third sandrock not found. Depth of driving-pipe, 32 feet. The company have a steam saw-mill in connection with their works, affording means for working up the timber on the property, thereby increasing the revenue to the company.

*Rice Farm.*—On the east side of Allegheny River, in Cranberry township. Contains 130 acres. Original owner, G. Rice. Present owners, Green Hill Oil Company and Allegheny Coal Oil Company. Boundaries not ascertained. Four miles from Franklin, fronting on river, which bears at this point to the southeast. A rough and rocky road leads along the river bank. Have two producing and seven non-producing wells, none flowing. Two pumping two barrels each per day. Amount of production not ascertained. Average depth of wells, 600 feet. Average cost, $5,000. Commenced producing in 1861. Are all on hillside, and are owned by the companies and lessees. Have three engines, stationary. Depth of first sandrock, 160 feet—35 feet thick; second sand, 355 feet—25 feet thick; third sand not found. Depth of driving-pipe, 45 feet. Narrow table along river, giving but little room for operations. Hills abrupt and high. There is a small island in the river, just opposite, called Porter Island, belonging to the Porter Oil Company, of Huntingdon County, on which there is one well, 365 feet deep, producing twenty barrels per day by pumping. Was struck in August, 1865. Sandrocks about same as those on main land.

*Hoover Farm.*—Located on west side of Allegheny River, Sandy Creek township. Contains 272 acres. Original owner, Robert Brandon. Present owner, Jas. P. & C. M. Hoover, Geo. H. Bissell & Co., and Cameron Petroleum Company. Boundaries not fully ascertained. Four miles from Franklin, in a southeast direction. There are fifteen producing and fifteen non-producing wells. River bears east of south, fronting farm on north and east. Average daily production, 150 barrels. The fifteen producing wells are all pumping, and yield from five to sixty barrels each. None flowing. The old Hoover well commenced producing in 1860, continued for about four years, then ceased. In cleaning it out the tools became fast. After losing different sets of tools, trying to get the first out, the old well was abandoned, and a new one drilled close beside it. The new one is now producing sixty barrels per day. Amount of oil farm has produced not ascertained. Average depth of wells, 460 feet. Average cost, $5,000. Commenced producing in 1860 and 1865. Wells all on flat and hillside. Owned by individuals and companies. Have fifteen portable and stationary engines. Depth of first sandrock, 300 feet—40 feet thick; second sand, 440 feet—25 feet thick; third sand not found. Depth of driving-pipe, 40 feet on lower part of farm. On upper end the bedrock came to the surface. There is a broad flat and bends back to hills, giving ample room for operations. Snake Island is in river opposite. Is owned by a Philadelphia company. Two wells on the island, one producing. Is 440 feet in depth. Amount of production not ascertained. Have one engine. Average cost of wells, $5,000. Sandrocks, &c. same as on main land.

*Pope Farm.*—On Allegheny River, west side, in Sandy Creek township. Contains ninety acres of land. Original owner, Widow Pope. Present owner, or lessee, Pope Farm Oil Company, of Philadelphia. Boundaries not fully ascertained. Fronts the river on east side. Five miles from Franklin in an east of south direction. Road leading to Franklin passes through it in northwest direction. Five producing and five non-producing wells on farm. None flowing. Five wells are pumping five to twenty-five barrels each. Daily aggregate production, fifty barrels. Average depth of wells, 400 feet. Average cost of wells, $5,000. Commenced producing in 1864 and 1865. All on flat. Owned by company and individuals. Have five portable engines, with ordinary machinery. Depth of first sandrock, 200 feet—20 feet thick; second sand, 400 feet—15 feet thick; third sand not found. Depth of driving-pipe, 36 feet. From surface indications this ought to be as good oil producing territory as the farm above it. Siefer's Run passes through in an eastern direction, emptying into the river. Hills broken and covered with timber.

*Smith Farm.*—On Allegheny River, west side, in Sandy Creek township. Contains —— acres. Original owner, S. Smith. Present owner not ascertained. Bounded on the north by Pope and river, east by river, south by Miller, west by Foster & Brown. River here bears east by south. Brown's Run passes through in an eastern direction. Also another small run, giving the hills a broken appearance. Have two engines. Nothing doing on property. No one to get items from. So far as developed, the territory has proved unproductive in paying quantities.

*Miller Farm.*—On Allegheny River, west side, opposite mouth of Sandy Creek, in Sandy Creek township. Contains —— acres. Bounded on the north by Smith, east by river, south by river and Foster, west by river. Original owner, D. Smith. Present owner, Excelsior Oil Company. There are eight producing and seven non-producing wells. One flowing twenty barrels per day. Daily production of the eight wells, about fifty barrels. Have two engines. Depth of first sandrock, 150 feet—20 feet thick; second sand, 350 feet—15 feet thick; third sand, 600 feet—15 feet thick. Depth of driving-pipe, 40 feet. Average cost of wells, $5,000. Amount of production not ascertained.

*Upper Two Mile Run.*—On Two Mile Run, above the Hiland's farm, in Sugar Creek township, the Philadelphia Oil Company have some six wells, all non-producing. Some four miles from the river are located several oil companies, represented as follows: Great Western Oil Company, Forest Shade Oil Company, Scott Farm Oil Company, Kunkel Oil Company, New York & Philadelphia Oil Company. On the Great Western and Forest Shade Oil Companies tracts there are six wells. Some of them have been sunk to more than the ordinary depth; one nearly 1,000 feet. All of them unprofitable. Some of them have produced oil, but not in paying quantities. The sandrocks are irregular, varying in depth and thickness, not having the appearance of regular oil-bearing rock.

*M'Cormick Farm.*—Located on French Creek, north and east side, Sugar Creek township. Contains 421 acres. Original owners, W. C. & R. S. M'Cormick. Present owners, M'Cormick heirs, Eclectic Oil Company, Beacon Oil Company, Pearson Petroleum Company, New York and Franklin Oil Company, P. F. Kelly, Cattaraugus and French Creek Petroleum Company. Purchased in 1864 and 1865. Bounded on the north by Blakely, east by Galloway, M'Calmont, and others, south by French Creek, west by French Creek and Dubbs. A portion of the territory lies in the limits of the borough of Franklin, and extends two miles up French Creek. The Creek has a bearing on upper part south, bends southeast, then northeast, and east at lower part of farm in borough limits. The Franklin branch of the Atlantic and Great Western Railway passes through on the line of the Creek. There are three producing and fourteen non-producing wells on the property. None flowing. Three pumping—two on Beacon Oil Company tract, three barrels each per day; one on the New York and Franklin Oil

Company tract, two barrels per day. Amount of production not ascertained. Average depth of wells, 350 feet. Some of the old wells have formerly produced; one reported to have flowed 100 barrels per day. One of the wells on the Beacon Oil Company tract, 800 feet in depth. Cost of wells from $3,000 to $6,000. Commenced to produce in 1861, while owned by M'Cormick's. Wells all on the flat land. Have six engines, portable and stationary, on the property. Depth of first sandrock, 265 feet—30 feet; second sandrock, 500 feet—28 feet thick; third sandrock not found. Driving-pipe, from 10 to 20 feet, according to locality. Less depth near Creek. Pearson Petroleum Company have 216 acres; have eleven wells drilled and in progress. None paying wells; some oil in all of them. The Ashland Company have half an acre of land close beside the railroad track, near to or in the borough limits, with two wells, of but little account. One has produced some oil formerly, claimed to be not thoroughly tested.

*Dale Farm.*—On south side of French Creek, joining Franklin borough. Contains 200 acres. Is in French Creek township, quarter of a mile from the junction of Creek with Allegheny River, extending half a mile on the Creek. Original owner, S. F. Dale. Present owner, Dale Farm Oil Company. Purchased in 1864. Capital stock of company, $1,000,000. Number of shares, 200,000; par value, $5 per share. Boundaries not ascertained. The Creek fronting it on the north, with a bearing due east. Road to Franklin leading along bluff. The Franklin and Jamestown Railroad runs through the property. The company have one producing well, pumping six barrels per day of lubricating oil of a very dark color. This is a new well. The old well has produced a considerable amount of oil. Average depth of wells, 466 feet. Cost, from $3,000 to $6,000. Both operated by one engine. One blower set to operate the old well, which commenced to produce in 1861 and 1865. Wells on bank of Creek under the hills. Owned by company. Have one engine, stationary, oscillating. Depth of first sandrock, 217 feet—not regular, in shells; second sand, 248 feet —12 feet thick; third sand not found. Driving-pipe, 10 feet. A white sandrock, nine feet thick, was found at the depth of thirty-nine feet. Oil found at 248 feet. Company have opened a vein of coal in the hills. Found it two feet four inches thick. Supposed to be thicker at a greater depth. There is but little flat territory. A narrow bench along the bank of the Creek. Hills high, but not so abrupt as at some other points.

*Hayes Farm.*—On south side and west side of French Creek township. Containing 272 acres. Original owner, Samuel Hayes. Present owner, Ohio Carbon Oil and Mining Company. Boundaries and time of purchase not ascertained. One mile from Franklin. A small ravine makes through the hills. In this ravine, tradition says, the French found lead ore. None has been found in later days. There are no producing wells. Four non-producing; one of them drilling. There has been no oil produced on the farm. Average depth of wells, 650 feet. Average cost, $5,000. Wells on the flat and hillside. Three owned by company and one by lessees. Have two portable engines. Depth of sandrocks not ascertained.

*Bowman Farm.*—On west side of French Creek and mouth of Trout Run, in French Creek township. Contains 200 acres of land. Original owner, Mrs.

Bowman. Present owner, Mammoth Oil Company. Boundaries not ascertained. Half a mile from Franklin. French Creek has a general bearing at the point south, Trout Run east. Road leading to Franklin, and Jamestown and Franklin Railroad crosses the farm on Creek side. There are two non-producing wells. No producing ones on farm. One has formerly produced some oil. Amount of production not ascertained. One of the wells 300 and the other 600 feet in depth. Average cost, $5,000. Both on flat, and owned by company. Two portable engines on the property. Depth of first sandrock, 80 feet—60 feet thick; second sand, 330 feet—14 feet thick; third sand not found. Here is a fine flat for operating, both on Run and Creek. Hills are broken and recede back on a more gentle elevation.

*Longwell Farm.*—On west side of French Creek, in French Creek township. Containing 106 acres. Original owner, R. C. Longwell. Present owner, Centralia Oil Company, of Philadelphia. Purchased in 1864. Boundaries not ascertained. One and three quarter miles from Franklin. French Creek fronting it on east, with a bearing south. Flat broad and well located for oil purposes, being convenient to operate upon. There is one producing well, pumping three barrels per day of lubricating oil. Has produced 125 barrels. There are three non-producing wells, and one in progress. One of the old, and now non-producing wells, has produced 10,000 barrels of oil. Average depth of wells, 250 feet. Average cost, $3,800. Commenced producing in 1862 and 1863. Have shipped 10,000 barrels. Wells all on flat. Owned by Centralia Oil Company, and Excelsior Oil Company. Have three portable and stationary engines. Depth of first sandrock, 80 feet—

70 feet thick; second sand, 330 feet—14 feet thick; third sandrock not found. Depth of driving-pipe, 42 feet. Hills recede gradually from flat.

*Sutley Farm.*—On southwest and south side of French Creek, in French Creek township. Contains 100 acres of land. Original owner, J. W. Sutley. Present owners, Sutley Lubricating Oil Company, Eureka Oil Company, Longwell Petroleum Company, M'Cormick & M'Kissick Oil Company. Purchased in 1864. Boundaries not fully ascertained. Two miles from Franklin. French Creek fronts it on the north, bearing southeast and south. Road to Franklin and Jamestown and Franklin Railroad passing through the farm. There are four producing and fourteen non-producing wells on the farm, and two wells in progress. One flowing six barrels per day, and three producing three barrels per day—whole daily product, fifteen barrels. Amount of oil produced not ascertained exactly, but several thousand barrels. Average depth of wells, 300 to 500 feet. Average cost, $3,000. Commenced producing in 1860, 1864, and 1865. Wells all on the flat. Owned by the companies. There are six portable and stationary engines on the property. Depth of first sandrock, 250 feet—25 feet thick; second sand, 600 feet—25 feet thick; third sand not found. Depth of driving-pipe, 10 feet to bedrock near the Creek; 60 feet back near the hill. There has never been a failure on this farm (so reported). All the wells drilled have produced some oil. Here we find a large scope of territory on the flat. A number of the wells were drilled in the early days of oil development, and abandoned in the days of cheap oil.

*Dubbs Farm.*—On north side of French Creek and

on Patchel's Run, in Sugar Creek township. Contains 125 acres of land. Original owner, J. Dubbs. Present owners, French Creek Lubricating Oil Company. Purchased in 1864. Bounded on the north by Blakely & Clark, east by Blakely & M'Cormick, south by French Creek, west by Newell. Two and a quarter miles from Franklin. French Creek at this point has a bearing south of east, Patchel's Run south and southwest. Road to Franklin, and Atlantic and Great Western Railway pass through the farm. Patchel's Run forming a junction here with French Creek affords ample room on flats of Creek and Run for extensive operations, and so far as developed has proved to be oil producing territory. There are one producing and five non-producing wells on the farm, and one in progress. The producing well is pumping five barrels of lubricating oil per day. Amount of production not fully ascertained, but over 7,000 barrels. Average depth of wells, 300 feet. One 650 feet. Average cost, $5,000; one $7,000. Commenced producing in 1860, and one since. Wells on flat and hillside. Owned by company and lessees. There are four portable and stationary engines on the farm. Depth of first sandrock, 90 feet—20 feet thick; second sand, 240 feet—25 feet thick; third sand not found. Depth of driving-pipe, 11 feet on the Creek, a greater depth back from the Creek toward the hills. On Patchel's Run, one mile from river about 40 feet.

On French Creek at this point and below, the bed-rock comes nearer to the surface on the borders of the Creek on both sides than those back toward the hills, are nearly or quite level, requiring the driving-pipe (which, as before described, has to be driven to the bed or surface rock, as it is called, and acts as a conductor for the drilling-tools, to ensure uniformity in the walls of the wells), to a depth of 10 feet next to the Creek, while back from 10 to 15 rods on either side of the Creek, and on Patchel's Run, requiring to be driven a depth of from 40 to 90 feet. Showing that the waters of French Creek are running upon a high ridge of rocks. Why this ridge should cling with such pertinacity to the tortuous course of a stream, and for so great a distance exhibit such uniformity of depth, is a phenomena of which we cannot essay a practical reason.

*Newell Farm.*—On north side of French Creek, in Sugar Creek township. Contains 160 acres. Original owner, H. Newell. Present owner, December Lubricating Oil Company. Purchased in 1864. Bounded on the north by Simmonds, east by Dubbs, south by French Creek, west by Brown. Two and a half miles from Franklin. French Creek bears east. A small stream comes from the hills, giving them a broken appearance. Wagon-road to Franklin and Atlantic and Great Western Railway passing through. There is one well producing very little by pumping, and ten non-producing wells. No oil been shipped. Average depth of wells, 400 feet. Average cost, $6,000. Are all on the flat, and owned by lessees. Six portable and stationary engines on the property. Depth of first sandrock, 50 feet—15 feet thick; second sand, 290 feet, so claimed, but is not distinct. Depth of driving-pipe, 30 feet. The number of non-producing wells, and small indications in the producing one, shows a poor return for the investment. Over $70,000 have been expended here without any return. This is like the name of the company—*Cold December*—cheerless comfort to those who have invested their time and money to induce mother earth to favor them with a flow of the liquid illuminator.

*Brown Farm.*—On north side of French Creek, French Creek township. Contains 220 acres of land. Original owner, W. Brown. Present owner, Venango Lubricating Oil Company, of New York. Purchased in October, 1864. Bounded on north by lands of Philadelphia Oil Company, east by Newell, south by French Creek, west by M'Elrath. Two and three quarter miles from Franklin. French Creek here bearing east. Road to Franklin, and Atlantic and Great Western Railroad same. There are no producing wells on the farm at the present time. Three non-producing. One of them formerly produced thirty barrels per day. Amount of production not ascertained. Wells all on the flat, and owned by the company. Have two portable engines on the ground. Depth of first sandrock, 40 feet—20 feet thick; second sand, 290 feet claimed, not distinct. Depth of driving-pipe from 20 to 50 feet. The territory here is ample for a large number of wells, but the developments are not very promising.

*M'Elrath Farm.*—On northeast side of French Creek and east side of Sugar Creek, at its junction with French Creek, in Sugar Creek township. Contains 174 acres. Purchased in Fall of 1864. Bounded on north by Overmire, east by Brown, south by French Creek, west by Brown and French Creek. Original owner, D. M'Elrath. Present owner, M'Elrath. Leased to Sugar Valley Oil Company, of Philadelphia. Three and a quarter miles from Franklin. French Creek and Sugar Creek both border on the territory, the former having a bearing southeast, the latter south. Wagon-road to Franklin, and Atlantic and Great Western Railroad crossing it. Lick Run borders on the northwest corner of the tract. There is a broad flat on this territory, affording ample room for oil operations. Oil found in small quantities, lubricating. There are two producing wells, pumping five barrels each daily, three non-producing, and two in progress. Amount of production not ascertained. Average depth of wells, 325 feet. Average cost, $5,000. Commenced producing in August, 1865. Wells on the flat and hillside, hillside wells the best. Owned by company and lessees Are seven stationary and portable engines on the property. Depth of first sandrock, 40 feet—25 feet thick; second sand, 285 feet—25 feet thick; third sand not found. Depth of driving-pipe, 9 feet. These wells are claimed not to be properly tested.

*Brown Farm.*—On north side of French Creek and Sugar Creek. Containing 130 acres of land. Original owners, F. & W. Brown. Present owners, Junction Oil Company and Brown Farm Oil Company. Purchased in the Spring of 1864. Bounded on the north and east by Sugar Creek, west by Roberts, south by French Creek. Three and a half miles from Franklin. French Creek bearing east, Sugar Creek passing through in a southeast direction. Road, and Atlantic and Great Western Railroad to Franklin, passing through in an eastern direction. There are four producing wells on the farm, as follows: two on land of Junction Oil Company are producing ten barrels daily; two on Brown Farm Oil Company land, producing daily twenty barrels. All the wells are pumping. There are five non-producing wells, all old ones, drilled three and four years since. They formerly produced some oil. Amount of production not ascertained fully. Reported, 750 barrels. Average depth of wells, 350 feet. Cost, $4,000. Commenced producing in 1865. Wells all on the flat. Owned by companies. Have four stationary and portable engines. Depth of first sandrock, 75 feet

—10 feet thick; second sand, 264 feet—30 feet thick; third not found. Depth of driving pipe, 10 to 38 feet.

On the opposite or south side of French Creek, at this point, are three non-producing wells. No information could be ascertained relative to them. Judging from outward appearances, they are an unprofitable investment. The Atlantic and Great Western Railway have a station on the land of the Junction Oil Company. The Brown Farm Oil Company have good and substantial machinery for pumping. Have one large engine which operates two wells by means of long belts connecting with pulleys attached to machinery, which drives by a direct attachment, by the use of a cross head and slides, making a very complete and substantial apparatus for pumping. There is a hotel and a few other buildings, giving this point the appearance of life and activity.

*Roberts Farm.*—On the north side of French Creek, reaching over and crossing Sugar Creek on the north, and containing 275 acres of land. Is in Sugar Creek township. Original owner, R. Roberts. Present owners, Indian Spring Oil Company, 93 acres; Philadelphia and Providence Oil Company, 100 acres; and 82 acres to individual companies. Bounded on the north by Shotwell, Homan & Brown, south by French Creek, west by Hayes and others. Purchased by the companies in December, 1864. Is four miles from Franklin. French Creek bears east and northeast; Sugar Creek southeast, passing through north part of the tract. Indian Spring Oil Company's territory is located on southeast portion of tract. The Philadelphia and Providence Company on the northeast corner, bordering on Sugar Creek. One producing well on the farm, producing five barrels per day; and seven non-producing, three of them not tested. Average depth of wells, 400 feet. Average cost, $5,000. Commenced producing, August, 1865. Wells on second bench of land back from Creek on Sugar Creek, owned by the Companies. Have five engines, portable and stationary. Average depth of first sandrock, 120 feet—12 feet thick; second sand, 320 feet—45 feet thick. No third sand found. Depth of driving-pipe, from 12 to 96 feet. Less on French Creek than on Sugar Creek.

*Hayes Farm.*—On the north side of French Creek. Containing 196 acres. Bounded on the north by White and Galvin, east by Roberts, south by French Creek and Lullman. Original owner, J. Hayes. Present owner, Hayes Farm Oil Company. Purchased in December, 1864. Four and a half miles from Franklin. French Creek here has a bearing north of east. Road and railroad to Franklin having an eastern bearing. Are no producing wells on farm—one testing and one drilling. One 415 feet in depth, the other to be drilled to depth of 600 feet. Average cost, $5,000. Wells on flat and hillside, and owned by company. Have two stationary engines. Depth of first sandrock, 120 feet—12 feet thick; second sand, 320 feet—45 feet thick; third sand not found. Depth of driving-pipe, 12 to 90 feet. Lesser depth on the Creek.

In passing up the Creek there are some developments, but not very promising. On the Hannah farm, on south side of Creek, there are one or two small producing wells. There are also some ten or more abandoned and in progress on this tract and above. Depth of wells and sandrocks not ascertained. Owned by the Hannah Goss Oil Company, and others. The developments to any material extent, upon French Creek ter minated here, at the time of our visit, and we ended our exploration of French Creek for the time.

*Blakely & Clark Farm.*—On Patchel Run, Sugar Creek township. Containing 313 acres. Original owners, Blakely & Clark. Present owner, Franklin Lubricating Oil Company. Purchased in July, 1864. Bounded on the north by Simmonds and others, east by Neely and others, south by M'Cormick & Dubbs. One mile from French Creek, Patchel Run passing through it in a southern course. Turnpike to Franklin passing through in a southeast course. There are two producing wells on the farm, not pumping any material amount, and three non-producing. Amount of production, about 400 barrels. Average depth of wells, 325 feet. Cost, $5,000. Commenced producing in 1864 and 1865. Are all on the flat land, and owned by the company. Have three portable engines. Depth of first sandrock, 130 feet—20 feet thick; second sand, 280 feet—25 feet thick. Depth of driving-pipe, 40 feet.

*Neely Farm.*—On Patchel Run, in Sugar Creek township. Original owner, J. Neely. Present owner, Silver Farm Oil Company. Leased by Development Oil Company. Bounded on the north by Dow and others, east by Galloway, south by Galloway & Bleakley, west by Bleakley & Dow. Patchel Run passing through in a south of west direction. There are three non-producing wells, one of them is 650 feet deep. Sandrocks about the same as on farm below. Depth of driving-pipe, 10 feet.

*Homan Farm.*—On northeast side of Sugar Creek and Foster Run. Containing 363 acres. Original owner, S. Homan. Present owners, Sugar Creek Petroleum Company. Leased by Development Oil Company in part. Is in Sugar Creek township. Purchased in 1864. Sugar Creek has a bearing southeast. Foster Run passes through in a due south course. Turnpike to Franklin crosses the tract northwest to Cooperstown. There are four non-producing wells, and one non-producing (old well). None flowing. Average daily product, sixty barrels. These wells are known as the Shippen wells. Total amount of production not fully ascertained, but estimated at 4,500 barrels. Average depth of wells, 325 feet; one 700 feet, not producing. Average cost, $6,000. Commenced producing in February and March, 1865. Wells all on flat, and owned by company. Have four portable and stationary engines. Depth of first sandrock, 100 feet—20 feet thick; second sand, 275 feet—60 feet thick; no third sandrock found. Depth of driving-pipe, 56 to 65 feet. These wells are known as the best on Sugar Creek. Here is a large scope of territory, broad flats, as yet unoccupied by oil operators. Hills broken and not so abrupt as below.

*Shotwell Farm.*—On east side of Sugar Creek, in Sugar Creek township. Containing 160 acres. Original and present owner, J. Shotwell. Report says this gentleman was offered $100,000 for his farm, but declined to sell. There are no developments. Sugar Dale Oil Company are located on the northern portion of this tract. West Sugar Creek forms a junction with the main Creek on the opposite side of the same, passing through the Hawthorn farm. Apparently good territory. On the next tract above this is the location of the New York, Philadelphia, and Baltimore Consolidated Petroleum and Mining Company, and the New York and Pennsylvania Oil Company. No developments.

*Russell Farm.*—On both sides of Sugar Creek and Foster Run. Containing 112 acres of land, in Sugar Creek township. Original owner, J. M. Russell. Present owner, Russell Farm Oil Company. Bounded on the north by Homan and others, east by Alexander, south by Day and others, west by township line. Four miles from French Creek. Sugar Creek passes through in a southern direction, making a short bend to the west, thence south again. Foster Run passes through the western part in a southeast course. Road from Franklin to Cooperstown, running west of north. One small producing well on farm, and two non-producing wells. The producing well is pumping one barrel per day. Average depth of wells, 400 feet. Cost, $3,000. Have one engine. Depth of sandrocks, &c., not ascertained. No operators on the ground to get items from. There is ample room on this tract for future development.

*Homan Farm.*—On both sides of Sugar Creek and Foster Run. Original owner, H. H. Homan. Present owners, Homan Oil Company, New York, Philadelphia & Baltimore Consolidated Petroleum and Mining Company. No paying developments.

*M'Calmont Farm.*—On both sides of Sugar Creek. Original owner, J. M'Calmont. Present owner, Rochester Petroleum Company, in part. Bounded on the north by Frazier and others, east by Brown, south by Homan, west by township line. Sugar Creek bears west by south. Road to Cooperstown running north. There is one well, when operated producing five barrels per day by pumping, and one non-producing. Have two engines. No operators on the ground. No items or details ascertained.

*Frazier Farm.*—On Sugar Creek, in Sugar Creek

township, we find the Breen, Howes & Co., Maple Shade, and M'Kenzie well, and in passsing up further the Osgood, the Kohler well, and the territory of the Maple Grove Lubricating Oil Company, of Baltimore. On the tract above is the Frothingham and Beech Tree well, Smith well, and Pennebaker well. None of these wells are at present being operated, and have never produced oil in paying quantities. No details could be obtained of this territory. No operators on the ground.

Passing on above Cooperstown we find a well belonging to the Rochester Petroleum Company; one of Olmstead, Chapman & Co.; H. P. Adams & Co., Forest wells (two). Further up, Phil. Sheridan Oil Company, six wells. Some of these wells have given signs of oil, and produced slightly, but not in very large quantities. The surface of the country and its general appearance would indicate fair producing territory but present developments do not bring forth favorable results, to those who have invested their means. The future explorer may open up here a rich mine of wealth. Here end the developments on Sugar Creek.

### OIL CREEK VALLEY.

*Graff, Hasson Farm.*—On the north side of Allegheny River, at the mouth of Oil Creek, and on east side of latter. Is in Cornplanter township, embracing the mouth of Cornplanter Run. Contains 400 acres, in the United Petroleum Farms Association purchase, the portion now being described. Bounded on the north by Woods, & McAboy, and Clapp farms, west by Haliday Run and Oil Creek, east by Hasson Reserve, south by river. The Creek bears southwest and south. Cornplanter Run empties into the Creek from the west side, having a southeast bearing. Oil Creek empties into the Allegheny on the east side, bearing west. A portion of the 400 acres was subsequently sold to Hoffman Petroleum Company.

In May, 1862, there were two producing wells, with an average daily product of twenty barrels, and twenty-eight non-producing. To that date the farm had produced 5,900 barrels of oil. Average depth of wells, 400 feet. Cost, $3,000. In the beginning of 1866 there were thirty producing wells on the farm, and fifteen non-producing, and none in progress. Of the producing wells eight were flowing and twenty pumping. Average depth of wells at that date, 400 feet. Cost, $4,000. Daily production, sixty-nine barrels. Wells commenced producing along from year to year, the greater number in 1865. Many are old wells, never drilled through the second sandrock. Those through third sand are better. Oil has been found principally in the second sandrock. Wells are located on the flat and table land, principally the former. Owned chiefly by lessees. Have twenty engines on the farm. Average depth of first sandrock, 200 feet—20 feet thick; second sand, 340 feet—25 feet thick; grey sand, 455 feet—25 feet thick; at bottom of this, slate and mud, 8 to 10 feet; third sand or pebble-rock, 482 feet—6 feet of hard pebble and 4 feet hard, fine sand—10 feet in all. Depth of driving-pipe, 27 feet. At this time there were seven refineries on the farm, as follows: J. Cornwall & Co., capacity, 200 barrels crude per week; Economy Oil Works, Hill & Thumm, 250 barrels; Union & Albany Oil Works, two, one on each side of the Creek, M'Cammon & Babcock, proprietors, capacity, one of 180 barrels, the other, 300 barrels per week; Anglo-Saxon Oil Works, Everson Bros., 360 barrels per week; Romeo Oil Works, Babb, 100 barrels; Standard Oil Works, 50 barrels. In all, 1,380 barrels per week. One other, Beebe & Combs, was in operation formerly, now abandoned. The wells on this farm are small, but noted for their longevity, some of them having produced for several years, and are still being successfully operated. The success has been uniform. The oil is generally of 42° gravity. The shipping facilities are unequalled, being by pipe, river, and railroad. The Reno and Pithole Railroad skirts the brow of a hill at a point on Cornplanter Run, some 200 feet above the Creek Valley, making a heavy down grade in passing up the Creek.

The lands of the United Petroleum Farms Association embrace the eastern portion of Oil City, including Cottage Hill, to the river, their line extending some distance beyond the borough limits. This comprises the chief portion of the city, including the railroad depot. The Atlantic and Great Western and Oil Creek and Allegheny River Railroads pass through the property, on the river front, and along the Creek Valley. Cottage Hill has an altitude above Creek level, of 275 feet. Most of the principal dwellings are here located. The view is magnificent from the hill, embracing the river, Creek, and surrounding country. During 1868 a number of new wells have been drilled on the flat, and two on the hill north of the Warren and Franklin Turnpike. Both of the hill wells are producing. Many of the old wells on the flat have been cleaned out, and are now yielding. The present number of producing wells at date of January 1, 1869, fifty-four. Daily production, 100 barrels. Forty non-producing wells. Thirty-two new leases have been given out on United Petroleum farms and on the Hoffman property. Total amount of oil sold by the United Petroleum Farms Association, since date of November 1, 1864, 14,271 barrels. Amount received for same, $73,511.73. Depth of sandrocks on the hill differ, so far as developed, according to altitude. Are the same in all other general respects. There will be a large development on the lands of the Petroleum Farms Association during present season. Several of the wells have been producing for eight and nine years, the present production of same averaging from one to five barrels. The company have sold a large portion of land for building purposes, but have still left a large amount of territory suitable for oil purposes, to which it is admirably adapted.

*Clapp Farm.*—Bissell & Plummer Tract, located on both sides of Oil Creek, in Cornplanter township. Contains 125 acres of land. Original owner, R. Clapp. Present owners, Geo. H. Bissell and Arnold Plummer. Purchased in 1859. Bounded on the north by Cornplanter tract of same farm, east by Wingar, south by Hasson Reserve and Petroleum Farms Association, west by Petroleum Farms Association. Half a mile from Oil City, two miles from M'Clintockville, and three miles from Rouseville. Oil Creek passes through the farm, bearing southwest. Glen Run comes across with a southeast bearing, emptying into the Creek. The road from Oil City to M'Clintockville, and on up the Creek, passes through; also the Farmers' Railroad, from Oil City to Petroleum Centre. The Creek flats are about thirty rods in width, lying mainly on the north side of the Creek. The Reno and Pithole Railroad passes along the hillside, 150 feet above level of Creek. Hills high and abrupt on east side, allowing only room sufficient for one tier of wells along the Creek bank. This farm is among the first upon which extensive operations were carried on, including the upper portion, or Cornplanter tract, and has been among the most prominent producing ones.

In May, 1862, there were four flowing wells, and ten pumping ones, with an average daily production of thirty-five barrels. There were then eighty non-producing wells. To that date the farm had produced by flowing, 7,474 barrels; by pumping, 17,826 barrels. From May, 1862, to date of 1865, the farm has produced upward of 92,000 barrels. There are now nineteen producing wells, all pumping from two to seventy-five barrels per day, and eighty-seven non-producing or idle wells. Daily production not regular. Average depth of wells, 530 feet. Average cost, $4,500. Commenced producing in 1861 and on to 1865. Wells are mainly on flat; some few on the first table of bluff. Located on both sides of the Creek, as follows: On west side of Creek, seven producing and seventy non-producing. On east side, twelve producing and seventeen non-producing. Owned in part by lessees, as follows: Plummer, Gibson, Grove & Co., Buckeye Company, Wolverine, Parker, Robson, Bliss & M'Kinley, Gilbert & Co, Boston Rock Oil Company, Adams, Pitker & Pool, Thompson & Christie, Williams well, Stanton wells, &c. There are thirty engines on the property, portable and stationary. Depth of first sandrock, 190 feet—30 feet thick; second sand, 330 feet—30 feet thick; third sand, 460 feet—25 feet thick. Depth of driving-pipe, 25 feet. There is one refinery on Glen Run, owned by Coope & Co., with a capacity of 200 barrels crude oil per week. The hills on west side are gradual in elevation, affording territory proportionally easy of access for operating upon.

Many of the old wells have been abandoned. New ones have been drilled with varied success. Not more than twenty of the non-producing wells were ever drilled to the required depth. All but about ten of them are now filled up with clay, sand, and cement, in the following manner: First, clay and sand to third sandrock; then cement in third sandrock; sand and clay to second sandrock; cement in rock; sand and clay to first sandrock, and cement in rock; and then sand and clay to surface, thus effectually closing them. The surface water is shut off in the balance by a seed-bag in the first sandrock. Preparations are making for casing remainder of wells by process described in preceding chapter of this work.

*Ham. M'Clintock Farm.*—On both sides of Oil Creek, in Cornplanter township, containing 350 acres of land. Original owner, Hamilton M'Clintock. Present owner, M'Clintockville Petroleum Company. Purchased in Spring of 1864. Bounded on the north by Buchanan farm, east by Tolles, south by Shaw and Cornplanter tract, west by Hood and others. Three miles from Oil City, and half a mile from Rouseville. Oil Creek passes through on east of south and south course. Loyd's Run comes through a ravine on east side of Creek in a south of west course, emptying into Creek. Also Hamilton Run, from same side of Creek, having a southwest course. Turnpike from Oil City to various points up the Creek, passes through. Also Reno and Pithole and Farmers' Railroad, now building. The Reno Railroad passes along the side hill to near the upper or northern line of this farm, crossing the Creek and valley at this point, on a high bridge and trestle-work, some forty feet high. This farm is famous as having been one of the localities from which oil was obtained for perhaps half a century or more previous to the oil development. The oil flowed up on the surface of springs and a portion of the Creek. The Indians collected oil from same locality, so far as memory runs. Each year the Senecas and other tribes of the six nations, held a yearly festival upon Oil Creek, and this farm is supposed to have been one of their meeting places. Montcalm, the French commander, gives a description of one of these festivals, in a letter to his government. As a fitting finale, at the close of their ceremonies, the oil on the surface of Oil Creek was set on fire, making a scene that filled the mind with awe, if not terror. The nature of the ceremonies, objects, etc., have unfortunately been unrecorded, so that we can give no definite idea of them. The oil was collected and used by the Indians both for medicine and as a cosmetic for adorning their persons, being used for mixing their war-paint. Paint rock—red—has been found at various points along the Allegheny. All that was required for the purposes specified was thus provided by nature.

In May, 1862, there were eight producing wells, all flowing and fifty-six non-producing. Average daily product at this time, 235 barrels. Amount of production up to that time, as follows: Flowed, 40,698 barrels; pumped, 41,700 barrels—total, 82,390 barrels. Average depth of wells, 500 feet. Average cost of wells, $3,000. All on flat, and owned as follows: Shirk, Curtiss, Hebbard, Noble & Co., Centre County, Haines & Foster, Willard, Tuft & Co., Hyde Town Oil Company, Wilkins & Shaw, Loomis & Co., G. W. Howard, Kuhns, Gates, Hasson & Co., Persan, Dewey & Co., Kintner & Co., Rolph, Tuft & Son, Long, Brewer & Watson, Hotchkiss, Garfield & Elliot, Whiting, and others. There are at the present time twenty producing wells—three flowing, seventeen pumping. There are ninety non-producing wells, and fifteen in progress. Present daily production, 400 barrels. Average depth of wells on the flat, 520 feet; on the hill, 610 feet; one well drilled 746 feet. Amount of production since 1862 not ascertained. The present owners know nothing of the production previous to their purchase, and seem disinclined, or something else, to run over their books and give us the desired information. Commenced producing in 1861, and continued on to the present time. Their wells are on the flat and hillside or table land of same. Owned generally by lessees. Seven are owned by company. Wells known as follows: M'Kinley wells, old and new; Hebbard, Greenback, Abbott, Jenkins, Gurney, Weaver, Ashton, Rood, Long, Paxton, Warehouse, Refinery, Page, Dunn, Cunningham, &c. There are forty engines, portable and stationary, on the farm. Depth of first sandrock on the hill, 190 feet; on the flat, 225 feet—20 feet thick; second sand, 330 feet on flat, and 450 feet on hill—25 feet thick; third sand, 490 feet on flat, and 600 feet on hill—15 feet thick. Depth of driving-pipe, 30 feet on the flat, and 18 feet on the hill. Two refineries on farm—Bowen & Clark, capacity 400 barrels crude per week; and the Wamsutta Oil and Mining Company, capacity, 425 barrels per week. Hills on the hillside abrupt and difficult of access. On the east side more gradual in ascent affording good table land for operating on. This is also the residence of the former owner of the farm. M'Clintockville, at one time a promising little town, is located principally on the flat. Has a hotel and store, &c. A toll-bridge across Oil Creek at this point.

*Buchanan Farms.*—Located on both sides of Oil Creek, three and a half miles from its mouth, in Cornplanter township. Containing 150 acres in the lower, or John Buchanan, farm, and 276 acres in the upper, or

Archie Buchanan, farm. Original owners, J. & A. Buchanan. Present owners, of J. Buchanan farm, Buchanan Royalty Petroleum Company. Purchased in 1865 of A. Buchanan by Haldeman, Hostetter & Co., and others. Purchased in September, 1864. Bounded as follows: north by J. M'Clintock, and J. W. Steele, east by Tolls, south by H. M'Clintock, and by Hayes. Is three and a half miles from Oil City, and four miles from Plummer Road, leading to Oil City in an east of south course. Kersey Railroad, from Creek through Cherry Run, crosses this tract. Also the Farmers' Railroad. The Reno and Pithole Railroad crosses the valley of Oil Creek on this farm, on a high trestle-work, supported by spiles driven into the ground, the whole structure reaching from the high banks of the Creek on the west side to the hillside or the east side of the Creek, near the point of the hill made by the valley of Oil Creek and Cherry Run, crossing the J. Buchanan farm in a diagonal northeast course. This trestle-work will average forty feet in height across the vale. It still remains as a monument of the energy and lavish expenditure of the projectors of this famed "mountain" railroad. It is an airy-looking structure, and has a heavy grade, well calculated to make the passenger nervous.

In May, 1862, there were on both farms five wells, all flowing, and thirty-nine non-producing. Average daily production, 165 barrels. The amount of production up to that time by flowing, 48,284 barrels; by pumping, 57,577 barrels—total, 105,861 barrels. Average depth of wells, 500 feet. Average cost, $3,000. Commenced to produce in 1861. Wells principally on the flat, and owned generally by lessees, a few of which we give: Haldeman and Murray, Burning Well Company, Rockwell and others, Brawley, Sherman, York, drilled with spring-pole, some with horse-power, and others by steam, any way in those days to get the well drilled and obtain the oil. Oil was also obtained on this farm by white people, and previous to them by the Indians, similar to that of the M'Clintock farm, long before any developments were dreamed of. At the present time there are, on the J. Buchanan farm, three producing and twenty-eight non-producing wells. On the A. Buchanan farm, eighteen producing and 110 non-producing. None flowing. Producing wells all pumping. Daily production of J. Buchanan farm, thirty barrels. Of A. Buchanan farm, 250 barrels. Amount of production since May, 1862, not ascertained. The property having passed through so many different hands, it is difficult to get reliable figures. Wells are now on both flat and bluff, owned by lessees, a few of whom are as follows: Alhambra well, Taylor & Abbott, Champion, Cherry Run, Smith, Charter Oak, Trundy & Patton, Taylor & Rockwell, U. S. No. 4, Sprague, Rouseville Oil Company, Shaft well, &c., on A. Buchanan farm. On J. Buchanan farm: Clark & Banks, Watson & Irwin, Haldeman & Murray, Trundy, Camp & Strong, Cunningham, Comet, Guernsey, Imperial, Allen Wright Oil Company, Burning well, &c. There are seventy-five engines on the farms. Sandrocks are as follows: Depth of first sandrock, 190 feet—30 feet thick; second sand, 340 feet—25 feet thick; third sand, 480 feet—18 feet thick. On the hills these are a greater depth, corresponding to the elevation of the location of the wells. The flats are broad at this point, and at the junction of Cherry Run with Oil Creek. Hills not as abrupt as below, giving a large scope of territory to operate upon both on the hills and flats. There are two refineries on the A. Buchanan farm—Brown, Mitchell & Co., capacity, 125 barrels crude per week; Horton, capacity, ninety barrels crude oil per week. The flourishing village of Rouseville is located on the A. Buchanan farm, on both sides of Cherry Run, at its mouth, extending to the table lands of the J. M'Clintock farm.

The Buchanan Farm Oil Company, Working Interest Oil Company, Oil Basin Petroleum Company, Haldeman & Co., and J. Alexander, are represented on these farms, in the leasehold of the oil interest. The Curtiss & Hickok well, on the A. Buchanan farm, is reported to have been the first flowing well on Oil Creek. Flowed a large quantity of oil at the depth of 196 feet. It was struck before the days of tubing wells to flow. Not having the experience of later days, much of the oil was lost, and the well injured by the surface water obstructing the flow. The Clark & Banks wells, and Haldeman & Murray wells, on the J. Buchanan farm, are among the most productive. The Burning well, on the J. Buchanan farm, is among the most noted from its tragical history, which is given in a preceding chapter.

*John M'Clintock Farm.*—On east side of Oil Creek, extending across the valley of Cherry Run, in Cornplanter township. Containing 200 acres; one-half or more of which is good territory for oil operations. Original owner, J. M'Clintock. Present owner, J. M'Clintock in fee. In leasehold, Bliven Oil Company, of New York, formerly B. R. Alden, seven forty-eighths; Working Interest Oil Company, twenty-four forty-eighths; Wallace Oil Company, formerly Curtiss, Haldeman & Fawcett, three thirty-seconds; H. R. M'Clintock, one sixteenth; Allegheny & Pittsburgh Oil Company, formerly John M'Clintock Reserve, one eighth, Kendrick Oil Company, one twenty-fourth; Dr.

A. Carey, one thirty-second. Bounded on the north by the Rynd farm, east by lands of the Union and Brevort Oil Companies, south by Buchanan, west by Oil Creek. Is four miles from Oil City, three miles from Plumer, and four miles from Petroleum Centre. Oil Creek here has a bearing south, Cherry Run southwest; where it crosses road to Oil City, south, Plumer northeast, Petroleum Centre north. Kersey Railroad crosses, passing up Cherry Run.

In May, 1862, there was but one producing well, the Anderson, flowing seventy barrels per day, and twenty-one non-producing wells. Amount of production up to that time from the farm by flowing, 12,100 barrels; by pumping, 250 barrels. Average depth, 500 feet. Average cost of wells, $3,000. Commenced producing in 1860. Wells principally on the flat, and owned or known as follows: Anderson well, Alden & Chase, Buttonwood wells, Pool well, Buell well, and others. At the present time there are thirty-two producing wells, ten flowing, and twenty-two pumping. Average daily product, 200 barrels. There are also fifty-three non-producing wells, and fourteen in progress. Average depth of wells, 500 feet. Commenced producing in 1863, 1864, and 1865. Wells both on flat and bluff. The best wells are on the bluff and table lands at the present time, and are known in part as follows: St. Nicholas No. 2, formerly Ennis, Sucker State, Whyte & Main, Wright & Hall, Main & Horn, Meade, Orr Bros., Bluff Cottage, Chamberlin & Babbit, Ennis, Babcock, all flowing. Seven others pumping. There are fifty engines, portable and stationary, with the ordinary machinery, on the farm. Depth of first sandrock on flat, 200 feet; on hill, 260 feet—25 feet thick; second sand, 340 feet on flat, 400 feet on hill—25 feet thick; third sand, 520 feet on flat, 560 feet on hill—12 feet thick. The difference in depth is the difference in elevation. Depth of driving-pipe, 30 feet.

There is one refinery—Auburn Oil Works, Orr Bros.; capacity, 200 barrels crude per week. Is located on the eastern part of the farm, on a small ravine making down into Cherry Run on the east side. This farm has been among the most productive on the Creek, and is well located for oil purposes. Many new wells have been drilled on the hills, with good success, and are now producing abundantly. The Excelsior well, in the bluff, has been among the best wells on the Creek near the M'Clintock reserve. Is now doing nothing. This farm was leased by Chase & Alden, in September, 1859, for one half the oil. In 1860, Messrs. Curtiss, Haldeman. Fawcett & Carey, one eighth. Other parties having smaller interests, M'Clintock reserving one eighth. Several of these interests have since changed hands, and are now owned by different parties. Amount of production since 1862 not ascertained.

*Steele Farm—(or Widow M'Clintock.)*—On the west side of Oil Creek, opposite J. M'Clintock farm, in Cornplanter township. Containing 100 acres. Original owner, Widow M'Clintock. Present owner, J. W. Steele, by will or inheritance, or will as an adopted son. He came into the possession of it in March, 1864. Bounded on the north by Rynd, east by Oil Creek, south by Buchanan, west by Moore and Hayes. Four miles from Oil City, and four miles from Petroleum Centre, Oil Creek bears south. Roads to Oil City south, to Petroleum Centre north.

In May, 1862, there were three producing wells, all flowing, and twenty-four non-producing. Average depth, 500 feet. Average cost, $3,000. Commenced producing in 1861 and 1862. Average daily product at that time, 350 barrels. Amount produced up to May, 1862, 89,000 barrels. Wells all on the flat, and owned in part as follows: Van Slyke, Eastman, Lloyd, Christie, Ocean well, Hayes & Merrick, Hernden & Co., and others.

At the present time there are twenty producing wells—two flowing and eighteen pumping. There are twenty-five non-producing wells, and five in progress. Amount of production since May, 1862, not ascertained. Average depth of wells, 520 feet. Average cost, $6,000. Commenced producing in 1863, and on since. Are mostly on flat. One, Little Giant, on bluff. A few others have been drilled on the hillside. The average daily production at the present time is 300 barrels. The names of the principal producing wells are as follows: Little Giant, Mammoth, Sterritt No. 2, Bumstead, Morrison, Lafayette, Blacksmith, Sterritt No. 4, Chase well, Hebbard & Chamberlin, Chamberlin, and nine other smaller ones. Depth of first sandrock, 160 feet—40 feet thick; second sand, 330 feet—21 feet thick; third sand, 473 feet;—30 feet thick. Depth of driving-pipe, 25 feet. On this farm is a machine shop, one store, and three hotels. There is a broad flat, well adapted for oil operations. The hills are more broken, and recede back from the Creek with a gradual slope. The wells, producing in 1862, are doing nothing now. Many wells drilled on the farm have proved unprofitable. Some have been large flowing wells. The great difficulty on this farm is the connection between the crevices in the rocks, letting the surface water from one flood the other. Many of the wells are now cased, and are producing more oil by having the surface water cut off. When this course is taken with all the wells, we may look for a large production.

*Rynd Farm.*—On both sides of Oil Creek, and mouth of Cherry Tree Run, in Cornplanter township. Consists of 300 acres. Original owner, J. Rynd. Present owner, Rynd Farm Oil Company, of New York. On the east side of the Creek are 196 acres. It extends over the hills, bordering on the valley of Cherry Run. On the west side there are 114 acres, embracing the mouth of Cherry Tree Run and Wykle Run. Bounded on the north by Blood, east by Smith farm, south by John M'Clintock and Steele farms, west by Ward. Purchased by the Rynd Farm Oil Company in the Summer of 1864. Bearing of Oil Creek south and southeast; Cherry Tree Run southeast; Wykle Run east. Roads to Oil City south; to Cherry Tree west of north; to Petroleum Centre northeast.

In May, 1862, there were four producing wells, all flowing. Daily production, 140 barrels. There were sixteen non-producing wells. Amount of production up to that time, 28,000 barrels. Average depth of wells, 500 feet. Average cost, $2,500. Commenced producing in 1861 and 1862. Wells all on flat at that time.

At the present time there are twelve producing wells—one flowing, eleven pumping. Average daily production, fifty-six barrels. There are forty-five non-producing wells, and four in progress. The flowing well has been flowing over four years, and is now producing six barrels per day; formerly flowed eighty barrels per day. Amount of production, since 1862, not ascertained. Since Rynd Farm Oil Company purchased, 18,339 barrels. Present average depth of wells, 550 feet. Average cost, $5,500. Commenced producing in 1863, 1864, and 1865. Wells principally on flat. Some on bluff and hill. The producing wells are owned as follow: Favorite Petroleum Company, of New York, Frost Petroleum Company, Northern Light

Petroleum Company, Rynd Farm Oil Company, National Oil Company, of Pittsburgh; Ozark Petroleum Company, Lumbermen Dealers Oil Company, Rochester and Cherry Run Oil Company, Syracuse Oil Company, Boston Oil Company, Story Farm Oil Company, of Philadelphia. Depth of first sandrock, 167 feet—30 feet thick; second sand, 345 feet deep—30 feet thick; third sand, 470 feet—25 feet thick. The depth of the sandrock increases with the elevation of the hills. Depth of driving-pipe, from 8 to 30 feet, according to locality. Is less on the hills. There are five refineries, as follows: Littlefield, Swift & Co., on Cherry Tree Run, capacity, 350 barrels crude per week; John Colby, capacity, 60 barrels; Masters & Abthorp, capacity, 100 barrels crude per week; Anderson & Dyer, capacity, 100 barrels; Eagle Spring Refining and Oil Company, on Wykle Run, capacity, 350 barrels crude per week. There is a broad flat convenient to operate upon on the east side of the Creek. On west side, embracing the mouth of Cherry Tree Run and Wykle Run, there is a large scope of territory, eligibly located for oil development. Hills abrupt on east side, forming a high ridge between Oil Creek and Cherry Run. On Cherry Tree Run and Wykle Run, hills are more broken, and not as difficult of access. Some wells have been drilled recently on the high lands with varied success. Some are paying wells, others are not.

There is an island in the Creek at the upper part of this farm and lower part of Blood farm, formerly known as Rynd Island. Now owned by the Island Oil Company. There are six wells upon the island. Some of them have formerly been productive, but are doing nothing now, and have the appearance of being unprofitable investments. Have two engines on the island. The wells on Cherry Tree and Wykle Runs have never been very productive.

*Blood Farm.*—On both sides of Oil Creek, in Cornplanter township. Containing 440 acres. Original owner, J. Blood. Present owners, Home Petroleum Company, of New York, of that portion lying on the east side of Oil Creek; Blood Farm Petroleum Company of that portion lying on the west side of Oil Creek. The Ocean Petroleum Company own five acres of the flat on the east side, on which there were formerly twelve flowing wells, all drilled in 1861. The Home and Blood farms purchased in May, 1864. Bounded on the north by Story, Tarr, and Ray farms, east by M'Fate Bros., south by Rynd, west by Ward and others. Is one mile from Rouseville, three miles from Petroleum Centre. Oil Creek bears south and southwest. Roads to Rouseville south, to Petroleum Centre north.

In May, 1862, there were fourteen producing wells, all flowing. Average daily product, 450 barrels. There were eight non-producing wells. Amount of production up to May, 1862, 151,000 barrels. Average depth of wells, 485 to 500 feet. Average cost, from $1,000 to $5,000. Commenced producing in 1861 and 1862. Wells chiefly on the flat on east side of Creek, and known in part as follows: Kelly well, Carnes & Co., Painter, Lauffer & Co., Crane & Stowe, Filkins, Hart & Bros., Burning well, Maple Tree Company, Wright & Co., .Collins, Rowley, Lehigh Company, Blood well, Reed well, and several others.

There are at the present time fifteen producing wells, all pumping, and sixty non-producing wells. Are located as follows: ten producing and forty non-producing on east side, Home Oil Company; five producing and twenty non-producing on west side, Blood Farm Oil Company. Present daily production, 200 barrels. Average depth of wells, 500 feet. Average cost of wells, $5,000. Commenced producing in 1863, 1864, and 1865. Amount of production since May, 1862, not ascertained. Wells on flat and bluff. Known in part as follows: Lady Washington, Bushnell & Co., Maple Grove. Have forty engines, portable and stationary, on the farm. Depth of first sandrock, 190 feet—25 feet thick; second sand, 315 feet—25 feet thick; third sand, 465 feet—25 feet thick. Depth of driving-pipe, 35 feet on the flat, and 15 feet or less on the bluff. One refinery—Barrows & Hazleton. Not in operation. Capacity not ascertained. There are but a few acres of the flat upon the east side, yet they produced in 1861, 1862, and 1863, large amounts of oil, and were at the time mentioned as the great producing centre of Oildom. The flat on west side is more extensive, but not so productive, never having been as extensively developed. The companies are now casing their wells, in anticipation of making many of them productive again. So far as they have progressed they have met with good success. Some of the wells having doubled their production.

The Pithole speculation and excitement drew the interest from this locality and materially retarded operations, as it did all along the Creek. This company are now offering extra inducements in the way of reduced royalty, which will be beneficial, and no doubt draw back many of the wanderers. At the Lady Washington and several other wells, small gas-pipes are extended from the wells to a height above the tops of the derricks, conducting off the surplus gas not used for the engines. The gas is ignited at the end of these pipes, and kept burning, illuminating the flats at night, giving them a very cheerful appearance. The Burning well, belonging to the Maple Tree Company, flowed nearly 2,500 barrels per day, but soon fell off to about 250 barrels per day, producing at this rate up to date of October, 1862, when fire communicated and destroyed the same, together with a number of other wells, and a large amount of oil. (See chapter on Disasters.) The territory burned over has not since proved as good as before.

*Tarr Farm.*—Located on the east side of Oil Creek, Cornplanter township. Contains 198 acres. Original owner, Jas. Tarr. Present owners, Jas. Tarr, one half; Clark & Sumner, one fourth; Tarr Farm Petroleum Company, one fourth. Clark & Sumner purchased their interest in August, 1861, for $20,000. The same parties offered $60,000 for the entire royalty, but Mr. Tarr refused the offer. Clark & Sumner resold one third of their purchase to Colonel Payne, Bernard Hughes, L. C. Spencer, and others, of Rochester, for $35,000. The rapid decline in the price of oil, made these parties regret their purchase, and they subsequently disposed of their interest back to the same firm at the buying price. The Tarr Farm Petroleum Company purchased their interest more recently. Particulars of purchase not ascertained.

Bounded on the north by the Story farm, east by Russell & Seymour, south by Blood farm, west by Oil Creek. Is five and a half miles from Oil City, two and a half from Petroleum Centre. The Creek bears southwest and south. Road to Oil City south, to Petroleum Centre north. In May, 1862, there were seven producing wells, all flowing, and three non-producing wells. Average daily production at that time, 1,200 barrels. Amount of production to May, 1862, 80,900 barrels. Average depth of wells, 480 to 505 feet. Average cost, $2,000 to $3,000. Commenced producing in 1861 and 1862. Are located principally on the flats.

At the present time there are twenty-six producing

wells. Average daily production, 1,000 barrels. There are also thirty-two non-producing; twenty-one in progress and testing. Two of the producing wells are flowing and twenty-four pumping. Average depth of wells on flat, 500 feet; on bluff, 580 feet. Average cost, $5,000. Wells are on flat and bluff. Amount of production, name of well, and date of commencement is shown by table below. Depth of first sand-rock, 233 feet—20 feet thick; second sand, 367 feet—25 feet thick; third sand, 507 feet—35 feet thick on first table land; less in depth as same descends to flat, more as ascends the hill. Depth of driving-pipe on the hill, or bluff, 40 feet; on the flats, 20 feet.

| Name of Well. | Time Struck. | No. Bbls. per Day. | Total Production. |
|---|---|---|---|
| Crescent | June, 1861 | 600 | 32,000 |
| Philips No. 1 | " " | 400 | 25,000 |
| " No. 2 | Sept., " | 4,000 | 300,000 |
| Globe No. 1 | Oct., " | 80 | 4,000 |
| Woodford | Dec., " | 3,000 | 150,000 |
| Elephant No. 1 | " " | 600 | 20,000 |
| Raymond | " " | 50 | 8,000 |
| Smith No. 2 | " " | 120 | 7,000 |
| Union | April, 1862 | 300 | 30,000 |
| Curwin | July, " | 75 | 20,000 |
| Eagle | Aug., " | 100 | 8,000 |
| Anaconda | " " | 80 | 8,000 |
| Janes | Oct., " | 75 | 3,500 |
| Weitzel | June, 1863 | 60 | 2,200 |
| Union No. 2 | " " | 40 | 2,000 |
| Crouch No. 1 | Oct., " | 45 | 8,000 |
| Monitor No. 1 | " " | 20 | 2,500 |
| Cornwall | Aug., 1864 | 120 | 6,500 |
| Sterling | Sept., " | 120 | 3,500 |
| Scare Cat, No 2 | Nov., " | 80 | 3,000 |
| Densmore | Dec., " | 40 | 600 |
| Spooner | May, 1865 | 50 | 600 |
| Crouch No. 2 | July, " | 50 | 600 |
| Monitor No. 2 | " " | 40 | 200 |
| Lot 253 | Aug., " | 50 | 700 |
| Lot 233 | Oct., " | 50 | 200 |
| Lot 252 | " " | 50 | 200 |
| | | | 646,300 bbls. |

The above is to October 26, 1865.

The names of non-producing and wells in progress, not ascertained in full. The larger portion of the farm lies upon the table land, hillside, and top of same. Flat narrow, rising gradual up to the first bench, where some of the best wells are located. Many of the old wells are now cased, and others closed, showing a perfect success in this manner of treating wells, from the fact that last of March, 1865, the farm was not producing a barrel of oil; now it is producing 1,000 barrels per day. After casing, the Philips and Woodford wells pumped six to eight weeks on water before commencing to produce oil. Gas is burned at several wells on this farm, as on the Blood. The Tarr farm boasts of a population of 1,000 inhabitants, and has quite a lively little town, with hotels, stores, dwellings, church building, school-house, machine-shops, &c.

When gold was worth $2.50, Mr. Tarr was offered $800,000 in cash for the fee-simple of the farm, which was equivalent to $2,000,000. He declined the offer. During the last season quite a number of good wells have been struck on the hillside, and have given a new impetus to speculation.

We give more extended information of some of these farms than most of others, from the fact that we find their business in good working-shape, and all items of information convenient to get at, everything having been kept so that those interested in the farm could know how their matters stood. Many, in fact a majority of the companies, either could not, or did not wish to furnish us with the information we desired.

Several wells are being pumped with machinery forming a direct action with slides and ways, with cross-head, &c., dispensing with working-beam, &c.

*Story Farm.*—On both sides of Oil Creek, in Cornplanter township. The larger portion being on the west side, extending nearly half a mile. The east side is principally bluff and hills, high and abrupt. Farm

PHILLIPS AND WOODFORD FLOWING WELLS—TARR FARM.

contains 500 acres. About 100 acres of flat and table land. Original owner, Wm. Story. Present owner, Columbia Oil Company, of Pittsburgh. Purchased in 1859, by Messrs. Ritchie, Hartje & Co., for about $30,000. The following year it was put into a stock company, having a nominal capital of $200,000, represented by 1,000 shares, at the par value of twenty dollars each. The company was incorporated by act of Legislature of May 1, 1861, under the title of the Columbia Oil Company. Bounded on the north by G. W. M'Clintock & Hayes; east by Oil Creek, Hayes & Sutley; south by Blood, Tarr, and Oil Creek; west by Sutley, Pierson, and others. Is six miles from Oil City, extending nearly to Petroleum Centre on the west side of the Creek. Oil Creek has a bearing south and southeast. Petroleum Centre northeast and north.

In 1862 there were twelve producing wells, all flowing, and two non-producing. Average daily production, 1,300 barrels. Amount of production to May, 1862, 60,375 barrels. Commenced producing in 1861 and 1862. Average depth of wells, 515 feet. Cost, from $1,000 to $4,000. Wells principally on the flat, and known in part as follows: Shreve & Co., Maloney, Painter, Wannamaker & Co., Hayes, Jack & Co., Hartje & Co., Dexter, Ladies' well, Shoup, Lloyd & Co., Eicholtz & Co., Breed Town, Reynolds, Floral Oil Company.

There are at the present time sixty producing wells —four flowing and fifty-six pumping. Average daily production, 700 barrels. There are sixty-two non-producing wells, and ten in progress. Amount of production, 805,587 barrels up to date of November 5, 1865. Average depth of wells at this time, 505 to 520 feet. Cost, $3,000 to $5,000. Commenced producing in 1863, 1864, and 1865. Wells are on flat and bluff.

Owned as stated below. Have seventy-five engines on the farm. Depth of first sandrock, 233 feet—20 feet thick; second sand, 367 feet—25 feet thick; third sand, 480 feet—25 feet thick. Depth of driving-pipe, 27 feet on the flat. Less on bluff. Wells now producing: Sheridan well; Guipner & Co., two wells; Manhattan, three wells; Titus Oil Company, two wells; Columbia No. 30, Eicholtz & Co.

The following is the amount some of these wells have produced: Guipner & Co., two wells, 36,000 barrels; Titus Oil Company, two wells, 91,000 barrels; Ladies' well, 60,000 barrels; Perry well, 25,000 barrels; Ramcat well, 40,000 barrels; Story Centre Oil Company, 36,000; Breed Town well, 35,000 barrels; Big Tank well, 40,000 barrels; Oil Creek and Cherry Run Company's well, 52,000 barrels; Western Pennsylvania Oil Company, Floral well, 58,000 barrels.

There are three refineries—Crystal Oil Works, capacity, 440 barrels crude per week, B. J. Kimball & Co., proprietors; Big Tank Oil Works, 240 barrels crude per week, Fish & Colvin; Croton Oil Works, capacity, 400 barrels crude per week, Scott, Crane & Co. The company have a machine shop of their own. There is a store-house, hotels, boarding-houses, shops, offices, &c., on the farm. We find here a new patent apparatus in use, working very successfully, styled Van Norman, Brown & Morrison's Patent Steam and Petroleum Burner. We will allow the inventor to tell his own story:

The patentees propose to burn steam in combination with Petroleum, as a cheap agent to generate steam in any description of boiler, for the purpose of driving engines. The machine receives the steam direct from the boiler, *in part*, the balance of the steam required for fuel is taken from the exhaust of the engine; the whole passing through a coil of iron pipe, in the top of the furnace, in which it becomes hot enough to effect the decomposition of the water or steam. From thence one part is admitted into the retort, which contains gasified Petroleum, which is nearly pure carbon. The combustion making carburetted hydrogen, which passes out of the retort, in small openings on its top, and another portion of the steam passing into a steam-jacket which surrounds the retort, and comes out of the top, immediately opposite to and close to the small streams of carburetted hydrogen, and at an angle of 40°, bringing the point of contact about two inches above the retort, at which point ignition takes place. The expansion of the gases during combustion is very great, burning with a strong blue and crimson blaze. The remaining portion of the steam is conducted into a coil at the bottom of the retort, which is submerged into the Petroleum, that converts it almost at once into gas, and from thence it passes out of the apparatus through pipes into a large hogshead, and through another coil for the purpose of preparing the residuum oil outside the engine-house for use.

*Hayes Farm.*—Principally on the east side of Oil Creek, the southwest corner crossing the Creek, giving nearly same amount of flat land on each side of the Creek. Is in Cornplanter township. Contains 108 acres of land. Original owner, —— Hayes. Present owner, Dalzell Oil Company. Purchased in 1864. Bounded on the north by J. M'Cray, east by Wm. M'Cray, south by Story, west by Story and Egbert & Hyde. Seven and a half miles from Oil City and half a mile from Petroleum Centre. Oil Creek bears southeast and south. Roads to Petroleum Centre north, to Rouseville south.

In May, 1862, there was one producing well, flowing. Daily production, twenty-five barrels. Amount of production to that date, 3,500 barrels. Depth, 460 feet. Cost, $2,000. Commenced producing in 1861. Well on the flat. At the present time there are two producing wells, pumping fifteen barrels each per day; fourteen non-producing wells, and five in progress. Amount of production since May, 1862, not ascertained. Average depth of wells, 520 feet. Average cost, $5,000. Commenced producing in 1863, 1864, and 1865. Wells on flat and bluff, three; on hillside, owned by the company, nine; by lessees, ten. The Chimney well has been the most productive on the farm, and is located on the southwest corner, near the line of the Columbia Oil Company's territory. There is an apparent lack of energy on the part of the company. From the surface appearance, this territory ought to be as productive as the adjoining Story & Egbert farms. The Petroleum Shaft and Mining Company, of Pennsylvania, are sinking a shaft on the northwest part of this farm, near the line of the Egbert & Hyde farm, on the hillside, in the rear of the Maple Shade and Jersey wells. Its location is about fifty feet above the level of the flats. The shaft is seven by seventeen feet. The foundations are built for the engines, which are now being put up for operations in the Spring. They intend to put up three engines, one of which is ninety horse-power, and two of fifty horse-power each. They propose to sink the shaft 500 feet, or to the third sandrock. The company expect to open up a new era in the process of mining for Petroleum.

There are at the wells on this property ten portable and stationary engines. Depth of first sandrock, 180 feet—25 feet thick; second sand, 360 feet—30 feet thick; third sand, 480 feet—20 feet thick. Hills high, but not as abrupt as at many points above and below.

*Hyde & Egbert Farm.*—On the east side of Oil Creek, Cornplanter township. Contains thirty-eight acres. Original owner, A. Davidson. Present owners, Egbert & Hyde. Purchased in 1859. Bounded on the north by Oil Creek, east by M'Cray and Dalzell, south by Story and Oil Creek, west by Oil Creek. Is eight miles from Oil City, and adjoins Petroleum Centre; two miles from Plumer. Oil Creek bears southwest and southeast. Roads, to Tarr farm south, Petroleum Centre north, to Plumer east.

In May, 1862, there were two producing wells, yielding thirty-five barrels per day of a production. No other wells had then been sunk on the farm. Depth, 500 feet. Cost, $2,000. Commenced producing in 1861; both on the flat. At the present time there are eighteen producing wells—one flowing, seventeen pumping. Average daily product, 800 barrels. Wells producing from 15 to 150 barrels each daily. Amount of production since May, 1862, not fully ascertained, but estimated at 1,000,000 of barrels. Average depth of wells, 550 feet. Cost, $500. Commenced producing in 1862, 1863, 1864, and 1865. Wells all on flat and bluff, principally on the flat. Owned by land interests and lessees. Have forty engines, portable and stationary, on the farm. Depth of first sandrock, 170 feet—20 feet thick; second sand, 320 feet—30 feet thick; third sand, 450 feet—40 feet thick. Depth of driving-pipe, 40 feet on the flat, and 90 feet on the bluff.

The present producing wells are Cole well, Bernheimer, Keystone, Tom Benton, Jersey Nos. 1 and 2, Maple Shade, Eagle, Olive Branch Nos. 1 and 2, Hollister, Coquette No. 27, Burlington, Bird, No. 34, Potts, Laurel Hill, &c. Twelve hundred barrels is the highest daily average of the farm. The Egbert &

Hyde is among the most productive of the farms along the Creek. The bluff wells are not as productive as those on the flats. Dr. M. C. Egbert has a fine residence on the farm. There is not a liquor saloon on the farm, which is more than can be said of many of those where so much business is transacted. In 1861 this farm was considered "dry territory." Subsequent developments proved the fallacy of previous predictions. The Maple Shade well commenced yielding August 5, 1863, and for some months flowed from 800 to 1,000 barrels per day, after which it fell off to 450 barrels per day, and gradually declined to about 100 barrels per day. Its yield, as estimated from August 5, 1863, to present time, has been about 190,000 barrels, selling at the well for upwards of $800,000, in addition to a considerable amount destroyed at the burning of this well, on March 2, 1864. Twenty-eight tanks of oil were burned, and the loss estimated at $100,000. The number of oil companies formed upon land and working interests of this farm, during the excitement of 1864 and 1865, were legion.

*G. W. M'Clintock Farm.*—Is located on the west side of Oil Creek, in Cornplanter township. Contains 207 acres. Original owner G. W. M'Clintock. Present owner, Central Petroleum Company, of New York, purchased in February, 1864. The Company first leased the property in November, 1863, and afterward purchased it. Bounded on the north by Lower M'Elheney and Benninghoff, east by Oil Creek, south by Oil Creek and Story, west by Sherman, Knox & Co. Is eight miles from Oil City, and eight from Titusville. Oil Creek has a bearing south and southwest. Bennehoff Run south-east, and makes into the Creek on the west side at the north line of the farm. Adjacent to the village of Petroleum Centre is Wild Cat Run, a circular ravine of about three fourths of a mile in length, running around a "hog-back"—a hill shaped similar to the name given it—of about 50 feet in elevation above the flat. Roads, to Shaffer northerly, to Tarr Farm south, to Cherry Tree westerly.

In May, 1862, there was one producing well, flowing 40 bbls. per day. There were nine non-producing wells. Amount of production up to that time, 1600 barrels. Average depth of wells, 440 feet. Average cost, $3,000. Commenced producing in August, 1861. Wells on flat. Owned by Brown, Catlin & Co. At the present time, there are 33 producing wells—10 flowing and 23 pumping. Average daily production, 900 barrels. Two wells are abandoned, and 40 in progress; three non-producing at present. Have produced formerly. Amount of production since May, 1862, 200,000 barrels. Average depth of wells, 550 feet. Average cost of wells, $6,000. Commenced producing in 1864 and 1865. Wells are on flat and hillside. The hillside wells are the most productive. Owned by the company and lessees. Have 65 engines, portable and stationary, with ordinary machinery, on the farm. Depth of first sandrock, 180 feet—20 feet thick; second sand, 340 feet—25 feet thick; third sand, 460 feet—25 feet thick. Depth of driving pipe, 40 to 70 feet. Wells known in part as follow: Company's 5-9-10-11-12-13-14. G. H. Clark's well, Meyer well, Anderson wells, Swamp Angel, No. 4, &c. There are five refineries, as follows: Crystal Spring, capacity 440 bbls. crude oil per week; Monitor, M. L. Bates & Co., 175 bbls. per week; Baker, Farnes & Co., 200 bbls. per week; Sweet, Moore & Co., 160 bbls.; Bartlett & Newton, 160 bbls. There are two abandoned refineries. This farm embraces a large extent of borable territory, taking in the larger part of Wild

Cat Hollow. It is the location of the flourishing town of Petroleum Centre. The new strikes on the Stevenson Farm and Bennehoff Run have given a fresh impetus to the place, and all the busy scenes of the various points on Oil Creek are witnessed here.

This territory was held in high estimation at the time of the speculative fever. The bonus of one hundred thousand dollars was offered in 1865, for ten leases on this farm, the royalty being one half. The offer was refused by the Company. It has been one among the best producing farms of the Oil Region, and from the commencement admirably managed.

*Stevenson Farm.*—Joins the G. W. M'Clintock Farm, on the north, on Stevenson Run and Tarr Run. Is in Cherry Tree township. Contains 153 acres. Original owner John Stevenson. Present owner, Ocean Oil Company, of Philadelphia. Purchased in November, 1864. Bounded on the north by R. Stevenson, east by Bennehoff Petroleum Company, south by G. W. M'Clintock, west by Drake, and others. Is half a mile from Petroleum Centre, not reaching to the Creek by a quarter of a mile. Stevenson Run bears southeast, and empties into Bennehoff Run a short distance from its mouth, Tarr Run bears east, and empties into Bennehoff Run above Stevenson Run. Road to Petroleum Centre south, to Cherry Tree northwest.

There are nine producing wells on the farm : six flowing and three pumping. Average daily production, 1,000 barrels. There are 34 wells in progress, five nearly completed. Average depth of wells, 690 feet. Average cost, $6,500. Commenced producing September, 1865. Amount of oil shipped, over 50,000 bbls. The wells are all on the hill-side, three quarters of a mile back from, and on an elevation of 170 feet above the level of, the

Creek. The derricks and necessary buildings about them give it much the appearance of a city set on a hill, and whose light will not be hid, judging from its present flow of illuminating oil. The company own two wells, Ocean and No. 3. The balance are owned by lessees, giving to company one half the oil as royalty. Have 43 engines, portable and stationary, on the farm. Depth of first sandrock, 350 feet—8 feet thick; second sand, 530 feet—9 feet thick; third sand, 650 feet—40 feet thick. Depth of driving pipe, 15 to 20 feet. The principal producing wells are the Ocean well, 300 bbls.; Arctic well, 300 barrels; No. 12—60 bbls.; No. 27—230 bbls.; No. 18—30 bbls.; No. 20—75 bbls.; No. 19—20 bbls.; No. 24—80 bbls.; No. 25—20 bbls. There is one refinery—M'Cool's, capacity 200 barrels crude oil per week. This farm shows conclusively that the oil is not confined to the valleys of the streams. The first well, the Ocean, commenced flowing September 1st, 1865. Since then, eight more good wells have been obtained, which looks well for the future of the territory.

*McCray Farm.*—On east side of Oil Creek, and on Rattlesnake Run. Original owner, Jas. M'Cray. Present owner, same. Bounded on the north by Boyd, east by Boyd & M'Cray, south by M'Cray—or Boston Petroleum Company—and Hayes, west by Egbert & Hyde and Oil Creek. In Cornplanter township, adjoining Petroleum Centre. Oil Creek has a bearing southwest.

There are no producing wells, six non-producing, one pumping—testing. One well over 700 feet in depth; the others, 600 feet deep, are on Creek bank, hill-side and top. Sandrocks same as on flat, depth to them varying according to difference of elevation, hills high and abrupt, difficult of access from Creek side, only a small portion coming to the Creek.

*Boyd Farm.*—On the east side of Oil Creek, Cornplanter township; contains 75 acres. Original owner, James Boyd. Present or principal owners, Woods & Wright, of New York. Time of purchase not ascertained. Bounded on the north by M'Elhenny, east by Phillips & Bro., south by M'Cray, west by Oil Creek. Is half a mile from Petroleum Centre, and three miles from Shaffer. There is one producing well on the farm, yielding when pumped, ten barrels per day. Oil Creek bears south at this point. Road to Petroleum Centre south, to Shaffer north. There are 21 non-producing wells—one or two of them have produced formerly, but not in paying quantity. Average depth, 480 feet. Cost not ascertained. Wells all on the flat, owned principally by lessees. Have five engines, portable and stationary. Depth of first sandrock, 165 feet—20 feet thick; second sand, 325 feet—25 feet thick; third sand, 455 feet—25 feet thick. Depth of driving pipe, 40 to 60 feet. There are three refineries, Patterson & Sons, capacity, 320 barrels of crude per week; Ensign's or Carbon Oil works, 88 barrels per week; M'Cool, capacity, 160 barrels per week. The last two not in operation. In the general history of the Creek, this farm is known as "dry territory." Future development of the hills may yet bring it on a par with the other farms adjoining it.

*Lower M'Elhenny Farm.*—On both sides of Oil Creek, in Cornplanter township, and Cherry Tree township. Contains 100 acres. Original owner, M'Elhenny. Present owners, Hussey, M'Bride, and L. Haldeman & Co. Purchased in the winter of 1859 and 1860, for the sum of $20,000. Bounded on the north by Bennehoff and Oil Creek, east by Oil Creek and Espy, south by Boyd & M'Clintock, west by Bennehoff

Petroleum company. Is one mile from Petroleum Centre, and two miles from Shaffer. Oil Creek is the only stream, and north of west, south and east of south. Road to Shaffer north, to Petroleum Centre south.

In May, 1862, there were seven non-producing, all flowing, and 23 non-producing. Average depth, 475 feet. Cost from $4,000 to $7,500. Commenced producing in 1861 and 1862. Amount of production up to that time, 153,680 bbls. Wells all on the flat. The Crocker well commenced to flow in January, 1861. Flowed 1,000 barrels per day for some time, afterward falling off, and ceased altogether in 1862. Empire well commenced flowing in September, 1861, 2,500 barrels per day, yielding 2,000 barrels per day for most of the winter, falling off during the summer to about 300 barrels, and ceasing in 1863, after flowing 80,000 barrels. Buckeye well commenced flowing in September, 1861, 800 barrels per day, falling off to 200 barrels per day, and ceased in 1862, after flowing 43,000 barrels. Funk well is among the best producing ones of those days, the Burtiss, Aiken Davis wells, and others. There are at the present time 14 producing wells, all pumping. Average daily production, 217 barrels. There are 65 non-producing wells. Amount of production since May, 1862, 225,000 barrels. Average depth of wells 480 feet. Average cost, $5,000. Commenced producing from 1862 to present time. Wells all on flat. Owned by lessees. Have 35 engines, portable and stationary, with ordinary machinery, on the farm. Depth of first sandrock, 153 feet—30 feet thick; second sand, 311 feet—32 feet thick; third sand, 446 feet—27 feet thick. Depth of driving-pipe, 27 to 45 feet in the middle of flat; less nearer the hills of bluff. The hill on the west side is very abrupt and high—on the east side gently elevating until near the upper portion, then more abrupt. There is a narrow flat at the upper end on the west side. The developments are confined to the low land on the Creek. A large portion of the farm is on the up-lands, and is as yet undeveloped. The highest part of this, or the upper farm, is not more than 150 feet above the level of the Creek. The present producing wells are, Hibbard No. 2, Empire No. 1, Hatch No. 2, Olmsted Nos. 1 & 4, Densmore Nos. 7 & 11, Gray No. 3, American No. 2, Lincoln, Genessee Valley, No. 2, Bluff well. These farms are among the earliest of the oil-producing ones, and have been highly profitable to both land and working interests. Quite a smart little town, called Funkville, has been built up here, and comprises a few stores, hotel, school-house, machines, residences, &c.

*Benninghoff Farm.*—On the north side of Oil Creek, at the bend in same, between lower and upper M'Elhenny Farms, the south line crossing the creek twice, embracing two one-fourths in the bend of the creek on the south side. Owned by the Benninghoff Reserve Oil Company. Is in Cherry Tree township. Contains 150 acres of land. Bounded on the north by Benninghoff, east by upper M'Elhenny, south by lower M'Elhenny, west by Stevenson. Original and present owner, John Benninghoff. Five acres leased on the creek and Pioneer Run, embracing the principal part of the oil bearing territory, the greater part of the farm being on the hills. Pioneer Run crosses the north-east corner. Is one mile from Petroleum Centre and two miles from Shaffer. Oil Creek bears to the west at this point. Road to Petroleum Centre south, to Shaffer north.

There are eight producing wells on the farm: three flowing, and five pumping. Daily production, 445 barrels. There are ten non-producing wells, and five on Pioneer

Run in progress. Four wells on the Reserve, on south side of Creek—one producing 15 barrels per day, two with tools fast, and one in progress. Amount of production not ascertained. Wells on flat and hill side. Average depth of those on the Reserve, 570 feet. Cost $6,000, and are owned by the company. Those on the north side of Creek and Pioneer Run average 460 feet in depth, and cost $5,000. Are owned by lessees. Fifteen engines, portable and stationary, on the farm. Depth of first sandrock, 152 feet—20 feet thick; second sand, 308 feet—25 feet thick; third sand, 440 feet—27 feet thick. Depth of driving-pipe, from 25 to 40 feet. One old abandoned refinery on Creek, owned by DeKalb Oil Company, capacity not ascertained, probably 200 barrels. The principal producing wells are; W. Jenkins, two wells, 40 barrels each; Huidekoper Oil Company, two wells, 150 and 50 barrels; DeKalb Oil Company, two wells, 30 barrels; Ocean well, Haskin's well, on Pioneer Run, 120 barrels. Hills high and abrupt near the south line, leaving only a narrow bench along the Creek. More gentle in their ascent as they break over on to Pioneer Run.

*Espy Farm.*—On south side of Oil Creek, Cornplanter township, containing 154 acres. Original and present owner, Geo. P. Espy. Bounded on the north by Oil Creek, east by Caldwell, south by Phillip's Petroleum Company, west by Lower M'Elhenny. Mile and from Shaffer and Petroleum Centre. Oil Creek bears west. There are two producing wells, one flowing and one pumping. Daily product, 22 barrels. Five wells non-producing, because not worked, and one in progress on the hillside, back of Buckeye well. The old Buckeye well formerly flowed 800 barrels per day, but is now doing nothing. Commenced producing in 1862 and 1864. Amount of production, not ascertained. Wells on flat and hill-side. Owned by lessees. Have six portable engines on property. Average depth of wells, 460 feet. Cost of wells, $4,500. Depth of first sandrock, 165 feet—15 feet thick; second sand, 316 feet—15 feet thick; third sand, 440 feet—30 feet thick. Depth of driving-pipe, seven feet. One refinery, Northrop Bros., capacity 150 barrels crude oil per week. The diamond drill was tested on this farm, and failed to meet the requirements. So far, oil has been obtained in every well drilled. The property has never been thoroughly tested.

*Upper McElhenny Farm.*—On north side of Oil Creek, and on Pioneer Run; contains 80 acres of land, and is in Cherry Tree township. Original Owner, M'Elhenny. Present owners, Hussey & McBride, and Haldeman & Co., purchased in the winter of 1859 and 1860. Bounded on the north by Foster's farm, east and south by Oil Creek, west by Benninghoff, one and a half miles from Petroleum Centre, and same distance from Shaffer. Oil Creek runs northwest and west, Pioneer Run, south, emptying into the creek. Road to Shaffer northwest, to Petroleum Centre west. There are eleven producing wells on the farm, all pumping. Average daily product, 310 barrels. There are 47 non-producing wells. Average depth of wells, 480 feet. Average cost, $5,000. Commenced producing in 1862, and up to present time. Amount of production estimated at 200,000 barrels. Wells are all on the flat, and owned by lessees as below. Twenty-five engines, portable and stationary, with ordinary machinery, on the farm. Depth of first sandrock, 151 feet—30 feet thick; depth of second sand, 305 feet—32 feet thick; third sand, 440 feet—27 feet thick. Depth of driving-pipe, from

27 to 45 feet, more in middle of valley. There are two refineries on Pioneer Run: Great Western, capacity 160 barrels per week; Pioneer, capacity 160 barrels. The hills are less abrupt, rising gradually back from creek, and on the lower part of Pioneer Run, with a flat next to Creek and mouth of Run, from 15 to 20 rods in width, well located for oil operations. The developments are principally confined to the low-lands on the Creek. A large portion of the farm is yet undeveloped. There is a reserve of twenty acres on which there are five wells; only one producing, yielding about four barrels per day. Farmer's well, on the Reserve, formerly produced forty barrels per day. The present producing wells are Mount Vernon Nos. 4, 5, and 6; Pioneer Nos. 1, 2, and 3,; Briggs, Forest City, Champion, &c. The Fertig well No. 1 was formerly the best well on this tract. Is now doing nothing.

*Caldwell Farm.*—On east side of Oil Creek, Cornplanter township, and on Bull Run, containing 220 acres. Original owner, Caldwell. Present owner, Caldwell Oil Company, of Philadelphia. Purchased in fall of 1864. Bounded on the north by Farrell farm, east by Patterson, south by Irwin & Espy, west by Oil Creek. Two miles from Petroleum Centre, and one mile from Shaffer. The bearing of Oil Creek is southwest. There are four producing wells, all pumping. Average daily product, 105 barrels. Five non-producing wells; five in progress and testing. Average depth of wells, 480 feet. Average cost of wells, $6,000. Commenced producing in March and June, 1865. Amount of production, 3,000 barrels. Wells on creek and the flat; on Bull Run, and on the table lands two wells, 600 feet deep. Owned by company and lessees. Have ten portable engines on the property. Depth of first sandrock, 165 feet—40 feet thick; second sand, 306 feet—35 feet thick; third sand, 445 feet—30 feet thick. Depth of driving-pipe fifteen feet on flat, more on bluff. The celebrated Caldwell well was originally on this farm, but was sold, with one and a half acres of land, in the spring of 1863, to the Noble Well Company, for the sum of $145,000. It has not since been operated, the purchase having been a measure of protection merely, there being a connection between the Caldwell and Noble wells in the crevices of the rocks below, that promised to interfere with the production. The flat is narrow, broader at lower end. Hills high, and difficult of access.

*Foster Farm.*—On west side of Oil Creek, opposite Caldwell and Farrell. Is in Cherry Tree township, and contains — acres. Original owner J. Foster. Present owners, Irwin Petroleum Company, of Philadelphia; have some 25 or 30 acres, with a front upon the Creek of about 2,500 feet, running back over the hill to Pioneer Run. Owner of balance not ascertained. Bounded on the north by Gregg, east by Oil Creek, south by upper McElhenny, west by Benninghoff. Two miles from Petroleum Centre, and one mile from Shaffer. Oil Creek has a bearing southwest. Roads to Shaffer north, Petroleum Centre south. In May, 1862, there was one producing well, the Sherman, flowing an average daily production of 900 barrels, and 12 non-producing. Average depth of wells at that time, 470 feet. Cost not ascertained. Amount of production, 75,000 barrels. Wells on the flat, bored by Sherman and others. The Sherman well was struck in March, 1862, and for some months it was one of the largest flowing wells on the Creek. It commenced with a flow of about 2,000 barrels per day, but after a few months ran down to about 600 barrels. It ceased flowing in February,

1864, since which time it has been pumping, yielding large quantities of oil, but is now doing nothing, having recently been burned, with all the tanks, derricks, engine houses, &c., around it. No producing wells on the farm at present time. There are fifty non-producing wells, nearly all bearing the appearance of having been abandoned. Amount of production since May, 1862, not ascertained. Depth of wells, 475 feet. Cost of wells, $6,000. Commenced producing in 1862 and 1864. Wells on flat. Owned by various companies and lessees. Have five engines, portable and stationary, on the property. Depth of first sandrock, 140 feet—15 feet thick; second sand, 294 feet—20 feet thick; third sand, 435 feet—30 feet thick. Depth of driving-pipe, 50 feet on the creek, and 10 feet on the bluff.

The flat here is quite broad and very well located for operating. Is low and swampy, making plenty of mud in wet weather. The Crocker is one of the noted wells on this farm. It was pumped for two months on clear salt water. Mr. Crocker received the jeers of his neighbors with coolness, politely informing them that he thought he knew his own business. Future developments proved that he did, for after pumping the time above specified, the water began to color with oil, and in a few days after was flowing at the rate of 500 barrels per day. But it was a short-lived well, there being so many wells around it, letting down the surface water, that it was soon flooded out.

Here, where two years since was one of the busiest places on the Creek, desolation now reigns. Doubtless if the same energy of development prevailed in this locality as on the Tarr farm, a large yield of oil would be obtained.

*Farrel Farm.*—On the east side of Oil Creek, and

on Bull Run, contains 36 acres. Original owner, J. Farrel. Present owner, J. Farrel and Commonwealth Oil Company. The Company purchased an undivided half in the tract in October, 1864. Bounded on the north by Beaty farm, south by Irwin & Espy, west by Oil Creek. One mile from Shaffer, and two miles from Petroleum Centre. Oil Creek runs west at this point, Bull Run north of west, emptying into the creek.

There is one producing well, the Norton, flowing 40 barrels per day, and twelve non-producing, two of them being tested. Average depth of wells, 515 feet. Average cost, $4,000. Commenced producing in 1863, 1864, and 1865. Estimated production from Noble & Delamater well, 1,000,000 barrels; from Craft well, 101,860 barrels. Wells on flat and on ravine of Bull Run. Owned by lessees. Have six engines, portable and stationary, on the property. Depth of first sandrock, 185 feet—10 feet thick; second sand, 345 feet—10 feet thick; third sand, 478 feet—36 feet thick. Depth of driving-pipe, 31 feet. Some of the principal wells besides the Noble and Craft, are the Mulligan, Commercial Oil Company, &c. The Noble & Delamater well commenced flowing in January, 1863, at the rate of 3,000 barrels per day, as estimated at the time, and ceased the 28th day of February, 1865. This well was the most productive, as well as the most remunerative one ever struck, making its owners, in this brief time, millionaires in a literal sense. The value of the product of this well has been variously estimated, at from *two to five millions of dollars!* The first investment in sinking the well was about $4,000. Is it any wonder that people became excited, and plunged into wild speculation, when a fact of such magnitude was apparent to their visual organs. This well proved to be one among the many, and demonstrated fully the immensity of the deposit that nature had hid away in the rock-ribbed recesses, for the use and benefit of mankind. The Craft well ceased from the failure of its seed-bag. The owners have experimented, and endeavored in every possible way to resuscitate it, but in vain. A sad accident occurred on this farm a short time previous to our visit. Messrs. Noble & Delamater had built a new office. Before it was quite finished, their book-keeper, James Bonner, jr., son of Rev. James Bonner, senior, an Episcopal clergyman of Ashtabula, Ohio, removed to it. Shortly afterward he was taken ill with typhoid fever. He soon became past all hope. On the seventh day of his illness the office, from some cause unknown, took fire, and was consumed. Young Bonner was carried by his father to the window, where by the help of those outside, placing boards from an engine house to the window, he was taken out; but he died in a very short time. His father fell by the window overcome with the heat and smoke, and was with difficulty rescued; the young man's mother had previously made her escape by jumping from the window. The elder Mr. Bonner was seriously injured by inhaling the heated air and smoke.

*Beaty Farm.*—On the west side of Oil Creek, and on Hemlock Run, in Allegheny township. Contains 160 acres. Original owner, R. P. Beaty. Present owner, Clinton Oil Company, of New York. Purchased in 1864. Bounded on the north by Sanney, east by Beaty, south by Farrel, west by Oil Creek. One mile from Shaffer, two miles from Petroleum Centre. Oil Creek bears south, Hemlock Run west. There are two producing wells, both pumping; daily production 50 bbls. Five non-producing wells, and three in progress on the Run. Average depth of wells on Creek,

500 feet; on the Run, 650 feet. Average cost of wells, $5,000. Commenced to produce in August, 1865. Amount of production not ascertained. Wells located in bluff on Creek, and on Hemlock Run. Have five engines, portable and stationary, on the farm. Depth of first sandrock, 170 feet on the Creek, and 280 feet on the Run—20 feet thick; second sand, 210 feet on Creek, 420 on Run—25 feet thick; third sand, 480 feet on Creek, 550 feet on Creek. Depth of driving-pipe, five feet. There is but little boring territory adjoining the Creek. Hills high and abrupt, covered with timber. Hemlock Run very narrow.

*Gregg Farm.*—On the west side of Oil Creek, in Cherry Tree township. Contains over 300 acres. Original and present owner, D. Gregg. Bounded on the north by Sanney, east by Oil Creek, south by Foster, west by Benninghoff. Half a mile from Shaffer; two and a half miles from Petroleum Centre. Bearing of Oil Creek south. No producing wells on the farm, forty non-producing. Two or three of them have formerly yielded some oil. Average depth, 500 feet. Cost not ascertained. Wells on flat. Owned by lessees. Have ten engines and ordinary machinery on the farm. Depth of first sandrock, 170 feet—10 feet thick; second sand, 210 feet—15 feet thick; third sand not found. Depth of driving-pipe, 45 feet. Reported third sandrock found in the Almer well, at depth of 700 feet, ten feet thick. Two refineries on the farm—Victoria Oil Works, capacity 160 bbls. crude per week; Continental Oil Works, capacity 150 bbls. crude per week. There is generally a very good show of oil in the second sandrock. The farm has been but slightly productive.

*Sanney Farm.*—On both sides of Oil Creek. Con

tains 125 acres. Original owner, Sanney. Present owners Potter & Nurse, the widow's one-third, and two heirs' interest; other parties the balance. Purchased in 1865. Bounded on the north by Shaffer, east by Tallman, south by Beaty and Gregg, west by Stevenson and others. Two and a half miles from Petroleum Centre, and half a mile from Shaffer Station. Oil Creek has a bearing southwest and south. Road to Shaffer north, Petroleum Centre south. No producing wells—eleven non-producing ones, and three in progress and testing. Two of the wells have produced a small amount of oil. Average depth of wells, 500 feet. Average cost, $4,000. Amount of production not ascertained. Wells all on the flat, and owned by lessees. Have six engines, portable and stationary, on the farm. Depth of first sandrock, 170 feet—15 feet thick; second sand, 210 feet — 15 feet thick; third sand not found. Depth of driving-pipe, 40 feet. Three refineries are located here—Continental Oil Works, capacity 150 bbls. crude per week; Hope Oil Works, capacity 150 bbls. per week—both owned by the same parties, Robison, Cone & Co.; Silver Rock Oil Works, 96 bbls. per week, Twing & Dorr, proprietors. This farm is one of the unproductive ones, and may be classed as "dry territory." Hills not abrupt on east side; higher and more difficult of access on west side.

*Shaffer Farm.*—On both sides of Oil Creek, in Cherry Tree and Allegheny townships. Contains 54 acres of land. Original owner, G. Shaffer. Present owner not ascertained. Bounded on the north by Miller and Fleming, east by Tallman, south by Sanney, west by Benninghoff. Six miles from Titusville, and fourteen from Oil City. Oil Creek bears southeast, south, and southwest. Road to Titusville northerly, to Petroleum Centre southerly. Oil Creek Railroad to Titusville. Shaffer is the present terminus of the Oil Creek Railroad, and has assumed considerable importance as a shipping point. A little town has sprung up as by magic, and hotels, stores, and saloons abound. All this prosperity will, we think, be of brief duration. In a short time the Oil Creek Railroad will make its way down to the natural terminus of the road, Oil City; then the enterprising denizens of Shaffer, finding their occupation gone,

> "Will fold their tents like the Arabs,
> And silently steal away."

There are no producing wells on the farm at the present time. Twelve non-producing wells, and twelve more have been partially drilled and abandoned. One well pumping. Sherman well testing. Some of the wells have formerly produced—the Rangoon well, owned by Samuel Downer; Brewer & Watson well, and others. Amount of production not ascertained. Average depth of wells, 650 feet. Average cost, $5,500, except the old, partially drilled ones. Wells all on flat, and owned by lessees. There are six engines, portable and stationary, on the property. Depth of first sandrock, 180 feet—19 feet thick; second sand, 230 feet—10 feet thick; third sand not found. Depth of driving-pipe, 40 feet.

The principal feature of this point is the railroad depot, and the thrifty town which has grown up within the past two years. The Oil Creek Railroad was completed to this place in July, 1864, running along the edge of the hills, and the flat, where ample landings, depot buildings, turnouts, &c., are constructed. The depot buildings occupy the western side of the track, the landings in front forming an elongated semi-circle of several hundred feet in length, a double-track passing around it, while on the Creek side are a number of warehouses. There are in the town four first-class hotels, about a dozen stores, several livery stables, a large number of dwellings, one good school-house, post-office, &c. Boasts of a population of 1,500. Everything is new and in a crude state. In July, 1864, when the railroad was completed to this point, there was a single frame dwelling and a small restaurant on the banks of the Creek. The continuation of the O. C. RR. to Petroleum Centre is now in progress, the grading being nearly completed. On the west side of the Creek, the flat is about thirty-five rods wide, in the rear of which the hills rise gradually westward, a ravine causing a break in the hill immediately back of the Railroad Station. On the east side the hills are abrupt and high, admitting of a mere passage along the Creek bank.

*Miller Farm.*—On both sides of Oil Creek and Hemlock Run, above Shaffer farm. In Cherry Tree and Allegheny townships. Contains 303 acres. Original owner, R. Miller. Present owner, Indian Rock Oil Company, of New York. Purchased in the fall of 1863. Bounded on the north by Fleming, east by Fleming, south by Shaffer, west by Jameson and Howe. One mile from Shaffer, five from Titusville. Oil Creek bears south-west, south, and south-east. Road to Titusville northerly, to Petroleum Centre southerly. Plank road to Pithole southeast. Hemlock Run has a bearing south-east. There are no producing wells on the farm—fourteen non-producing ones, and six abandoned. Lincoln well formerly produced 60 bbls. per day; Boston well, 50 bbls. Bobtail well has been a producing one. Amount of production, 10,000 bbls. Average cost of wells, $4,000. Commenced producing in 1863.

Wells on the flat. Owned by company and lessees. Average depth of wells, 600 feet. Have 10 engines, portable and stationary, on the property. Depth of first sandrock, 143 feet—20 feet thick; second sand, 290 feet—25 feet thick; third sand not found. Depth of driving-pipe, 40 feet. Names of some of the wells as follows: Phillips, Barnsdall, Hemlock, Kerosene, Irrawaddy, Hoyt, Downer, New Bedford, Webb, Van Vleck, Railroad, &c.

Here is the terminus of the Pithole and Miller Farm Transportation Company. They have tankage for 20,000 barrels of oil, convenient to the railroad, and have large platforms on which they fill the barrels by means of pipes from the tanks, and roll the barrels on the cars from platforms. The other end of their works are located on the Thos. Holmden farm, on Pithole Creek, where the oil is received in tanks from pipes extended from the different wells, and from barrels, when hauled to platforms erected for the purpose. The oil is taken from these tanks by force-pumps driven by steam engines, and forced in two-inch pipes over the hills, and discharged into the tanks on the Miller farm. Length of pipe, five and a quarter miles. Size of pipe, two inches. Have four ten-horse power engines. The first is located near the tanks on the Holmden farm, and forces the oil up one half mile from the tanks. The second engine forces it three quarters of a mile over to West Pithole, near the Paxton House. The third engine forces it up one mile further. The fourth engine forces the oil to the summit of the hill, from which it runs down to the Miller farm. Capacity of transportation, 1,500 barrels per day. The Company are laying a second string of pipe, and adding four more engines, thus doubling the present capacity of their works. The oil is forced in a continuous stream by the action of the engines and pumps, working in concert along the whole line. They have a telegraph wire running along the whole length of the pipe, with an instrument and operator at each end, that instant communication can be had along the whole line, and in case of any accident to the machinery, it can be known at once, and the damage repaired. Barnsdall Oil Company are on Shreve tract, west side of Creek, and on Hemlock Run. There is one refinery—Sunshine Oil Works, Geo. Bartlett; capacity 120 bbls. crude oil per week. Hills on west side abrupt, high, and broken. On east side is a broad and flat hillside, rising gradually to the east. There are here two or three new hotels, several stores, offices, dwellings, railroad stations, &c., making the commencement of the new town of Meredith. They boast of a population of 800. The location is good, being on a gentle elevation of ground, facing to the westward.

*Fleming Farm.*—On the west side of Oil Creek, in Cherry Tree township. Number of acres not ascertained. Original and present owner, —— Fleming. Bounded on the north by Henderson, east by Oil Creek, south by Miller, west by Shugert. Is one and a half miles from Shaffer, and four and a half from Titusville. Oil Creek bears south and southeast. Road to Titusville north, to Shaffer south; railroad crosses on Creek bank. There are ten non-producing wells on the property, all on the flat. Owned by lessees. Have five engines on the farm. Depth and cost not ascertained. Sandrocks same as on Miller farm. One of the wells formerly flowed oil in considerable quantities. The operator thinking he could increase the production, drew up his tubing, changing the locality of the seed-bag. The result was a stoppage of the well entirely, it having produced nothing since. Thus showing that it is better to let a well alone when it is producing in paying quantities.

*Jones Farm.*—On east side of Oil Creek, and on Trout Run, in Allegheny township. Original owner, Widow Jones. Bounded on the north by Fleming, east by Cecil and others, south by Miller, west by Oil Creek. The Creek has a bearing south and southeast. Trout Run west. Fifteen non-producing wells on the farm. Depth and cost not ascertained. Operations have entirely ceased on the farm. There is ample room on the flat and table lands to operate upon, but this tract has the appearance of dry territory.

*Henderson Farm.*—On the west side of Oil Creek, in Cherry Tree township, only one corner coming to Creek. Original and present owner, Henderson. Bounded on the north by Stackpole, east by Fleming, south by Fleming, west by Lacy. Creek bears southeast at this point. There are no developments here. Railroad runs along the west bank of Creek, in the bluff. There is a point here, called "Night Cap," in a bend in the Creek and hills. Hills high and abrupt.

*Fleming Farm.*—On west side of Oil Creek. Owned by Mrs. Fleming. Bounded on the north by Oil Creek, and Stackpole, east by Oil Creek, south by Oil Creek. Creek has a bearing here southeast, south, and southwest. Road to Titusville northerly, to Shaffer southerly. Railroad crossing. No developments.

*Stackpole Farm.*—On east side of Oil Creek, and on Hubbard's Run. Original owner, J. Stackpole. Present owners, Henderson, Brewer, Watson & Co.

Bounded on the north by Shreve, east by Smith, south by Cecil and others, west by Oil Creek. Creek bears southwest, Hubbard's Run westerly. There are ten old non-producing wells, and one engine on the property. No developments at this time.

*Shreve Farm.*—On east side of Oil Creek, and on Shreve Run, in Allegheny township. Original owner, S. Shreve. Present owner, Great Western Consolidated Oil Company. Bounded on the north by Potts, east by J. Shreve, south by Stanford, west by Oil Creek. Bearing of Creek southeast, Shreve Run south of west. All quiet. No developments.

*Pott Farm.*—On east side of Oil Creek, Allegheny township. Owned by D. Pott. Bounded on the north by Stackpole, east by Shreve & Bros., south by Shreve, west by Oil Creek. Creek has a bearing south. No developments on this tract.

*Stackpole Farm.*—On west side of Oil Creek, in Cherry Tree township. Owned by Stackpole & Fletcher. Bounded on the north by Griffin, east by Oil Creek, south by Henderson, west by Curry. Creek bears south. Oil Creek Railroad crosses. There are ten non-producing wells on the property. All on flats. Depth of sandrocks not ascertained. Have three engines, portable and stationary, on the farm. Two refineries on the property — one abandoned; the other, Knight, Clement & Co., not now in operation. Capacity, 120 barrels crude per week. Flat is broad and well adapted for Oil operations, but is deserted at the present time.

*Stackpole Farm.*—On east side of Oil Creek, in Allegheny township. Original owner, O. Stackpole. Present owner, Northern Light Oil Company, Brewer, Watson & Co. Bounded on the north by Bissell & Co., east by Henderson, south by Pott, west by Oil Creek. Creek has a bearing south. Two old wells and one engine on the property. Nothing doing. Forsaken by oil operators. On the Creek at this point is the lower saw-mill. The dam raises the water so that it covers nearly the whole flat for some distance above. This is out of the dams which are cut in making the Pond Freshets.

*Bissell Farm.*—On both sides of Oil Creek, in Cherry Tree and Allegheny townships. Owned by Geo. H. Bissell & Co. Bounded on the north by Conley, east by Henderson, south by Stackpole, west by Griffin. Oil Creek bears south. There are thirty-four leases laid out on the west side, and seventeen on the east side of Creek. Two old wells, non-producing, on the property. The flats are covered by water raised by the dam. Flat narrow. Hills high and abrupt on west side, more gradual in ascent on east side. Railroad on west side. Wagon road on east side.

*Conley Farm.*—On east side of Oil Creek, in Allegheny township. Owned by Conley. Bounded on the north by Thompson, east by Carter, south by Bissell & Co., west by Oil Creek and Griffin, only southwest corner coming to the Creek. Bearing of Creek southeast. Six old wells on the farm, none in operation. Water from mill-dam forces back upon this tract.

*Griffin Farm.*—On both sides of Oil Creek, in Cherry Tree and Allegheny townships. Original owner, B. Griffin. Present owners, New York and Pennsylvania Petroleum Company. Bounded on the north by Bissell & Co., and Lang. East by Oil Creek and Conley, south by Stackpole, west by Lang and others. Bearing of the Creek southeast. Railroad crossing on the Creek bank. There are three producing wells, all pumping, but not largely. Average depth, 600 feet. Average cost $5,000. Commenced producing in 1861, and on since. Amount of production not ascertained. Wells all on the flat. Have three portable engines on the property. Average depth of wells 600 feet. Depth of first sandrock, 145 feet—59 feet thick; second sand, 415 feet--14 feet thick; third sand not found. Depth of driving-pipe, 45 feet. The Creek flats are narrow. Hills on west side high and abrupt; on east side more gradual.

*Bissell Farm.*—On both sides of Oil Creek, in Cherry Tree and Allegheny townships. Contains eighty acres. Original owner, Geo. H. Bissell & Co., present owner, Original Petroleum Company. Bounded on the north by Weisman, east by Thompson, south by Griffin, west by Davidson. Two miles from Titusville and four from Shaffer. Bearing of Creek east of south. Railroad crossing on west side of Creek. There are six producing wells on the farm, all pumping. Average daily product eighty barrels. Average depth of wells, from 200 to 600 feet. Average cost $5,500. Commenced producing in 1861, and in July, 1865. Amount of production several thousand barrels—not fully ascertained. Have ten portable engines on the farm. Wells all on the flat. Owned by the company and lessees. Depth of first sandrock, 145 feet deep—20 feet thick; second sandrock, 415 feet—59 feet thick; third sand, 471 feet —14 feet thick. Depth of driving-pipe, nine feet on the Creek, 45 feet back by bluff.

Here are about 100 of the pits, described in a former chapter. Oak-trees, standing in some of these, on being cut, show 400 rings or grains, indicating that they are at least 400 years old. The pits must have been dug sometime previous to the deposit of the acorns from which these trees sprang. Nor have we any evidence of the present being the first growth of timber upon them. This point presents the most activity in oil operations of any farm above the M'Elhenny and Farrel farms.

*Watson Petroleum Company.*—On both sides of Oil Creek, next above Bissell & Co., in Cherry Tree and Allegheny townships. Contains thirty-six acres. Original owner, Hibbert. Present owner, Watson Petroleum Company. Purchased in 1864. Bounded on the north by Pierce & Lock, east by Thompson, south by Original Petroleum Company, west by Davidson. One and a half miles from Titusville, four and a half miles from Shaffer. Bearing of Creek south. There are four non-producing wells on the farm; have formerly produced largely; one well in progress. Amount of production not ascertained. Average depth of wells, 400 feet—some greater. Average cost $5,000. Commenced producing in 1859, 1861, and since. Wells on flat and island formed by tail-race to saw-mill. Owned by the company. Have three stationary engines on property. Depth of first sandrock, 150 feet—10 feet thick; second sand, 370 feet—55 feet thick; third sand, not found. Depth of driving-pipe, 40 to 60 feet. The wells, Drake, 480 feet in depth; White well, 101 feet; Riley well, 454; Polhemus, 545 feet.

The Drake well was the first drilled in Venango County. It averaged thirty-two barrels per day for over two years, and would produce oil at the present time if operated. The engine which operates the Drake well, operates the Polhemus well also. The derrick and engine house of the Drake well unite under cover, the derrick being boarded up close up to the top. The history of this famous well has been given in a previous chapter.

Here is the location of what is called the upper saw-mill, just below Titusville. The owners of the mill have two wells drilled under the roof of their mill, both worked by the same power that drives the mill. The wells are producing some oil. Amount of production not ascertained. This mill dam is one that is cut in making the pond freshets.

*Kingsland Farm.*—On both sides of Oil Creek, in Cherry Tree and Allegheny townships. Number of acres not ascertained. Original owner, Bexford Pierce. Present owner, Original Petroleum Company. Purchased in 1864. Bounded on the north by county line and Watson Flats, east by Lock, south by Watson Petroleum Company, west by Davidson. One mile from Titusville. Oil Creek bears south-southwest and south. Road to Titusville and the railroad running northerly. There is one producing well on the farm, yielding twelve barrels per day. There are ten non-producing wells, and four in progress. Most of the wells yield some oil when pumped. Amount of production not ascertained. Wells all on the flat. Owned by company and lessees. Have six engines, portable and stationary, on the property. Depth of first sandrock, 150 feet—10 feet thick; second sand, 370 feet—53 feet thick; third sand-not found. Depth of driving-pipe, 45 feet. The flat begins to widen out as they extend up the Creek. Crossing the county line upon the Watson Flats in Crawford County. The hills are less abrupt than below.

*Watson Flats.*—On west side of Oil Creek and opposite the mouth of east Oil Creek, (or Pine Creek, as it is termed.) Contains ninety acres. Original owner, Jonathan Watson. Present owners, Oil Creek Petroleum Company of New York and Pennsylvania, and Oil Creek Petroleum Company of Philadelphia. Purchased in summer of 1864. Bounded on the north and east, by Oil Creek, south by Kingsland. Is half a mile below Titusville. There is a broad flat, well located for oil operations. Oil Creek bears east and south. There are at this time six producing wells, all pumping—average daily production thirty barrels—and fifty-four non-producing wells. Average depth of wells, 450 to 525 feet. Average cost, $5,000 to $7,500. Commenced producing in 1861, and on since. Amount of production, upward of 73,000 barrels. Wells all on flat. Eight owned by companies, balance by lessees. Twenty engines, portable and stationary, on the property. Depth of first sandrock, 150 feet—10 feet thick; second sand, 365 feet—55 feet thick; third sand, not found. Depth of driving-pipe, 30 to 70 feet. The Deep, or Barren well, is 1,200 feet in depth. Its name gives its history. There is one well on W. Palmer's Reserve, owned by the Mountain Oil Company, producing some eight barrels of oil per day. Among the wells which have formerly, and are now producing, are the Little Hope, Wood King, Continental, Kats, Baker, Grant, Drum, Eagle, Ingersoll, Goodrich, Augusta, Eureka, Ellenville, Finch, Palmer, and others. There has been a large amount of money expended here in developing this territory, and it has formerly been a point of much activity in the oil business. It is now very quiet. But few engines at work. A portion of these flats are within the borough limits of Titusville. Developments were made here in the earlier part of the oil excitement.

*Parker Farm.*—On the north side of Oil Creek, joining and in the Borough of Titusville, in Crawford County. Contains about 150 acres. Original owner, — Parker. Present owners, Parker Farm Petroleum Company of Philadelphia. Purchased in 1863. Fronts on the Creek nearly one fourth of a mile, extending back across the flats. There are twelve producing wells, all pumping. Daily production not ascertained, nor whole amount. Average depth of wells, 500 feet. Average cost, $4,000. Commenced producing in 1861, and on since. Wells all on the flat. Owned by company and lessees. Names of some of them as follows: Bunker Hill Petroleum Company, Canfield & Funk, Philadelphia Oil Company, Tremble & Phelps, Oceanic Company, Artesian Company, Firth & Abbott, and others. Have twenty engines, portable and stationary, on the farm. Depth of first sandrock, 150 feet—15 feet thick; second sand, 365 feet—50 feet thick; third sand not found. The bearing of Oil Creek is south of east, and south. Depth of driving-pipe, 30 to 80 feet. There is a large scope of territory conveniently located for oil operations. The Bunker Hill Company have three wells, all worked with a large engine. Plank road to Pithole and road to Pleasantville pass through this farm. Adjacent to the road, is the refinery of H. Hinkley, capacity 400 barrels crude per week; and between the road and creek is the Parker Farm Refinery, capacity 400 barrels crude per week. Formerly great activity prevailed at this point. At present, oil operations are limited.

The city of Titusville has been described in a previous chapter of the work. As stated, there are twelve refineries in the Borough limits. Two, the Hinkly and Parker Farm are described in account of Parker farm. The others are as follows: White Rose Oil Works,

Curtiss & Co., capacity 140 barrels crude per week Spring Hill Oil Works, 400 barrels; Mowbray & Co. 300 barrels; A. A. Bennett, 120 barrels; J. Brannon, 140 barrels; Bunker Hill, 640 barrels; Wm. Barnsdall, 240 barrels; Crosby, Young & Co., 120 barrels; Bartlett, 120 barrels; Rice, 120 barrels.

Jonathan Watson, Esq., is sinking a well in the Borough limits which is now at a depth of 1,600 feet. The third sandrock has not yet been found. He proposes to go 2,000 feet in search of it, unless sooner found. At the depth of 1,500 feet, oil was brought up in the sand-pump. This being (in the general term used by oil operators) a good show.

Passing above Titusville, there are developments at various points, but not of a successful character. Whether it is from the fact of its being outside of the oil basin, or from failures to strike the lucky vein, we leave for the theorizing of others. The explorer for oil has left his mark at various points all along the valley. From the appearance of the various localities, we think oil might be found. The speculative genius of the age has victimized not a few along the upper portion of Oil Creek. Tracts of land at various points bear the honorable name of some oil company, whose prospectus and stock have figured conspicuously in the eastern cities. Mills and mill-ponds are more frequent than productive oil operations. Passing Hydetown and its developments, we come to Clappville. Here the New York & Oil Creek Petroleum Company have some territory. Also the Sun Oil Company. Tyronville and Centreville, with some developments, are above. Passing on up the Creek, we find Riceville, Lincolnville, and other points, with here and there developments, until we reach Oil Creek Lake, about thirty-five miles above Oil City, the headwaters of the renowned Oil Creek, whose classic name and fame, coupled with Venango and Petroleum, is now known wherever civilization exists.

There are about one hundred wells above Titusville, nearly all of them non-producing, or, more properly speaking, dry holes. Some oil has been found at various points, but not in paying quantities. What the future of this territory will be, time alone can solve.

East Oil Creek, or Pine Creek, is a branch of Oil Creek, forming a junction with Main Oil Creek in Crawford County, near the Venango County line. At its confluence, is the John Watson farm, lying principally on the north and east side of the bend of Main Oil Creek, as it turns from an east to a south course. This farm contains 200 acres, and is owned by John Watson. Leased in part by Erie & Allegheny Oil Company. Bounded on the north by ——, on the east by Guild, south by Oil Creek and East Branch, west by Parker farm, joining the Borough of Titusville. Oil Creek bears south, East Branch southwest. Plank road to Pithole crossing easterly. There are two producing wells on the farm, both pumping—daily product, five barrels—and eighteen non-producing wells. Average depth, 500 feet. Cost of wells, $3,000. Commenced producing in 1861 and since. But few of the non-producing wells have formerly produced oil. Amount of production not ascertained. Wells all on flat. Owned by company and lessees. Five engines, portable and stationary, on the property. Depth of first sandrock, 150 feet—50 feet thick; second sand, 350 feet—25 feet thick; third sand not found. Depth of driving-pipe, from 50 to 80 feet on bluff, and from 80 to 90 feet on the flat. The flat lands at this point are low, and covered with hemlock and pine timber.

*Guild Farm.*—On north and west side of East Oi. Creek. Contains 200 acres. Original owner, James O. Guild. Present owners, Watson & Howland, Morrill Petroleum Company, and others. Creek bears south and west. Plank road to Titusville and Pithole crosses the tract. There are no producing wells. Sixteen non-producing wells, and two in progress. Some of them have produced formerly. Average depth of wells, 500 feet. One 930 feet. Cost not ascertained. Wells all on flat. Owned by lessees. Eight engines, portable and stationary, on the property. Depth of first sand-rock, 160 feet—13 feet thick; second sand, 440 feet—30 feet thick; third sand not found. Depth of driving-pipe, from 40 to 60 feet. Guild Island is in the Creek opposite, containing a few acres. The flats are broad and low, and were formerly covered with pine timber. The oil developments have caused the forest trees to disappear, and give place to oil well derricks and their machinery.

*Brewer Farm.*—On east and south sides of East Branch, crossing on to the north side. Number of acres not ascertained. Original owner, Dr. Brewer. Present owners, Hopeman & Co., (good name for oil operators). There is one non-producing well, and one engine on the property, also a grist-mill and saw-mill. Depth of well, 700 feet, dry-hole. Territory has never been sufficiently developed to know whether it is productive or not. Bearing of Creek is south and west.

There are no developments above this point of any account, until we reach Enterprise. The Creek makes a very winding way through a range of broken hills in a course nearly eastward for about ten miles, thence northerly. The external appearance of the land would indicate the existence of oil. A large portion of the territory has been taken up and put into oil stock companies, but the developments are yet to come.

On the Doris farm, some fellow has been sufficiently courageous to erect a derrick, but his heart failed before he got his engine on the ground. At Enterprise, there are some promising developments. Three wells are producing oil. Reported as producing ten to forty barrels per day each. They promise fair for new wells in a locality so far distant from other paying interests.

On Church Run, above Titusville, the Atlantic and Great Western Petroleum Company have put down a well, which is reported to be producing fifty barrels per day, pumping, with a good flow of gas. From this point, we will pass down Oil Creek, and give the results of the developments on its several tributaries.

The first of these is Bull Run, a tributary of Oil Creek, emptying into it on the east side, on the Farrell farm.

*Patterson Farm.*—Above Farrell. Three producing wells, one flowing (Johnson well) thirty barrels per day. Two pumping, with Crocker Blowers (M'Kinney well, ninety barrels per day; Railroad well, fifteen barrels per day.) Average depth of wells, 625 feet. Average cost, $6,000. Commenced producing in April, 1865, and later. Amount of production not ascertained. Five portable engines on the property. Depth of first sandrock, 255 feet—20 feet thick; second sand, 415 feet—30 feet thick; third sand, 705 feet—30 feet thick; fourth sand, 825 feet—40 feet thick. Depth of driving-pipe, 30 feet. This farm lies on an average of 125 feet above the level of the Creek.

*Skinner Farm.*—On Bull Run, in Cornplanter township, containing 216 acres. Original owner, J. B.

Skinner. Present owner, M. C. Egbert. Purchased in February, 1865. There are three producing wells, all pumping. One forty-five, the others fifteen barrels each daily. Average daily product, seventy-five barrels. Depth of wells, 760 to 860 feet. Cost of wells, $8,000. Commenced producing in June, 1865, and later. Amount of production not ascertained. Wells all on hill. Owned by Dr. Egbert and lessees. Seven portable engines on the property. Four wells in progress. Depth of first sandrock, 30 feet—20 feet thick; second sand, 280 feet—25 feet thick; third sand, 440 feet—30 feet thick; fourth sand, 730 feet—30 feet thick; fifth sand, 850 feet—40 feet thick. This farm lies on the hill or dividing ridge, between Oil Creek and the headwaters of Cherry Run. Well No. 109 is 370 feet above the level of Oil Creek, 860 feet deep, and is producing fifteen barrels per day. No. 5 is 270 feet above the level of the Creek, 760 feet deep, and is producing forty-five barrels per day. No. 4 is 270 feet above the level of Oil Creek, producing fifteen barrels per day. One well, drilling 350 feet above level of Creek, has a good show of oil at 330 feet from the surface, and gas in considerable quantities.

*Benninghoff Run.*—Beginning at the mouth, the territory formerly belonged to G. W. M'Clintock. In 1861, he sold off the land immediately on the Run, reaching from the west side at the brow of the hill across the valley of the Run, and over the narrow ridge, between the Run and Oil Creek. It was sold in tracts ten rods wide, embracing two acres at the mouth; wider as the Run ascends.

The first is Bartlett & Newton. One refinery, capacity 160 bbls. crude per week. Four non-producing wells. Average depth, 500 feet. Second, Col. Chambers, two acres, with two non-producing wells. Third, Lesler & Co., two acres, three non-producing wells—have formerly produced some oil. Average depth, 600 feet; Sandrocks same as on M'Clintock farm. Fourth, Thorn & Prindle, two and a half acres; two non-producing wells; one refinery, capacity 120 bbls. crude per week. Fifth and sixth, two and a half acres each; two non-producing wells, and one old abandoned refinery; owners not ascertained. Seventh, Cornwall Oil Company, four and a half acres; three non-producing wells, and one in progress on hillside. Average depth, 550 feet. Depth of first sandrock, 225 feet—25 feet thick; second sand, 360 feet—20 feet thick; third sand, 485 feet—50 feet thick. One well formerly produced ten barrels per day. Eighth, A. G. Wilcox & Co., four and a half acres; no wells; one refinery, Rock Spring, capacity 120 bbls. crude per week. Ninth, Orris, Hall, and others, one well in progress on Run, one producing on Creek; have four and a half acres. E. Olmsted, four and a half acres; two non-producing wells. One refinery, Olmsted Oil Works, capacity 120 bbls. crude per week. There is a large iron tank near the refinery, constructed for the use of the same. It has never been used. The lot is called Iron Tank lot. This tract is the first above the line of the M'Clintock farm, embracing the valley of the Run and the hillsides. Next comes the Brewer & Wilson tract, now Benninghoff Petroleum Company, eleven acres, no wells. Next is the Titus and Drysdell tract, 13 acres, three of which are now owned by Newton & Clark, with one refinery; capacity 160 bbls. crude per week. Two wells on the tract of Titus & Drysdell, one non-producing, and one in progress.

At the mouth of Tarr Run, on a tract of three acres belonging to Titus & Drysdell, there are two producing wells, one flowing, the other pumping; daily product.

each 50 and 25,total 75 barrels. Commenced producing in spring of 1864. Amount of production not ascertained. There are four non-producing wells. The valley of the Run will average about fifteen rods in width. At the mouth of Tarr Run it is broader. Hills high, and in many parts rugged.

*Warner Farm.*—On Benninghoff Run, both sides, at and above the mouth of Tarr Run, in Cherry Tree township, containing thirty-seven acres. Original owner, George Warner; present owner, Benninghoff Petroleum Company, of New York. Purchased in September, 1864. One mile from Oil Creek, which is reached by wagon road and horse railroad. There are fifteen producing wells—twelve flowing and three pumping; three non-producing, and ten in progress. Amount flowed, pumped, or shipped, not ascertained. (Superintendent refused to tell, saying that was the Company's private business, &c.) Average depth of wells, 600 feet. Average cost, $4,500. Commenced producing March 22d, 1865. Wells on flat and hill-side. Three owned by company, balance by lessees. Twenty engines, portable and stationary, on the property. Depth of first sandrock, 270 feet—20 feet thick; second sand, 380 feet—in shells and irregular; third sand, 525 feet —54 feet thick. Depth of driving-pipe, 15 feet on flat and 27 feet on the bluff. The producing wells are the Coddington, 80 bbls.; Haskins, 105; Getty, 250; Clark, 240; Beach, 80; Lady Harmon, 70;. Western Union Telegraph, 280; Rencie, 200; California, 60; Philadelphia, 100; Bodine, 40; Spence, 180; Story, 60; No. 19, 40 bbls.; No. 18, 10; No. 15, 15; No. 16, 4 bbls. The reported daily production making a total of 1,721 bbls., which we think is an over-estimate. This is a very busy point, and any of the producing wells hereabouts we should consider very desirable property,

*Weiss Farm.*—Next above Warner. Owned by Cornwall Company. Contains from 80 to 100 acres. Some five or six wells drilled. All non-producing. Next above is the Widow Foster farm, M'Corns, owner; twelve acres; three wells drilled, all non-producing. Next is Martin Benninghoff farm, owned by Benninghoff Petroleum Company and Martin Benninghoff Petroleum Company. Two wells—one pumping, but producing nothing to pay expenses of working; the other blowing large quantities of gas, and about four bbls. of oil per day. Hills are very broken, and rise more gradually from the valley.

*Wild Cat Hollow.—Widow Stevenson Farm.*—On west side of Wild Cat Hollow. Original owner, Widow Stevenson. Present owner of thirteen acres, Petroleum Centre Company, of Philadelphia. Purchased in spring of 1864. There are five producing wells, all pumping; daily production, 25 bbls. per day. There are four non-producing wells. Total amount of production not ascertained. Average depth of wells, 550 feet. Cost, $4,000. Wells all on the flat. Owned by lessees. Six portable and stationary engines on the property. Depth of first sandrock, 200 feet—20 feet thick; second sand, 380 feet—30 feet thick; third sand, 480 feet—25 feet thick. Depth of driving-pipe, 70 feet. The Coldwater Oil Company have eight leases immediately on the Run, from the mouth up, on which there are six wells, at the average depth of 550 feet; none of them producing, from not being operated.

M'Calmont tract, on north-west side of Wild Cat Hollow, containing thirty-six acres. Original owner, Judge M'Calmont. Present owners, Knox & Co. Purchased in 1864. There are four producing wells, one flowing 3 bbls. and three producing 6, 6, 15 bbls. each. Daily production, 30 bbls. Ten non-producing wells, and two in progress. Average depth, 550 feet. Cost $5,000. Commenced producing in 1864–5. Amount of production not ascertained. Wells on flat and hillside. Owned by lessees. Ten engines, portable and stationary, on the property. Depth of first sandrock, 195 feet—20 feet thick; second sand, 245 feet, 30 feet thick; third sand, 480 feet—50 feet thick. Depth of driving-pipe, 70 feet on the flat, and 40 feet on the bluff. Names of some of the wells: Bradbury well, Wild Cat well, Forrest well, Parmer well.

*Wykle Run.*—This is a small stream emptying into Cherry Tree Run, a short distance from its confluence with Oil Creek, on the Rynd Farm. The Hap farm and Gibson farm are on this Run, above Rynd farm. Is one mile from Oil Creek. The Run bears south-east. There are eight non-producing wells on these tracts. Average depth, from 500 to 700 feet. Cost not ascertained. Wells principally on flats. Owned by a Philadelphia Company, and lessees. Four portable engines on the property. Depth of first sandrock, 170 feet—30 feet thick; second sand, 350 feet—30 feet thick; third sand, 475 feet—25 feet thick. The valley is about ten rods wide. Hills gently elevating, and not very high. There is some good operating territory on this Run, but the wells thus far have proved non-producing. The Whiting well blowed strongly with gas—could be heard a considerable distance—it was all gas and no oil. The Paxton Petroleum Company drilled one dry hole on this territory. At the mouth of the Run there are hotels, stores, saloons, &c., the beginning of a small town.

*Cherry Tree Run.*—A tributary of Oil Creek, emptying in from the west side, on Rynd farm, four miles from Oil City.

*Tarr Farm.*—First above Rynd, in Cornplanter township, containing twenty-five acres. Original owner, George Tarr. Present owner not ascertained. Seven acres leased to Barnes. Bounded on the north by Ward, east by Blood, south by Rynd, west by Gibson. One and a half miles from Rouseville. The Run has a bearing east of south. Road to Cherry Tree north, to Rouseville south. Two non-producing wells. Have produced formerly a small amount of oil, and are from 500 to 750 feet deep. Average cost, $5,000. Are on the flat. One portable engine on the property. Depth of first sandrock, 240 feet—20 feet thick; second sand, 372 feet deep—12 feet thick; third sand, 500 feet—22 feet thick. Depth of driving-pipe, 14 feet. The valley of Cherry Tree Run will average about twenty-five rods in width. The hills in some portions of it are very high and steep, but generally have a gradual elevation, the stream crossing the valley many times from side to side on its winding way.

*Ward Farm.*—On Cherry Tree Run, in Cornplanter township, containing twenty-two acres. Original owner, Robt. Ward. Present owner not ascertained. Bounded on the north by Boyd, east by Blood, south by Geo. Tarr, west by Gibson. One and a half miles from Rouseville, five miles from Cherry Tree. General bearing of Run, south. There are three non-producing wells, averaging in depth froom 600 to 758 feet. Average cost, $5,000. Wells on flat. Have two portable engines on property. Depth of first sandrock, 243 feet—40 feet thick; second sand, 385 feet—35 feet thick; third sand, 525 feet—25 feet thick. Depth of driving-pipe, 12 feet. Flat about twenty-five rods in width. Hills steep. Developments not very healthy or encouraging to owners.

*Boyd Farm.*—On Cherry Tree Run, in Cornplanter township, containing thirty acres. Original owner, J. R. Newman; present owners, Kingston & Randolph Petroleum Company, and Plymouth Rock Oil Company. Purchased in 1864. Five acres by Plymouth Rock Oil Company, and ten acres by Kingston and Randolph Oil Company. Bounded on the north by Newman, east by Pearson & Ray, south by Tarr, west by Gibson. Two miles from Rouseville, four and a half from Cherry Tree. Roads to Cherry Tree and Rouseville. Three non-producing wells; one owned by K. & R. Company, other by P. R. Oil Company. Depth of wells, 550 to 666 feet. Average cost, $6,000. Wells on flat. Two portable engines on property. Depth of first sandrock, 390 feet—40 feet thick; second sand, 465 feet—35 feet thick; third sand, 628 feet—25 feet thick. Depth of driving-pipe, six and half feet. There is a difference in the drillers' report of sandrocks from those above or below, each claiming to be right. At 250 feet they found sand and slate, which is probably the first sand. Ten acres of the tract belong to some parties whose names we failed to ascertain. One well, non-producing. No one on the premises to get items from.

*Brown Farm.*—On Cherry Tree Run, in Cornplanter township, containing 22 acres. Original owner,—— Brown. Present owner, Metropolitan Oil Company, of New York. Bounded on the north by Brown, east by Pearson & Ray, south by Ward, west by McElvane. Two and a half miles from Rouseville, four from Cherry Tree. The Run bears south and east of south. Road to Cherry Tree north, to Creek south. One well will produce six barrels per day when worked. Has produced 300 barrels. Depth of well, 600 feet. Cost, $6,000. Commenced producing in 1865. Well on flat, and own ed by company. One portable engine on the property. Depth of first sandrock, 248 feet—43 feet thick; second sand, 405 feet—22 feet thick; third sand, 537 feet—20 feet thick. Depth of driving-pipe, 12 feet. The Seneca Oil Company have one well on a three acre reservation on this lot. Drilled by water power. Is one of the unprofitable investments.

*Sutley Farm.*—On Cherry Tree Run, two and three quarter miles from Rouseville, and three and three quarter miles from Cherry Tree, in Cornplanter township, containing 75 acres. Bounded on the north by Black, east by Pearson & Ray, south by Newman. There is one producing well, pumping twelve barrels per day. One non-producing well. Depth 590 to 655 feet. Average cost, $6,000. Commenced producing in July, 1865. Are on the flat, and owned by the company. Have one engine. Depth of first sandrock, 265 feet—40 feet thick; second sand, 407 feet—22 feet thick; third sand, 573 feet—20 feet thick. Depth of driving-pipe, 17 feet. The flat is broader than below. Hills on a gentle slope. The producing well shows there is oil here, although not in paying quantities, so far as developments have gone.

*Black Farm.*—On Cherry Tree Run, in Cornplanter township, contains — acres. Original owner,——— Black. Present owners, Big Tank Oil Company, on east side; and Pennsylvania Oil Company, on the west side. Bounded on the north by Dempsey, south by Brown, east by Sutley, west by Elvane. Bearing of the Run at this point is east of south. Road to Cherry Tree, north, to Rouseville east of south. There are two non-producing wells, and twelve in progress, but all quiet now. Depth of the wells drilled, one is 600 feet, the other 700. Average cost, $6,000. Wells are on the flat. Owned by company and lessees. Have five engines, portable and stationary, on the property. Depth of first sandrock, 265 feet—40 feet thick; second sand, 407 feet—22 feet thick; third sand, 575 feet—20 feet thick; fourth sand, 700 feet—10 feet thick. Depth of driving pipe, 10 feet. The 700 feet well made a fine show when first tested. Broke the sucker-rods after pumping half an hour. Has not been tested any since.

*Dempsey Farm.*—On Cherry Tree and Cain's or Little Cherry Run, in Cornplanter township, containing 160 acres. Owned by T. Dempsey. Leased to different parties. Bounded on the north by Baney & Lichtenthaler, east by Thomas Tarr, south by Black, west by Mallory. Three and a half miles from Rouseville, and three and a half from Cherry Tree. Bearing of Cherry Tree Run, south, of Little Cherry Run, east. Roads to Cherry Tree and Oil Creek. There are two producing wells, pumping each three and four barrels daily. six non-producing, and twenty in progress; operations having been commenced on this extensive scale, but suspended apparently for want of confidence on the part of those interested. Average depth of wells drilled 600 feet. Average cost, $6,000. Commenced producing in 1865. Amount of production not ascertained. Wells on flat and owned by lessees. Ten engines on the property. Depth of first sandrock, 284 feet—20 feet thick; second sand, 450 feet—22 feet thick; third sand, 573 feet—32 feet thick. Depth of driving pipe, 10 feet.

*Mallory Farm.*—On Cain's, or Little Cherry Run, above Dempsey farm, there are four wells drilled, all non-producing, and twenty more commenced. On next farm above, Chichester & Co. have one well, over 1,000 feet in depth; Clark & Co., one 800 feet. There are also five more wells drilled 600 feet, all dry holes. Comment, under the circumstances, would be superfluous. Average cost of wells, $6,000 to $8,000.

*Baney Farm.*—On Cherry Tree Run, in Cornplanter township, containing 125 acres. Original owner, — —Baney. Present owners, Day & Co., of New York, and others. Purchased in 1865. Bounded on the north by Pierson, east by Stevenson and Lichtenthaler, south by Dempsey, west by Mallory. Four and a half miles from Rouseville and two and a half from Cherry Tree. The Run bears in a southeast course. On this farm is one producing well, the Williams, pumping five barrels per day, and four non-producing. Average depth, 600 feet. Cost of wells, $5,000. Commenced producing in 1865. Wells on the flat and owned by lessees. Three portable engines on the property. Depth of first sand-rock, 285 feet—20 feet thick; second sand, 450 feet—22 feet thick; third sand, 575 feet—32 feet thick. Depth of driving pipe, twelve feet. There are thirty-five leases on the Creek, mostly on the east side—a small tract sold in fee on lower end and adjoining the leases. The valley here is broad, affording a fine field for oil operations, but has been very slightly tested.

*Pierson Farm.*—On Cherry Tree Run, in Cornplanter township. Number of acres not ascertained. Original owner, J. Pierson. Present owner, M'Clintockville and Cherry Tree Run Petroleum Company. Bounded on the north by Reed & Kilgore, east by Stevenson, south by Baney, west by Chambers. Five miles from Rouseville, and one and a half from Cherry Tree. There are two non-producing and four wells in progress on the farm. One of the wells is testing. Depth of wells, 485 to 736 feet. Average cost, $5,000. Located on the flat. Two portable engines on the property. Depth of first sandrock, 290 feet—35 feet thick; second sand, 450 feet--35 feet thick; third sand, 595 feet--25 feet thick. Depth of driving pipe, 15 feet. The flat is broad. Hills gently undulating. Oil has not yet been found in paying quantities. Developments are meagre, not sufficient to test the territory. Above this tract is a strip of undeveloped territory. Adjoining and above this is the Stewart farm, containing 230 acres, owned formerly by J. B. and J. Stewart. Present owners, Claremont Oil Company of Baltimore, and Potomac Oil Company of Washington. Purchased in the spring of 1864. One mile from Cherry Tree and five miles from Oil Creek. Bounded on the north by Drake & Irwin, east by Chambers, south by Pierson, west by J. Stewart. Bearing of the Run is south. Roads to Cherry Tree, and Oil Creek. There are four non-producing wells, two to each company—two of them testing. Average depth 828 feet. Average cost, $6,000. Wells located on flat and hill-side, one nearly on the top, on east side of Run. Owned by the companies. Four portable engines on the property. Depth of first sandrock, 330 feet —30 feet thick; second sand, 435 feet—40 feet thick; third sand, 725 feet—17 feet thick. This record is of the well on the hill. The difference of depth is the difference of elevation. Depth of driving pipe, seven feet on the hill; 54 on the flat. This territory has been but slightly developed, and is covered with heavy timber on flat and hill sides.

*Irwin Farm.*—On Cherry Tree Run, in Cherry Tree township, containing 119 acres. Original owner, S. N. Irwin. Present owner, Dervan Oil Company of Chicago. Purchased in spring of 1865. Bounded on the north by Disile, east by Stewart, south by Stewart, west by Tarr. Six miles from Oil Creek. There are two non-producing wells on the farm. Both were tested at a depth of 600 feet. Are now drilling them deeper under a contract for a depth of 1,000 feet. Average cost of wells so far, $7,000. Wells located on the flat and owned by the Company. Depth of first sandrock, 340 feet—45 feet thick; second sand, 590 feet—25 feet thick; third sand, 750 feet—30 feet thick. Depth of driving-pipe, 17 feet on the east side of the Run, and 45 feet on the west side. We here found the most, complete record of the rocks drilled through, yet met with. The driller has boards prepared by boring one-inch holes into them in rows, has his boards or planks about eighteen inches wide and two and a half feet thick fitted into a box to receive it,—has some of the drillings taken out after each reaming, washed and put into these holes, thus forming a complete record of the rocks, showing all the various grades and colors, from the surface, down. The holes are all numbered and filled in regular rotation. On the lower part of this tract is a custom grist mill, with the necessary buildings about it.

Above, this point we come to the village of Cherry Tree. There are three or four wells drilled in and around the village, all non-producing. Above the village are a few wells drilled, of the same sort, dame Nature, seemingly turning aside from this locality with her supply of oleaginous fluid. This has not partaken of the fell spirit of speculation that has visited some of its neighbors. It is located principally in the valley. There are a few fine dwellings, and a church on the hill, as well as the requisite buildings for its population and business. It is an old town, and bears the marks of age. No developments worthy of notice above this point.

## CHERRY RUN.

Cherry Run is one of the main tributaries of Oil Creek, and forms a junction with the same at Rouseville.

*John M'Clintock Farm.*—On both sides of Cherry Run, in Cornplanter township, containing 138 acres. In 1859 this tract was sold by Mr. M'Clintock to New York parties by the name of Curtiss, Macumber, and others. In April, 1864, these parties sold one-half the tract to the Brevoort Petroleum Company of New York, and stocked the other half in the Union Petroleum Company of New York, dividing the tract diagonally across the centre from corner to corner. Half a mile from Rouseville, three miles from Plumer. General bearing of the Run is southwest. Road to Rouseville southwest, to Plumer, northeast. Bounded on the north by J. Rynd, east by Knapp, south by Buchanan, west by J. M'Clintock.

The Union Petroleum Company have the south-western half, and have eighteen producing wells, three flowing and fifteen pumping, with an average daily product of 120 barrels. Three non-producing wells, and six in progress. Average depth of wells, 610 feet. Average cost $6,000. Commenced producing in June, 1864. Amount of production over 15,000 barrels. Wells located on flat and hill-side. Owned by the Company, in fee, and by lessees, for a royalty of one-half the oil. Have twenty-one engines, portable and stationary, on the property. The Union Petroleum Company organized with a capital stock of $125,000. In June, 1864, they declared a dividend of sixty per cent, and expended as much more.

The Brevoort Oil Company own the north-eastern

half, and have nine producing wells, one flowing and eight pumping, with an average production of 190 barrels, per day. Two non-producing wells, and six in progress. Average depth of wells, 594 feet. Average cost, $6,000. Commenced producing in 1865. Amount of production 14,560 barrels. Wells located on the flat and hill-side. Wells owned by Company in fee, and by lessees for one-half royalty, after paying a heavy bonus. The Company have received bonus on eight three-quarter acre leases, $80,000, and one-half the oil. Fourteen engines, portable and stationary, on the property. Sandrocks, on this and the Union tract, are as follows; Depth of first sandrock, 240 feet on flat, and 311 feet on bluff—30 feet thick; second sand, on flat, 370 feet—on bluff, 453 feet—20 feet thick; third sand, on flat, 535 feet, on bluff, 600 feet—25 feet thick. Depth of driving pipe, 13 feet on the flat, and 25 feet on hill. The rocks are very near on a level on the flat and hill, the difference in depth being the difference in the elevation. The Brevoort Oil Company organized with a capital stock of $500,000. Have been expending about $15,000 per month. Are now paying monthly dividends on the capital stock of two and a half per cent. A small run and ravine cross the tract, and empties into Cherry Run, on the south east side, called Brevoort Run.

*Knapp Farm.*—On Brevoort Run, about Brevoort Company's tract, in Cornplanter township, containing fifty acres. Original owner, ——— Knapp. Present owner St. Nicholas Oil Company of Philadelphia, purchased in 1864. Bounded on the north by Cherry Run Petroleum Company, east by Balliett & Co., south by Tolls, west by Brevoort. One mile from Rouseville, two and a half from Plumer. Bearing of the Run is west. There are three producing wells, all pumping, with a daily average of fifty barrels, twelve wells in progress, and twenty-five more leases taken to operate upon. Average depth of wells, 760 feet. Cost of wells, $6,000. Commenced producing in September, and November, 1865. Amount of production 900 barrels. Wells located on ravine, and owned by lessees. Twelve engines, portable and stationary, on the property. Depth of first sandrock, 730 feet--39 feet thick; second sand, 565 feet—41 feet thick; third sand, 730 feet—25 feet thick. Depth of driving-pipe, 20 feet. This tract is estimated to be 150 feet above the level of Cherry Run.

*Tolls Farm.*—On Brevoort Run, in Cornplanter township, containing eighty-seven and a half acres. Original owner ——— Tolls. Present owner, Murray Oil Company. Bounded on the north by Cherry Run Petroleum Company, east by Balliett and Tolls, west by St. Nicholas Company. Two and a half miles from Plumer, and two from Rouseville. There is one producing well, pumping very little, and twelve in progress. Average depth, 800 feet. Cost of well, $7,000. Wells located on Ravine flats. Estimated to be 200 feet above the level of Cherry Run. Owned by lessees. Eight portable engines on the property. Depth of first sandrock, 450 feet—39 feet thick; second sand, 600 feet–41 feet thick; third sand, 730 feet—25 feet thick. Depth of driving-pipe, 20 feet.

*Tract from Rynd Farm.*—On Cherry Run, in Corn planter township, containing five acres. Originally owned by John Rynd. Purchased from him by Mr. Duff, and afterwards Mr. R. Criswell purchased two acres and eight perches. Present owners of Criswell tract Mingo Oil Company of Philadelphia, and Wadsworth and Wynkoop. One and a quarter miles from

Rouseville, and two miles from Plumer. There are on this tract nine producing wells, all pumping, producing 200 barrels per day. Five non-producing wells and six in progress. Total amount of production not ascertained. Average depth of wells 600 feet. Average cost, $6,000. Commenced producing in 1862 and 1864. Wells located on flat and hill-side, owned by lessees. Known as Reed wells, Nos. 1, 2, 3 and 4. Criswell wells, Nos. 1 and 2, Phillips and Egbert, mountain well, Bradly, Moore, Coyle, Sweeny and McCauley. Fifteen engines, portable and stationary, on the property. Depth of first sandrock, 240 feet—35 feet thick; second sand, 365 feet—25 feet thick; third sand, 550 feet—25 feet thick. Depth of driving-pipe, 12 feet on flat, and 28 feet on hill-side. The old Reed well is pumping about twenty barrels per day, Mountain well very little, Criswell wells, lightly. Reed well, No. 4, is pumping very well.

This tract has been highly productive. The Reed wells, four in number, all located on a single acre, produced upwards of 100,000 barrels of oil. The old Reed well was struck July 18th, 1864, commencing with a production of 280 barrels per day, by flowing. As before stated, R. Criswell owned some two acres here. The purchase was made in the first years of Petroleum development. Mr. Criswell drilled one well on his property to a depth of about 700 feet. This well only produced by pumping three or four barrels per day. In the days of cheap oil, this amount of production only served to impoverish the operator. After struggling along for several years, Mr. Criswell joined Mr. Wm. Reed who owned an interest near by, in sinking the Reed well. During its progress, Mr. I. N. Frazer took an interest in it. After the well had been sunk to the proper depth, the appearances were not very favorable.

After several days had been spent in testing it, the well commenced to flow at the rate stated above, and continued to produce for several years. Mr. Criswell sold his interest—one-fourth of the land, soon after the well was struck, to the Mingo Oil Company of Philadelphia, for $280,000, having previously realized about $30,000 from the oil produced by the well. After realizing about $75,000 from the well, Mr. Wm. Reed sold his one-half interest of all the oil, to Bishop, Bissell & Co., for $200,000. The interest of Mr. Frazer, consisting of one-fourth in original tract, or one-third of the working interest of the well, was sold about the same time to other parties, for $100,000, making a total cash income realized by the fortunate owners, of $685,000, in a few months after its commencement. The purchasers, it is fair to presume, realized handsomely on their first investment. The Reed well was a "fat thing" even in the land of Oildom.

*Part of Rynd Farm.*—On Cherry Run, in Cornplanter township, containing 36 acres. Original owner, J. Rynd. Present owners, Curtin Oil Company, five-eighths; Johnson, one-eighth; North American Petroleum Company, two-eighths: purchased in June, 1864. Bounded on the north by Blood, east by Smith, south by Brevoort, west by Rynd. One and a half miles from Rouseville, and two miles from Plumer. Bearing of Cherry Run, southwest. Road to Plumer northeast, to Rouseville southwest. Twenty-one producing wells on tract, six flowing and fourteen pumping. Daily product 200 barrels. Six non-producing wells, and seven in progress. Have produced over 30,000 barrels. Average depth of wells 600 feet. Average cost $6,000. Commenced producing in 1861 and 1865. Wells located on flat and bluff. One on top of hill, 746 feet in depth, flowing

75 barrels per day. Owned by lessees. Twenty-five engines, portable and stationary, on the property. Depth of first sandrock on the flat, 245 feet; on the bluff 430 feet—39 feet thick; second sand, on flat 390 feet; on bluff 565 feet—41 feet thick; third sand on flat 530 feet; on bluff 710 feet—36 feet thick. Depth of driving pipe, 18 feet on the flat, and 25 feet on the bluff. Hill high and abrupt.

*Smith Farm.*—On Cherry Run, in Cornplanter township, one and a half miles from Rouseville, and two miles from Plumer. Contains 57 acres. Original owner, Wm. Smith. Present owners, Cherry Run Petroleum Company of Connecticut. Purchased in 1863. Bounded on the north by Blood, east by Cherry Run Petroleum Company, south by Brevoort Oil Company, west by Curtin Oil Company. Bearing of Cherry Run southwest. Roads to Plumer and Rouseville, and railroad from Rouseville to Plumer passing through property. There are twenty producing wells on the farm, seven flowing and thirteen pumping. Average daily product 300 barrels. There are seven non-producing wells, and thirteen in progress. Have produced over 150,000 barrels. Average depth of wells, 600 feet. Average cost of wells, $7,000. Commenced producing in 1864 and 1865. Wells located on flat and bluff, and owned by lessees. Known in part as follows: Yankee well, Gruninger, No. 11; S. S. Watkins, Fry, No. 34; No. 47, Auburn well, Bradley, Weaver, Hathaway, Wade; Faunce & Auburn well, Mackie well, &c. Thirty-five engines, portable and stationary, on the property. Depth of first sandrock, 270 feet—40 feet thick; second sand, 405 feet—40 feet thick; third sand, 550 feet—40 feet thick. Depth of driving pipe, 10 feet on the flat, and 20 feet on the bluff, varying according to elevation.

This has been one of the best tracts for oil production of its size in the oil region. Royalty one-half the oil. Average bonus on leases, $4,000 each. Several of the former producing wells are now doing nothing. The old Wade well, which flowed very largely for some time and then ceased, had a set of tools in the bottom. After it had ceased flowing, efforts were made to extricate them, but without success. A new well was drilled three feet from the old one, and is pumping about ten barrels per day. This is fair evidence that they are not both on the same vein. The working interest of the old Wade well, flowing about 100 barrels per day, was sold to a New York party for $300,000 cash.

This farm was purchased in 1858 by Mr. Smith from H. M'Clintock, for a yoke of oxen. Mr. Smith sold it in 1862, for $1,000. The present owners paid $6,500 for it. The farm has since been valued at several millions of dollars. Such are the chances in this favored land of oil. What one loses another gains.

*Cherry Run Petroleum Company of Philadelphia Lands.*—On Cherry Run, in Cornplanter township. Contains 401 acres of land. Is two miles from Rouseville, and one mile from Plumer. Original owners, John M'Fate, Joseph M'Fate, and D. B. Knapp. Present owners, Cherry Run Petroleum Company of Philadelphia. Purchased in 1863. Bounded on the north by Russell and M'Fate, east by Seymour, south by Murray and St. Nicholas Oil Companies, west by Smith and Blood. Bearing of Cherry Run, west, M'Fate Run, south. Roads to Plumer and Rouseville passing through. There are on the property twenty-five non-producing wells. Some of them have produced, but not largely. Average depth of wells 625 feet. Average cost $6,000. Commenced producing in 1863. Wells are located on the flat, owned by the company in part, and by lessees. Ten engines, portable and stationary, on the property. Depth of first sandrock, 285 feet—15 feet thick; second sand, 425 feet—15 feet thick; third sand, 570 feet—14 feet thick. Depth of driving pipe, 15 feet. The wells drilled have tested the flats effectually, or nearly so. But the hills and ravines are as yet untested, and we look for good wells to be obtained some day on the uplands of this farm. The hills are not very abrupt, of easy access. It seems strange that this tract should prove unproductive, when such good territory as the Smith farm adjoins it. But Dame Nature plays some strange freaks, leaving her admirers often surprised and bewildered.

*Widow M'Fate Farm.*—Adjoining the Cherry Run Petroleum Company's lands on the south side. Owned at present by Balliett & Co.; is a tract lying on the table-lands back from the valley of Oil Creek, on which there are several wells in progress, none of which have yet been tested. There are four engines on the property. Depth of sandrocks, &c., not ascertained.

*M'Fate Farm.*—On the north side of the Cherry Run Petroleum Company's lands, bordering on the valley of Cherry Run, containing 160 acres. Original owner, Joseph M'Fate. Present owner, Oil Creek and Cherry Run Petroleum Company. Purchased in 1864. Bounded on the north by Russell, east by Lamb and Knapp, south by Cherry Run Petroleum Company, west by Blood. One and a half miles from Plumer and one mile from Rouseville. Roads to Rouseville and Plumer over the hills. On this tract are twenty-one non-producing wells. Some have been tested, but proved unproductive. Average depth of wells, 625 feet. Average cost, $6,000. Wells located on hill-sides and ravines. Twelve engines, portable and stationary, on the property. Depth and thickness of sandrocks similar to those on the Cherry Run Petroleum Company's lands. Nineteen of the wells are on the border of the valley of Cherry Run, the others are on the ravine on upper part of farm. This farm has the appearance of dry territory.

*Hogg Farm.*—Originally a portion of the Lamb farm, contains 25 acres, now owned by the Springfield Petroleum Company, and several parties who purchased in fee, one and two acres each. On the tract are several wells drilled and others in progress, now abandoned, none of those drilled proving paying investments. The Springfield Petroleum Company have a small refinery on this property, not now in operation. Capacity not ascertained.

*Humboldt Petroleum Company.*—Have 24 acres off from same farm, bearing on Cherry Run, and lying across the valley. Was purchased by Messrs. Bruns and Ludovici, of New York, and the Humboldt Refinery built in 1862. Present owner, Humboldt Mining and Refining Company of New York. Purchased in 1865. The whole tract is enclosed with a high board fence. Capacity of works, 20 stills of 40 barrels each, capable of refining 1500 barrels of crude oil per week. Is half a mile below Plumer. Have one non-producing well in their enclosure over 700 feet in depth. Depth of first sandrock, 325 feet—40 feet thick; second sand, 507 feet—37 feet thick; third sand, 623 feet—34 feet, thick. Depth of driving pipe, 30 feet.

The Company have a machine shop in which they do all their own machine work and fitting, also, an extensive barrel factory of their own. Have storage for 8,000 barrels of crude oil. This is the second largest refinery in the oil region. They claim that they can turn out more refined oil than Downer's works, at Corry. Their principal shipping point heretofore has been Walnut Bend, on the Allegheny river. The completion of the Cherry Run Railroad will enable them to ship direct by rail. The railroad runs directly by their works.

*Lamb Farm.*—On west side of Cherry Run, a small portion coming to the Run mostly on the hill side. Contains 230 acres. Original owner, J. Lamb. Present owner, Cherry Run Central Oil Company. Purchased in 1864. Bounded on the north by M'Fate and Lamb, east by Turner and Anderson, south by Cherry Run Petroleum Company, west by M'Fate Brothers. Half a mile from Plumer, and two and a half miles from Rouseville. The bearing of Cherry Run is south and southwest. On the east side of Cherry Run, joining the Springfield Petroleum Company's lands, and the Humboldt works, is the Robert M'Fate farm, containing 80 acres, on which there are several non-producing wells. Name of present owner not ascertained. No one operating it. It is in the hands of some stock company. The Sleuts and Turner farms join the Lamb and M'Fate on the north, several wells having been sunk on each, all non-producing. On these several tracts, embracing the Springfield company, Hogg farm, Lamb farm, M'Fate, Turner, Slents and Humboldt Company, &c., all lying between the Cherry Run Petroleum Company's lands and the village of Plumer, there are a large number of stock companies represented, many leases having been sold out in fee to parties operating. It is difficult, nay impossible, to get the names of the different companies, as there is no one operating, except the Cherry Run Central Petroleum Company.

There are on these tracts below the Humboldt Petroleum Company's lands, sixty-four derricks standing; from the works, reaching up to the village of Plumer, ninety-four derricks. Of this number, 154 in all, there are not more than forty wells drilled, varying in depth, but on an average 600 feet. The engines, twenty-four in number, are portable and stationary, as in other localities, with the ordinary machinery. One of Button's Patent Drilling Apparatus commenced the drilling of a well with flying colors, but failed to get down to the required depth. This territory looks desolate, indeed. There are here reared many monuments of departed. hopes, marking the spot where is buried, safe from moth, rust and corruption, and from the thief who prowls in darkness, the "Working Capital" of Stock Companies and of individuals, 600 feet below the earth's surface. It may yet prove productive.

Depth of sandrocks, as generally found, first sand, of flat, 100 feet, on bluff 163 feet—42 feet thick; second sand, on flat, 336 feet; on bluff 400 feet—37 feet thick; third sand, on flat 500 feet, on bluff 525 feet—34 feet thick. In the Kane well, on the Lamb farm, the fourth sandrock, was found 650 feet deep—10 feet thick; fifth sand, at 684 feet—14 feet thick; sixth sand, 908 feet, 12 feet thick; seventh sand, 973 feet—4 feet thick. At the last or seventh sandrock, was the first show of oil. The well is 1007½ feet in depth. Is now being tested. Have pumped for four weeks on salt water. Have subsequently learned that no great amount of oil was obtained, and operations ceased. The Cherry Run Central Petroleum Company are testing two wells. Thus far all their wells have proved non-producing. Average cost $8,000, are located on both hill and bluff. Hills at this point are gently undulating, and not as high as at points below.

The village of Plumer is located on the M'Calmont

tract, seven miles from Oil City, and 3½ miles from Rouseville. It is one of the mushroom creations of Petroleum, has revelled in bright anticipations for a brief time, and went into rapid decay. The Osceola and Warren Refineries were the principal features of the place, and to them it owed its chief importance. Both are abandoned. The land about the village is principally owned by Prather & Duncan. A grist mill is located here.

*M'Calmont Farm.*—On Cherry Run, above and joining the village of Plumer, containing forty-five acres. Original owner J. N. M'Calmont. Present owners, Rochester Petroleum Company. Purchased in 1864. Bounded on the north by Prather, east by First National Petroleum Company, south by Turner, west by Seymour, seven and a half miles from Oil City. Cherry Run passes through it, bearing south. Warren & Franklin turnpike, northeast. Leased to Duncan & Prather, and sub-leased by them to various parties. No developments.

*Prather Farm.* —On Cherry Run, in Cornplanter township, containing 282 acres. Original owner, A. C. Prather. Present owners, Prather heirs, M'Aboy and Cherry Run Petroleum Company, Rochester Petroleum Company, and Liberty Oil Company. Bounded on the north by Prather, east by First National Petroleum Company, south by M'Calmont, west by Seymour. Seven and a half miles from Oil City. Bearing of Cherry Run south. Warren Turnpike, passing through, northeast.

On Liberty Petroleum Company's tract, fifty acres, are twelve non-producing wells. On Rochester Petroleum Company's tract, 100 acres, no developments. On M'Aboy and Cherry Run Company's tract, seventy-one acres, are seven non-producing wells. On Prather's heirs' tract, eleven acres, are five non-producing wells. Average depth, 300 feet. All on the flat. Have ten engines, portable and stationary, on the property. On the east of this tract and Plumer village, on the J. M. Hodge farm, are the lands of the First National Petroleum Company and the Reliance Petroleum Company. No paying developments. Joining the lands of the Prather heirs, on a portion of the valley of Cherry Run, off from the main valley, is the territory of the Philadelphia, Lancaster & Cherry Run Oil Company, containing 175 acres. Purchased in 1864. One well drilling, now 980 feet deep, being little indications of oil. One more well in progress. Cost of one now drilling up to present time, $7,000. Have two portable engines. Depth of first sandrock, 237 feet—18 feet thick; second sand, 465 feet—20 feet thick; third sand, 552 feet—23 feet thick; fourth sand, 606 feet—30 feet thick; fifth sand, 840 feet—10 feet thick. Depth of driving-pipe, 22 feet.

*Prather's Farm.*—On Cherry Run, Cornplanter township, containing 112 acres. Original and present owners. Prather Bros. Leased to various parties in 1864. No paying developments. Only one well drilled. Bounded on the north by Bloss & Barringer, east by Hodge and others, south by heirs of Prather, west by Wilson and others. Is one mile from Plumer. Bearing of Cherry Run south. The eighty acres lying just above, is known as the Hatch farm. No wells or developments. Cherry Run is a small stream at its upper waters. Flats are narrow and broad, varying in different localities. Hills gently undulating, with good tillable land bordering on either side at various points.

*Rickets Farm.*—On Cherry Run, in Cornplanter township. Number of acres not ascertained. Original owner, J. Rickets. Present owners, J. Rickets' heirs, and Plumerville Oil Company. Bounded on the north by Patterson, east by Seely and others, south by Prather and west by Davidson. One and a half miles from Plumer, Cherry Run bears southwest and south. There are no producing wells. Seven non-producing, two in progress. Average depth, 650 feet—one 837 feet. Average cost not ascertained. Probably $6,000. Located on flat. Owned by Companies and lessees. Have four portable engines on the property. Depth of first sandrock, 150 feet—35 feet thick; second sand, 275 feet—20 feet thick; third sand, 500 feet—55 feet thick; fourth sand, 740 feet—8 feet thick. Depth of driving-pipe, 50 feet.

On Morrison farm, just above, is one well over 700 feet deep. Are now testing; no oil yet. Between Morrison farm and Rattlesnake School house, are three wells. Above the road and below Stowell farm, are two more, all non-producing.

*Stowell Farm.*—On Cherry Run, in Cornplanter township, containing ——— acres. Original owner, ——— Stowell. Present owner not ascertained, Titus Oil Company have six acres. Two miles from Plumer. Bounded on the north by Allegheny township, east by Bates, south by Seely, west by Patterson. One producing well, the Benton, pumping twenty-five barrels per day. One well drilling. Depth of Benton well, 740 feet. Cost $7,000. Commenced producing in 1865. Wells located on the flat. Owned by the Companies. Have five portable and stationary engines on the property. Depth of first sandrock, 150 feet—40 feet thick; second sand, 275 feet—20 feet thick; third sand, 500 feet—60 feet thick; fourth sand, 740 feet—8 feet thick. Depth of driving-pipe, 53 feet.

On the tract above Stowell farm, the Great Republic Oil Company have two producing wells, both pumping—one five, the other fifteen barrels—twenty barrels in all, One well drilling. Average cost of wells, $7,000. Average depth of wells, 750 feet. Commenced producing in Nov. 1865. Are on flat and owned by Company. Have three portable engines on property. The fifteen barrel well belongs to the Lyons Oil Company. These wells are on the upper waters of Cherry Run, between Oil Creek and West Pithole Creek, one and a half mile from each.

*Huidekoper Farm.*—On Cherry Run, above Great Republic Oil Company's lands. Contains seventy-six acres. Original owner ——— Huidekoper. Present owner, Crescent City Oil Company. Purchased in 1864. Three and a half miles from Plumer. No producing wells. Five in progress—none tested. Are to be drilled 750 feet deep. Present cost, $5,500. Located on the table lands. Owned by lessees. Have five portable engines on the property. Depth of first sand-rock, 125 feet—25 feet thick; second sand, 275 feet—60 feet thick; third sand, 512 feet—15 feet thick. Depth of driving-pipe, 36 feet. One well on branch of Cherry Run, east of Crescent City Oil Company's land, and east of the Titusville road, called the Green well, producing twenty barrels per day, pumping. Sand-rocks about same as those on the land of the Crescent City Oil Company.

## CORNPLANTER RUN AND CABOOSE RUN.

Cornplanter Run is a tributary of Oil Creek, and

empties into the same at the Clapp farm, just above Oil City. Caboose Run empties into Cornplanter Run a short distance above its mouth

*Nevins Farm.*—In Cornplanter township. Contains 385 acres. Original owner, J. Nevins. Present owners, Woods, M'Aboy & Co., 185 acres; Chamberlin & Kasson, 200 acres. Purchased in 1863. Bounded on the north by Hayes, east by Hood, south by Haliday & Kasson, west by Foster and others. One mile from Oil City. Cornplanter Run bears southeast, emptying into Oil Creek one mile above its mouth. Caboose Run bears east of south, emptying into Cornplanter Run. One producing well, flowing about three barrels per day when operated. Nine non-producing wells—five on Chamberlin tract, two on M'Aboy, and two on Kasson. Average depth of wells, 600 feet. Average cost, $4,000. Commenced producing in 1863. Wells located on the flat. Owned by the Companies. Have three engines, portable and stationary, on the property. Depth of first sandrock, 190 feet—20 feet thick; second sand, 330 feet—40 feet thick; third sand, 460 feet deep—15 feet thick; fourth sand, 560 feet—10 feet thick. Depth of driving-pipe, 12 feet. Cornplanter Run is a small stream, with narrow flats. Hills high in many places. The Reno & Pithole Railroad winds its way in a circuitous route over the hills into the valley of Oil Creek.

*Shaw Farm.*—On the uplands between the H. M'Clintock and Clapp Farms, on Oil Creek, and the Keyser and Lamb farms, on the Allegheny River, near Walnut Bend. Contains 402 acres. Original and present owner, R. Shaw. Bounded on the north by H. M'Clintock, Tolls & Keyser, east by Keyser and Lamb, south by Lamb, Winger and Clapp, west by Clapp and M'Clintock. Is three miles from Oil City, on the Warren & Franklin Turnpike. Five wells are being drilled on the farm. One being tested. Average depth of wells, 630 feet. Average cost of the one now testing, $8,000. There are eighteen leases given out on the farm to various parties, who are now operating. Have five engines on the property. Depth of sandrocks and driving-pipe about the general average of same on the Lolls farm adjoining.

*Downing Farm.*—On the south side of Allegheny river, opposite Oil City, on Sage Run, in Cranberry township. Contains —— acres. Original owner, T. G. Downing. Present owners, Imperial Oil Company, and Sage Run Oil Company. Purchased in 1864. Bounded on the north by Allegheny river, east by E. G. Clapp, south by Crammond and others, west by Bastian and others. Sage Run has a bearing at this point north, emptying into the river on the south side. There are several wells on the tract, and most of them have produced largely in former years. The old Phillips & Frew wells, near the ferry landing, were the first producing wells on the Allegheny River, and yielded at the rate of forty barrels per day. Were drilled in 1860. Several other wells produced some oil at about the same time. This farm is the location of a portion of Venango City, and for about the first time oil operations have given place to human habitations.

*Graff, Hasson & Co's Farm.*—On north side of Allegheny river, in Cornplanter township. Contains 600 acres. Original and present owners, Graff, Hasson & Co. Purchased in 1856. Thirty-six acres on southeast corner on river, sold to different parties in small lots. Bounded on the north by Clapp and Winger, east by Siverly, south by river, west by United Petroleum Farms Association, and joining Oil City Borough. Bearing of Allegheny river west. There are on the farm three producing wells. These, when operated, produced about four barrels each per day, and one non-producing well. All old wells. Average depth, 442 feet. Average cost, $2,500. Commenced producing in 1861. Wells located on flat or river side. Owned by Buffalo Oil & Mining Company, Kier & Co., Seneca Oil Company, and others. Have three stationary engines on the property. Depth of first sandrock, 180 feet—30 feet thick; second sand, 240 feet—30 feet thick; third sand, 475 feet—25 feet thick. Depth of driving-pipe, 15 feet. On the thirty-six acres sold to the various parties, is one well flowing, belonging to the Riverside Oil Company, as stated in connection with their lands on river above. There is only a narrow bench along the river to operate on, and on this is the wagon-road to Oil City and the line of the Warren & Franklin Railroad. Hills high and rocky.

*Siverly Farm.*—On the north side of Allegheny river, in Cornplanter township. Contains 118 acres. Original owner, P. H. Siverly. Present owner, J. W. Knox. Purchased in 1864. Bounded on the north by Graff, Hasson & Co., east by Alcorn, south by river. One mile from Oil City. River has a bearing north-west. Siverly Run south-west, emptying into river. There are four producing wells all pumping, one non-producing, and twenty in progress. The Gardner well is producing twenty barrels per day; Coyle well, thirty barrels; others twenty barrels. Daily production, seventy barrels. Whole amount produced, over 2,000 barrels. Average depth of wells, 530 feet. Average cost, $5,000. Wells on the flat and owned by lessees. Have ten engines portable and stationary, on the property. Depth of first sandrock, 250 feet—30 feet thick; second sand, 370 feet—40 feet thick; third sand, 490 feet—30 feet thick. Depth of driving-pipe, 60 feet. The flat and table lands are broad, affording a fine field for operating. On this property is located the village of Siverlyville. It is a small place, and has a highly desirable location. Cornplanter Island is a small island in the river opposite this and the Graff, Hasson Farm, belonging to the Riverside Petroleum Company. On it are two wells, each 520 feet deep. Both non-producing. Have one engine. Sandrocks same as on the shore.

*Clapp Farm.*—On south side of Allegheny river, in Cranberry township. Contains 164 acres. Bounded on the north by river, east by river and Alcorn, south by ———, west by Downing. Original owner, Ellen G. Clapp. Present owners, Ocean Oil Company, 100 acres; Sage Run Oil company, sixty-four acres. Purchased in summer of 1864. Half a mile from Venango City. River bears north-west. Sage Run northwest. A high and narrow ridge lying between the two. Covered with heavy timber. There is only a narrow bench upon which to operate. Hills high and abrupt. There is one non-producing well on the Ocean Company's tract, 600 feet deep. Has produced some seventeen barrels of oil. Cost $6,000. Located on river bank, and owned by Company. Have one engine on property. Depth of first sandrock, 240 feet—30 feet thick; second sand, 350 feet—25 feet thick; third sand, 500 feet—15 feet thick. Depth of driving-pipe, 40 feet.

*Crammond Tract.*—On Sage Run, one and a half miles from the mouth, in Cranberry township. Contains 126 acres. Original owner, ——— Crammond. Present owners, Ogden Oil Company, 26 acres; Sage Run Oil Company, 100 acres. Sage Run, the only stream crossing, has a bearing northwest. Road to Venango City, same bearing. On the Ogden Oil Company's tract, there are two producing wells, both pumping—each one barrel of heavy oil per day, two non-producing wells, and two in progress. Have produced in all forty barrels. Average depth, 600 feet. Average cost, $6,000. Commenced producing in 1865. Wells located on flat. Four portable engines on property. The wells are thought by company not to be deep enough. Depth of first sandrock, 190 feet—12 feet thick; second sand, 420 feet—33 feet thick; third sand, 552 feet—16 feet thick. Depth of driving-pipe, 10 feet. Sage Run Oil Company have two wells on their territory, which lie above the Ogden Company. The wells are 650 feet deep. Others in progress. None yet producing. Have not been thoroughly tested.

Forward Oil Company, on tract farther up the Run, have some wells in progress. One testing at a depth of 800 feet. The valley of Sage Run will average about fifteen rods in width. Hills on east and north side steep and high; on the opposite side not as abrupt. The valley presents a good field for oil operations. Its value will depend upon future developments. There was formerly a tram railroad passing up the valley, for the purpose of bringing iron ore from the banks above to the Allegheny river, from whence it was taken to the furnace then in operation on the Hasson farm, at the mouth of Oil Creek, on the east side.

*Alcorn Farm.*—On north and west side of Allegheny river, Cornplanter township, containing 175 acres. Owned originally by R. Alcorn. Present owners, Alcorn Oil Company, and Congress Oil Company. Purchased in 1864. Bounded on the north by Winger, east by Downing and river, south by river, west by Siverly. One and a half miles from Oil City. Bearing of river at this point southwest. Road to Oil City and Oil City & Pithole Railroad crossing it. There are two producing wells, both pumping, yielding but very little. There are two non-producing wells. Average depth, 576 feet. Average cost, $5,000. Commenced producing in Oct. 1865. Amount of production, small. Wells located on the flat. Have one portable engine on the property. Depth of first sandrock, 130 feet—30 feet thick; second sand, 250 feet—25 feet thick; third sand, 510 feet—10 feet thick. Depth of driving-pipe, 28 feet. On western portion of farm, hills gently elevated—eastern portion, hills abrupt and difficult of access. The river at this point, between the R. Alcorn farm and the W. P. Alcorn farm, on the opposite side of the river, makes a short bend, changing from a southwest to a northwest direction. There is a small island in the river here, belonging to Campbell, Lamberton & Co., formerly known as the Alcorn Island. On it there is one well over 600 feet in depth. Never produced any oil. A new one commenced, but never drilled. Failed in driving the pipe, and was abandoned in consequence.

*Alcorn Farm.*—On south and east side of Allegheny river. Contains 300 acres. Original owner, W. P. Alcorn. Present owner, Seneca Oil Company. Purchased in 1864. Bounded on the north and west by river, east by Eaker and others, south by ———. Is one and a half miles from Venango City. One producing well, pumping. Has produced about 1,000 barrels of oil. One non-producing well. Average depth, 556 feet. Cost of wells not ascertained. Commenced producing Sept. 1864. Wells located on the ravine, a few rods back from the river, and owned by the Company. Have one portable engine on the property. Depth of first sandrock, 150 feet—30 feet thick; second sand, 300 feet—25 feet thick; third sand, 540 feet—10 feet thick. Depth of driving-pipe, 30 feet. The river bank is higher than at many points. The table lands are well located for oil operations. Hills not as abrupt as on the opposite side of the river.

*Downing Farm.*—On north and west side of Allegheny River. Contains ——— acres. Original owner, N. Downing. Present owners, Baltimore Petroleum Company, and W. B. Wier, of Philadelphia. Purchased in 1863. Bounded on the north by Cary, east by Cary and the river, south by river, west by Alcorn. Two and a half miles from Oil City. Bearing of river southwest. On this farm are two producing wells, both pumping two barrels each daily, and five non-producing wells. Amount of production, 1,100 barrels. Average depth of wells, 560 feet. Average cost, $3,000. Commenced producing in September, 1864. Wells located on river bank, and owned by company. Have two portable engines on property. Depth of first sandrock, 213 feet—28 feet thick; second sand, 350 feet—28 feet thick; third sand, 475 feet—10 feet thick. Depth of driving-pipe, 28 feet. Hills high and abrupt, narrow table along the river, with horse and foot-path along the bank and railroad track. There are some old derricks standing here, as monuments of the past.

*Eaker Farm.*—On east and south side of Allegheny

River, in Cranberry township. Contains 120 acres. Original owner, E. R. Eaker. Present owner, Riverside Petroleum Company. Purchased in December, 1864. Is bounded on the north by river and Evans, east by West, south by Alcorn, west by river. Two and a half miles from Venango City. Bearing of river at this point southwest. Have one producing well, pumping very little, and two non-producing wells. Average depth of wells, 550 feet. Average cost of wells, $6,000. Commenced producing in 1864. Wells located on river bank, and owned by company. Have two engines on property. Depth of first sandrock, 160 feet—20 feet thick; second sand, 380 feet—26 feet thick; third sand, 485 feet—30 feet thick. Depth of driving-pipe, 28 feet. Hills gently elevating back from river for thirty rods. Are considerably broken. Oil found in the second sandrock.

*Evans Farm.*—On the south and east side of the Allegheny River, and on Horse Creek. Contains 2,190 acres. Original owner, E. Evans. Present owner, Horse Creek Oil and Manufacturing Company of Philadelphia. Purchased in 1863. Bounded on the north by river, east by Blakely and others, south by Law and others, west by river. Three miles from Venango City. The bearing of the Allegheny River is southwest. Horse Creek northwest. At the mouth of Horse Creek is a little settlement, one store. Was formerly the location of a blast furnace. One of the old stacks is still standing. Here is a good tract for oil operations on the river and on both sides of Horse Creek. There is one producing well, pumping, but not very much, and two non-producing wells. Average cost of wells, $5,000. Average depth, 500 feet. Wells located on flat, and owned two by company, and one by Arren &

Bannon. Have three portable and stationary engines on the property. Depth of first sandrock, 200 feet—30 feet thick; second sand, 300 feet—25 feet thick; third sand not found. Depth of driving-pipe, 20 feet.

This property extends up Horse Creek for a long distance. The Philadelphia Oil Company, Venango Petroleum Company, and Allegheny Petroleum Company are represented in leases on the river flats. Coal is found on the tract, but has never been worked in the banks to ascertain the extent of the supply. It is thought to be valuable, being an under vein. The company have a new saw-mill about one mile and a half up the Creek. In one well now in progress on Horse Creek, was found a white sandrock, 75 feet below the surface, which is 100 feet thick.

*Cary Farm.*—On the north side of Allegheny River, in Cornplanter township. Number of acres not ascertained. Original owner, H. Cary. Present owner, Rockwood Oil Company, twenty-five acres on river front and table lands, the American Illuminating Oil Company have the balance, lying on the hill tops. Four miles from Oil City. Bounded on the north and east by Lamb, south by river and Downing, west by Downing and Wingar. Allegheny River bears southwest. Railroad crossing on river bank. There are three producing wells, all pumping. Daily product, thirty barrels. One non-producing well. Whole amount of production, 20,000 barrels. Average depth of wells, 312 to 510 feet. Average cost, $4,000. Commenced producing in 1861 and 1865. Wells located on river bank, and owned by the Wellsville Oil Company. Have three portable engines. Depth of first sandrock, 220 feet—30 feet thick; second sand, 350 feet—25 feet thick; third sand, 485 feet—10 feet thick, not firm but shelly. Depth of driving-pipe, 10 feet. The American Illuminating Oil Company have nine non-producing wells on the hills, some 200 feet above the level of the river. Have three engines.

*Lamb Farm.*—On north side of Allegheny River, and on Lamb Run, in Cornplanter township. Number of acres not ascertained. Original owner, S. Lamb. Present owners, Rogers & Macy, New York, Philadelphia and Baltimore Petroleum Company, and Philadelphia, Lancaster, and Cherry Run Petroleum Company. Purchased in 1864. Bounded on the north by Shaw, east by Keyser and others, south by river, west by Cary. Four and a half miles from Oil City. River bears southwest, Lamb Run south. Roads to river and Oil City. There are four producing wells, all pumping from five to twenty-five barrels each per day. Daily product, fifty barrels. Six non-producing wells. Amount of production not ascertained. Average depth of wells, 380 feet. Average cost, $5,000. Wells located on the flat, and owned by the companies. Have four portable and stationary engines. Depth of first sandrock, 190 feet—7 feet thick; second sand, 298 feet—15 feet thick; third sand not found. Depth of driving-pipe, 12 feet. The Tarr Homestead Oil Company are represented on this farm. Hills high and rocky. Only a narrow bench on river for operating upon.

*Rennoff Farm.*—On north and west side of Allegheny River, in Cornplanter township. Original owner, S. Rennoff. Present owners, M'Clintock & Cornwall Petroleum and Mining Company, and Ross Oil Company. Purchased in 1864. Bounded on the north by Conver, east and south by river, west by Lamb and others. Five miles from Oil City. Bearing of river east of south and south of west. There are two producing wells, pumping three barrels per day, and eighteen non-producing wells. Total amount of production not ascertained. Average depth of wells, 500 feet. Average cost, $5,000. Commenced producing in 1860 and 1861. Wells are all on flat, and owned by various companies, as follows: Carbon Oil Company, three; Woods & Wright, two; Grant & Stewart, one; Ravenna Oil Company, two; Lockhart & Frew, two; Lay, Rathbone & Co., six. Have six portable and stationary engines on the property. Depth of first sandrock, 270 feet—28 feet thick; second sand, 340 feet—28 feet thick; third sand not found. Some of the wells on this tract have formerly produced quite largely, but are now doing nothing The wells to some extent interfere with each other.

*Blakely Farm.*—On south and east side of Allegheny River, and on Panther Run, in President township. Original owner, J. Blakely. Present owners, Harrington Oil Company, Star Oil Company, and others, owning different tracts. Purchased in 1864 Bounded on the north by Ulrick & Foster, east by Blakely and Dale, south by Evans, west by Evans and river. Five miles from Venango City. Bearing of river, southwest; Panther Run, west. Roads to Venango City south, to Walnut Bend north. There is one producing well, pumping fifteen barrels per day, owned by Harrington Company, and six non-producing wells. The non-producing wells are owned as follows: Harrington Oil Company, two; Star Oil Company, two. Average depth of wells, 500 feet. Average cost, $6,000. Commenced producing, September, 1864. Amount of production not ascertained. Wells located on river side bench. Owned by the companies. Have four portable engines on the property. Depth of first sandrock, 210 feet—15 feet thick; second sand, 350 feet—25 feet thick. Depth of driving-pipe, 50 feet. No third sand found. Hills high. Covered with hemlock timber. Panther Run is a small stream, at the mouth of which are four wells. None of them in operation, apparently abandoned. There is a thick growth of hemlock on the Run. A good place for panthers, we should judge. Solitary looking place.

*Nellis Farm.*—On east side of Allegheny River, in President township. Containing——acres. Original owner, James Nellis. Present owners, V. Ulrick, Panther Run Oil Company, and Clapp & Lindermann. Bounded on the north by Bruner, east by Foster, south by Blakely, west by river. Five and a half miles from Venango City. Bearing of river east of south. Road to Walnut Bend north, to Venango City east of south. There are three non-producing wells, and one drilling. Average depth, 600 feet. Average cost, $6,000. The wells have formerly produced some oil; are located on the flat, and owned by companies. Have three portable engines. Depth of first sandrock, 210 feet—15 feet thick; second sand, 350 feet—25 feet thick; third sand not found. Depth of driving-pipe, 40 feet. The flat affords a fine territory to operate upon. Is narrow at lower end. Various oil companies are here interested in the working interest.

*Krotzer Farm.*—On Allegheny River, east side, and in Walnut Bend, President township. Contains ninety-five acres. Original owner, Philip Krotzer. Present owner, Continental Oil Company. Purchased in 1864. Bounded on the north by Nevins, east by Nellis, south and west by river. Six miles from Oil

City. Bearing of river, southeast. Road to Walnut Bend north, to Horse Creek, southeast. There are seven producing wells on the farm, all pumping from six to fifteen barrels each. Average daily production, sixty barrels. Total amount of production not ascertained. Average depth of wells, 450 feet.. Average cost, $5,000. Commenced producing in 1864 and 1865. Wells located on the flat, and owned by the Continental Oil Company, Walnut Bottom Company, and New York and Walnut Bend Oil Company. Have seven portable and stationary engines on the property. Depth of first sandrock, 170 feet—30 feet thick; second sand, 260 feet—20 feet thick; third sand not found. Depth of driving-pipe, 45 feet. This is good territory for operations in oil, and good land for cultivation.

*Dille Farm.*—On east side of Allegheny River, and in Walnut Bend, President township. Contains ninety-five acres. Original owner, Henry Dille. Present owner, United States Petroleum Company. Purchased in 1864. Bounded on the north by Jones & Fry, east by Fry & Bruner, south by Krotzer, west by Krotzer and river. Six miles from Oil City. Bearing of river south. Road to Horse Creek south. On this tract are three old abandoned wells; one formerly produced about four barrels per day. Average depth, from 200 to 500 feet. Are located on the flat; one back near bluff. Depth of first sandrock, 165 feet—40 feet thick; second sand, 265 feet—16 feet thick; third sand, 315 feet—12 feet thick. Depth of driving-pipe, 40 feet.

*Dotson Farm.*—On east side of river, in Walnut Bend, President township. Contains twenty-five acres. Original owner, Alexander Dotson. Present owner, New World Petroleum Company. Purchased in 1864

Bounded on the north by Dille, east by Fry, south by Nevins, west by river. Six and a half miles from Oil City. Bearing of river south. There are three non-producing wells. Two formerly yielded some oil, one as high as forty barrels per day. Average depth of wells, 488 feet. Average cost, $4,500. Commenced producing in 1862 and 1865. Total amount of production not ascertained. Wells on flat, and owned by the company and lessees. Have two portable engines. Depth of first sandrock, 170 feet—45 feet thick; second sand, 270 feet—16 feet thick; third sand, 320 feet—12 feet thick. Depth of driving-pipe, 50 feet.

*Barr Farm.*—On south side of Allegheny River, in President township. Containing thirty and a half acres. Original owner, Jacob Barr. Present owners, Walnut Bend Oil Company, Shoe and Leather Oil Company, and Star Oil Company. Purchased in 1864. Seven miles from Oil City and seven from President. One producing well, pumping five barrels per day, on property. Ten non-producing wells, and one in progress. One of the Star Company's wells has produced about 11,000 barrels; and another one, 2,000 barrels. Average depth of wells, 500 feet. Average cost, $5,000. Commenced producing in 1861 and 1865. Wells located on the flat, and owned by companies. Four engines, portable and stationary, on the property. Depth of the first sandrock, 175 feet—40 feet thick; second sand, 275 feet—16 feet thick; third sand, 325 feet—12 feet thick. Depth of driving-pipe, 58 feet. The Star well flowed, when first struck, at the rate of 200 barrels per day; diminished gradually to forty barrels per day, and pumped afterwards for two years. The other one pumped at first seventy barrels per day; afterward ran down to twenty barrels per day, averaging this amount of production for eight months. Is regarded as good oil territory hereabouts.

*Krotzer Farm.*—On Allegheny River, south side, in Walnut Bend, President township. Contains fifty acres. Original owner, Henry Krotzer. Present owners, Krotzer Farm Oil Company, and Gibson Oil Company. Purchased in 1864. Bounded on the north by river, east by Bruner, south by Fry, west by Barr. Seven and a half miles from Oil City. Bearing of river west. Road to Henry Bend east. There are here two producing wells, both pumping, one nine barrels; one forty-eight barrels per day; and eight non-producing wells. Average depth, 550 feet. Average cost, $5,000. Commenced producing in July and October, 1865. Amount of production, 1,500 barrels. Wells located on the flat and owned by companies. Have six engines, portable and stationary, on the property. Depth of first sandrock, 185 feet—40 feet thick; second sand, 285 feet—16 feet thick; third sand, 335 feet—12 feet thick. Depth of driving-pipe, 60 feet. The flat and table land here and below, embracing the greater portion of the territory in Walnut Bend, is well located for oil, and is also good land for agricultural purposes.

*Tolls Farm.*—On west side of Allegheny River, in Cornplanter township. Number of acres not ascertained. Original owner, A. Tolls. Present owners, Brooklyn Petroleum Company, Walnut Bend and Cherry Run Petroleum Company, Allegheny and Walnot Bend Petroleum Company, Munhall & Co., M'Clintock & Kinter, Philips & Van Arsdall, and Brown & Co. Bounded on the north by Conner, east by river, south by Rennoff, west by Shaw. Six miles from Oil City, and four from Pithole Creek or Oleopolis. Bearing of river southeast. Road to Warren and Franklin Turnpike northwest. On this farm are two producing wells, and nine non-producing. The two producing wells are pumping one two barrels, and the other eight barrels per day. Average depth of wells, 500 feet. Average cost, $5,000. Wells located on bench or hillside, and owned by the various companies. Have three portable and stationary engines on the property. Depth of first sandrock, 150 feet—40 feet thick; second sand, 360 feet—25 feet thick; third sand, 480 feet, not distinct. Depth of driving-pipe, 30 feet. Narrow bench along river bank, giving only room for one tier of wells. The Oil City and Pithole Railroad passing over the same. Hills high and abrupt.

*Conner Farm.*—On west side of Allegheny River, in Cornplanter township. Original owner, S. Conner. Present owner, Plainfield Petroleum Company in part. Bounded on the north by Knapp, east by Kintzler, south by river and Tolls. Five and a half miles from Oil City, and four and a half from Oleopolis. Bearing of river south. On this farm are twelve non-producing wells. Average depth of wells, 500 feet. Average cost, $5,000. Located on the river bank. Owned by the company and lessees. Have four portable and stationary engines on the property. Depth of first sandrock, 150 feet—25 feet thick; second sand, 300 feet—30 feet thick; third sand, 400 feet—20 feet thick. Depth of driving-pipe, 20 feet. No operators on the ground. All quiet as a sleeper. Hills high and abrupt. Narrow bench along the river bank.

*Kintzler Farm.*—On the north side of the Allegheny River, in Cornplanter township. Contains 116 acres, Original owner J. Kintzler. Present owner,

Humboldt Petroleum and Mining Company. Purchased in 1864. Bounded on the north by Seymour, east by Russell, south by river, west by Conner. Seven miles from Oil City, and three miles from Oleopolis. Bearing of river west. Road to Oleopolis east, to Plumer north. Railroad crosses on river bank. On the farm is one producing well, pumping twenty-five barrels per day, and six non-producing wells. Several of them have formerly produced some oil. Amount not ascertained. Average depth of wells, 400 feet. Average cost, $4,000. Commenced producing in 1861, 1862, and 1864. Wells located on flat and river bank. Have two engines, portable and stationary, on the property. Depth of first sandrock, 160 feet—18 feet thick; second sand, 280 feet—25 feet thick; third sand, 340 feet—20 feet thick. Depth of driving-pipe, 12 feet. Kintzler Run empties into the river at this point, and near the Run is the Humboldt Company's Ferry across the Allegheny.

*Russel Farm.*—On north side of Allegheny River, at Walnut Bend in Cornplanter township. Contains 100 acres. Original owner, Archibald Russel. Present owner, Lamb Bros., and H. M'Clintock. Purchased in 1862, Lamb Bros.' part; M'Clintock's in 1864. Seven and a half miles from Oil City, and two and a half from Oleopolis. Road to Plumer north, to Oleopolis east. Three non-producing wells. Depth, 200 feet. Cost, $2,000. Wells located on flat and river bank. Owned by Lamb Bros. & M'Clintock. Have one stationary engine on property. Depth of first sandrock, 160 feet —10 feet thick; second sand, 289 feet—25 feet thick. Depth of driving-pipe, 20 feet. Hills high and rocky. Difficult of access.

*M'Fate Farm.*—On north side of Allegheny River,

and on M'Fate Run, at Walnut Bend, in Cornplanter township. Containing twenty-five acres. Original owner, Samuel M'Fate. Present owner, Lamb Bros. Two and a half miles from Plumer. Bearing of river west, M'Fate Run south. Road to Plumer north, to Oleopolis, east. There are four producing wells, all pumping, each six barrels daily. Average daily production, twenty-four barrels. Eight non-producing wells. Some of these have formerly yielded a considerable amount of oil. Amount of production, 5,000 barrels. Average depth of wells, 350 feet. Average cost, $4,000. Commenced producing in 1861, 1862, and 1865. Located on river side table land, and owned by Lamb Bros. Have four engines, portable and stationary, on the property. Depth of first sandrock, 160 feet—11 feet thick; second sand, 280 feet—25 feet thick; third sand, 340 feet—15 feet thick; shelly. Depth of driving-pipe, 12 feet.

On M'Fate Run, above this farm, on the Hogg farm, there are five wells in progress. None of them producing. Average depth, 400 feet. Sandrocks about same as on M'Fate. Have five engines.

*M'Mahan Farm.*—On M'Mahan Run, and north side of Allegheny River. Containing 300 acres. Original owner, J. M'Mahan. Present owner, M'Mahan Farm Oil Company. Purchased in 1864. Bounded on the north by Caldwell, east by Anderson, south by river, west by M'Fate. Two miles from Oleopolis. Road to Oleopolis southeast, to Plumer north. Bearing of river northwest, M'Mahan Run west of south. One producing well, pumping fifteen barrels per day, known as the Warren well. Has produced fifty barrels per day. Is now owned by the Oil Creek, Cherry Run and Allegheny Petroleum Company, of

Syracuse, N. Y. There are ten non-producing wells, three of them formerly yielded some oil. Average depth of wells, 400 feet. Average cost, $3,000. Commenced producing in 1864. Amount of production, over 7,000 barrels. Wells located on river bank and ravine. Owned by company and individuals. Three portable and stationary engines on the property. Depth of first sandrock, 180 feet; second sand, 280 feet—30 feet thick. Depth of driving-pipe, 12 feet. M'Mahan Run is a small spring stream, making from the hills to the river. Warren & Bros. formerly had a landing at this point, and oil from their works at Plumer, pumping it through pipes over the hills, and shipping it from this place by river.

*Bruner Farm*—On Allegheny River, south side, in President township, containing 160 acres. Original owner, W. Bruner. Present owners, Bruner Oil Company, Almeda Oil Company, Collins Oil Company, and Bruner Farm Oil Company. Purchased in 1864. Bounded on the north by river, east by Foster, south by Bruner, west by Krotzer and others. Eight miles from Venango City, and two from Oleopolis. On this tract are seventeen non-producing wells, one of which formerly flowed thirty barrels per day. The whole amount of production, 6,000 barrels. Depth of wells, from 300 to 600 feet. Average cost, $5,000. Commenced producing in 1861 and 1863. Wells located on hillside and table lands. Owned by companies. Have five engines, portable and stationary, on the property. Depth of first sandrock, 130 feet—30 feet thick; second sand, 230 feet—30 feet thick; third sand not found. Depth of driving-pipe, 40 feet. On the farm is one small refinery, owned by Mr. Collins; capacity, 42 barrels crude per week. Hills high and rugged, with only a narrow table next to river.

*Lower Walnut Island*—Is in the river, between the Barr farm on south side, and Kintzler farm on the north side. Roads to it in all directions by skiffs. There is one producing well on the island, pumping five barrels per day, reported to have formerly produced seventy-five barrels per day. Original owner, —— Downing. Present owners, Farley, Hodgkiss, and others. Purchased in 1860. Full amount of production not ascertained. Well owned by the company. Have one portable engine. Depth of first sandrock, 165 feet—40 feet thick; second sand, 168 feet—15 feet thick; third sand, 315 feet—12 feet thick. Depth of driving-pipe, 40 feet.

*Middle Walnut Island*—Is in Allegheny River, a few rods above the first, between the Krotzer farm on the south side, and M'Mahan farm on north side, and contains seven acres. Owned originally by Wm. Bruner. Present owners, Whitmore & Hull. There are three old wells, non-producing. One of them has formerly yielded some oil. The flood of March, 1865, swept the works all away. Average depth of wells, 500 feet. Depth of first sandrock, 174 feet—20 feet thick; second sand, 250 feet—20 feet thick; third sand not found. Depth of driving-pipe, 40 feet.

*Upper Walnut Island*—In Allegheny River, a short distance above the Middle Island, between Krotzer and Bruner farms on the south side, and M'Mahan and Anderson farms on north side. Eight miles from Oil City and two miles from Oleopolis, containing twenty acres. Originally owned by Wm. Bruner, by whom it was sold to Dr. L. Haldeman, C. Curtiss, S. Q. Brown, and others; sold by them to Upper Walnut Island Oil Company, who are the present owners. There are four producing wells, all pumping from three to twelve barrels each per day. Daily product, thirty-five barrels. Total amount of production, 5,200 barrels. Average depth of wells, 400 feet. Average cost, $5,000. There are six non-producing wells, and three in progress. Wells commenced producing in August, 1864. Are owned by company. Have nine portable engines on property. Depth of first sandrock, 174 feet—20 feet thick; second sand, 256 feet—20 feet thick; third sand not found. Depth of driving-pipe, 40 feet.

*Anderson Farm.*—On north side of Allegheny River, in Cornplanter township. Containing 200 acres. Originally owned by J. Anderson. Present owners, Great Western Consolidated Petroleum Company, Anderson Petroleum Company, of New York, First National Petroleum Company (leased). Purchased in 1864. Bounded on the north by Caldwell, east by M'Calmont, south by river, west by M'Mahan. Bearing of river northwest. Road to Plumer north, to Oleopolis southeast. There are four producing wells, all pumping from 5 to 140 barrels per day. Average daily production, 200 barrels. Fourteen non-producing wells; five of these are reported to have formerly yielded considerable—200 barrels per day. Total amount of production not ascertained. Estimated at 10,000 barrels. Average depth of wells, 400 feet. Average cost, $3,000. Commenced producing in 1860, 1861, and 1865. Wells located on the table lands along the river sides. Owned by individual companies and lessees. Have eight engines, portable and stationary, on the property. Depth of first sandrock, 180 feet—30 feet thick; second sand, 230 feet—30 feet thick; third sand not found. Depth of driving-pipe, 40 feet. A portion of the town of Oleopolis is located on this farm.

*M'Calmont Farm.*—On north side of Allegheny River, and on both sides of Pithole Creek, after leaving the river a few rods. Contains 170 acres. Extends one and a half miles on the Creek. Original owner, M'Calmont estate. Present owner, Pithole Oil Company, of Philadelphia. Purchased in 1864. Ten miles from Oil City, three miles from Plumer, six miles from Pithole City, and two miles from Eagle Rock. Bearing of river west, of Pithole Creek south. Road to Plumer north, to Eagle Rock east. Oil City and Pithole Railroad crossing, having shipping platforms for branch railroad and river. There are three producing wells on Pithole Creek, one and a half miles up from river. All pumping from thirty-five to forty barrels each per day. Eight old non-producing wells on the river. Average depth of wells, 335 feet on the Creek and 500 feet on the river. Average cost of wells, $4,000. Commenced producing in October, 1865. Amount of production not ascertained. Wells located on flat and table lands. Owned by the company and lessees. Have three engines on the property. Depth of first sandrock, 110 feet—20 feet thick; second sand, 230 feet—22 feet thick; third sand, 320 feet—24 feet thick. Depth of driving-pipe, 10 feet. On river sandrocks same as on Anderson farm. The new town of Oleopolis is principally located on this farm, fronting on the river with a southern slope toward river, hills gently elevating, affording a fine location for a town. The Pennsylvania Pipe and Transportation Company have their tanks located on this tract on Pithole Creek, and on the Anderson farm on the river. The town of Oleopolis has already assumed considerable magnitude, comprising dwellings, stores, shops, hotels, also a steamboat landing, railroad station, and showing all the energy of a new and flourishing business point.

Upon the hill on this farm, is one of the deep pits from which Pithole Creek takes its name. The opening is about sixteen inches wide and six feet long, running in about a northeast and southwest course. It is a separation in the rocks, and seems to have been occasioned by some violent upheaval or disturbance of internal earth. Its depth has never been ascertained. Another pit has been found on what was originally the Widow Shaw tract. This opening is larger and longer than the other. The earth has evidently washed into this pit, making an oblong basin before coming to the rock of about fifteen feet wide, and twenty or more feet long. The opening in the rocks is some eighteen inches in width, and about twelve feet long, increasing some in size as it goes below the surface. Depth of these pits not ascertained. Stones thrown into these openings, roll and tumble as long as they can be heard. Parties have been in for some distance, or as far as the purity of the air would allow. The openings are now nearly filled with logs and brush, to keep animals from falling in. Snow never remains about the openings—melting away at once. Throw a handful of leaves over the openings, and the current of air at once lifts them up and carries them away. That these mysterious openings have some bearing upon the Petroleum formation, we have no doubt, but the evidence so far as presented is too vague and unsatisfactory for illustration. A scientific and thorough exploration of these pits might possibly enlighten mankind greatly as to many of the seeming mysteries that surround the wonderful product known as Petroleum.

*New York, Philadelphia & Baltimore Petroleum Company's Tract.*—This is a triangular tract of ten acres on the Allegheny river and at the mouth of Pithole Creek. Original owner, M'Calmont estate. Present owners, New York, Philadelphia & Baltimore Petroleum Company. Ten and a half miles from Oil City, six miles from Pithole City, and two miles from Eagle Rock. Bearing of river west, Pithole Creek south; road to Pithole City north, to Henry Bend, east. Warren & Franklin Railroad now building to this point. There are two non-producing wells. Depth 400 feet. Cost $4,000. Wells located on the flat, and owned by the Company. Have two engines on the property. Depth of first sandrock, 170 feet—26 feet thick; second sand, 290 feet—15 feet thick; third sand not found. Depth of driving-pipe 40 feet. The Pennsylvania Pipe & Transportation Company have a shipping dock at this point and their principal office. They have also another office at Pithole City. The length of the pipe is seven and a half miles, it is cast iron, and six inches in diameter. The head of the pipe is at the United States well. The oil is received from branch pipes running to the different wells. It is first measured in the tanks, and then let into the main pipe for transportation to the receiving tanks at Oleopolis. These tanks are made of boiler-iron, holding some 15,000 barrels. They are located about a quarter of a mile from the river, up the Creek. The oil is conveyed in a six-inch discharge-pipe to the shipping platform on the bank of the river. At this platform is a cross-head pipe, of same calibre, forty to fifty feet long. In this, at convenient distances for filling barrels, are inserted pieces of two-inch service-pipe, to each of which is attached a stop-cock and rubber hose, with goose-neck for filling barrels. Pipe can be attached to these for running the oil into bulk boats. The Company can transport 6,000 barrels of oil per day through the pipe, if required. The pipe is laid six feet under ground and is not affected by frosts. It has proved a successful mode of transportation, but has been highly detrimental to the growth of Pithole, and annihilation to teamsters.

*Foster Farm.*—On the south side of Allegheny river, in President township. Owned by S. D. Ulrick. Bounded on the north by Allegheny river, east by Heydrick, south by Blakely, west by Ulrick and others. No oil developments of consequence here. Two wells were commenced at one time, but never finished. At one, the engine house was burned down, and the other well was "kicked" down with a spring-pole. There are three abandoned derricks. Hills high and abrupt. Only a narrow rocky beach next to river. Not very inviting for oil men. No road, only a narrow foot-path along the river bank. Is opposite Oleopolis.

*Heydrick Farm.*—On south side of Allegheny river, and on upper waters of Panther Run. Owned by C. W. Heydrick. Two miles from Walnut Bend, and two miles from Henry Bend. On this farm are two abandoned wells. A derrick erected for another is also abandoned. Nothing doing in the way of development. No road. Narrow foot-path along the river bank. Hills high and rocky, covered with timber. There is a small island in the river here, directly opposite the mouth of Pithole Creek and Oleopolis.

*Howe Farm.*—On south side of Allegheny river, in President township, containing 133 acres. Original owner,—Howe. Present owner, Hazleton Petroleum Company. Purchased in 1864. Bearing of river west, road to Henry Bend, east. Two producing wells on the farm, both pumping, one twelve barrels, the other 20 barrels per day. Have produced over 7,000 barrels of oil.

Average depth of wells, 350 feet. Average cost, $5,000. Commenced producing in 1864 & 1865. Wells located on the flat and a small ravine. Have five non-producing wells. Owned by the company and lessees. Have four engines on the property. Depth of first sandrock, 165 feet—15 feet thick; second sand, 210 feet—30 feet thick. Depth of driving-pipe, 36 feet. Here is a good flat for operating. Is narrow at the lower end, where the best wells are located. Well No. 2 is pumping with a good, substantial, direct attachment rig, all arranged for durability and convenience.

*Alexander Farm.*—On south side of Allegheny river. Containing forty-four acres, originally part of the Heydrick tract. Present owners, Pithole Creek and Allegheny river Petroleum Company. Purchased in 1864. Is two and a half miles from Walnut Bend. Bearing of river northwest. Road to Walnut Bend northwest, to President southeast. There are two producing wells, both pumping one barrel each per day, when worked; two non-producing wells, and one in progress. Average depth of wells, 400 feet. Average cost, $5,000. Wells located on river flat and Oil Run. Owned by the Company. Have three engines on the farm. Depth of first sandrock, 165 feet—15 feet thick; second sand, 210 feet—30 feet thick. Depth of driving-pipe on Oil Run, 16 feet; on river, 36 feet. Here is some fine territory for operating. Only a few wells drilled. Oil not yet found in paying quantities.

*Heydrick Farm.*—On Allegheny river, southeast side, in Henry Bend. Contains 194 acres. Original owners, P. C. & C. W. Heydrick. Present owners, Heydrick Brothers, and Heydrick Oil Company. Bounded on the north by river, east by President Petroleum

Company, south by Alexander, west by river. Three miles from Walnut Bend. Bearing of Allegheny river southwest. Road to Walnut Bend northwest. Have one producing well, pumping 15 barrels per day. Known as the Starrow well. Twenty-five wells non-producing and in progress. Average depth, 375 feet on Excelsior lease, on Heydrick 300 to 500 feet. Commenced producing in 1861 and 1865. Total amount of former production 36,000 bbls. Latterly not ascertained. Wells located on flat and table land. Owned by Companies and Excelsior Oil Company lessees. Have five portable and stationary engines on property. Depth of first sandrock, 175 feet—20 feet thick; second sand, 350 feet—30 feet thick. Depth of driving-pipe, 70 feet on lower part of tract, 18 feet at old Heydrick well. This well commenced producing in 1860. Flowed ten days at the rate of 1,800 bbls. per day, then stopped. Would flow afterward only when agitated with the pump. Pumped 100 barrels per day for several months. Is only 280 feet deep. Are reaming out the hole, hoping to open the vein.

*Shaffer Farm.*—On north side of Allegheny river, and on Pithole Creek, containing 400 acres. Original owner — M'Kissick. Present owners, Barrows, Hazleton & Co. Bounded on the north and east by M'Calmont tract, south by river, west by lands of Pithole Oil Company. Ten miles from Oil City, four miles from President. Bearing of river southwest, Pithole Creek east of south. Road and railroad to Pithole City northwest and north. No producing wells. Fifteen non-producing and five old abandoned wells. The Irvine well flowed quite largely when first struck, but was never chambered or the oil saved. Was plugged up in consequence of an objectionable barrel contract, and the cheap price of oil. Average depth of wells, 400 feet, one 700 feet. Average cost, $4,000. Commenced producing in 1862. Wells located on flat and hillside. Have six engines, portable and stationary, on the property. Wells owned by lessees. Depth of first sandrock, 170 feet—20 feet thick; second sand, 290 feet—15 feet thick. Depth of driving-pipe, 50 feet on river at lower end, and 25 feet on creek. On the upper end of farm they drive 15 feet.

*M'Calmont Farm.*—On Allegheny river, on north side, and on Pithole Creek, in President township. Contains 291 acres. Original owner, Judge M'Calmont. Present owners, Farrar Oil Company, of Boston. Purchased in 1861. One non-producing well on the farm, 650 feet in depth. Cost $6,000. Located on the riverside or bank. Owned by the company. Have one portable engine. Depth of first sandrock, 125 feet—12 feet thick; second sand, 245 feet—25 feet thick. Depth of driving-pipe, 35 feet. This farm fronts on the river, running back and crossing Pithole Creek in the form of an L. The company have surveyed out leases on the Creek, inviting capitalists to invest.

*Armstrong Farm.*—On north side of Allegheny river in President township, containing 175 acres. Original owners, S. & J. Armstrong. Present owners, Winona Petroleum Company. Hussey & M'Bride purchased the farm of the original, owners, and sold same to Winona Petroleum Company in 1865. Bounded on the north by Wood & M'Kissick, east by Henry, south by river, west by Farrar Oil Company. Three miles from President and three from Plumer. Bearing of river west, road to Plumer, northwest, to President, east. Have eight non-producing wells on the farm. One well formerly flowed 300 barrels per day. The Hussey & Mc-

Bride well flowed one month, when it was plugged, in consequence of an unfortunate barrel contract. [To the uninitiated reader, this phrase needs explanation. In the earlier years of the Petroleum development, many of the lessees were bound by the terms of their lease to furnish the royalty or landowners portion of the oil obtained, in barrels, thus furnishing both barrels and oil. When oil sold for less per barrel than the cost of the package or barrel, it was inconvenient, nay, impossible in the majority of instances, for the lessee to fulfil his obligations to the landowner. The average price of barrels ranging from three and one half to four dollars each, while the oil could only command a market value of from twenty to thirty cents per barrel. Many valuable wells were plugged up or closed, and their production lost, from this cause. The celebrated Phillips well, on the Tarr Farm, was a notable instance of this kind. It flowed when struck 4,000 barrels per day. The owners of the well being lessees, and having a clause in their lease requiring them to furnish the royalty in barrels, refused to comply, and after a costly law suit, the matter was compromised or settled]. Average depth of wells, 300 feet. Average cost $4,000. Amount of production not ascertained. Wells located on flat and hillside. Have four engines on the property. Depth of first sandrock, 125 feet—10 feet thick; second sand, 250 feet—30 feet thick. Depth of driving-pipe, 45 feet. This farm has a river front of 60 to 70 rods. Hills high and abrupt at lower corner on river. There is a small flat at the upper part on the river, the location of the Hussey & M'Bride well.

*Henry Farm.*—On north side of Allegheny river, in President township, containing 150 acres. Original owner, J. Henry. Present owners, Hussey and M'Bride.

Purchased in 1860. Bounded on the north by M'Kissick & Ludlow, east by Culbertson, south by river, and west by Armstrong. Four miles to Plumer, and three to President. Bearing of river northwest. There are six non-producing wells on the farm, some of them so in consequence of not being worked, and one well in progress. The Orchard well formerly flowed 75 barrels per day, and the Walnut Tree well pumped 15 barrels per day. Doing nothing now. Average depth of wells 300 feet. Cost of same $4,000. Oil is obtained at a depth of from 240 to 300 feet. Commenced producing in 1861 and 1862, and produced several thousand barrels. Wells located on flat and owned by company. Have four engines, portable and stationary, on property. Depth of first sandrock, 125 feet—10 feet thick; second sand, 250 feet—30 feet thick; third sand not found. Depth of driving-pipe on river, 45 feet, deeper back toward the bluff. Warren & Brother are pumping oil from Pithole over the hills in pipes, and have their tankage on the farm. They contemplate removing their refinery from Plumer to this place.

A new town is springing up called Henryville, from being on the Henry Farm, and at Henry Bend. It already boasts of hotels, stores, shops, dwellings, &c. Is a good shipping point, for both river and rail. The flat is broad, presenting a fine location for a town, and is well located for oil purposes. The Warren and Franklin Railroad passes through, and intend building a station at this point.

*Culbertson Farm.*—On north side of Allegheny river, and on Muskrat Run, in President township, containing 150 acres of land. Original owner, H. Culbertson. Present owner, Beekman Oil Company. Three and a half miles from Walnut Bend, two and a half from

President. Bearing of river northwest, Muskrat Run southwest. Road to President east, to Plumer northwest. There are four non-producing wells, and one in progress. One formerly pumped five barrels per day for four months, and then ceased. Average depth of wells 350 feet, Cost of same $5,000. Commenced producing in June, 1864. Total production, 1,500 barrels. Wells located on the flat, and owned by the company. Four portable and stationary engines on the property. Depth of first sandrock, 125 feet—10 feet thick; second sand, 225 feet—30 feet thick. Depth of driving-pipe 45 feet—less near river. Muskrat Run here empties into the Allegheny. On this Run alone are eight wells, all non-producing — on Ludlow and J. M. M'Cabe tracts, part of them belonging to a Michigan Company, and part to a Boston Company.

*Culbertson Farm.*—On Allegheny river, and on Muskrat and Culbertson Runs, in President township. Contains 190 acres. Original owner, A. Culbertson. Present owner, Commercial Oil Company. Purchased in 1864. Bounded on the north by Ludlow, east by M'Crea, south by river, west by Culbertson. Two miles from President, five from Plumer. Bearing of river northwest, Muskrat Run, southwest, Culbertson Run, southwest. Road to Plumer over the hill northwest, to President southeast. Have seven non-producing wells, —two on Culbertson Run. Some of these wells would doubtless produce some oil if worked, but not in material quantity. Average depth of wells, 300 feet. Average cost of same, $4,000. Wells located on flat, and owned by lessees and Company. Have four portable and stationary engines on property. Depth of first sandrock, 130 feet—12 feet thick; second sand, 260 feet —25 feet thick. Depth of driving-pipe, 50 feet.

This property was purchased of A. Culbertson by C. Curtiss, and stocked by him in Commercial Oil Company, of Philadelphia. There are about 60 acres of flat on the river, and some fine leases on Culbertson Run, also some excellent territory on Muskrat Run. Railroad crosses on the flats. Hills gently elevating.

*M'Crea Farm.*—On north and west side of Allegheny river, and on Culbertson Run, contains 404 acres. Original owners, H. & J. M'Crea. Present owners, Eagle Rock Oil Company. Purchased in 1864. Bounded on the north by M'Calmont and Griffin, east by Griffin and river, south by river, west by river and Culbertson. Is in President township. One mile from President, extending up the river opposite, and five miles from Plumer. Bearing of river south, west, and northwest, making a short bend. Road to Plumer northwest, to Pithole City northwest, to President east and north. There are three wells on the farm, which will produce when worked an aggregate of 20 barrels per day; seven non-producing wells, and three in progress. Commenced producing in 1861. Average depth of wells, 550 feet. Average cost, $6,000. Wells located on flat, and one on Culbertson Run. Owned by the company and lessees. Have eight engines on the property. Depth of first sandrock, 178 feet—14 feet thick; second sand, 275 feet—30 feet thick. Depth of driving-pipe, 50 feet. Here is a broad flat, good territory for oil operations, and good land for cultivation and for building purposes. The young and flourishing city of Eagle Rock is being built upon it. This property was purchased by C. Curtiss, and stocked by him in the Eagle Rock Oil Company, of Philadelphia. This point is more generally known as M'Crea Landing, one of the best on the river, and has been the principal shipping point for the Pithole oil. There are 215 houses already built, comprising dwellings, stores, hotels, warehouses, &c. The Warren and Franklin Railroad crosses, and have a station and depot built here. Hills gently elevating on lower part, on upper portion high and abrupt. This is the location of Elliott's Ferry, crossing the river to the village of President.

*Clapp Farm.*—On south and east side of Allegheny river, and on M'Crea Run and on Hemlock Creek, in President township, containing 8,400 acres. Original owner, R. Clapp. Present owner, President Petroleum Company. Purchased in 1864. Bounded on the north by river and Elliotts, east by Tionesta township, south by Pine Grove, west by Blakely and others. Fourteen miles from Oil City, and six from Tionesta. Bearing of river south, west, and northwest, M'Crea Run, north-east, Hemlock Creek, northwest. Road to Tionesta, north, to Venango City southwest. There are fifteen wells, all non-producing, some are in progress of drilling, and others are idle. Average depth of wells, 500 feet. One well drilled by water-power, at mouth of Mc-Crea Run, is over 900 feet deep. Average cost of wells, $4,000. Wells located on the flat and owned by the company. Have five engines. Depth of first sandrock, 124 feet—11 feet thick; second sand, 230 feet—10 feet thick. In Water-wheel well, a hard sandrock was found at the depth of 700 feet, 24 feet thick.

On the lower end of the property there are four old wells, one called the Hemlock Shade well, belonging to Jesse Heydrick (leased), producing when pumped about five barrels per day. The M'Crea and Cherry Run Petroleum Company have territory back from the river—one well 600 feet deep, non-producing, not operating at present. The President Company's land extends some five miles up Hemlock Creek, and some four miles up Porcupine Creek. Coal is supposed to be in the hills, but no bank has yet been opened. There is a large amount of good operating territory on this property. The larger portion of the land lies on the hills, on which there is a large amount of timber, pine, hemlock, and other varieties.

*Harper Farm.*—On south and east side of Allegheny river, in President township, containing 100 acres. Original owner, R. E. Harper. Present owner, Great Western Oil Company. Purchased in 1864. Bounded on the north by river and Clapp, east and south by Clapp, west by Clapp and river. Fourteen miles from Oil City, six miles from Tionesta. Bearing of river southwest and west. Roads to President and Tionesta east and north. Have two producing wells when operated, pumping three barrels each daily; two non-producing, one formerly flowed at first, but was of brief duration. Average depth, 450 feet. Average cost, $5,000. Wells located on flat, and owned by the Company. Have three engines on property. Depth of first sandrock, 124 feet —11 feet thick; second sand, 230 feet—10 feet thick; third sand not found as a distinct rock. Depth of driving-pipe 25 feet. This tract lies on the flat below the mouth of Hemlock Creek, the Clapp farm bounding it on all but the river side. There is a small island in the river opposite the mouth of Hemlock Creek, on which is one well. Is producing, but not in paying quantities. Owners name not ascertained.

*President.*—This is a small village on the Allegheny river and Hemlock Creek, at the confluence of the latter with the former, and contains several dwellings, one excellent hotel, two stores, one grocery, a custom grist-mill, one water, and one steam saw-mill. Population, about 300. There was formerly a blast-furnace in operation here on the Creek, supplied with ore taken from the President Company's property, on which it is found in great quantity. Both white and yellow pine timber, also Hemlock and oak, are in abundance on same; a larger portion of the village is on the Elliott tract.

*Elliott Farm.*—On east side of Allegheny river, and north side of Hemlock Creek, containing 200 acres. Original owner, R. P. Elliott. Present owner, Farrar Oil Company, of Boston. Purchased in 1864. Bounded on the north by Elliott, east by Clapp, south by Clapp, west by river. Six miles from Tionesta. Bearing of river south, Hemlock Creek northwest. Road to Tionesta, north. Have three non-producing wells. Average depth, 483 feet. Average cost, from $3,000 to $5,000. Wells located on flat and owned by Company. Have two portable engines on property. No regular oil bearing sandrock found here, the only kind met with being red sandstone, varying in depth and irregular. Four acres were sold out to Dr. Smiley and others, on which are three wells drilled ; four more derricks where wells were commenced but not completed. One about 300 feet in depth. Second sandrock, 300 feet deep. Depth of driving-pipe, nineteen feet.

Between this and the President Petroleum Company's property, are some seven derricks, on a tract formerly sold to Dr. Smiley. There are only three wells on the tract. Is now owned by Messrs. Brognard & Richardson. Depth of wells not ascertained, there being no one operating on the place. The table lands at this locality extend back fifty rods or more on a gentle slope to the west and south, affording ample room for operating, and is well located for cultivating. It seems to be dry territory so far as yet tested. The future may possibly prove it different in point of results.

### PITHOLE CREEK.

*Caldwell Farm.*—On Pithole Creek, in Cornplanter township. Contains seventeen acres. Original owner, Harrison Caldwell. Present owner or owners not ascertained. Purchased in 1864. Two and a half miles from Oleopolis, and two and a half from Plumer. Bearing of Pithole Creek west of south. Have three non-producing wells. Average depth, 350 feet. Average cost, $4,000. Wells located on flat, and owned by company. Have two portable engines. Depth of first sandrock, 10 feet—20 feet thick; second sand, 230 feet—22 feet thick; third sand, 320 feet—24 feet thick. Depth of driving-pipe, 10 feet.

Here we find another distinct and separate sandrock, from and at about the regular distance above the first sandrock, at the mouth of the Creek. This gradually grows deeper as we ascend the Creek, until we reach the Prather tract, at the mouth of Allender Run, where we find another distinct sandrock, and this becomes the second sandrock. The one we strike at the mouth of Allender Run, at ten feet from the surface, is the same which is found at a depth of ninety-five feet on Rooker & Holmden farms, at Pithole City, and constitutes the first sandrock of that locality. Consequently, we find that the third sandrock on the Holmden farm is the first sandrock at the mouth of the Creek. The grade of the Oil City and Pithole Railroad is 313 feet in six and a half miles, the distance from Oleopolis to the Holmden farm. The difference in depth of the first sandrock at the river and the third at the Holmden farm, is 270 feet, according to the report of drillers, or record of wells kept by them at the different points.

This is forty-three feet less than the grade of the railroad. This difference may be accounted for in the dip of the rock, or discrepancies in the record of the wells. The latter is as likely to be the case as the former. In descending the Creek from the Holmden farm, we lose one sandrock at or near the mouth of Allender Run, and another at or near the upper line of the Pithole Oil Company's territory, on the M'Calmont farm. So that if this theory is correct, the oil which is found in the second sandrock on the river, is in the same strata of sandrock which constitutes the fourth sandrock at the Holmden farm, and may account for the not finding a third sandrock on the river. Here can doubtless be found the solution of the vexed question raised by the different operators in relation to the seeming geological "fault."

*Shaw Tract.*—On Pithole Creek, in Cornplanter township. Contains ten acres. Original owner, E. Shaw. Present owners, Dill Oil Company, Hope Oil Company, Penn Yan Oil Company, International Oil Company, and Baily. Purchased in 1864. Each company have from one to two acres. Is two and a half miles from Plumer. Bearing of Creek southwest, road to Plumer north. Have six non-producing wells, and one in progress. All have been tested, but yielded no oil. Average depth of wells, 650 feet. Average cost, $6,000. Wells located on the flat, and owned by the companies. Have one portable engine. Depth of first sandrock, 50 feet—20 feet thick; second sand, 240 feet; third sand, 360 feet—25 feet thick each. Depth of driving-pipe, 12 feet.

*Shaw Farm.*—On Pithole Creek, in Cornplanter township, containing 100 acres. Original owner, John

Shaw. Present owner, Syracuse Oil Company. Purchased in 1865. Two and a half miles from Plumer. Bearing of Creek southwest. Road to Plumer north, to Oleopolis southwest. Have one non-producing well, 700 feet deep. Cost, $7,000. Located on the flat, and owned by the company. Have one engine on property. Depth of first sandrock, 50 feet—20 feet thick; second sand, 240 feet—25 feet thick; third sand, 360 feet—25 feet thick. Depth of driving-pipe, 12 feet.

*M'Kissick Tract*—On Pithole Creek, in President township, containing 100 acres. Original owner, —— Keyser. Present owners supposed to be Baltimore Petroleum Company, so reported to us in the absence of any one knowing. Bearing of Creek southwest. To Plumer two and a half miles. Road to Plumer north, to Oleopolis south. One non-producing well, located on the flat, and owned by the company. Have one engine. Sandrocks, &c., not ascertained. The territory appears to be of little value for oil purposes. Future developments may dispel these appearances.

*Wood Farm*—On Pithole Creek, in President and Cornplanter township, containing 139 acres. Original owner, Luther Wood. Present owners, Venango Region Oil Company of Chicago, and Excelsior Oil Company of New York, Cleveland Company, and others, in small tracts. Purchased in 1864. Bounded on the north by Rickets, east by Ludlow, south by Armstrong, west by Shaw. Three miles from Oleopolis, three and a half from Pithole City, and two and a half from Plumer. Bearing of Pithole Creek south and west. Road to Plumer west, to Pithole City north, to President east, railroad south. No producing wells; ten non-producing on Excelsior, two on Cleveland, two in progress on Cleveland, and one on Venango Region Oil Company's tract. Average depth of wells, 600 feet, some of them 700 feet. Average cost, $6,000. Wells located on flat, and owned by the companies. Have six engines on the property. Depth of first sandrock, 260 feet—20 feet thick; second sand, 390 feet—25 feet thick; third sand not distinct, loose and shelly. Depth of driving-pipe, 15 feet. There is a mill-dam and saw-mill at this point, a fine water power—known as L. Wood's mill. This mill has furnished large quantities of lumber to the oil operators. On a reserve of two acres, is the residence of Mr. Wood, a store, and barn. The flat varies in width from ten to thirty rods. Hills on each side high and rugged.

*Rickets Farm*—On Pithole Creek and on Hemlock Run, in Cornplanter and President township, containing 160 acres. Original owners, J. & A. Rickets. Present owners, Faunce & Auburn, Rickets Bros., and Wadsworth & Smith. Purchased in 1864. Bounded on the north by Wood, east by Prather, south by Wood & M'Kissick, west by Hodge. Two and a half miles from Plumer, and same distance from Pithole. Bearing of Pithole Creek south. Road to President east, Plumer west, Pithole City north, and railroad south. There are seven non-producing wells on the farm. Average depth of wells, 600 feet. Average cost, $6,000. Wells located on flat and bluffs of Hemlock Run. Owned by companies and lessees. Have six portable engines on the property. Depth of first sandrock, 270 feet—20 feet thick; second sand, 300 feet—25 feet thick; third sand, 425 feet—not distinct. Depth of driving-pipe, 12 feet. Apparently dry territory.

*Wood Farm*—On Pithole Creek, containing 150

acres. Original owners, L. Wood and S. M. Hodge. Present owners, Phillips Petroleum Company and New World Petroleum Company. Purchased in 1864. Bounded on the north by Rickets, east by Prather, south by Rickets & Shaw, west by Hodge. Two miles from Pithole City, and four miles from Oleopolis. Bearing of the Creek south. Road to Plumer west, to Pithole City north. Nine non-producing wells on the farm—five on the Phillips Company, and four on New World Company's land. Average depth, 700 feet; one over 1,000 feet, drilling for a test well. Wells located on flats, and owned by companies. Seven portable engines on the property. Depth of first sandrock, 270 feet—20 feet thick; second sand, 300 feet—25 feet thick; third sand, 405 feet—not distinct—shelly.

*Prather Farm*—On Pithole Creek and on Allender Run, in Cornplanter and Allegheny township, containing 100 acres. Original and present owners, Prather Bros. Bounded on the north by Woods and others, east by M'Kinley, south by Creek, west by Hatch and others. One and a half miles from Pithole City, and three miles from Plumer. Bearing of Pithole Creek south, west, and north of west, Allender Run southeast. Road to Oleopolis south, Pithole City north. No producing wells; three non-producing—one on corner by mill and two on Creek on east side on the bluff. Have two engines. Prather's saw-mill is located on the Creek just below the mouth of Allender Run, propelled by both water and steam. Depth and cost of wells not ascertained. Depth of first sandrock, 10 feet—30 feet thick; second sand, 280 feet—20 feet thick; third sand, 400 feet—25 feet thick; fourth sand, 525 feet, not distinct.

*Lake Farm.*—On Allender Run, in Cornplanter

township. Contains sixty acres. Original owner, —— Lake. Present owner, New York Star Petroleum and Mining Company. Purchased in the fall of 1864. Two and a half miles from Plumer, and one and a half from Pithole City, and three quarters of a mile from Prather's mill, at the mouth of the Run. West end of the farm half a mile from west Pithole Creek. Have two wells drilled, ready for testing. Average depth, 750 feet. Average cost, $7,500. Located on flat, and owned by company. Have two portable engines. Depth of first sandrock, 36 feet—84 feet thick; second sand, 235 feet—18 feet thick; third sand, 574 feet—20 feet thick; fourth sand, 640 feet—15 feet thick. Depth of driving-pipe, 36 feet on centre of tract; 21 feet on west end of farm.

*Balliet Farm.*—On Allender Run, in Cornplanter township. Contains forty acres. Original owner, —— Reynolds. Present owners, Lehigh Oil Company of Lehigh County. Purchased in 1863. Two and a half miles from Plumer, one and a half miles from Pithole City. There are two non-producing wells, and three in progress. The two wells drilled are testing. Average depth, 740 feet. Average cost, $8,000. Wells located on flat and owned by lessees. Five engines, portable and stationary, on the property. Depth of first sandrock, 25 feet—54 feet thick; second sand, 233 feet—20 feet thick; third sand, 455 feet—35 feet thick; fourth sand, 548 feet—25 feet thick. Depth of driving-pipe, 25 feet. Drillers record shows a *fifth* sandrock at the depth of 613 feet, and a *sixth* sandrock at the depth of 710 feet. Above this tract and below the territory of the Central Basin Oil Company, are four wells begun. None producing. Have the appearance of being abandoned.

*Reynolds Farm.*—On Allender Run, in Cornplanter township. Original owner, Central Basin Oil Company. Two and a half miles from Plumer, two and a half from Pithole City. Have one non-producing well, and one in progress. Depth, 883 feet. Cost, $7,000. Located on flat. Owned by company and lessees. Have two engines. Depth of sandrocks, &c., not ascertained accurately. Some three or four more wells were partially drilled by lessees and abandoned.

*Rickets Farm.*—On Pithole Creek, west side. Originally owned by Rickets' heirs. Present owners, Phillips Bros. and Prather Bros. Purchased in 1864. Bounded on the north by Woods, east by Creek, south by Prather, west by Hatch. Bearing of Creek, south. No developments on the tract. Covered with timber. Hills rugged. Railroad crosses, passing down the Creek.

*Wood & Copeland Farm.*—On Pithole Creek. Contains 160 acres. Originally owned by L. Wood and H. Copeland. Present owners, Phillips Bros. and American Illuminating Oil Company. Purchased in 1864. Bounded on the north by Blackmer & Rooker, east by Reynolds, south by Prather, west by Creek. Five miles from river, three miles from Plumer. Bearing of Creek south. Road to Pithole City north, to Oleopolis south. No producing wells. Seven non-producing and five in progress. Average depth, 630 feet. Average cost, $5,000. Wells located on the flat, and owned by lessees. Have six portable and stationary engines on the property. Depth of first sandrock, 60 feet—80 feet thick; second sand, 309 feet (grey)—5 feet thick; third sand, 342 feet—40 feet thick; fourth sand, 436 feet—40 feet thick; pebble-rock, 595 feet—3 feet thick. Depth of driving-pipe, 16 feet. The United States Oil Company have alternate leases on the Woods tract.

On the east of this tract, bordering on the valley of the Creek, is the Reynolds farm. Original owner, —— Reynolds. Present owners, Garden City Oil Company, Corn Exchange Oil Company, and Philpot, Sherman & Co. Eleven wells drilled on the farm, all non-producing.

*Blackmer Farm.*—On Pithole Creek, and on west Pithole Creek, at the mouth of the latter, where it empties into main Pithole. Contains 104 acres. Originally owned by —— Blackmer. Present owners, Prentice, Clark & Seely, of New York. The United States Oil Company have one fourth of the oil by virtue of a former lease. Purchased by the former parties in February, 1865. Bounded on the north by Holmden, east by Rooker, south by Wood & Copeland, west by Ellison and others. Bearing of Creek west and south, west Pithole east of south. Have two wells drilled, both non-producing, and eighteen in progress. Average depth, 650 feet. Average cost, $6,000. Wells located on the flat. Two owned by the company, and the balance by lessees. Have ten engines on the property. Depth of first sandrock, 180 feet—40 feet thick; second sand, 350 feet—30 feet thick; third sand, 460 feet—30 feet thick; fourth sand, 605 feet—13 feet thick. Depth of driving-pipe, 12 feet. The fourth sand as found here, is irregular and shelly.

Oil City and Pithole Railroad have a shipping platform on this tract, and the Reno, Oil Creek, and Pithole Railroad intend to have one when their road is finished.

*Rooker Farm.*—On Pithole Creek, in Cornplanter

township. Contains 100 acres. Original owner, Wm. S. Rooker. Present owners, Bonta & Bates, Drew, and others. Purchased June 4, 1865. Bounded on the north by Thos. Holmden, east by Hyner, south by Reynolds & Copeland, west by Blackmer. Three and a half miles from Plumer, twelve miles from Titusville, and four from Shaffer. Bearing of Creek southwest and west. Road leading to Titusville north of west, Henry Bend southeast—stone road. There are seven producing wells on the farm, three flowing and four pumping; six non-producing wells, and forty-five in progress. The producing wells are: 110, J. R. Johnson and others, 800 barrels; No. 15, Pratt & Sumner. 400 barrels; No. 18, Andrews, Hart & Co., 200 barrels; 108, Perkins, Dunshee and others, 400 barrels; No. 16, Sumner & Pratt, 200 barrels; No. 4, Williams lease, 30 barrels; No. 149, 200 barrels. Daily product, 2,230 barrels. Average depth of wells, 615 feet. Average cost, $6,000. Commenced producing in August, September, and October, 1865. Amount of oil produced, over 200,000 barrels. Wells principally on the flat. Some in progress on the hillside. Owned principally by lessees. Have fifty portable engines and the ordinary machinery on the property. Depth of first sandrock, 95 feet—35 feet thick; second sand, 340 feet—30 feet thick; third sand, 440 feet—18 feet thick; fourth sand, 595 feet—14 feet thick. Depth of driving-pipe, 23 to 25 feet. Average of bonus paid on leases, $3,000. On the farm are three machine shops, one extra. The Pithole Petroleum Company have one well in operation, and claim the flat-lands of the farm. Is in litigation at present time. One of Button's Patent Drilling Apparatus is in operation on the farm. The location of Prather City is on the east side of the Creek. From the office of the company is a fine view of the flats and numerous wells on this and the Holmden and other farms in the famous valley of Pithole The valley of the Creek is about twenty-five rods in width. Hills on either side not very high or abrupt, somewhat broken. Prather City is of few days, but is flourishing in growth. Prather City boasts of one excellent hotel, the Bonta House, built in modern style, and has all the latest improvements, with accommodations for 175 guests. The entire establishment cost about $80,000. Its location is admirable, commanding a fine view of the Pithole oil district. Warren Bros. Pipe Company are pumping oil from this farm over the hills to Henry's Bend, on the Allegheny River. Size of pipe, two inches, with a capacity of 2,000 barrels per day. Oil is also run in pipes to the receiving tanks of the Pennsylvania Transportation Company, who have the six-inch pipe to Oleopolis.

*Holmden Farm*—On Pithole Creek, Cornplanter township, embracing both sides of the Creek, and containing 160 acres. Original owner, Thos. Holmden. Present owners, Wright & Chittenden. United States Oil Company leased the farm in the first place from Holmden, for a term of twenty years. Duncan & Prather purchased the farm in fee from Holmden afterward, and sold the same to Wright & Chittenden, September 4, 1865. The flat and hillsides are divided into 152 half-acre leases. Bounded on the north by Copeland, east by Hyner, south by Rooker, west by Walter Holmden. Four miles from Plumer, twelve miles from Titusville, seven miles from Shaffer. Bearing of Pithole Creek southwest, plank-road to Titusville north, railroad to Oleopolis south, road to Plumer southwest. There are twenty producing wells on the farm—four flowing and sixteen pumping; and 115 wells in progress.

The producing wells are: Frazier well, owned by the United States Oil Company, commenced producing Jan. 7, 1865, flowing at the rate of 650 barrels per day, ceased to flow November 10, 1865; No. 47, Pool, Perry & Co., September 15th, flowing 400 barrels per day; Nos. 1 and 2, Twin wells, Kilgore, Keenan & Co., Jan. 17th and 19th, 800 barrels per day; No. 54, Pool, Perry & Co., August 28th, 800 barrels per day; No. 19, Grant well, August 2d, 450 barrels per day; No. 77, August 1st, 150 barrels per day, E. Deshler; No. 63, July 20th, 35 barrels per day, J. B. Fink; No. 37, August 4th, Robinson & Co.; No. 73, September 5th, Ogden, Curtiss & Co., 100 barrels; No. 35—flowing—September 10th, J. Long; Nos. 104, 106, 107, 108, 72, 79, testing. Daily production of farm, 3,685 barrels. Average depth of wells, 612 feet. Average cost, $6,000. Commenced producing as above. Have produced over 360,000 barrels. Wells principally located on the flat. Several in progress on the hillside. Owned as above and by others in large number. One enterprising operator on this farm is reported to have sold *seventeen* sixteenths in his producing well. Seventy-five portable and stationary engines on the property. Depth of first sandrock, 95 feet—35 feet thick; second sand, 340 feet—30 feet thick; third sand, 440 feet—18 feet thick; fourth sand, 595 feet—14 feet thick. Depth of driving-pipe, 23 to 25 feet. Eureka well, No. 148, commenced flowing November 22d, 600 barrels per day. Had flowed 150 barrels per day for several days previous. On drawing the sucker-rods, the above-stated increase took place. Is 618 feet deep. No. 76 produced 250 barrels per day, and No. 95, 150 barrels. Several of the wells named have ceased, and all have fallen off materially. Other new ones have been struck and taken their places. Pithole City is located on a portion of the farm. Its rapid growth and similar decadence was of the first a marvel, and of the second a natural consequence attendant upon seasons of brief and unhealthy excitement and speculation. It is mentioned fully in another part of this work.

The Pennsylvania Pipe Transportation Company have their receiving tanks and shipping platform on this farm.

A curious phenomenon manifested itself at Pithole, about the last of January or first of February, 1866. A fire occurred, and in drawing the water to extinguish the flames, it was discovered that it only added fuel to the flame. An examination of the well from which the water was obtained was made, and it was found that oil was running into it. On further examination, other wells were found in like condition. The excitement consequent upon this discovery was very great. The wells were visited by thousands, and large prices offered for them. It was not confined to the wells alone. Two or three springs, in the vicinity of the water wells mentioned, were found to be covered with oil. Great was the demand for real estate in Pithole, and the excitement among the holders of leases thereof bordered on the ludicrous. To their view a new Oil Dorado had opened itself up to the light of day, getting weary of the slow advance of the miner's drill. The first well in which oil was discovered, belonging to a widow lady, is only sixteen feet in depth, and from this over fifty barrels of oil were taken with a common pump. Another, and the second, is higher up on the hillside, twenty-three feet deep, belonging to a gentleman named Hill. From this well were obtained over 100 barrels of oil. These wells produced at the rate of five to twenty barrels per day when operated. From the springs near them, the owners dipped several barrels per day. Some parties sank wells to a moderate depth, and in cases obtained oil. One of these, only a few feet from the widow's well, struck a crevice in the surface-rock, at a depth of about twelve feet from the surface of the earth. The crevice, or opening, was small, but there was a fine little stream of oil running out from it. Another well, a short distance from Mr. Hill's, was similarly affected, the water coloring with oil. The well of Mr. Hill was the most productive. The yield from these wells was of brief duration, and speculators on a small scale were badly bitten. The location of these wells is about 150 feet above the level of the Creek, on the second bench or table land, half a mile from the Creek, where the producing wells are located. Many theories were advanced as to the cause of this freak of nature, but the short life of the new producing wells spoiled most of them. It was doubtless an accidental affair, or Dame Nature may have been in a jocular mood, and intended the surprise as a joke. The discovery of oil at Pithole came soon after the thousand or so of oil companies had been formed, and when the bubble was fully inflated. The success of the first operators was more uniform than in any previous locality. The people of the country indulged in a mania, and Petroleum was king. The new discovery added an impetus to oil stocks, and every thing connected with Petroleum. The speculators flocked to Pithole by thousands, and wild and high the excitement grew. Sixteenths in producing, non-producing, and in wells that were to, but never did go down, were bought up eagerly, and changed hands as rapidly as any of the standard stocks on Wall street. Many fortunes were made by the lucky ones. One lease on the Holmden farm was speculated on until the total of the bonus paid amounted to $24,000, and the well on the same is yet to be drilled. Dealers in interests swarmed in all parts of the locality like the locusts of Egypt. Two, five, and even eight thousand dollars were paid for sixteenth interests in wells, and in exceptional cases even higher sums were obtained. After a short but brilliant career, the bright prospects of Pithole waned, and has left but a sorry-looking wreck behind to remind us of its former glory.

*Hyner Farm*—On Pithole Creek and on Simmonds Run, in Allegheny township, containing 150 acres. Original owner, R. Hyner. Present owners, Botolph Oil Company, of Boston, and Hyner Oil Company. Purchased in July, 1864, by the Botolph Oil Company. Hyner Oil Company purchased in February, 1865. Bounded on the north by M'Kinney, east by Manross and M'Cammon, south by ———, west by Holmden & Rooker. Bearing of Pithole Creek southwest, Simmonds Run northwest. On the Botolph Oil Company's tract there are seven producing wells—four flowing and three pumping; two non-producing, and seven in progress. The producing wells are the Homestead well, formerly flowed 500 barrels of oil per day, now pumping five barrels per day; No. 2, Stevenson, flowing about 175 barrels per day; No. 9, flowing about 50 barrels; No. 5, Arletta, flowing 250 barrels; No. 14, pumping 25 barrels; No. 15, pumping 50 barrels; No. 23, flowing 25 barrels. Kate and Annie well, non-producing. Estimated daily product, 500 barrels. Average depth of wells, 615 feet. Average cost, $6,500. Commenced producing from June to November, 1865. Have produced over 50,000 barrels. Wells located on flat and hillside. Owned by company and lessees. Have ten portable engines on the property.

On the Hyner Oil Company's tract there are no

producing wells as yet. Have ten wells testing with good indications. Average depth of wells, 670 feet. Average cost, $6,000. Wells located on the flat and hillside. Owned by company and lessees. Have ten portable engines. The first sandrock is the surface-sand, so-called, 50 feet thick; second sand, 100 feet deep on the flat, and 145 feet on hillside—from 60 to 100 feet thick; third sand, 500 feet deep on the flat, and 505 feet on hillside—from 20 to 40 feet thick; fourth sand, 565 feet deep on flat, 600 feet on bluffs—from 10 to 68 feet thick. Depth of driving-pipe, 10 feet on hillside and 18 feet on flat. Adamsville, a new town—one of the mushroom creations of the oil development, is located on Hyner Oil Company's tract, lying back from the Creek, on Simmonds Run—the Botolph Oil Company's tract lying on Pithole Creek, and embracing the mouth of the Run. The flats are narrow; the hills gently elevated and broken, not very high. The Botolph is better known as the Boston Oil Company—Botolph being Boston in modern English. The Homestead well is their first well. Commenced producing in June, 1865. Ceased flowing August 30, 1865. Only a small portion of the farm lies on the Creek, the northwest corner crossing, embracing the ground occupied by the Homestead well.

*Copeland Farm*—On Pithole Creek, in Cornplanter township, containing 106 acres. Original owner, P. S. Copeland. Present owners, A. G. Mowrey and D. H. Burtiss. Copeland has a reserve of five acres on the Creek flat. Purchased in 1865. Bounded on the north by Dunham, east by M'Kinney, south by S. Holmden, west by Lyons. Bearing of Creek west of south. Four miles from Plumer, twelve from Titusville, six miles from Shaffer. There are eight producing wells—

four flowing and four pumping, as follows: Rice well, No. 1, flowing 300 barrels; No. 2 well, flowing 150 barrels; Clara well, flowing 300 barrels; No. 84, Burtiss well, flowing 800 barrels; Satterlee well, pumping 60 barrels per day; No. 185, pumping 25 barrels; No. 145, pumping 20 barrels; No. 139, pumping 25 barrels. Estimated daily product, 1,600 barrels. On west side are five wells, doing nothing. Including these, there are sixty-two wells in progress. Average depth of wells, 625 feet. Average cost, $6,000. Bonus paid for leases, in addition to the royalty of one half the oil, from $500 to $4,000. Wells commenced producing in October, 1865. Located on flat and hillside. Owned as above stated, and by numerous other parties, lessees. There are sixty portable and stationary engines on the property. Depth of the first sandrock on the flat, 120 feet, on hillside, 170 feet—60 feet thick; second sand on flat, 359 feet, on hillside 385 feet—30 feet thick; third sand, 440 feet on flat, and 480 feet on hillside—25 feet thick; fourth sand, 587 feet on flat, and 604 feet on hillside—20 feet thick. Depth of driving-pipe on flat, 27 feet, on hillside, from 30 to 45 feet. Twenty of the wells in progress are on the reserve of five acres belonging to Copeland, on the flat. Pithole Creek Petroleum Company have leased the reserve, giving Copeland one fourth of the oil as royalty. The Morey House, one of the largest and best kept hotels in the Oil Region, is located on the table lands of this farm.

*M'Kinney Farm*—On Pithole Creek, in Allegheny township, containing seventy-five acres. Original owner, W. M'Kinney. Present owner, Second National Petroleum Company, of New York. Purchased in February, 1865. Bounded on the north by Dawson, east by Ball, south by Morey, west by Copeland.

Bearing of Creek west of south. Have five producing wells, all pumping—No. 6, 35 barrels per day; No. 10, 35 barrels; No. 11, 200 barrels; Island well, 50 barrels; and one other well producing a small amount. Estimated daily production, 140 barrels. There are fifty wells in progress, some with the usual success—drilling-tools fast in the bottom. Average depth of wells, 625 feet on the flat, and 660 feet on the bluff. Average cost, $7,000. Commenced producing in the summer of 1865. Wells located on flat and hillside, or bluff. Owned principally by lessees. Have thirty portable and stationary engines on the property. Depth of first sandrock, 158 feet on the bluff, on the flat not ascertained—52 feet thick; second sand, 404 feet on the bluff—24 feet thick; third sand, 472 feet on bluff—27 feet thick; fourth sand, 633 feet—12 feet thick. Depth of driving-pipe, 25 feet on the flat, and 54 feet on the bluff. At the depth of 520 feet, there is a red sandstone, 12 feet thick. The Island well formerly flowed 250 barrels per day. Ceased flowing November 21, 1865. Hill and table lands on this tract are gently elevating, affording a fine field for oil operations. There is one of Atwood's Patent Drilling Machines on the farm, professing to do away with derricks, &c. Like all the other apparatus devised for this purpose, it has so far met with but partial success.

*Ball Farm*—On Pithole Creek, east side, in Allegheny township, containing 106 acres. Original owner, C. M. Ball. Present owners, S. Zents, J. G. Dale, M. Henry, and J. T. Sawyer. Purchased in June, 1865. Pithole Petroleum Company have a lease of four acres, comprising all the bottom land of the farm, on which there is one well pumping twenty-five barrels per day. Bounded on the north by Blank, east by

Stewart, south by Hyner & Manross, west by M'Kinney and Creek. Five miles from Plumer, ten miles from Titusville Bearing of Creek west of south, plank-road to Titusville west of north, road to Eagle Rock east of south. Is four miles from Shaffer. Seventeen wells in progress, none tested; have good indications. The wells, two in number, one on the Pithole Petroleum Company's, and the other on the Sawyer tract, are the only producing ones on the farm. Average depth of wells, 650 feet. Average cost, $6,000. Wells located on hillside and owned by lessees. Have fifteen portable engines on property. Depth of first sandrock, 124 feet—24 feet thick; second sand, 357 feet deep on the flat, and 400 feet on hillside—40 feet thick; third sand, 457 feet on flat, 510 feet on hillside—30 feet thick; fourth sand, 495 feet on flat, and 610 on hillside—20 feet thick. Depth of driving-pipe, 30 feet. On Sawyer's tract are nine wells in progress—none at the proper depth to test thoroughly, one is pumping about fifteen barrels per day. This tract embraces eighty acres off from the north side of the farm.

Ball Town is a new and flourishing village, located on the table lands of this farm, on the road leading to Eagle Rock City, on the Allegheny River. The town was surveyed into town-lots in July, 1865. In December there were 100 houses, with a population estimated at 1,000, having stores, shops, and all the appearances of like villages.

Dawson Centre, located on the Sawyer tract, is a young and flourishing village of few days' growth, having stores, hotels, shops, dwellings, &c., giving a fine show of life and energy, and will compare well with some of its neighbors.

*Dawson Farm.*—On Pithole Creek, in Allegheny

township. Contains 175 acres. Original owner, J. Dawson. Present owners, W. B. Tuell & Co., Miner & Holmes, Burtiss, Hart & Co., and J. T. Sawyer. Purchased in 1865. Bounded on the north by Blank, east by Ball, south by M'Kinney, west by Copeland. Five miles from Plumer, ten miles from Titusville. Bearing of Creek west of south. Road to Plumer southwest. Plank-road to Titusville west of north. One producing well, the Hoosier, on Tuell tract, pumping thirty barrels per day. Eight wells in progress on the Burtiss tract, thirteen in progress on Tuell's tract, five in progress on Sawyer tract. There are ten non-producing wells on the farm. Average depth of wells, 630 feet. Average cost, $5,500. Commenced producing in August, 1865. Has produced over 500 barrels. Location of wells on flat and table land. Owned principally by lessees. Hoosier well is owned by Tuell & Co. Have twenty portable engines on the property. Depth of first sandrock, 124 feet—30 feet thick; second sand, 357 feet—40 feet thick; third sand, 457 feet—30 feet thick; fourth sand, 595 feet deep—20 feet thick. Depth of driving-pipe, from 30 to 40 feet. The valley at this point is about twenty rods wide. Hills gently sloping, affording a fine location for wells, buildings, and all the various operations of the oil business. The plank-road to Titusville is now completed to this point, also the plank to Miller farm, on Oil Creek.

*Blank Farm*—On Pithole Creek, in Allegheny township, containing 106 acres. Original owner, Samuel Blank. Present owners, Irvine Petroleum Company, of New York, and Frank W. Allin. Purchased in March, 1865. Bounded on the north by J. Blank, east by Pithole Creek, south by Dawson, west by Dunham. Bearing of Creek south. Two non-producing wells—one pumping and one testing. Average depth, 688 feet; one to be drilled 750 feet. Average cost, $7,500. Wells located on flat and on hillside Owned by companies. Have three engines on property. Depth of first sandrock, 130 feet—12 feet thick; second sand, 400 feet—8 feet thick; third sand, 448 feet —24 feet thick; fourth sand, 645 feet—6 feet thick. Depth of driving-pipe, 18 feet on the flat, and 65 feet on the table lands.

*Blank Farm*—On Pithole Creek, in Allegheny township, containing eighty acres. Original owner, J. Blank. Present owner, Pithole Petroleum Company. Bounded on the north by Haworth, east by M'Caslaw, south by Ball, west by Pithole Creek. Ten miles from Titusville, five miles from Plumer. Bearing of Creek southwest. Four wells in progress, none tested. Average depth, 650 feet. Cost, $7,000. Located on flat, and owned by company. Have four portable engines. Depth of first sandrock, 123 feet—12 feet thick; second sand, 380 feet—10 feet thick; third sand, 443 feet—22 feet thick; fourth sand, 500 feet— 15 feet thick; fifth sand, 598 feet—12 feet thick. Depth of driving-pipe, 30 feet. On the tract between this and the Haworth farm, are two wells in progress, not ready for testing. Here is a large scope of territory unoccupied, a broad flat, covered with cherry and other woods, and under-brush; in appearance much like the Rooker, Holmden, and Mowrey farms. Seems to be a good field for the oil seeker to invest his money in.

*Haworth Farm.*—On Pithole Creek and on Dunham Run. Contains thirty-five acres. Original owner, —— Haworth. Present owners, G. L. & J. W. Haworth. Purchased in 1859. Bounded on the north by Young, east by Anderson, south by Blank, west by Blank. Bearing of Creek west, Dunham Run south. Have two non-producing wells, one in progress, and two testing. Have two engines. Depth of first sandrock, 125 feet—25 feet thick; second sand, 365 feet—45 feet thick; third sand, 455 feet—25 feet thick; fourth sand, 510 feet—25 feet thick. Depth of driving-pipe, 9 feet on the flat, 27 feet on the bluff. This is the locality of Haworth's mill, formerly a grist-mill, now converted into a machine shop. Plank-road crosses; to Titusville eight miles.

*M'Caslaw Farm*—On Pithole Creek valley, in Allegheny township, containing 126 acres. Original owner, G. L. Haworth. Present owner, Genesee & Venango Oil Company, of Batavia, N. Y. Purchased in January, 1865. Bounded on the north by Anderson, east by Siggins, south by Culver, west by Blank. Two and a half miles from Pithole City; ten and a half miles from Titusville by plank-road. There are two non-producing wells, and six in progress. Average depth of wells, 625 feet. Average cost, $5,500. Wells located on the flat. Owned by company and lessees. Have seven portable engines on property. Depth of first sandrock, 125 feet—20 feet thick; second sand, 365 feet—45 feet thick; third sand, 455 feet—25 feet thick; fourth sand, 510 feet—25 feet thick. Depth of driving-pipe, 9 feet. Hills at this point grow less in height. Land more rolling, and in better shape for cultivating, provided it was good for any thing in that line.

On the Vanderlin tract, above the Haworth lower mill, are two non producing wells. Developments not very extensive, the territory so far having shown but slight indications of oil.

*Stewart Farm.*—On Pithole Creek, in Allegheny township. Original owner, —— Stewart. Present owners not ascertained. Bounded on the north by Haworth, east by M'Kinley, south by Vanderlin and Anderson, and west by Young. Bearing of Creek west and southwest. Four non-producing and wells in progress. Average depth, 625 feet. Average cost, $6,000. Wells located on flat and hillside. Have two portable engines. Sandrocks not ascertained. Is the location of Stewart's mill.

*Haworth Farm.*—On Pithole Creek, in Allegheny township. Owned by —— Haworth. Bounded on the north by Siggins & Dawson, east by Dawson and others, south by Marshall, west by Stewart. Bearing of Creek southwest. There are eight non-producing wells, some of them testing. Average depth, 620 feet. Average cost, $7,000. Wells located on flat and hillside. Owned by lessees. Have eight portable engines. Depth of first sandrock, 130 feet—25 feet thick; second sand, 370 feet—45 feet thick; third sand, 460 feet—25 feet thick; fourth sand, 515 feet—25 feet thick. Depth of driving-pipe, 27 feet. On this tract are sixteen derricks erected, called the "sixteen derrick clump." Eight are merely derricks, without development.

*Conley Farm.*—On Pithole Creek, in Allegheny township. Original owner, —— Conley. Present owners, Rochester Oil Companies. Bounded on the north by Clark, east by Hayes and others, south by M'Kinley, west by Siggins. There are five non-producing wells and five in progress. Some are testing. Average depth, 600 feet. Average cost, $5,000. Located on the flat. Owned by companies and lessees. Have seven portable engines. Sandrocks, &c., not ascertained.

*Siggins Farm.*—On Pithole Creek and M'Custin Run. Contains fifty-three acres. Original owner, J. Siggins. Present owner, Geneva Oil Company. Purchased in January, 1865. Bounded on the north by Clark & Brown, east by Dawson, south by Stewart, west by Haworth. Bearing of Creek south. Have one well, producing very largely of gas, and but little oil. Is called the Henry Clay, or Gas well, and two non-producing wells, both testing. Average depth, 631 feet. Cost of wells, $8,000. Wells located on the flat, and owned by Northern New York Oil Company, Kipp & White, Working Interest Oil Company, of Oil City, and others. Have two portable engines. Depth of first sandrock, 232 feet—45 feet thick; second sand, 398 feet—45 feet thick; third sand, 488 feet—25 feet thick; fourth sand, 602 feet—25 feet thick. Depth of driving-pipe, 18 feet. This tract joins the Geneva & Venango Oil Company's territory, and lies about fifty feet higher.

*Pratt Farm.*—On Pithole Creek, in Allegheny township. Original owner, —— Pratt. Present owner not ascertained. Bounded on the north by Brown, east by Haworth, south by Dawson, west by Brown. Bearing of Creek south, branch southwest. Three non-producing wells. One, the Pratt well, has produced some oil. Average depth of wells, 600 feet. Cost, from $5,000 to $6,000. Wells located on the flat. Owned by —— Chichester. Rochester Petroleum Company, Michigan Company, and Pomeroy have ten acres. Two engines on property. Sandrocks, &c., not ascertained.

*Van Wykle Farm.*—On Pithole Creek, at the forks, in Allegheny township. Original owner, —— Van

Wykle. Present owner not ascertained. Bounded on the north by Pratt and others, east by Haworth, south by Clark, west by Beebe. Bearing of Creek east by south. Have two non-producing wells, over 600 feet deep. Cost of wells, $6,000. Located on the flat. Have one portable engine. Depth of first sandrock, 203 feet; second sand, 375 feet; third sand, 450 feet; fourth sand, 600 feet. Just above, on the west branch of Pithole, is the upper Conley farm, in Allegheny township. Has one non-producing well, over 600 feet deep. Cost, $6,000. Located on the flat. A portion of this tract belongs to the Brooklyn Petroleum Company. Above this is the M'Garrah farm. Has one non-producing well, over 600 feet deep, located on the flat. Sandrocks same as on the Van Wykle farm. Hills gently elevated. Judging from present indications, this land is better for cultivation than for oil purposes.

*Tyrrell Farm.*—On east branch of Pithole Creek, above the forks. Has one non-producing well. Above this tract, on the Carson farm, are two wells, one reported as producing some oil, the other as non-producing. Above the Carson is the Griffin farm, with one non-producing well. Three engines on the three tracts. Here is a large scope of territory, reaching up to about Nealsburg. Much of the land is good for farming purposes, and is valuable, even if oil is not found on it in paying quantities.

WEST, OR LITTLE PITHOLE CREEK.

*Widow Holmden Farm.*—On Little Pithole, containing 120 acres. Original owner, Walter Holmden; present owners, Widow Holmden and heirs by inheritance. The United States Petroleum Company have a lease for a term of years of the farm. Bounded on the north by Tyrrel, east by Holmden, south by Blackmer, west by Smith & Reynolds. The streams are Main Pithole Creek, with a bearing of southwest; Little Pithole, east of south. There are no producing wells. Two non-producing, drilled by United States Petroleum Company. Average depth, 615 feet. Average cost, $6,000. Located on the flat. Have two engines on the property. Depth of first sandrock, 95 feet—35 feet thick; second sand, 340 feet—30 feet thick; third sand, 450 feet — 18 feet thick; fourth sand, 500 feet—13 feet thick. Depth of driving-pipe, 23 feet. A part of Pithole City is located on this farm. The fourth sandrock found on this farm is not the regular one, being shelly, or in thin shells. The property is in possession of the heirs. The United States Oil Company, beyond drilling the two wells, both of which were failures, have done nothing; no development at the time of our visit.

*Lyons Farm.*—On West Pithole and on Holmden Run, containing 110 acres. Original owners, Widow Lyon and heirs. Present owner, Webster Petroleum Company. Purchased in 1864. Bearing of Creek south-east. Have three wells, two testing, and one has the tubing dropped into it. The two wells testing are getting some oil, with very good indications. Average depth, 700 feet. Average cost, $8,000. Located on flat. Webster well and Columbia well are on Holmden Run, and are owned by the Company. Bliss well No. 2 is on West Pithole. Have three portable engines. Depth of first sandrock, or surface, 12 feet—75 feet thick; second sand, 350 feet—3 feet thick; third sand, 450 feet—12 feet thick; fourth sand, 560 feet—13 feet thick. Depth of driving-pipe, 12 feet. The above record is taken from the Bliss well No. 2, on West Pithole.

*Clark Farm.*—On Little Pithole. Owned originally by——Clark. Present owner, Pittsburgh & Philadelphia Petroleum Company. Contains 200 acres. Bearing of Creek southeast. Have four wells all in progress. Average depth, said to be 715 feet. Record of sandrocks not ascertained. Depth of driving-pipe, 12 feet.

*Steen Farm.*—On Little, or West Pithole Creek, containing 107 acres. Original owner, L. R. M'Kissick. Present owner, Paxton Petroleum Company. Purchased in 1864. Three miles from Plumer, and one mile from Pithole City. Bearing of the stream south. There are five producing wells, all pumping. Average daily product 55 barrels. Average depth of wells, 720 feet. Average cost, $6,000. Commenced producing in Sept. 1865, and later. Have shipped over 700 barrels of of oil. Wells located on flat and owned by lessees. Have nineteen portable engines. Depth of first sandrock, 245 feet—35 feet thick; second sand, 550 feet—22 feet thick; third sand, 608 feet—25 feet thick; fourth sand, 713 feet—13 feet thick. Depth of driving-pipe, 10 feet. The Lyons Oil Company's well on the tract has the following record. Depth of first sand, 229 feet—25 feet thick; second sand, 459 feet—8 feet thick; third sand, 540 feet—16 feet thick; fourth sand, 595 feet—23 feet thick. The wells that have been drilled through the fourth sandrock, have none of them done as well as those stopped in the rock. This territory is well located, and gives promise of being productive.

*Turner Farm.*—On Little Pithole Creek, containing 106 acres. Original owner, A. Turner. Present owner, Geo. Cathrall, seven twelfths; Dr. Egbert, one twenty-fourth; S. Fell, one sixteenth; Beacon Oil Company of Boston, one sixteenth; Wm. Penn Oil Company's lease embraces five acres. Bearing of Creek south. No producing wells. Seven wells in progress—none tested. Average depth, 735 feet. Cost, $6,000. Wells located on flat and table lands. Owned by lessees of the various companies. Have seven portable and stationary engines. Depth of first sandrock, 101 feet, in some of the wells 280 to 320 feet—26 feet thick; second sand, 583 feet—25 feet thick; third sand, 653 feet—20 feet thick; fourth sand, 745 feet—12 feet thick. Depth of driving-pipe from 10 to 40 feet, according to the location. Geo. Cathrall & Co. control the land interest.

*Vose Farm.*—On Little Pithole Creek, containing 106 acres. Original owner, Wm. Vose. Present owner, West Pithole Creek Petroleum Company. Purchased in June, 1865, for $100,000. There is one producing well, called the Green well. Has formerly produced fifty barrels per day—doing very little at the present time, and are experimenting on it. The lease is in litigation, which may account for the diminished production. There are six wells in progress. The Green well pumps and flows alternately when operated. Average depth of wells, 750 feet. Average cost, $7,000. Commenced producing, August, 1865. Wells located on the flat and hillside. Owned by lessees and companies. Have seven portable engines. Depth of first sandrock, 45 feet—40 feet thick; second sand, 456 feet—40 feet thick; third sand, 610 feet—25 feet thick; fourth sand, 710 feet—15 feet thick. Depth of driving-pipe, 45 feet on flat, 25 feet on hillside. The Green well struck a crevice of five feet in the fourth sand. We can see no reason why the well would not produce as much as formerly, if operated rightly. A Saginaw Company are drilling with pole tools, such as are used in sinking salt wells. Hills are not high, gently elevating, forming a good field for operating on. The indications here are good for oil production.

*Brumegin Farm.*—On West Pithole Creek, containing 134 acres. Original owner R. Brumegin. Present owners, Brumegin Oil Company of Michigan. Purchased in Sept., 1865. On this tract are four wells in progress, none drilled to the proper depth, and fourteen more are to be put down by Jonathan Watson, of Rochester, N. Y. Heydrick Oil Company have the tract joining this and the Vose farm, with one non-producing well on the same.

*Austin Farm.*—On West Pithole Creek, containing 107 acres. Original owner, Sylvester Austin. Present owners, Davis & Parker. There are three wells in progress. Average depth 700 feet. Average cost, $6,000. Wells on the flat. Owned by Fritz & Co., Finch & Co., and Kinkaid & Co. Have three portable engines. Depth of first sandrock, 54 feet—this is surface sand, and 28 feet thick; the first sand proper, is 160 feet deep —20 feet thick; second sand, 598 feet—23 feet thick; third sand, 648 feet—not through it. Depth of driving-pipe 52 feet.

On the Mill Farm, above this, there is one—the Olive well—producing some oil, but not in large quantities, and four more in progress. The Olive well is 801 feet deep, and belongs to the Minerva Oil Company of New York. The third sandrock is split by a strata of slate, and is 735 feet deep—15 feet thick; fourth sand, 760 feet—12 feet thick. Both rocks have pebbles in them.

The Second National Oil Company of New York, two miles from Pleasantville, and two and a half miles from the Paxton House, have one well, 806 feet deep. Has four sandrocks. A crevice of three feet was struck at the depth of 708 feet. The well is pumping 100 barrels per day.

At Pleasantville, half a mile toward Enterprise, there is one well 891 feet deep, producing 75 barrels per day, and one well in progress, 750 feet deep—still drilling. No material developments further in this direction. Here our explorations ended, and we resumed the same on the Allegheny river.

*Elliott Farm.*—On Allegheny river, in President township. Original owner, J. S. Elliott. Present owners, Walter Scott Petroleum Company, of Boston, and Hemlock Petroleum Company, of Pittsburgh. Purchased in 1864 and 1865. Contains 150 acres. The Walter Scott Petroleum Company own 100 acres, and the Hemlock, 50 acres. The Walter Scott Company purchased in 1864, and the Hemlock in 1865. Bounded on the north by Brewer, east by Van Geisen, south by Farrar Oil Company and Fox, west by river. Half a mile from President, and six from Tionesta. Bearing of Allegheny river south. Road to Tionesta south and east, to President south. No producing wells on Scott Company's tract. One producing very little, and two non-producing wells, on the Hemlock Company's tract. Depth of wells, one 1,050 feet, one 530 feet, and one 600 feet deep. Average cost of wells, $1,500 to $5,000. Wells located on flat. Owned, two by company, and one—lease—by Grant Company, of Brookville. Have three engines. Depth of first sandrock, 300 feet—10 feet thick; second sand, 340 to 560 feet—five to sixteen feet thick. Depth of driving-pipe, 40 feet.

*M'Crea Farm.*—On west side of Allegheny river, on the hill, in President township. Original owner, M. M'Crea. Number of acres not ascertained. Present owner, Culbertson Oil Company. Purchased in 1864. Bounded on the north by McCalmont, east by Griffin, south by Griffin and Eagle Rock Oil Company, and west by same. Half a mile from President and one mile from Eagle Rock City. Bearing of Allegheny river south, Culbertson Run southwest. Road to Eagle Rock City south and west, to Tionesta north and east. No producing wells. One non-producing well over 600 feet deep. Have one engine on property. This farm does not come to the river.

*Griffin Farm.*—On Allegheny river, west and north side, in President township. Original and present owners, Widow Griffin and heirs. Bounded on the north by M'Calmont, east by M'Calmont, south by river, west by Culbertson Run Oil Company. One and a quarter miles from Eagle Rock City; three quarters of a mile from President. Bearing of Allegheny river southwest. Road to Eagle Rock City southwest, to Tionesta northwest. No developments on this farm. Hills rise abrupt from river bank, and are very high—barely room next to river for a road. The Warren & Franklin Railroad crosses on river bank.

*M'Calmont Farm.*—On Allegheny river, north side, containing 100 acres. Original owner, J. Nevins. Present owners, Fryburgh Oil Company, of Philadelphia. Purchased in 1864. Bounded on the north by M'Calmont, east by Smith, south by river, west by Griffin. One and a half miles from Eagle Rock City, and five and a half from Tionesta. Bearing of Allegheny river south of west. Road to Eagle Rock City south of west, to Tionesta north of east. No producing wells. Two non-producing wells. Have formerly produced a little oil. Depth of wells, 548 and 400. Average cost, $3,000. Commenced to produce in 1863 and 1865. Amount of oil shipped not ascertained. Wells located on flat next to river, and owned by Company. Have two stationary engines. Depth of first sandrock, 220 feet—15 feet thick; second sand, 340 feet—40 feet thick; third sand, 440 feet—80 feet thick. Third sand not distinct—doubtful. Depth of driving-pipe, 40 feet. One old well produced oil formerly in second sandrock. There is but a small flat on upper corner, next to river. The balance of river front is bluff. Hills rising high and abrupt. Barely room for a road next to river. Warren & Franklin Railroad crosses on river bank. This farm has 80 rods of river front.

*Smith Farm.*—On Allegheny River, north side, and on Johnson's Run, in President township. Original owner, D. Smith. Present owners, Oceanic Oil Company, 120 acres; Mercantile Oil Company, 12 acres; Farrar Oil Company, of Boston, 118 acres. Purchased by the different companies in 1863 and 1864. Bounded on the north and east by M'Calmont, south by river, west by Nevins. Two miles from Eagle Rock City and five from Tionesta. Bearing of Allegheny River south of west, Johnson's Run south. Road to Tionesta north of east, to Eagle Rock City south of west. Two non-producing wells, belonging to Mercantile Oil Company, produced eighty-five barrels, and one sixty barrels. Formerly nine wells pumping on this tract. Farrar Company had formerly one. Average depth of wells: Mercantile Company, of New York, one 547 feet and one 468 feet; Oceanic Oil Company, one 650 feet and one 630 feet; Farrar Oil Company, one 670 feet and one 460 feet. Average cost of wells, $5,000. Commenced producing in April and October, 1865. Mercantile Company have shipped 185 barrels; Farrar Oil Company, 3,000 barrels. Have six engines. Wells located on flat, and owned by companies. Depth of first sandrock, 172 feet—45 feet thick; second sand, 287 feet—8 feet thick; no third sandrock. Depth of driving-pipe, 31 feet.

*M'Calmont Farms*—On Allegheny River and on Stewart's Run, in President township, containing 200 acres. Original and present owners, H. P. & H. C. M'Calmont, owning 100 acres each. Purchased in 1816, or thereabouts. Bounded on the north by Haworth, east by M'Calmont, south by river, west by Smith. Two and a half miles from Eagle Rock City, four and a half from Tionesta. Bearing of river west, Stewart's Run south; road to Eagle Rock City west, to Tionesta east. No developments. Have one old well, drilled in 1860, 360 feet. Never produced any oil. Neither of these farms belongs to any oil company.

*M'Calmont Farm.*—On Allegheny River, and on Pine and Stewart's Run, in President township. Contains 212 acres. Original owner, Thos. M'Calmont. Present owners, Farrar Oil Company, of Boston, and Samuel Duff, of Pittsburgh. Have one well, pumping and flowing two barrels per day of very heavy oil. Belongs to the Van Dwyre Oil Association. The Farrar Oil Company have one non-producing well. The Van Dwyre well commenced producing in January, 1866. Average depth of wells, 450 to 500 feet. Cost, $5,000. Wells located on flat. One owned by company, and others by Van Dwyre Oil Association, the latter having a lease of five acres. Have two engines. Depth of first sandrock, 140 feet—12 feet thick; second sand, 230 feet—15 feet thick; third sand, 430 feet—40 feet thick. Depth of driving-pipe, 8 feet.

*Brewer Farm*—On Allegheny River, south side, containing seventy-four acres. Original owner, J. Brewer. Present owner, Germania Oil Company; Progress Oil and Mining Company have seven and a half acres. Purchased in 1864. Bounded on the north by river, east by Van Giesen, south by Elliott, west by river. One mile from President and five from Tionesta. Bearing of Allegheny River south of west. Road to President south of east, to Tionesta north of east. Have two wells in progress, one belonging to each company. Depth of Germania Oil Company's well, 600 feet; Progress Oil Company's well, 840 feet. Cost, from $7,000 to $8,000. Wells located on flat. Owned by companies. Have two portable engines. Depth of first sandrock, 215 feet—15 feet thick; second sand, 295 feet—32 feet thick; third sand not found as yet. Depth of driving-pipe, 48 feet.

*Myres Farm.*—On Allegheny River, south side, in President township. Original owner, James Myres. Present owner, Wabun Oil Company. Contains fourteen acres. Purchased in 1864. Bounded on the north by river, east by J. Van Giesen, south by Van Giesen, west by lands of Germania Oil Company. One and a half miles from President, four and a half miles from Tionesta. Bearing of Allegheny River west, Van Giesen Run northwest. Road to President west, to Tionesta east. Have one old well, bored for salt many years ago. No further developments.

*Van Giesen Farm*—On Allegheny River, south side, and on Van Giesen Run, containing 116 acres.

Original owner, J. Van Giesen. Present owner, C. Curtiss. Purchased in 1864. Bounded on the north by river, east by Sager, south by President Petroleum Company, west by lands of Germania and Wabun Companies. One and a half miles from President and four and a half from Tionesta. Bearing of Allegheny River west, Van Giesen Run northwest. Road to Tionesta east, to President west. No developments on this tract. Hills gently sloping, forming a good table land for operating both on river and Van Giesen Run, the portion lying upon the Run is said to be the best.

*Sager Farm*—On Allegheny River, south side, containing 120 acres. Original owner, A. Sager. Present owner, Caledonia Oil Company. Purchased in 1864. Bounded on the north by river, east by Van Giesen heirs, south by President Petroleum Company, west by Van Giesen. Two miles from President, four from Tionesta. Bearing of Allegheny River west. Road to Tionesta east, to President west. Have three non-producing wells, and one in progress. Two of the wells are over 600 feet deep. Oil not found in paying quantities. Have three engines. Intend to put the wells down to greater depth. Sandrocks, &c., not ascertained. No one on the ground operating.

*Van Giesen Farm.*—On Allegheny River, south side, and on Painter Run. Owned by Van Giesen heirs. Bounded on the north by river, east by Keeler, south by President Petroleum Company, west by Sager. Two miles from President, four from Tionesta. Bearing of Allegheny River west. Road to Tionesta east, to President west. Have three non-producing wells and three engines. No developments going on. The wells are apparently drilled to the proper depth; one in the bank next to river. No one on the ground.

*M'Calmont Farm*—On Allegheny River, north side, and on Stewart's Run, containing 400 acres. Original owner, J. S. M'Calmont. Present owners, Onondaga Oil Company, Baum Tract, Cicero Oil Company, of Syracuse, N. Y. Purchased in 1864. Bounded on the north by Haworth, Church & Herbert, east by Holeman, south by river, west by Farrar Oil Company. Bearing of Allegheny River west, Stewart's Run southwest. Four miles from Tionesta, and three miles from Eagle Rock City. Road to Tionesta east, to Eagle Rock City west. Have five non-producing wells—one on river, one at mouth of Pine Run, two at or near mouth of Sugar Camp. Have one well pumping on Stewart's Run, above Sugar Camp. Depth of wells, from 750 to 300 feet. Average cost, $6,000. Have no producing wells as yet. Wells located on flat, owned by company and lessees. One well belongs to Cicero Oil Company, of Syracuse. Have three portable engines on the property.

*Holeman Farm*—On Allegheny River and Holeman Run, containing 475 acres. Original owner, Alexander Holeman. Present owner, C. Curtiss. Purchased in the Spring of 1864. Bounded on the north by the Philadelphia Oil Company, east by Pierson, south by river, west by Onondaga Oil Company. Three miles from Tionesta, four from Eagle Rock City. Bearing of Allegheny River west, Holeman Run southwest. Road to Tionesta east, to President west, to Eagle Rock west. Have one non-producing well on Holeman Run. Depth of well, 650 feet. Cost, about $6,000. Located on flat, and owned by C. Curtiss. Have one portable engine. Depth of first sandrock, 16 feet—200 feet thick; second sand, 260 feet—20 feet thick; third sand, 564 feet—86 feet thick. Depth of driving-pipe, 14 feet. At the depth of 145 feet, the drillers had a small show of gas and oil, also at the depth of 265 and 564 feet. There is one well near the lower part of the farm, on a lease of four and a half acres on the river, owned by a Mr. Marshall, of New York. Is non-producing. Had a very good show of oil, but was not very thoroughly tested.

*Holeman Island.*—In Allegheny River, on east side of President township, opposite Holeman farm. Original owner, Alexander Holeman. Present owner, C. Curtiss. Contains sixty acres. Purchased in Spring of 1864. Bounded principally by river. Three miles from Tionesta and four from Eagle Rock City. Bearing of Allegheny River west. No oil development. Is excellent farming land, both island and Holeman farm, and affords an ample field for extensive oil operations.

*Maple Islands.*—Three islands, in Allegheny River, below Holeman Island. Original owners, Thos. M'Calmont & R. Clapp. Present owners, Dr. Harrison & Mockeridge, of Philadelphia. Purchased in 1864. Contains four acres in all. Bounded by river. Two miles from President, and four from Tionesta. Bearing of Allegheny River west.

These islands are fast disappearing by the floods of the Allegheny River. One well on the middle island has formerly produced some oil, but the flood in the Spring of 1865 washed away every thing but the engine, which now lies in a wrecked condition.

*Keeler Farm.*—On Allegheny River, south side, in President township. Original owner, —— Keeler. Present owner, R. Clapp. Contains ninety acres.

Purchased in 1864. Bounded on the north by river, east by Holeman, south by President Petroleum Company, west by Van Giesen. Three miles from Tionesta, and three from President. Bearing of Allegheny River west. Road to Tionesta east, to President west. No developments.

*Holeman Farm.*—On Allegheny River, south side, in President township. Original owner, Alexander Holeman, jr. Present owner, Phœnix Oil Company, of Philadelphia. Contains 154 acres. Purchased in spring of 1864. Bounded on the north by river, east by Clapp, south by Clapp & President Petroleum Company, west by Keeler. Three miles from Tionesta, three miles from President. Bearing or course of Allegheny river west. Road to Tionesta east, to President west. No producing wells. Have two non-producing wells, one an old well, drilled some time previous to the purchase of property by the company. Depth of wells, 750 and 450 feet. The well in progress is on the flat, and owned by company. Have one portable engine. Depth of sandrocks, &c., not ascertained.

*Clapp Farm.*—On Allegheny River, south side, President township. Owned by J. M. Clapp. Bounded on the north by river, east by Tionesta township line, south by Hanna & Gilfillan, west by lands of Phœnix Oil Company. Is two and a half miles from Tionesta, and three and a half from President. Bearing of Allegheny River southwest and west. Road to Tionesta east and northeast, to President west. No development. Two old wells, drilled at the time of the first oil excitement on farm, both non-producing.

*Pierson & Elder Farm*—On Allegheny River,

north side, in Tionesta township, containing 316 acres and allowance. Original owners, Williams & Eli Pierson and M. Elder. Present owner, C. Curtiss. Purchased in spring of 1864. Bounded on the north by township line, east by Hunter, south by river, west by Holeman. Two and a half miles from Tionesta, and five and a half from Eagle Rock City. Bearing of Allegheny River west. Holeman Run west and southwest. Road to Tionesta east, to Eagle Rock City west. Have one non-producing well, 930 feet deep. Cost, $5,000. Located on flat, and owned by C. Curtiss. Have one portable engine. First sandrock, 20 feet thick, was found at the depth of 183 feet. No regular sandrocks afterward.

*Hunter Farm.*—On Allegheny River, north side, Tionesta township. Original owner, D. Hunter. Present owners, Illinois Petroleum and Mining Company, of Chicago. Contains 100 acres. Purchased in summer of 1864. Bounded on the north by Allegheny township line, east by Hunter heirs, south by river, west by Elder. Two miles from Tionesta, six from Eagle Rock. Bearing of Allegheny River west, Holeman Run southwest. Road to Tionesta east, to President west. Have four non-producing wells. Three of them are 1,000 feet, and the other 600 feet deep. Cost, $5,000. Located, one on flat next river, and one on Pierson Run back of hill, and owned by company. Have three portable engines. Depth of first sandrock, 125 feet—4 feet thick; second sand, 260 feet—30 feet thick; third sand, 310 feet—30 feet thick. No regular sands. Depth of driving-pipe, 50 feet.

*Hunter Farm.*—On Allegheny River, north side, and on Dustin Run, in Tionesta township. Owners, Hunter heirs. Bounded on the north by township line, east by Dustin, south by river, west by Hunter. Two miles from Tionesta, six from Eagle Rock City. Bearing of Allegheny River west. Road to Tionesta east, to President west. No development.

*Dustin Farm.*—On Allegheny River, north side, and on Dustin Run, in Tionesta township. Original owner, A. Dustin. Present owners, Dustin Farm and Cherry Run Oil Company, of Philadelphia. Bounded on the north by township line, east by Watson, south by river, west by Hunter. One and three quarter miles from Tionesta, six and a quarter from Eagle Rock City. Bearing of Allegheny Run west, Dustin Run southwest. Road to Tionesta east, to President west. Have one non-producing well, 450 feet in depth. Cost, $7,000. Located on flat, and owned by company. Have one engine. Sandrocks, &c., not ascertained.

*Watson Farm.*—On Allegheny River, north side, in Tionesta township. Original owner, Wm. N. Watson. Present owner, Security Petroleum Company, of New York. Contains fifty acres. Purchased in 1864. Bounded on the north by township line, east by Holeman, south by river, west by Dustin. One and a half miles from Tionesta, six and a half from Eagle Rock City. Bearing of Allegheny River west. Road to Tionesta east, to President west. Have three non-producing wells. Depth of wells, from 630 to 830 feet. Cost, $7,000. Located on flat, and owned by company. Have three portable engines. Depth of first sand, 35 feet—10 feet thick. No regular sandrocks. Depth of driving-pipe, from 35 to 40 feet.

*Holeman Farm.*—On Allegheny River, northwest

side, in Tionesta township. Original owner, Chas. Holeman. Present owners, Security Petroleum Company. Contains 125 acres. Purchased in 1864. Bounded on the north by M'Cray, east by M'Clatchey, south by river, west by Watson. One and a half miles from Tionesta, six and a half from Eagle Rock City. Bearing of Allegheny southwest. Roads to Tionesta north, to President southwest. Have two non-producing wells, and two old wells on the river. Depth of new or old wells not ascertained. New wells located on table lands, and owned by company.

*M'Clatchey Farm*—On Allegheny River, northwest side, in Tionesta township, containing 116 acres. Original owner, Wm. M'Clatchey. Present owners, Security Petroleum Company. Purchased in 1864. Bounded on the north by ——, east by M'Cray, south by river, west by Holeman. One and a quarter miles from Tionesta, six and three quarter miles from Eagle Rock City. Bearing of Allegheny River southwest. Road to Tionesta northeast, to President southwest. Have one non-producing well, 750 feet in depth. Cost, $5,000. Located on the flat of Dustin Run, and owned by company. Have one engine. No regular sand-rocks.

*M'Cray Farm.*—On Allegheny River, northwest side, in Tionesta township. Original owner, S. M'Cray. Present owners, Hamlin & Moore. Bounded on the north by ——, east by Hamlin, south by river, west by M'Clatchey. One mile from Tionesta, seven from Eagle Rock City. Bearing of Allegheny River southwest, Tionesta northeast, President southwest. Have one old well, shallow depth, and non-producing.

*Towner Farm.*—On Allegheny River, south side,

and on Little Tionesta Creek. Original owner, A. Towner. Present owner, —— Lanier, of New York. Contains 125 acres. Purchased in 1864. Bounded on the north by river, east by Hilands, south by Hilands, west by Clapp. Two miles from Tionesta, and four from President. Bearing of Allegheny River southwest, Little Tionesta Creek northwest. Road to Tionesta northeast, President southwest. Have one old well with the tools fast in, 500 feet in depth. Had a fine show of oil previous to tools becoming fastened. Here is a good flat for operating on, with an excellent saw-mill on the same, built by Messrs. Towner & Brett; not now in operation. Is like much of the property of stock companies—allowed to remain idle when it could be used to advantage in raising a revenue from which to declare dividends, &c.

*Hilands Farms.*—On Allegheny River, southeast side. Original owner, W. C. Hilands. Present owner, J. A. Dale. Contains 200 acres. Purchased in 1865. Bounded north by river and Johnson; other boundaries not ascertained. One and a half miles from Tionesta, four and a half from President. Bearing of Allegheny River southwest, Little Tionesta Creek northwest. Road to Tionesta northeast, President southwest. No developments. Hills high and abrupt. A horse and foot-path along the river.

*Dustin Farm.*—On Allegheny River, west side, in Tionesta township. Original owner, Ananias Dustin. Present owner, Perry Oil Company, of Pittsburgh. Contains twenty-five acres on mainland, and twenty acres in island, opposite. Bounded on the north by Hilands, east by Dale and M'Bride, south by Hilands, west by river. Bearing of Allegheny River southwest.

One mile from Tionesta, five from President. Road to Tionesta northeast, President southwest. Have two non-producing wells—one on the main land and one on island. Depth of first, 200 feet; of well on island, 600 feet. Owned by company. No machinery on the ground at present. Was washed away and destroyed by flood in Spring of 1865. The well on shore was drilled by J. Hawkins, of the Burning well, Blood farm, on Oil Creek.

*Hulings Farm*—On Allegheny River, west side, in Tionesta township, containing 100 acres. Original owner, W. Hulings. Present owners, Philadelphia, Lancaster and Cherry Run Oil Company. Purchased in March, 1865. Bounded on the north by Shreeves, east by Noble, south by Dale & Johnson, west by river. Half a mile from Tionesta, five and a half from President. Bearing of Allegheny River southwest. Road to Tionesta northeast, to President southwest. Have one non-producing well, 600 feet in depth. Has produced some oil in former years. Cost of well, $5,000. Located on flat, and owned by company. Have one engine. No regular sand.

*Shreeves Farm*—On Allegheny River, southeast side, and on Tionesta Creek, containing eighty-four acres. Original owner, J. Shreeves. Present owners, Shreeves Farm Oil Company, of New York. Purchased in November, 1864. Bounded on the north by Tionesta Creek, east by Noble, south by Hilands, west by river. Seven miles from President, and fourteen from Tidionte. Bearing of Allegheny River west of south, Tionesta Creek north of west. Road to President west of south, to Tionesta east of north. Have three non-producing wells. Depth of wells, 910, 555, and 535 feet. Average cost, $8,000. Located on the flat, and owned by the company. Have three engines. Depth of first sandrock, 245 feet, white—23 feet thick; second sand, 560 feet, grey—11 feet thick; third sand, 890 feet, white—40 feet thick.

The wells are on the flat, south side of Tionesta Creek, at the mouth. Good territory for oil operations or for building purposes.

*Watson Farm.*—On Allegheny River, northwest side, in Tionesta township. Original owner, James Watson. Present owners, Hamlin & Moore. Contains 100 acres. Purchased in 1864. Bounded on the north by Hulings, east by Hulings and river, south and west by M'Cray. No development.

*Hunter Island.*—In Allegheny River, opposite mouth of Tionesta Creek. Owned originally by Wm. Hunter. Present owners, Cascade Oil Company, of Providence, R. I. Contains forty acres. Purchased in May, 1865. Have one well in progress, 370 feet deep; owned by company. Have one engine. First sandrock was met with at the depth of 240 feet—14 feet thick. Depth of driving-pipe, 50 feet. The Cascade Oil Company have one well on Siggins farm, above the mouth of west Hickory Creek, 880 feet in depth. Is non-producing and abandoned.

*M'Calmont Farm.*—On Allegheny River, northwest side. Original owners, heirs of A. M'Calmont. Contains 100 acres. Present owners, Bissell & Co. Purchased in 1862. Bounded on the north by Shriver, east by river, south by Hamlin. Half a mile from Tionesta, seven and a half from Eagle Rock City. Bearing of Allegheny River west of south. Road to

Tionesta east of north, to Eagle Rock City west of south. No developments. Hills high and abrupt. Warren and Franklin Railroad crosses on river bank.

*M'Calmont Farm.*—On Allegheny River, north-west side, in Tionesta township. Owned by heirs of A. M'Calmont. Contains 600 acres. Bounded on the north by Hunter, east by Hunter and river, south by Hulings, west by——. Eight miles from Eagle Rock City, and seven from west Hickory. Bearing of Allegheny River west of south. Road to Eagle Rock City west of south, to west Hickory east of north. No development. Hills high and abrupt. Railroad crosses on river bank.

*Hunter Farm.*—On Allegheny River, west side, and on Hunter's Run. Original owner, Wm. Hunter. Present owners, Pennock, Ball & Co., of Pittsburgh. Contains 200 acres. Purchased in 1864. Bounded on the north by M'Calmont, east by river, south by Hulings. Eight miles from Eagle Rock City, five from West Hickory. Bearing of Allegheny River west of south, Hunter's Run southeast. Road to Eagle Rock City west of south, to West Hickory east of north. Have two non-producing wells, drilled by horse-power. Depth, 545 feet. Located on flat, and owned by company. Depth of surface sandrock, 10 feet—40 feet thick; first regular sand, 280 feet—15 feet thick; second sand, 530 feet. Depth of driving-pipe, 50 feet. One of the wells was drilled and operated by the water-power of adjacent mill, formerly pumped twenty-five barrels per day. Produced a large quantity of oil. Is doing nothing now from not being operated.

The village of Tionesta is located here, a description of which will be found in the chapter on Towns. It is on the east side of the Allegheny, and at the mouth of Tionesta Creek.

*May Farm*—On Allegheny River, east side, Tionesta township, containing 100 acres. Original owner, Helen May. Present owner, C. Curtiss. Purchased in summer of 1864. Bounded on the north by heirs of May, east by ———, south by Tionesta borough, west by river. Joins Tionesta borough. Course of Allegheny River west of south. Seven miles from Hickory. Road to Hickory east of north, to Tionesta east of south. No developments.

*May Farm.*—On Allegheny River, west side, and on Tubbs Run, and known as the Saddle-bags tract. Original owners, heirs of May. Contains 425 acres. Present owner, C. Curtiss. Bounded on the north by Brown & Co., east and south by———, west by river. Six and a half miles from Hickory, one half mile from Tionesta. Bearing of Allegheny River west of south, Tubbs Run west. No developments.

*Hunter Farm.*—On Allegheny River, west side, in Tionesta township. Original owner, G. S. Hunter. Two acres leased to S. N. Hill & Co. Contains fifteen acres. Have one non-producing well on the lease of two acres, 530 feet deep, and drilling. Will cost from $7,000 to $10,000. Located on river side, and owned by S. N. Hill & Co. Have one portable engine. Depth of first sandrock, 280 feet—15 feet thick; second sand, 530 feet. First or bedrock, grey sand, 40 feet.

*M'Calmont Farm*—On Allegheny River, west side, Tionesta township, containing 200 acres. Original owner, A. M'Calmont. Present owner, Bissell & Co. Purchased in 1862. Bounded on the north by Sowers, east by river, south by Hunter, west by ———. Half a mile from Tionesta, and four and a half miles from West Hickory. Bearing of the Allegheny River west of south. Road to West Hickory east of north, to Tionesta west of south. No developments.

*Sowers Farm*—On Allegheny River, west side, containing 100 acres. Original owner, C. Sowers. Present owners, E. Fawcett & Co., Pittsburgh, Pa. Purchased in 1865. Bounded on the north by Jamison, east by river, south by Bissell & Co., west by ———. Half a mile from Tionesta, four and a half from West Hickory. Bearing of Allegheny River west of south. Road to West Hickory east of north, to Tionesta west of south. Have ten non-producing wells. The old Sowers well formerly yielded twenty-five barrels per day, at the depth of 150 feet; oil 35° gravity. Average depth of wells, from 150 to 500 feet. Wells located on flat and ravine, and owned by company. Have two portable engines. The company own the island in the river opposite. Has one well on it, but was nearly destroyed by the flood of 1865. Depth of sandrocks, &c., not ascertained.

*Ensign & Jamison Farm.*—On Allegheny River, west side, and on Jamison Run. Contains 162 acres. Original owners, C. Ensign & R. Jamison. Present owners, Denver Petroleum Company, of New York. Purchased in 1864. Bounded on the north by Jamison, east by river, south by Sowers, west by ———. One mile from Tionesta, four from West Hickory. Bearing of Allegheny River west of south. Road to West Hickory east of north, to Tionesta west of south. Have one producing well, called the Butternut Shade well. Amount of production not ascertained. Have shipped but a few barrels. Have thirteen old wells, only one tested. Depth of wells, from 250 to 730 feet. Cost, $5,000. Wells located on flat, and owned by company. Have two portable engines. Depth of first sandrock, 101 feet—35 feet thick; second sand, 149 feet—14 feet thick; third sand, 206 feet—14 feet thick. The best show of oil has been found on this farm at the depth of 268 feet. Oil has been found in all the wells. But one well tested. Oil not found in paying quantities.

*Jamison Farm*—On Allegheny River, west side, and on Jamison Run, containing sixty-nine acres. Original owners, R. & G. Jamison. Present owners, Jamison Oil Company, of Philadelphia. Purchased in 1864. Bounded on the north by Dawson, east by river, south by Jamison, west by ———. Four miles from West Hickory, one mile from Tionesta. Road to West Hickory northeast, to Tionesta southwest. Bearing of Allegheny River southwest, Dawson Run east. Have one producing well, called the Butternut Shade well, and one non-producing well, tools fast in it. The producing well has pumped very little. Depth of wells, one 408, the other 345 feet. Average cost, $5,000. Wells located on flat, and owned by company. Have one portable engine. Depth of first sandrock, 101 feet —35 feet thick; second sand, 149 feet—14 feet thick; third sand, 206 feet—14 feet thick. Depth of driving-pipe, 13 feet. There were formerly two old refineries on this farm, both small, now abandoned. There is a broad river flat on this property, furnishing an excellent location for oil operations.

*Jamison Farm.*—On Allegheny River, west side.

The Gilbert Jamison farm contains eighty acres, and the John Jamison farm 250 acres. Original owners, G. & J. Jamison. Present owners, St. Clair & Co., Pittsburgh. Bounded on the north by Jamison, east by river, south by Jamison. There is one old well on the Gilbert Jamison farm; and five wells, none of them tested, on the John Jamison farm. Four and a quarter miles from West Hickory, and one and a quarter from Tionesta. Bearing of Allegheny River southwest. Road to West Hickory northeast, Tionesta southwest. Average depth of wells, 557 feet. Cost not ascertained. Wells located on flat and owned by company. The flat on these farms is not so broad as on those below.

*Johnson Farm*—On Allegheny River, east side, Tionesta township, containing sixty acres. Owned by Henry Johnson. Bounded on the north by Dale, east by——, south by May, west by river. One mile from Tionesta, six miles from Hickory. Bearing of Allegheny River southwest. Road to Hickory northeast, to Tionesta southwest. No developments.

*Dale Farm*—On Allegheny River, east side, containing 250 acres. Original owners, J. Dale's heirs. Present owner not ascertained. Purchased in fall of 1864. Bounded on the north by township line, east by Woolcot south by Brown & Co., west by river. One and a half miles from Tionesta, and five and a half from Hickory. Bearing of Allegheny River southwest. Road to Hickory northeast, to Tionesta southwest. Have two old wells, 200 feet deep, on farm, not tested.

*Dawson Farm.*—On Allegheny River, west side, and on Dawson Run, in Harmony township. Owned by heirs of J. G. Dawson. Contains 300 acres.

Bounded on the north by Baldwin, Wilder & Co., east by river, south by township line and Jamison. Two miles from Tionesta, and three miles from West Hickory. Bearing of Allegheny River southwest, Dawson Run southeast. Road to West Hickory northeast, Tionesta southwest. One well has been drilled 160 feet; tools fast.

*Stave Factory Tract*—On Allegheny River, west side, and on Dawson Run, in Harmony township, containing 200 acres. Original owners, Dawson & Gordon. Present owners, Dawson Run Oil Company. Purchased of Gordon and leased of Dawson, in 1863. Bounded on the north by Gordon, east by river, south by Dawson, west by Stillwagon. Two miles from West Hickory, three from Tionesta. Bearing of Allegheny River southwest and west, Dawson Run southeast. Road to West Hickory northeast and east, to Tionesta southwest. Have five non-producing wells, from 500 to 650 feet in depth. Average cost, $6,000. Three of the wells are located on the river flat and two on Dawson Run, and are owned by company. Have four stationary engines. Depth of first sandrock, on Dawson's Run, 220 feet—16 feet thick; second sand, 489 feet—30 feet thick; third sand not found. On river, first sandrock, 121 feet—12 feet thick; second sand, 164 feet—9 feet thick; third sand, 780 feet. Depth of driving-pipe, 6 feet on Dawson Run, and 12 feet on river flat.

*Gordon Farm*—On Allegheny River, north side, in Harmony township, containing 437 acres. Original owner, J. Gordon. Present owner, C. Curtiss. Purchased in Fall of 1864. Bounded on the north by Hunter, east by Green, south by river, west by Baldwin,

Wilder & Co. Half a mile from West Hickory, and four and a half from Tionesta. Bearing of Allegheny River west. Road to West Hickory east, Tionesta west. Have one well, producing one and a half barrels per day, pumping. Have three non-producing wells. Amount produced, 175 barrels. Depth of wells, 315, 400, and 812 feet. Commenced producing in September, 1865. Wells located on flat. Owned by Alden & Co., and Wright & Co. Have four portable engines. Depth of first sandrock, 180 feet—30 feet thick; second sand, 387 feet—48 feet thick; third sand, 780 feet—30 feet thick. Depth of driving-pipe, 44 feet. This farm has a large tract of flat-land on the river, good territory to operate on. The wells already drilled have a good show of oil. Have not been thoroughly tested. Oil heavy lubricating, 35° gravity.

*Green Farm.*—On Allegheny River, northwest side, in Hickory township. Original owner, J. Green. Present owners, Hickory Farm Oil Company, and West Hickory Oil and Mining Company, of New York. Purchased in 1864. Three hundred acres owned by Hickory Farm, and 500 acres by West Hickory Oil and Mining Company. Five miles from Tionesta, and two from Hickory village. Bounded on the north by Philadelphia Oil Company, east by Siggins, south by river, west by Gordon. Bearing of Allegheny River west, West Hickory Creek east. Two producing wells on property of Hickory Farm Oil Company, and two on property of West Hickory Oil and Mining Company's property. Four non-producing wells, and twenty-five wells in progress on tract of former, and twenty-five wells in progress on tract of latter company. Wells of Hickory Farm Company producing ten and fifteen barrels per day each. One well of the West Hickory

Company, thirty barrels per day. Depth of wells on Hickory Farm tract, 111 and 240 feet; on West Hickory Company's tract, 420 feet. Oil struck at 320 feet. Wells commenced producing on Hickory Company's tract, April 8, 1865, and January 23, 1866; on West Hickory Company's tract, November 23, 1865. Wells of Hickory Farm Company located on flat, and owned by company. West Hickory Oil and Mining Company's, on Creek, one mile from river, belonging to New York and Philadelphia Petroleum Company. Have seven engines on the property. Depth of first sandrock, 13 to 31 feet—15 feet thick; second sand, 75 to 83 feet; third sand, 128 to 154 feet; fourth sand, 409 to 425 feet. Depth of driving-pipe, 12 to 20 feet. No regularity in the thickness of sandrocks; appear like broken, mixed-up rock. There is a new town springing up at this point, called West Hickory. A number of new buildings are erected, and more are in progress.

*Sibbald Farm.*—On Allegheny River, east side, in Hickory township. Original owner, Dr. J. Sibbald. Contains 100 acres. Present owners, Sinclair & Co., of Cleveland, O. Purchased in fall of 1864. Bounded north and east by river, and Ball south and west. Two miles from Tionesta, and five from Hickory village. Bearing of Allegheny River southwest. Road to Tionesta southwest, Hickory northeast. No developments.

*Woolcott Farm*—On Allegheny River, east side, in Hickory township, containing 100 acres. Original owner, Wm. Woolcott. Present owner, G. S. Loney. Purchased in 1865. Bounded on the north and east by river and Bennett, south and west by Woolcott and river. Two and a half miles from Tionesta, and four and a half from Hickory. Bearing of Allegheny River southwest. Road to Tionesta southwest, to Hickory northeast. No developments.

*Ball Farm*—On Allegheny River, east side, in Hickory township, containing fifty acres. Original owner, Nelson Ball. Present owner not ascertained. Changed hands in 1864. Bounded on the north by river, east by Dewoody, south by ——, west by Ball. Three miles from Tionesta, and four miles from Hickory. Bearing of Allegheny River southwest and west. Road to Tionesta southwest, Hickory northeast and east. No developments.

*Bennett Farm.*—On Allegheny River, south side, in Hickory township. Original owner, Luther Bennett. Contains fifty acres. Bounded on the north by river, east by Allen & Co., south by ——, west by Bennett. Three and a half miles from Tionesta, and same distance from Hickory. Bearing of Allegheny River west. Road to Tionesta west, to Hickory east. Have one old well. Had a good show at depth of 250 feet. Abandoned by lessee in first oil excitement.

*Riddles Farm.*—On Allegheny River, south side, in Hickory township, mouth of Little Hickory. Original owner, M. K. Riddles. Present owners, Forrest Landing, Pittsburgh and Philadelphia Oil Company. Bounded on the north by river, east by Walter, south by ——, west by Dewoody. Four and a half miles from Tionesta, and two and a half from Hickory. Bearing of Allegheny River west, Little Hickory Creek north of west. Road to Tionesta west, to Hickory east and northeast. Have four non-producing wells. All have produced a little oil. Two of them are old wells on the river, below the mouth of Creek. Depth of wells, from 175 to 500 feet. Average cost, $4,000. Located on the flat and river side. Have two portable and stationary engines. Depth of first sand-rock, 74 feet—36 feet thick; second sand, 171 feet—36 feet thick. Depth of driving-pipe, 18 feet. This property is well located, being at the mouth of Little Hickory Creek, extending up the same. Oil is found in all the wells, but not in paying quantities.

*Walter Farm.*—On Allegheny River, southeast side, in Hickory township. Original owner, M. Walter. Present owners, Euston, Allen & Co. Bounded on the north by river, east by Richardson, south by ———, west by Riddles and river. Three and a half miles from Tionesta, and same distance from Hickory. Bearing of Allegheny River southwest, Little Hickory west. Road to Tionesta southwest, to Hickory northeast. No developments.

*Richardson Farm.*—On Allegheny River, southeast side, in Hickory township. Original owner, C. Richardson. Bounded on the north by Hunter and west by river. Five miles from Tionesta, two from Hickory. Bearing of Allegheny River southwest, road to Tionesta southwest, to Hickory northeast. One well on the property, 300 feet deep, tubed and tested. Oil not obtained in paying quantity. The flat on east side of river and south side of East Hickory Creek here commences and increases in width as it approaches the Creek. Have one engine on the property.

*Hunter Farm.*—On Allegheny River, southeast side, in Hickory township. Owner, S. Hunter. Bounded on the north by Prather and west by river. Six miles from Tionesta, and one mile from Hickory. Bearing of Allegheny River south. Road to Tionesta south, Hickory northeast. One well, 300 feet deep, and one engine. Well located on the flat below the town. Has produced some oil, but is not now in operation. Here is a broad flat, with a good territory to operate upon as the West Hickory flats. The sandrocks below the first are very irregular and broken up, and when found at all, are at about the following depth: First sandrock, 96 to 100 feet; second sand, 130 to 150 feet. The first is a hard grey sand, from 50 to 75 feet thick.

*Prather Farm.*—On Allegheny River, southeast side, and on East Hickory Creek. Original owner, T. H. Prather. Present owners: a Philadelphia company have the upper end of the farm, and G. S. Long the part on the river. Boundaries not entirely ascertained. Seven miles from Tionesta and seven from Tideoute. Bearing of Allegheny River south, East Hickory Creek north. Road to Tionesta south, to Tideoute north. Have three old wells, two engines. Two wells on flat of Creek, and one at the mouth. One of the wells is 700 feet in depth. All non-producing.

*Hickory Village.*—Located on the Ball farm at the mouth of East Hickory Creek, north side, and on east side of Allegheny river. There is one hotel, several stores, &c. This an important shipping point for lumber which is obtained in large quantities from the lands up the Creek, and transported to this point by wagons on a plank road owned by the lumbermen on the Creek. Here is also an extensive boat dock, with all the necessary apparatus, where a large number of the boats are built that are used on the Allegheny river and Oil Creek for the transportation of oil, &c.

There is one old well on the Ball farm, with ample room on the flats for extensive oil operations.

*Ball Farm.*—On Allegheny river, southeast side, in Hickory township. Owned by J. Ball. Bounded on the north by Green, east by ——, south by Prather, west by river. Half a mile from Hickory, and six and a half from Tidionte. Bearing of Allegheny river southwest, East Hickory west. Road to Hickory southwest, to Tidionte northeast. The village of East Hickory is on this farm.

Between this point and Tideoute there are thirty-two wells, on both sides of the river and in the ravines. The majority are old abandoned wells—some few are new and operating. As yet, all are non-producing. We found it impossible to obtain any definite information in relation to them, there being no one on the ground to give it. Most of these wells appear to be of shallow depth, and imperfectly tested.

*Green Farm.*—On Allegheny river, southeast side, in Hickory township. Owner, J. Green. Bounded on the north by river and Stowe, east by ——, south by Ball, west by river. One mile and a half from Hickory, and five and a half from Tideoute. Bearing of Allegheny river southwest. Road to Hickory southwest. Have two non-producing wells, 500 feet deep, and two engines.

*Siggins Farm.*—On Allegheny river, northwest side, in Harmony township. Original owners, W. & G. S. Siggins. Present owners not ascertained. Bounded on the north by Siggins, east by Siggins and river, south by Green, west by Manross. Half a mile from West Hickory, one and a half miles from Hickory. Bearing of Allegheny river southwest. Road to West Hickory southwest, to Tideoute northeast. Have nine non-producing wells, one of which is 250 feet in depth, and has the tools fast in it. On these farms is a broad flat or table land, good for agricultural purposes. Has on it good dwellings and barns, and is under very good cultivation. There has been but little development for oil.

*Siggins Farm.*—On Allegheny river, west side, in Harmony township, containing 100 acres. Original owner, J. C. Siggins. Present owners, New York & Philadelphia Petroleum Co. Purchased in 1864. Bounded on the north by Fleming, east by river, south by Siggins, west by Manross. One mile from West Hickory and one mile from Hickory. Bearing of Allegheny south, of West Hickory Creek southeast. Road to West Hickory south, to Hickory north. Have three old abandoned wells, 300 feet deep. Flat or table land like the farms below, but not so broad, merging into hill lands at upper end. Depth of first sandrock, 100 feet—25 feet thick; second sand, 200 feet—40 feet thick; third sand not found. Depth of driving-pipe 14 feet.

The Company have thirty acres on Hicks Run, a tributary of West Hickory. No developments.

*Fleming Farm.*—On Allegheny river, west side, in Harmony township. Original owners, R. & J. and Mrs. Fleming. Present owners not ascertained. Bounded on the north by Hawthorn, east by river, south by Siggins, west by ——. Two miles from West Hickory and seven from Tideoute. Bearing of Allegheny river south. Road to West Hickory south, to Tideoute north. The flat here is lost in the hills, which rise high and abrupt and are covered with timber. No developments.

*Hawthorn Farm*—On Allegheny river, west side, in Harmony township. S. Hawthorn, owner. Bounded on the north by Jones, east by river, south by Fleming, west by ——. One mile from Hickory and six from Tideoute. Bearing of Allegheny river west of south. Road to Hickory west of south, Tideoute east of north. Have one old well, which produced some oil; but is now abandoned. Hills abrupt and high.

*Jones Farm.*—On Allegheny river, west side, in Harmony township. Original owner, J. Jones. Present owner, E. Ferrero. Bounded on the north by Scott, east by river, south by Hawthorn, west by ——. One and a half miles from Hickory, five and a half from Tideoute. Bearing of Allegheny river west of south. Road to Hickory west of south. Tideoute east of north. No developments.

*Scott Farm.*—On Allegheny river, west side, in Harmony township. D. Scott, owner. Bounded on the north by Jones, east by river, south by Jones, west by ——. Two miles from Hickory, five miles from Tideoute. Bearing of Allegheny river south. Road to Hickory south, to Tideoute north. Have 4 non-producing wells, two drilling and two abandoned. Depth of wells, two 500 feet, one 200, one 400 feet. Hills not abrupt as below. Broken by ravines.

*Jones Farm.*—On Allegheny river, west side, Harmony township. Original owners, E. Jones and E. L. Jones. Present owners, E. Ferrero and others.

Number of acres not ascertained. Bounded on the north by Jones, east by river, south by Scott, west by ———. Two and a half miles from Hickory, four and a half from Tideoute. Bearing of Allegheny river south. Road to Hickory south, to Tideoute north. Have some old wells, apparently abandoned, on the farm.

*Jones Farm.*—On Allegheny river, west side, in Harmony township. Owned by heirs of J. Jones. Bounded on the north by Smith, east by river, south by Jones, west by ———. Three miles from Hickory, four from Tideoute. Bearing of Allegheny river south. Road to Hickory south, Tideoute north. No paying developments.

*Smith Farm.*—On Allegheny river, south side, in Harmony township. Original owner, S. Smith. Present owner, G. S. Long. Bounded on the north by Warren County line, east by river, south by Jones, west by ———. Three and a half miles from Tideoute and some distance from Hickory. Bearing of Allegheny river south. Road to Hickory south, to Tideoute north. On this farm are some wells and one engine. None of the wells in operation.

*Prather Farm.*—On Allegheny river, east side, in Hickory township. Owned by T. H. Prather's heirs, or Woodford. Bounded on the north by Winson, east by Stowe, south by Green, west by river. Two miles from Hickory, five from Tideoute. Bearing of Allegheny river west of south. Road to Hickory west of south, to Tideoute east of north. Have one old well drilling, 250 feet deep, and one engine on the property.

*Brown Farm.*—On Allegheny river, east side, in Hickory township. Original owner, Geo. Brown.

Present owners, Hopewell & Jennings. Purchased in 1865. Bounded on the north by Long, east by Stowe, south by Prather, west by river. Two and a half miles from Hickory, and four and a half from Tideoute. Bearing of Allegheny river west of south. Road to Hickory west of south, to Tideoute east of north. No developments.

*Alcorn Farm.*—On Allegheny river, east side, in Hickory township. Original owner, James Alcorn. Present owners, Hopewell & Jennings. Purchased in January, 1865. Bounded on the north by Hunter, east by Stowe, south by Winson, west by river. Three miles from Hickory, four from Tideoute. Bearing of Allegheny river west by south. Road to Hickory west of south, to Tideoute east of north. No developments.

*Hunter Farm.*—On Allegheny river, east side, in Hickory Township. Original owner, W. Hunter. Present owners, Hopewell & Jennings. Purchased in January 1865. Bounded on the north by Warren County line, east and south by Alcorn, west by river. Three and a half miles from Hickory, and same distance from Tideoute. Bearing of Allegheny river south. Road to Hickory south, to Tideoute north. Have one non-producing well, 280 feet deep. Belongs to Messrs. Anderson, together with the acre of land the well is on. Depth of first sandrock, 96 feet,—second sand, 130 feet. Depth of driving-pipe 14 feet.

WARREN COUNTY.—DEERFIELD TP.

*Bozer Farm.*—On Allegheny river, west side, in Deerfield township. Owned by H. & W. Bozer. Bounded on the north by Kortman, east by river, south by Venango County, west by ——. Four miles from Hickory, three miles from Tideoute. Bearing of Allegheny river east of south. Road to Hickory east of south, to Tideoute west of north. No developments.

*Kortman Farm.*—On Allegheny river, west side, in Deerfield township, containing several hundred acres. Original owners, J. & I. Kortman. Present owners, Grove Farm Oil Company, of Pittsburgh. Purchased in January, 1865. Bounded on the north by Kortman, east by river, south by Bozer, west by ——. Four and a half miles from Hickory, two and a half from Tideoute. Bearing of Allegheny river east of south. Road to Hickory east of south, to Tideoute west of north. No developments.

*Kortman Farm.*—On Allegheny river, west side, in Deerfield township. Original owners, P. Kortman, H. Kortman, & J. Kortman. Present owners, Grove Farm Oil Co., of Pittsburgh. Contains several hundred acres. Purchased in 1865. Bounded on the north by Spangler, east by river, south by Kortman, west by ——. Five miles from Hickory and two miles from Tideoute. Bearing of Allegheny river east of south. Road to Hickory east of south, to Tideoute west of north. Company have several wells—none producing. Are drilling some on West Hickory.

*Dunn Farm.*—On Allegheny river, east side, Limestone township. Original owner, J. Dunn. Present owners, Hopewell & Jennings. Purchased in January, 1865. Bounded on the north by Magee, east by ——, south by county line, west by river. Four miles from Hickory, three from Tideoute. Bearing of Allegheny river east of south, Dennis Run west. Road to Hickory east of south, to Tideoute west of north. Have three old wells,—are drilling one, and have reached a depth of 300 feet. One engine on the property.

*Magee Farm.*—On Allegheny river, east side, in Limestone township. Original owner, Mrs. Magee. Present owners, Hopewell & Jennings. Purchased in Jan., 1865. Boundaries not accurately ascertained. Four and a half miles from Hickory, and two and a half from Tideoute. Bearing of Allegheny river east of south. Road to Hickory east of south, to Tideoute west of north. No developments.

*Dale Farm.*—On Allegheny river, east side, in Limestone township. Owned by William Dale. Bounded on the north by Hautter, east by ——, south by Magee, west by river. Five miles from Hickory, two from Tideoute. Bearing of Allegheny river east of south. Road to Hickory east of south, to Tideoute west of north. No developments. Bald Eagle Island is in the river, opposite this farm. One well on the island, which had a good show of oil when first struck. Is 600 feet deep. Not operating at present.

*Hunter Farm.*—On Allegheny river, east side, in Limestone township. Owned by M. Hunter. Bounded on the north by Richardson, east and south by ——, west by river. Five and a half miles from Hickory, and one and a half from Tideoute. Bearing of Allegheny river east of south. Road to Hickory east of south, to Tidionte west of north. Adjoining this tract is seven acres off the Richardson farm, belonging to R. R. Roberts. No developments.

*Richardson Farm.*—On Allegheny river, east side, in Limestone township. Containing 120 acres. Original owners, Mrs. Richardson and L. D. Richardson.

Present owners, R. A. Wilder and others. Purchased in 1864. Bounded on the north and west by river, east by Ralston, and south by ——. Six and a half miles from Hickory, and half a mile from Tideoute. Bearing of Allegheny river east of south and west. Road to Hickory east of south, to Tideoute west of north and east. Have six wells—five old and one new one. Average depth of wells 60 to 500 feet. Located on flat and ravine. Owned by R. A. Wilder and others. Have one stationary engine. No account of sandrocks could be obtained. Depth of driving-pipe, 6 to 12 feet.

*Tipton Farm*—On Allegheny river, east and south side, containing 100 acres. Original owner, Shadwick Tipton. Present owners, Tipton Farm Oil Company, of Philadelphia. Purchased in 1864. The Enterprise Mining and Boring Company, of New York City, are sinking a shaft upon this farm. They have a lease of 15 rods square of ground. They began one shaft, sank it to some depth, and, after expending some $10,000, abandoned it and commenced a new one a few rods distant from the first. Are now down 90 feet. The shaft is 7 by 9 feet in size, cribbed with six-inch plank to the bottom of the first sandrock, and caulked to keep the water out. Depth to the first or bed rock, 30 feet; next, 10 feet—shale rock; next, 10 feet of hard gray sandrock; next, 40 feet of slate and soap-rock. Some indications of oil found in the shale above the sandrock, and in the slate and soapstone below, but none in the sandrock. The sandrock is very hard and broken, with small crevices all through. The larger portion of the water came in through the crevices in the sandrock. As the water rises in the river, it rises in or about the shaft, requiring the cribbing to be kept tightly caulked above it. Some gas was found in the soft rock below the sand, sufficient to light it in jets. The location of the shaft is about 30 rods from the river, and about 30 feet above the level of the same. The company have a 20 horse-power engine to work their machinery, and a ten-inch pump to exhaust the water. The shaft was afterwards abandoned at the depth of 160 feet, owing to accidental explosion of accumulated gas, one man being killed by same.

The obtaining of the above information was a striking exemplification of the saying, "knowledge under difficulties." The works were inclosed by a high board fence, intended to shut out inquisitive mortals. Over a small door, apparently the only entrance, was painted, in large letters "No Admittance," a Yankee phrase, signifying "Come in;" accordingly we went in, but were soon accosted by a person evidently in authority, who requested us to "go out." Being hard of hearing, we asked for the Superintendent of the Institution, and were told he was in town, where we might see him. Our business was made known in a few words, that we should be pleased to incorporate some items regarding the Shaft, in our history. This request met with a peremptory refusal. Said we would have to get permission from C—— M——, as he was opposed to giving any thing of the kind. We then asked permission to look at and go down into the merits of the thing. But this faithful guardian was too old a bird to be caught by such chaff. He eyed us suspiciously, as though he were fearful we would pocket his big hole in the ground, or whisk away his 20 horse-power engine, and was proof against all our entreaties. We failed to call on C. M., having obtained all the requisite information from a thoroughly reliable source. If he or his subordinate will call upon us, we promise to enlighten them fully upon the subject. A large number of wells were sunk on this farm in 1860 and 1861, to a shallow depth, but are now abandoned. There are 25 old wells visible, most of them have derricks, or fragments of same standing. No wells producing. One well drilling—or fishing for tools. Have four engines on the property, besides the one sinking the Shaft. Many more wells were sunk, but not a vestige of same is now to be found.

*Cohell Farm.*—On Allegheny river, south side, in Limestone township, opposite lower portion of Tideoute. Original owners, D. & J. Ralston. Contains 100 acres. Present owners, Ralston and others, and a Philadelphia Oil Company. Purchased in 1864. Bounded on the north by river, east by Economy Oil Company, south ——, west by Richardson. Seven miles from Hickory, and opposite Tideoute. Bearing of Allegheny river west. Road to Hickory west, to Irvine east.

A large number of wells were drilled on this farm in 1860 and 1861. Most of them are now abandoned. There are now standing ten derricks—all over shallow wells. Some of these have produced considerable oil. There are eight more conductors in view, showing where operations have been carried on in former days. Have only three engines on the property now. The depth of the old wells average about 200 feet.

*Economy Oil Company Lands.* — On Allegheny river, south side. Original owner: Wm. Davidson. Present owners, Economy Oil Company. Purchased in 1858 or 1859. Contains 6,000 or more acres. Bounded on the north by river, east and south by ——, west by Ralston. Opposite Tideoute, and 14 miles from Irvine. Bearing of Allegheny river southwest and west. Road to Hickory west, to Irvine east and northeast. Have five producing wells, all pumping. Daily average yield, 70 barrels. Five non-producing wells. One of these formerly flowed. Commenced producing in December, 1860. The wells are producing about the same now as when first drilled. Have produced about 10,000 barrels. Depth of wells, 135 feet. Average cost, $500. Wells located on the table lands on the river side. Owned by the Society. Have six engines on property. Depth of sandrock, 107 feet—44 feet thick. Had ninety-two feet of soapstone and slate. Depth of driving-pipe, 15 feet.

The Economists to whom this tract belong are a German Association, who some fifty years ago founded Harmony, in Butler County. Lived there for many years. They afterward removed to Indiana, on the Wabash river. At their new home they lost quite a number of their members, and were quite unfortunate in many other respects. In a short time, they returned to Pennsylvania, and located fourteen miles below Pittsburgh, in Beaver County, and there built their present abiding place, or town, called Economy. There are now about 300 members. They formerly numbered 800. A large majority of the present members are quite aged. They are very wealthy, and own all things in common, neither marry nor give in marriage. Their business is transacted by two trustees. They have a council of twelve members, for the transaction of all important business. They are extensively engaged in the cannel and bituminous coal business, in Beaver County. They formerly manufactured various articles, such as, silks, &c. Are now large wine manufacturers, their vineyards being quite extensive, and their wine is noted for its excellence and purity. They have one of the most beautiful locations on the Ohio river, and several hundred acres of land under a high state of cultivation. The Society are noted, in connection with their wealth, for their morality and strictness in their religious principles. They are of the Lutheran persuasion. Their business outside of their home is carried on by men whom they employ. None of the members go outside of their home for the transaction of any business, except the two Trustees, to oversee the same. All wear a plain dress, similar in style, of blue or grey home-made cloth. Females have the common female dress, after a similar pattern, and eschew the modern fashions, *paniers*, Grecian bend, &c., and in fact dress like women of good common-sense, comfortable and decently. The present Trustees are R. L. Baker and Jacob Henrici. Davidson, the former owner of this property, became involved financially, and the Company, or rather Society, loaned him money. He failed, and the Society, in order to save themselves, bid in the property at the Sheriff's sale. They afterward gave Davidson a lease of it for two years. In this time, he became able to discharge all his other obligations, and had a handsome competency besides.

*Henry Farm.*—On Allegheny river, west and northwest side, and on Dennis Run and Grove Run, in Deerfield township, containing 300 acres. Original owner, Robert Henry. Present owner, New York & Allegheny Oil Company. Purchased in 1864. Bounded on the north by Tideoute and Warren County, east by river, south by Kortman. Fifteen miles from Irvine, ten from Garland, ten from Enterprise. Bearing of Allegheny river southwest and south. Dennis Run southeast, Grove Run northeast. Road to Hickory south, to Tideoute north and northeast. Have five producing wells, as follows: Bennett well, 15 barrels per day; Shugart well, 70 barrels; Porter well, 30 barrels; Puroleum well, 70 barrels; Slocum well, 180 barrels. All pumping. Bonus paid for leases, $500 to $3,000. Have five old wells non-producing. Average depth of wells, 325 feet. Average cost of same, $3,000. Commenced producing in November, 1865. Have produced 10,000 barrels. Have twenty-five wells in progress on Dennis Run, and thirty more to be commenced. Wells located on hillsides of ravine, on Dennis Run and Grove Run. Owned by lessees. Company have a royalty of one half the oil. Have thirty portable and stationary engines. Depth of first sand-rock, 70 feet—18 feet thick; second sand, 215 feet—20 feet thick; third sand, 325 feet—40 feet thick. Depth of driving-pipe from four to ten feet.

The company was organized in the year 1864. Sunk five wells on one acre on the river, and, after operating about one year, and expending some $15,000 to no profit to the company, were about to abandon the property as valueless. But owing to the perseverance of the business manager and a few members of the company, they held on, and finally extended their operations on to Dennis Run, where, in November, 1865, they succeeded in striking their first well, and are now on a good basis. They have room for a large number of wells on their territory, and the Warren & Franklin Railroad passes through the property on the river bank. The company are about erecting two large iron tanks on their property fronting on the river, convenient to the railroad, so they can ship by rail or river, and are laying pipes from their wells to same, thus saving the expense of teaming for a mile and a half. The business here is carried on by Mr. James Parshall, the business agent, and who is also one of the directors of the company, owning one fifth interest in same, and is a resident of the place. Office of the company, 134 Maiden Lane N. Y. City. President, S. R. Trembly; Secretary, W W. Phelps.

The company took this property into their own hands in June, 1864, and are now operating successfully. Intend to make many improvements the coming season. Have nearly one mile on the river opposite Upper Tideoute. They have a 6,000 barrel iron tank for storage when oil is cheap; their own carpenter, cooper and blacksmith shops, and every thing arranged to do their own work. Large bodies of timber on their property.

*Parshall Farm.*—On Allegheny river, and on Gordon and Dennis Runs. Original owner, Samuel Parshall. Present owner, W. W. Wallace. Contains 400 acres. Tideoute & Warren Company leased the farm in the spring of 1859. Company's lands extend about three quarters of a mile up Gordon Run. Bearing of Allegheny river west and south, Dennis Run northeast Have two wells on the river, producing in the aggregate ten barrels per day. Have 16 producing wells, all pumping—two of them on river. Have 26 old wells on the river, 14 old derricks, and 14 wells in progress on Dennis Run. Have a total of 20 producing wells, in part as follows: Combs well, 25 bbls.; Niel well, 25 bbls.; Cincinnati well—testing; Shaw well, 40 bbls.; Pilgrim well, 40 bbls.; McKiel first well, 20 bbls.; Babcock well, 10 bbls.; Barker well, 25 bbls.; Gilson well, 25 bbls.; Eckart & Ralston well, 50 bbls.; McKiel well No. 2, 30 bbls.; Mix well—testing, 20 bbls.; Mosier well, 30 bbls.; Biggs well, 70 bbls. Average depth of wells, 210 to 220 feet. Average cost of same, $3,000 to $5,000. Commenced producing in July, 1865, and continued to February, 1866. Amount of production not ascertained. Have 34 portable and stationary engines on the property. Depth of first sandrock on the river, 125 feet—50 feet thick. No other one found. Depth of first sandrock on Dennis Run, 70 feet—18 feet thick; 2d sand, 215 feet—20 feet thick; 3d sand, 325 feet—40 feet thick. Depth of driving-pipe, 4 to 10 feet.

As the ground rises in ascending the Run, the rock rises ten feet for every 200 feet of elevation.

*State Land.*—On Dennis Run. Patented by E. Jones, in 1865. Present owners, New York & Dennis Run Oil Company, of New York. Purchased in 1865. Number of acres not ascertained. Have no producing wells. Have twelve wells in progress. Some have just commenced drilling and are making preparations. Have eight engines on the property. The wells are located on the upper portion of Dennis Run.

*Tideoute Borough.*—Is located on the property of the Tideoute & Warren Oil Company, formerly Parshall; Purdy property, formerly Kemier; Ralston & Co., formerly Arter property; McGuyr Petroleum Company, formerly McGuyr property, and the Anthony Courson property. It extends two miles along the river on the table lands. A description of the town will be found in another portion of this work.

Tideoute Island, McGuyr Island, McGill & McGee Island, and Church Island, are all in the river at this point. Are unoccupied at present. Some developments were made in former years, but oil was not obtained in paying quantities. Tideoute Island had four producing wells on it, at the time of the great flood in the spring of 1865, three engines in operation, and three more not set in place. All were swept away, involving a loss of $6,000. The company have done nothing since. Two of the wells are flowing a small quantity, which is now running into the river. The company contemplate resuming operations during the coming summer. It belongs to the Philadelphia & Tideoute Island Oil Company, of Philadelphia. There are along the river bank, Tideoute Creek, and McGuyr Run six or eight old wells to be seen, and four on the islands above Tideoute Island.

There has been a large number of wells drilled along the river, but the derricks, &c., being removed, nothing but the holes remain to tell of the toil of other days, and the investments of disappointed fortune seekers. The ground being covered with snow at the time of our visit, even the holes were not visible.

The refineries of this locality are as follows: Star Works, Pinney, Williams & Co., capacity, 100 bbls. crude per week; O. Osborn & Son, capacity, 150 bbls. per week; Capt. Dingley, 180 bbls. per week. Have one well drilling on the property 425 feet deep. Aurora Oil Works, J. W. Porter, 175 bbls. per week; Rising Sun Oil Works, White & Stilwell, 450 bbls. per week; Nonpareil Works, Brown, Babbett & Co., 150 bbls. Have six engines at these works.

*McGuyr Run.*—A tributary of the Allegheny river, emptying into it in Upper Tideoute. One well in progress, back from the river above the town, on the property of the McGuyr Run Petroleum Company. Have one engine. The Oil City Petroleum Company have forty acres, off the former Parshall tract, three quarters of a mile from the river, on which is one well, over 600 feet in depth, non-producing. On the Henry farm is 300 acres, now owned by the Midas Oil Company, of New York and Pittsburgh. Commenced four wells. No engines on property at present. The Brown tract, now Minturn & Bro., contains 600 acres, and extends for three miles on the Run. Purchased in Jan.,

1865. Have two wells—one of them drilling. One of the wells is 600 feet, and the other 500 feet, with a good show of oil at 250 feet. Have two engines, one of which drives the saw-mill on the property. Depth of first sandrock, 120 feet—42 feet thick; 2d sand, 210 feet—30 feet thick; 3d sand, 320 feet—13 feet thick. Depth of driving-pipe, six to ten feet. The property is from two and a half to three miles from the river by the Run. Are arranging to drill three more wells on the property during coming spring.

*Tideoute Creek.*—A tributary of the Allegheny river, emptying into the same between Upper and Lower Tideoute. There are six wells and three engines on the Kinnear property, now Tideoute Association and Empire Oil Co. of New York. Some of the wells are in progress, others apparently abandoned. No operations at present. Grist mill on the property, not in operation. Other estate on east side of Creek, has one well, not in operation. One refinery, not in operation, and one engine. Adjoining this estate, are four wells, all non-producing. On the Denning property are no developments. Next to this is the Bristol & Preston property, now Tideoute Creek Oil and Lumbering Company of New York. Contains 1,000 acres. Purchased in Jan., 1865. Have a steam saw-mill. Is two and a half miles from river by Run. Have two wells in operation. One is operated by the engine of the mill. Are 200 and 500 feet deep. Next is the Shaw tract, no developments. Then the Thompson property, now Oil City Petroleum Company. Contains 300 acres. Purchased in November, 1864. Have one well, 600 feet in depth, non-producing, and one engine. Next is the property of the Arcade Oil Company of Rochester, New York. Have two wells, one abandoned—one drilling. Have one engine.

*Gordon Run.*—A tributary of the Allegheny river, emptying into it near the lower end of Lower Tideoute. The west Branch, or Dennis Run, empties into Gordon Run about thirty rods from the river. (For account of oil interest, &c., of same, see Tideoute & Warren Oil Co., New York & Allegheny, and New York & Dennis Run Oil Co's. Above the Tideoute & Warren Oil Co.'s lands come the lands of the Gordon Run Association. No developments.

Next is a tract of 117 acres, formerly owned by C. Roupe. Now owned by the Oil City Petroleum Company. Have one well in progress, and one engine. Intend to drill to depth of 600 feet, unless sooner stopped by obtaining oil in paying quantities. Purchased in 1864. Is two and a half miles back from the river, by the stream. General bearing of the stream east. There are no producing wells on Gordon Run at present. Several parties are erecting tanks above and below the mouth of the West Branch for the purpose of storing oil. Preparations are making to put down a number of wells on the Run in the Spring. The flat is narrow near the river, broader at the mouth of West Branch, and above West Branch very narrow, affording barely room for the stream and road, by crowding most of the wells up to the bluffs. On Gordon Run, upon the lands of Oil City Petroleum Company's property, the sandrocks are as follows: first sand, 125 feet deep—15 feet thick; second sand, 350 feet—20 feet thick. Depth of driving-pipe, 8 feet.

*East Hickory Creek.*—Above the Ball and Prather farms, at the mouth of the Creek, is the C. Brown farm. There is one well on this farm, 802 feet in depth. Known as the Heth well, and two wells belonging to the Fremont Oil Company, all non-producing. Next to this is the lumbering tract of H. & H. H. Stowe, containing some 5,000 acres. Have on property one large gang-mill, capable of cutting 30,000 feet of lumber per day. The other is a single mill. This is a valuable tract of land, one mile from the river, with a good plank road to same, for the transportation of lumber for shipment on river. Have one well, 200 feet deep, drilled in 1861. Was never tested. Had some oil, but was thought to be a non-paying well, on account of the low prices then prevailing. On the Bennett tract above, owned by the Beaver Mining and Lumbering Company of Philadelphia, formerly Holbrook, Allen & Co's. property, is one well drilled in 1865, 600 feet deep. Produced some little oil, but not in paying quantities. This is also a lumbering tract.

On the Tyndall property, seven miles up the Creek from the river, is one well, 600 feet in depth. Had a large show of oil, one engine and all the machinery for operating, now idle. Parties intend to test it more fully when the Spring opens.

*Manross Farm*—Is on West Hickory Creek, first above the Green farm. Original owner, —— Manross. Present owner, West Hickory Mining Association of Philadelphia. Contains 1,000 acres. Purchased in June 1865. Is a mile and a half from the river. Bearing of West Hickory Creek, southeast. Road to river southeast, to Pithole northwest. Have seven producing wells, and fifteen in progress. Amount of production not ascertained. Average depth of wells, 125 to 240 feet. Average cost, $3,000 to $4,000. Commenced producing in January and Feburary 1866. Wells located on flat of Creek and McArthur Run. Owned by lessees. Have twenty-two portable and stationary engines on the property. Depth of first sandrock, 60 to 90 feet—10 to

45 feet thick; second sand 78 to 147 feet—3 to 33 feet thick; third sand, 109 to 190 feet—10 to 18 feet thick. Depth of driving-pipe, 10 feet.

The sandrocks vary so much on this farm that we will give the record of a few of the producing wells.

*Bull Well.*—First sandrock, 60 feet—10 feet thick; second sand, 86 feet—9 feet thick; third sand, 116 feet—10 feet thick; fourth sand, 162 feet—12 feet thick. Cook well—First sand, 61 feet—8 feet thick; second sand 78 feet—3 feet thick; third sand, 109 feet—11 feet thick; fourth sand, 143 feet—23 feet in rock. Lawrence well, on McArthur Run—First sand, 96 feet—35 feet thick; second sand, 147 feet—18 feet thick; third sand, 190 feet—6 feet in rock.

Seventy-five acres of this farm was sold to the Second National Company. On this are two wells, one producing 50 barrels per day, and the other testing with a good show.

The following are the producing wells, in part: Corson well, light oil, 25 barrels per day. Bull well, do., 50 barrels. Cook well, do., 40 barrels. Percy well, heavy oil, 50 barrels. Reserve well, do. 60 barrels. Three other wells have formerly produced oil. Are not in operation at present. The old Manross well commenced producing some two years ago. Is doing nothing now. The sandrock is very irregular, and the oil varies in gravity in wells not one hundred feet distant from each other.

*White Farm*—On West Hickory Creek and on Buckhorn Run. Original owner, John White. Present owner, Hickory Creek Petroleum Company. Contains 87 acres. Purchased in 1864. Two miles from river. Bearing of West Hickory Creek southeast, road to river southeast, to Neillsburgh northwest. Have four producing wells, all pumping. Twenty-five wells in progress, and fourteen more leases given out. The producing wells are as follows:—Union well, 20 barrels; No. 19, 15 barrels; No. 40, 30 barrels; No. 38, —— barrels. Gravity of oil obtained, 27°. Have produced some 2,000 barrels. Average depth of wells 400 feet. Average cost $5,000. Commenced producing in November, 1865, and January, 1866. Wells located on flat and owned by lessees. Have twenty-nine engines on the property. Depth of first sandrock, 115 feet—90 feet thick; second sand, 120 to 180 feet—10 to 12 feet thick. Depth of driving-pipe 6 to 20 feet.

On this farm is a new town laid out, called Hickory City. A number of buildings, consisting of hotels, stores, dwellings, machine shops, &c., have been erected, and the place presents a brisk appearance.

*Fleming Farm.*— Sub-Division No. 1.— On West Hickory Creek. Original owner, J. Fleming. Present owner, Cherry Run & Trout Run Petroleum Company of Philadelphia. Contains 133 acres. Purchased in 1865. Two and a half miles from the river. Bearing of West Hickory Creek, southeast. Road to river southeast, to Neillsburgh northwest. Have four wells in progress, and four engines. On this tract are being laid out 48 one-acre leases.

*Fleming Farm.*—Sub-Division No. 2. On West Hickory Creek. Original owner, J. Fleming: containing 160 acres. Present owners, Pennock, Bagaley & Co., of Pittsburgh. Purchased in 1864. Three miles from the river. Bearing of West Hickory Creek, southeast. On this tract are two wells in progress, and fifty-six leases. On sub-divisions 2 and 3, are sixty-three leases taken by different parties, many of which are being operated.

*Fleming Farm.*—Sub-division No. 3.—On West Hickory Creek. Original owner, J. Fleming. Present owners, Pennock, Bagaley & Co., of Pittsburgh. Contains 219 acres. Three and a half miles from the river. Bearing of West Hickory Creek, southeast. Road to river southeast, to Neillsburgh northwest. There are eleven wells in progress on this and Sub-division No. 4—fifty-three leases on No. 3. Two wells drilling—five engines. Depth of first sandrock, 85 feet—12 feet thick; second sand, 155 feet—7 feet thick. Depth of driving-pipe, 10 feet.

*Fleming Farm.*—Sub-division No 4. On West Hickory Creek. Original owner, J. Fleming. Present owners, Pennock & Co., and —— Davis, at Hickory Centre. Contains 219 acres. Purchased in 1864. Four miles from the river. Bearing of Hickory Creek southeast. Road to river southeast, to Neillsburgh northwest. A part of the wells mentioned in account of Sub-division No. 3, are on this farm. Sandrocks are similar, so far as drilled.

*Fleming Farm.*—Sub-division No. 5—on West Hickory Creek. Original owner, J. Fleming. Present owners, Pennock, Bagaley & Co., of Pittsburgh. Contains 269 acres, making in the five sub-divisions a total of 1,000 acres. Purchased in 1864. Four and a half miles from river. Bearing of West Hickory Creek southeast. Road to river southeast, to Neillsburgh northwest. Have five wells in progress—none tested, and five engines. Depth of wells from 200 to 700 feet. On this sub-division are ninety-four leases. Depth of first sandrock, 125 feet—10 feet thick; second sand, 250 feet—15 feet thick; third sand, 390 feet—15 feet thick.

*Giles Farm.*—On West Hickory Creek. Original

owner, R. H. Giles. Present owners, Shaw, Giles, Howe & Eastman. Contains 100 acres. Four and a half miles from river. Bearing of Creek southeast. Road to river southeast, to Neillsburgh northwest. Have one small saw-mill on property. No developments.

*Gorman Farm.*—On West Hickory Creek, containing 350 acres. Original owner, W. Gorman, Leased. Have two wells—one old—400 feet deep, with tools fast in it; and one well testing 400 feet deep. Depth of first sandrock, 85 feet—12 feet thick; second sand, 200 feet—15 feet thick. Depth of driving-pipe 18 feet.

*Miles Farm.*—On West Hickory Creek. Original owner H. Miles. Present owners, Warren & Venango Oil Company. Contains 6,000 acres. 300 acres were leased to Duncan & Brittanbaugh. Five miles from river. Bearing of West Hickory Creek southeast, to Neillsburgh northwest. Have one well, producing ten barrels per day. Depth of well 367 feet. Commenced producing in September, 1865. Oil is 40° gravity. Eighteen feet to bed-rock. Depth of sandrocks not ascertained. Two saw-mills on the property. An effort is being made to build up a town. This is a very good lumbering point.

*Gorman Farm.*—On west branch of West Hickory. Original owner, Michael Gorman. Contains 185 acres. Present owners, Adnah Neyhart, and Salem Oil & Lumbering Company, of Salem, Ohio. Purchased in 1864. Two miles from Neillsburgh. Bearing of west Branch of West Hickory south of east. Road to Neillsburgh southwest, to Tideoute northeast. No producing wells. Have two engines on property. Have two wells not tested, and several leases given out. The Neyhart hotel is located here. Mr. Neyhart purchased the whole tract, afterwards sold fifty acres of same to the Salem Oil & Lumbering Company, who also own a one-fourth interest in the lands of the West Hickory Oil & Lumbering Company, on the Main Creek and a branch of same.

Next to this is a reserve of a few acres with one well drilling—400 feet deep and one engine. Next to this is the property of D. Beatty, 421 acres, on both branches at the forks of the Creek. Have one well drilling, and another in progress; two engines, and two old abandoned derricks; have one sawmill on the property.

The next property is that of the West Hickory Oil & Lumber Association. Contains 200 acres, and is a lumber tract. Are building a steam sawmill. Have one large engine. The property lies on branches of Creek.

Next is the property of John and Benjamin Gorman, on Main Creek. No developments.

Next is the property of the Fremont Oil Company, of Philadelphia, being ninety-four acres off from Michael Gorman farm. Have one well drilling and one engine.

One well drilling on the Dow property, joined by Mr. Bonier, owner of twenty acres off the same tract, on lower end.

On West Branch, J. Pierce purchased of Wm. Wallace and Wm. S. Alcorn, 285 acres. Sold twenty-four acres to Cherry Run & West Hickory Petroleum Company. One well on Pierce tract, and one on Wallace tract—two engines. Steam sawmill run by J. S. Pierce & Bros.

On Cranberry Run, a tributary of West Hickory, is the property of R. C. Scott, containing 100 acres. He sold a one-half interest to a Chicago Oil Company. Have four wells drilling, and four engines.

Adjoining, J. R. Smith & Co., have fourteen acres, on which are two wells, nearly completed, and two engines. A steam sawmill on the property, run by J. R. Smith & Co.

On East branch of West Hickory is thirty acres belonging to Capt. Birch of Waterloo, New York. No developments.

Next on same branch, is the property of the Grove Farm Oil Company, of Pittsburgh, containing 950 acres. Have two wells drilling—nearly completed, and two engines. This tract has a large river front, below the property of the New York & Allegheny Oil Company of New York City. Parshall Run passes through property.

Next is seventy-eight acres, owned by W. Wallace —one well non-producing.

Next is the property of the Triumph Oil Company of Pittsburgh, 200 acres from off the Wallace tract. Have two wells in progress, and two engines.

## LITTLE HICKORY.

*Walter Farm.*—On Little Tionesta Creek, east side, and on the Allegheny river. Original owner M. Walter. Present owners, Frazee & Co., of Rochester, N. Y. Contains 600 acres. Have two wells, one 250 feet deep, and the other 560 feet. Oil found in both, but not as yet in paying quantities. Have one engine pumping the deep well.

Next to this is the Walter Barnes farm, containing 300 acres. Is now owned by a Pittsburgh Company.

Have one well 500 feet deep, non-producing, and one portable engine.

Next are the Barnes & Church farms, now owned by the Little Hickory Oil Company, of Philadelphia and New York. Contains 460 acres. Purchased in 1864. Have one well drilling, which is now 400 feet deep. Had good shows of oil, but don't find it in paying quantities by testing.

Next is the Hatch Farm, now owned by the Cincinnati and Pittsburgh Oil Company, containing 833 acres. Have two wells, one 760 feet, the other 1,160 feet. No oil. Average cost, $8,000. Depth of first sandrock, 255 feet—75 feet thick; second sand, 310 feet—6 feet thick, from 1,004 to 1,010 feet—hard grey sand. There is a sawmill on the property, not in operation.

*Jamison Farm*—On Jamison Run, a tributary of the Allegheny river, on west side. First farm back from river, containing 40 acres. Owned by Fahnestock Bros., of Pittsburgh. Purchased in 1864. Have one well, 715 feet deep, tools fast; no oil. Next to this is the Henry Sutley farm, one and a half miles from river, owned by a Buffalo Oil Company. Number of acres not ascertained. Have one well, 440 feet deep.

### TIONESTA CREEK.

First above Tionesta Borough is the G. G. Sickles farm, one well on same, belonging to Hughes & Wickes, 600 feet deep. Depth of first sandrock, 245 feet; second sand, 560 feet—23 feet of white and 11 feet of grey sand; third sand, 40 feet of white sand. Depth of driving-pipe, 30 feet. These parties have an improved plan for drilling—using a cam arrangement instead of the walking-beam, and turning the tools by the same machinery. Have another well on same property, called the Shirley well,

600 feet deep, owned by lessees. Next come farms in the following order: Noble's farm, 80 acres; Dale Bros. farm, 60 acres; Towner & Brett farm, 130 acres; Keyser farm, 80 acres; Dale Bros. farm, 150 acres; known as the C. J. Lloyd purchase,—New York parties. Dale Bros., south side, 200 acres—now Rose & Griffith; Dale & Bro., 108 acres; D. Stow, north side, 75 acres—now Pittsburgh, New York & Tionesta Oil Company. Purchased in 1864. Have two wells, one 600 feet deep, and the other, 1,000 feet, one old well 200 feet deep. Have two engines. Depth of first sandrock, 300 feet deep—40 feet thick; second sand, 900 feet deep—60 feet thick. Depth of driving-pipe, 36 feet, wells not tested.

Conroy farm, on north side of Creek, containing 50 acres; has one well, known as the Conroy well. Is 600 feet deep; the oil from this well is peculiar, different from any other discovered in this region. It is of a beautiful straw color, but is not obtained in paying quantity. Are now drilling the well deeper. This has been a remarkable well; the drillers claim to have drilled through a vein of copper. Rock all hard. Have one engine.

J. A. Dale farm, north of the Creek, 120 acres; no developments. John Noble farm, 58 acres, no developments, now owned by Capt. Brett & Co. May & Bros., north side of Creek, 300 acres; now owned by Morris & Co., Fredonia, N. Y. Have one well, 500 feet deep, non-producing, and one old well. May & Bros., north side, 120 acres, now owned by S. Q. Brown, of Pleasantville; leased by May Farm Oil Company of Pottsville, Pa. Have one well 713 feet deep, not tested, with a good show of oil and gas. Flowed water and oil for some time. Difficulty between the lessees and land interest. Have stopped operations. Stow Blakely farm, north side of Creek, 400 acres. No developments.

Now owned by Wm. Richardson & Co. Hyner farm, north side, 100 acres. Now owned by Richardson & Brognard. No developments. Henry Pierson farm, on north side, 500 acres. No developments. Now owned by Rose, Griffith & Co. Stephen Smith farm, south side, 100 acres. Now owned by Richardson & Brognard. No developments.

Tionesta Lumbering & Mining Company, formerly Hall & Lacy—called Lacy Town. Population, 150. Have here a gang saw-mill, grist-mill, hotel, store, tannery, dwellings, &c. 5,000 acres in the tract. Have three wells, not tested. One has the tools fast in, and one is being drilled. Depth from 300 to 700 feet. Have three engines. Is on south side of Creek. Wm. White farm, on both sides of Creek, containing 1,000 acres; owned by Woolf & Co., and Eastman & Co. No developments on south side. Two wells on north side. One is 300 feet in depth, and has the tools fast in it. The other is 400 feet deep. Have two engines. Amasa Purdy farm, lies on both sides of the Creek. Now owned by Frank Allin & Co. Contains 1,000 acres. Harrington farm, on both sides of the Creek, contains 200 acres. No developments. Green & Gordon farm lies on both sides of the Creek, containing 1,800 acres; now owned by Spooner, Mowry & Co. Have a saw-mill, and a number of tenement houses. Is a good lumbering location. No oil developments.

Newtown mill property lies on both sides of Creek, known as the H. Stow property formerly, now as the Stow, Wheeler & Co. property. Contains 6,000 acres. Is twelve miles from the mouth of Creek. This is an excellent lumbering tract, containing a large amount of pine, hemlock, cherry and oak timber. No oil developments on the property. Have two large saw-mills, capable of cutting out millions of feet of lumber per year.

Have eight miles of a tram railway for the transportation of logs to the mill, using horses for the motive power. Surface indications here are good for oil.

Next is the Johnson farm, on both sides of the Creek, containing 960 acres. Now owned by the Tionesta Creek Oil Company of Buffalo, N. Y. Have four wells on the property, one 1074 feet in depth. Known as the deep well of Tionesta Creek—flows about one quart of oil per week. When first struck was supposed to be a good well. The hole is rough and crooked, preventing it from being chambered. If reamed out straight would probably prove productive. Are now drilling a well near it, which has reached the depth of 570 feet. Oil is obtained from the depth of 254 feet from this by means of sand pump, being just beneath a white conglomerate rock. There is one well on Fork Run, a branch of Tionesta, 800 feet deep, non-producing, and one on Salmon Run, 650 feet deep, tubed, but not tested.

Joining the above tract is the property of the Salmon Creek Petroleum Company, on Salmon Creek, containing 500 acres. This is a portion of a 7,000 acre tract belonging to the Marcy Oil, Lumbering & Mining Company, of Philadelphia. Have two wells on the 500 acre tract, one 650 feet deep, non-producing, and one 650 feet deep, now testing. Have plenty of salt water. Company intend to drill to depth of 1,000 feet. The sandrocks of this locality are close, gray and hard. Depth of first sand, 103 feet—6 feet thick; second sand, 248 feet—3 feet thick. Sand and shale rock alternate all the way down. Salmon Creek is a tributary of Tionesta Creek, emptying into it fourteen and a half miles from the mouth of same at Allegheny river.

Next is the May farm, known as the Minister Eddy property, now owned by Schooley & Co., of Philadelphia. Purchased in 1865. Lies on both sides of the

Creek, containing 1,600 acres of land. Is a fine lumbering tract. Have one steam and one water mill. Timber, pine, hemlock, white oak and cherry.

Next is the W. W. Crook property, containing 6,600 acres. Lies on both sides of the Creek. Is a good lumbering tract. Abundance of pine, hemlock, oak and cherry timber. Mr. Crook at his death willed this tract to the Pennsylvania Hospital and Blind Asylum. This property reaches to and beyond the county line.

### TUBS RUN.

On the Dale & Irwin farm, now owned by A. B. Kelly & Co., one and a half miles from the creek; there is one well 600 feet deep, pumping some oil, but not in paying quantities. On the Ford property, one mile further up, is one well, with the tools fast in a mud vein.

### HUNTER RUN.

*Hunter Farm.*—On Hunter Run. Original owners, Wm. & John Hunter. Present owners, Banner Oil Company of Pittsburgh. Contains 200 acres. Purchased in January 1865. Near the river, opposite Tionesta. Bearing of Hunter Run, south of east. Four non-producing wells. Depth of wells from 355 to 800 feet. Average cost, $5,000. Wells located on the flat and owned by the Company. Have four portable and stationary engines. Depth of first sandrock, 200 feet—10 feet thick; second sand, 310 feet—20 feet thick; third sand, 342 feet—13 feet thick; fourth sand, 370 feet—10 feet thick; fifth sand, 420 feet—25 feet thick. We find the sandrock very irregular here. Depth of driving-pipe, 4 to 8 feet.

One well on next farm above, owned by a New York Oil Company, 600 feet deep. Had some indications of oil. Non-producing.

LITTLE TIONESTA CREEK.

*Irvine Farm.*—On Little Tionesta Creek. Original owner, Richard Irvine. Present owners, John Anderson Petroleum Company of New York. Contains 1,500 acres. Purchased in 1865. Half a mile from the river, and two miles from Tionesta Borough. Bearing of Little Tionesta Creek, east of south. Road to Tylersburgh east of south. Have four non-producing wells. Depth of wells, 600 to 700 feet. Average cost of wells, $7,000. Located on the flat and owned by the company. Have four portable engines. Depth of first sandrock 100 feet —8 feet thick; second sand, 250 feet—15 feet thick. Depth of driving-pipe, 5 to 17 feet. Well No. 1 at the depth of 240 feet struck a crevice, and had a fine show of gas and oil. This was so strong that for three nights no lamps could be lighted in the derrick without danger of igniting.

*Edmonson & Sligo Farms.*—On Little Tionesta. Original owner, J. Edmonson and G. Sligo. Present owners, Little Tionesta Petroleum Company. Purchased in December, 1864. Contains 213 acres. Bounded on the north by Neeley, east by Proper, south by Dale, west by ———. Four miles from the river, and six miles from Tionesta Borough. Bearing of Little Tionesta Creek north of west. Road to Tylersburgh east of south, to Hemlock south. Have two wells, one of them not tested, one flowing water; and two testing. Depth of wells, 750 to 865 feet. Average cost, $6,000. Wells located on the flat, and owned by the Company. Have two portable engines. Depth of first sand, as follows: to bed-rock, 35 feet, to regular sand, 103 feet.

The Proper Farm, next above, was purchased by J. T. Hegg, of New York. No developments.

### STEWART'S RUN.

Beginning at the mouth of this Run, is the Thomas M'Calmont farm, now owned by the Farrar Oil Comdany of Boston, and Samuel Duff of Pittsburgh; each owning an undivided half. For a more full description consult the beginning of this chapter. The J. S. M'Calmont farm, which has also been previously described, is the next adjoining, fronting on the Allegheny river, reaching back and across Stewart's Run, at the mouth of Pine Run, and Sugar Camp Run. Hills are high and abound with timber. The flat is about twenty rods in width.

*Taylor Farm.*—On Stewart's Run, in President township. Original owner, —— Taylor. Present owner, S. Q. Brown. Have two non-producing wells. One is 200 feet deep, and has the tools fast. The other is an old well. Have two engines.

*McKinley Farm.*—On Stewart's Run. Original owner, —— M'Kinley. Present owners, Church & Herbert. Number of acres not ascertained. Have four non-producing wells, only one of which has been tested. Depth of wells, from 600 to 700 feet. Cost not ascertained. Wells located on the flat and hill-side. Have three portable engines. This property adjoins the Hood farm on the side of the hill.

*Heckard Farm.*—On Stewart's Run. Original owner, Jonathan Heckard. Present owner, ——— Hutchins. The United States Oil Company has a lease of thirty-six acres. Contains 200 acres. Have one producing well, yielding three barrels per day. Is reported to have produced twenty-five barrels, when first struck. Have one well drilling. Total production, reported, 300 barrels. Doubtful. Average depth, 650 feet. Cost,

$8,000. Commenced to produce in September, 1865. Wells located on the flat and owned by the United States Oil Company. Have two engines. Depth of first sandrock, 94 feet—40 feet thick; second sand, 344 feet—30 feet thick; third sand, 455 feet—20 feet thick; fourth sand, 602 feet—19 feet thick; fifth sand, 635 feet—8 feet thick; sixth sand, 659 feet—23 feet thick; seventh sand, 700 feet—3 feet thick. Depth of driving-pipe 40 feet.

*Fair Farm.*—On Stewart's Run, and at the mouth of Johnson's Run. Original owner, ——— Fair. Present owners, Fair Farm Oil Company. Contains 120 acres. Purchased in the fall of 1864. Have one non-producing well, and one in progress. Average depth of wells, 704 feet. Average cost, $8,000. Wells located on the flat and owned by Company. Have two engines. Sandrocks same as on Heckard farm. Depth of driving-pipe, 40 feet.

*Dawson Farm.*—On Stewart's Run. Original owner, James Dawson. Present owners, Pittsburgh & Stewart's Run Oil Company, of Pittsburgh. Contains 300 acres. Purchased in February, 1865. Bounded on the north by Abbott, east by Savage, south by Butterfield, and west by Roorer. Three miles from Tionesta, three miles from Pithole, three from Howarth, and four from Tyrrel's. Bearing of Stewart's Run, south. Road to Tionesta east, to Pithole west, to Howarth and Tyrrell's northwest. One well has produced a small quantity—100 barrels in all. Have two non-producing wells—one not tested, and the other a failure. Average depth of wells 625 feet. Average cost, $8,000. Commenced producing in November, 1865. Wells located on flat, 80 feet above the level of the United States well at Pithole. Owned by the Company. Have three portable engines. Depth of first sandrock, 119 feet (grey)—18 feet thick; second sand 231 feet (grey)—7 feet thick; third sand, 374 feet (grey)—16 feet thick; fourth sand, 393 feet (grey)—22 feet thick; fifth, 431 feet—19 feet thick; sixth, 455 feet—11 feet thick; seventh sand, 473 feet—(white) 12 feet thick; eighth sand, 508 feet—5 feet thick; ninth sand,—grey and red—524 feet—20 feet thick; tenth, 614—grey—5 feet thick. When first struck, well No. 3 blew gas very strong for twelve hours, when it commenced to produce oil. Hills gently sloping. Not as high as below on the Creek.

*Abbott Farm.*—On Stewart's Run. Original owner, M. Abbott. Present owner, Stewart's Run Petroleum Company, of New York. Contains 180 acres. Purchased in January, 1865. Bounded on the north by M'Kinley, east by Ball, south by Dawson, west by W. J. M'Kinley. Bearing of Stewart's Run, south. Have three non-producing wells. Have tested two. Depth of wells 600 to 700 feet. Average cost, $8,000. Wells located on flat, and owned by company. Have four portable engines. Depth of first sandrock, 119 feet—18 feet thick; second sand, 213 feet—7 feet thick; third sand, 374 feet—16 feet thick; fourth sand, 393 feet—22 feet thick; fifth sand, 431 feet—19 feet thick; sixth sand, 455 feet—11 feet thick; seventh sand, 473 feet—12 feet thick; eighth sand, 508 feet—5 feet thick; ninth sand, 524 feet—20 feet thick; tenth sand, 614 feet—5 feet thick. Depth of driving-pipe, 22 feet.

*M'Kinley Farm.*—On Stewart's Run. Original owner, John M'Kinley. Present owner, Chicago Oil Company. Contains 87 acres. Purchased in January, 1865. No developments.

The Wm. M'Kinley Farm joins on the Run. Property of Ironsides Petroleum Company, and Meadville Company, located here. Have one well drilling. No wells down. Have one engine. Next to this is the Range Farm. No developments.

*Dawson Farm.*—On Stewart's Run. Former owner, J. H. Dawson. Present owners, Queens County Association, of Long Island. Contains 100 acres. Purchased in April, 1865. Have one well, the Remsen, testing. Depth, 606 feet. Cost, $5,500. Wells located on flat and owned by lessees. Have one portable engine. Depth of first sandrock, 102 feet—16 feet thick; second sand, 241 feet—26 feet thick; third sand, 553 feet—70 feet thick. Mixed with slate. Depth of driving-pipe, 58 feet.

*Green Farm.*—On Stewart's Run. Original owner, Robt. Green. Present owners, Jocelyn Oil Land Association, of New York. Contains 350 acres. Purchased in 1865. Bounded on the north by Kelly, east by Rynd & Copeland, south by Siggins & Dawson, west by ———. Four miles from Tionesta, four from West Hickory, and five miles from Pithole. Bearing of Stewart's Run, west of south. Road to Tionesta southeast, to Pleasantville northwest. Have two wells now producing, two testing, and one in progress. Depth of wells 680 and 730 feet. Oil found in both at the depth of 580 feet. Cost $6,000. Wells located on the flat, and owned by lessees. Have three portable engines. Depth of first sandrock, 112 feet—10 feet thick; second sand, 145 feet—16 feet thick; third sand, 309 feet—22 feet thick; fourth sand, 580 feet—white—40 feet thick; fifth sand, 636 feet—22 feet thick; sixth sand, 676 feet—grey and red—34 feet thick.

These wells are on the upper waters of Run. Hills gently sloping. Here are a few buildings, the commencement of a new town, called Jocelyn City. Have three hotels, one church, stores, shops, school-houses, &c.

JOHNSON'S RUN.

*Johnson Farm.*—On Johnson's Run. Original owner, John Johnson. Present owner, Boston Oil Company; Capt. Pray, agent. Own 30 acres of farm. Other owners not ascertained. Purchased in 1865. Three miles from East Pithole, and three-fourths of a mile from Stewart's Run. Bearing of Johnson Run, south. Road to Pithole west. Stewart's Run and Tionesta east. Have one non-producing well, 600 feet deep—tools fast. Cost $7,000. Wells located on the flat and owned by Company. Have one portable engine. Depth of first sandrock, 100 feet—22 feet thick; second sand, 244 feet—22 feet thick; third sand, 584 feet—40 feet thick. Depth of driving-pipe, 18 feet.

*Leedham Farm.*—On Johnson's Run—table lands. Original owner, John Johnson. Present owners, Northern New York Petroleum Company. Purchased in February, 1865. Contains 67 acres. Bounded on the north by Dr. Egbert, east by lands of Providence Petroleum Company, south by ———, west by Switzer. Three miles from Pithole east; three-fourths of a mile from Stewart's Run. Bearing of Johnson's Run south. Road to Pithole City west—to Tionesta east. Have two wells, both blowing gas in large quantities, and one well in progress. Depth of wells 683 feet, and 708 feet. Are to be drilled deeper when the gas can be exhausted. Wells located on table lands above the Run. Owned by the Company.

Have two portable engines. Depth of first sandrock, 146 feet—22 feet thick; second sand, 290 feet—22 feet thick; third sand, 630 to 684 feet—17 feet of hard sand, and 23 of soft rock. Depth of driving-pipe, 18 feet.

These wells are on the table lands, west side of Johnson's Run, and known as the great Gas Wells. They are located 250 feet above the level of Pithole Creek, at the United States well. The first well was drilled to the depth of 630 feet and tested, with but slight indications of gas or oil. It was then drilled through the next hard rock to the depth of 680 feet. Tested it for two days, getting water with but little gas. The sucker-rods were then withdrawn to repair the sucker. After taking out a few rods, the gas began to come out, causing the water to bubble and boil in the tubing. This agitation increased as the sucker, or pumping rods were withdrawn. When the rods were all removed, the water spouted above the top of the derrick, and continued until all the water was exhausted. Then a powerful volume of gas burst forth with a deafening roar, which was heard a distance of five miles, jarring the ground about the well like an earthquake. A pipe being put into the well, and projecting above the derrick, with branches leading off in different directions, carried off the gas. This was ignited, and at night could be seen a distance of twelve miles. The second well, drilled a short distance from the first, is similar in all respects. The company contemplate drilling both deeper when they can overcome the gas so it will be safe to operate them. They are drilling another well 1100 feet distant, and run all the machinery by the gas obtained from these wells, besides lighting up and warming the engine-house and derrick. When the pipes at the well are opened, the gas makes a loud report, like the sudden letting off of steam from the safety-valve of a steam engine under a powerful head of steam. According to early ideas of oil miners, where so much gas is found, there should be a corresponding quantity of oil. But in all subsequent wells of this kind, but little oil has been found, with perhaps a few exceptions. We don't propose to theorize on the subject.

### SUGAR CAMP RUN.

*Vandine Farm.*—On Sugar Camp Run. Original owner, ——— Vandine. Present owner, Oak Ball Oil Company, of Philadelphia. Purchased in the fall of 1864. No developments.

William Willings, adjoining, on Sugar Camp Run and on Stewart's Run. Present owners not ascertained. Contains sixty-six acres. Purchased in the fall of 1864. Have one non-producing well 630 feet deep. Cost $5,000. Wells located on flat and owned by Company. Have one stationary engine. Depth of first sandrock, 130 feet—18 feet thick; second sand, 390 feet—30 feet thick; third sand, 520 feet—40 feet thick. Depth to rock, which is sand—of driving-pipe, 14 feet.

*Hoffman Farm.*—On Sugar Camp Run. Original owner, J. Hoffman. Present owners, Oak Ball Oil Company, of Philadelphia. Contains 100 acres. Purchased in the fall of 1864. Have four non-producing wells. Depth of same, 450 to 563 feet. Cost $5,000. Located on flat and owned by Company and lessees. Have four engines on property. Wells are named as follows: Turing well, Wilcox wells, Nos. 1 & 2, and Oak Ball well. Depth of first sandrock, 170 feet—18 feet thick; second sand, 460 feet—30 feet thick; third sand, 520 feet—40 feet thick. Depth of driving-pipe 14 feet to bed-rock—a gray sand, 20 feet thick.

*Young's Farm*—On Sugar Camp Run. Original owner, ——— Youngs. Present owners, Oak Ball Oil Company, of Philadelphia. Contains 100 acres. Purchased in the fall of 1864. Have one non-producing well, partially tested, 500 feet deep. Cost $5,000. Located on the flats and belong to Company. Have one stationary engine. Depth of first sandrock, 170 feet—18 feet thick; second sand, 460 feet—30 feet thick; third sand, 520 feet—40 feet thick. Depth of driving-pipe, 14 feet.

Switzer Farm, on Sugar Camp Run, owned by ——— Switzer, has four non-producing wells on same. Present owners not ascertained.

### PINE RUN.

At the mouth of this Run, where it empties into Stewart's Run, is one well on the J. S. M'Calmont farm—now Onondaga Oil Company. On the Brown tract is one well 750 feet deep, belonging to the Cicero Oil Company of Syracuse, N. Y. Is non-producing. On the Thos. M'Calmont farm, now Farrar Oil Company, is one well, 450 feet deep, non-producing. Belongs to the Vandyne Oil Association, of Newark, N. J. Pumping but very little oil. Reported out of the region as producing forty-one barrels per day, but only has to be seen to know to the contrary. One well about one mile above, owned by the Pithole & Oil City Oil Company, over 550 feet deep. Are drilling but very little. Had a fair show of oil, at the depth of 520 feet; are 150 feet above level of Stewart's Run. One well still above, belonging to the same Company, 450 feet deep, with tools fast. The company own 1,000 acres or more of land, and have made but little development. The valley of Pine Creek Run is narrow, hills high, growing less in height as it reaches back from Stewart's Run. Empties into Stewart's Run one mile from the river. Two engines above the Vandyne Company's wells.

The new oil development of the two last years, 1867 and 1868, have been less general than in former years, being confined to localities. The principal among these have been Tideoute and vicinity, Shamburg and Pleasantville. We have concluded to give such sketches of these localities as we think will prove of general interest.

### TIDEOUTE AND VICINITY.

*Economy Oil Company.* The large tract of land owned by this company has been previously described. The old wells produced for seven years from the time of their drilling. The Company leased a considerable part of their land on the bluff in the early part of 1867. This tract has since proved highly productive. Total of producing wells at date of Jan. 1, 1869, twenty-two; daily production, 600 barrels. Number of non-producing wells, twenty-two. The bluff is very abrupt, and some 500 feet above the level of the Allegheny river. Upon the top of this bluff, or hill, are the present wells located. The average depth of wells is from 430 to 590 feet, and cost about $3,000; all the wells are put down by contract at $1.25 per foot. All the wells are owned by the Company save those on Hiland's ten acre lease, and part of Ralston & Eckert's. Total production of farm to date 210,000 barrels. Depth of first sandrock, 165 feet—10 feet thick; second sand, 312 feet—18 feet thick; third sand, 408 feet—23 feet thick. Wells all cased to first sandrock, and some of them below. The Company have declined to lease any more at present, and intend to put down a number of wells on their own account in the Spring. The Company have purchased during the last season lease given by them to Fisher Bros., consisting of twenty acres, for $80,000; Eckert's portion of the

Ralston & Eckert lease, twenty acres, being one half of the same, for $60,000.

There is a great discrepancy manifest here regarding the depth at which the sandrocks are found. The old wells on the flats average about 150 feet in depth ; on the bluff, some 500 feet above the river, the third sandrock found at an average depth of 430 to 575 feet. We have noticed no similar feature between bluff and flat wells,in any other locality. Have thirty engines on the farm.

*Noble Tract*, or *Irwin's*—Located on the Allegheny river, adjoining lands of Economy Oil Company, opposite Tideoute. Original owner, ——— Davidson. Contains 800 acres. The property is now in litigation between Noble and Irwin, Noble in possession. Have four pumping wells on the farm, yielding a daily production of sixty barrels, and three non-producing wells. Depth of wells, 500 feet. Average cost, $3,500. Wells located on bluff, owned by Company and lessees. Five engines on property. Sandrocks, same as those on farm of Economy Oil Company. A portion of this farm fronts on the river, making a broad flat.

*Tideoute & Warren Oil Company*—Located on Dennis Run. Since our previous report, some forty wells have been drilled. The new development commenced in August, 1866. Average daily production, since that time, 300 barrels. Twenty-one wells are now producing. Daily production, five hundred barrels. The lands of the Company are being rapidly developed. The lands of the Company originally cost them $10,000. The stock of the Company is owned by ten stockholders. In 1867, the dividends from the property amounted to $15,000 per share. In proportion to capital invested, this is probably the largest dividend ever declared by any Oil Company.

Recently some excellent wells have been obtained on the Jason farm, adjoining the lands of the Tideoute & Warren, and development is progressing rapidly. The lands of the Tideoute & Warren Oil Company lay on both sides of Dennis Run.

*New York & Allegheny Oil Company*—Located just above Tideoute & Warren Company's. Have been previously described. The development and production since 1866 has been considerable. The Superintendent being absent in New York, we failed to get the information so much desired. The territory of this Company is excellent, and when fully developed, will doubtless prove highly productive.

*Triumph Oil Company's Tract.*—Located on headwaters of main branch of West Hickory Creek, and of Dennis Run. Former owner, W. W. Wallace. Present owners, Triumph Oil Company, of Pittsburgh. Contains 239 acres. Twenty-five acres have been sold off to various parties. Purchased in 1864. Have drilled altogether 44 wells. Of these 31 were producing. During a portion of the summer of 1867, the daily production of this farm was 2,000 bbls. per day. Only lasted for about two months. Average depth of wells 700 feet. Average cost $4,000. Total production on main tract 170,000 bbls. On the 25 acre tract, 60,000 bbls. Commenced producing in Nov. 1866, and on to date. Wells are on the hill. Owned by Company & lessees. Depth of first sandrock, 450 feet—20 feet thick; second sand 500 feet—25 feet thick; third sand, from 50 to 60 feet in the rock. Total thickness unknown. Company are taking up the old leases as fast as forfeited, and grant no new ones. Are now sinking a well through the third sand. The wells make no show or sign before being tested. The largest well produced when first struck, by pumping, 400 bbls. per day. About 50 acres of the property have been developed. Only about half a dozen wells are now being operated. Daily production 100 bbls. per day. The highest point of elevation on this farm above the Allegheny river is 610 feet. The city of Triumph now consists of but a few houses, and is a mere semblance of its former greatness. The lands on the hill are rolling, and afford ample room for oil operations. Triumph City is two miles by Run, from Tideoute. Road to Titusville, north of east; to Tideoute, south of west. Bearing of Dennis Run, north and south of east.

At the time of our visit, the weather was extremely cold, and some two feet of snow covered the surface of the ground. As a consequence, operations were inactive. For some time previous, operations had been limited. The excitement that prevailed in this neighborhood in 1867, has left its signs, in the shape of scores of deserted derricks. The wells from the mouth of Dennis Run to the top of Triumph Hill were clustered too thickly together, and flooded each other by dozens. The new fields of Shamburg and Pleasantville having been opened up about this time, a general exodus of the Triumph operators took place. The landowners having made a handsome profit from their royalty, are content with the reversion of their leases, so fast as they become forfeit. By plugging up a portion of the wells, effectually shutting off the surface water, they can obtain a number of excellent wells at a moderate cost. This plan is being fully carried out by the land owners in all former large producing localities. The operator, or one who leases, is content to skim over the surface of the new oil fields that are being opened up from time to time.

On the road to Triumph City, the traveler passes

through the once noted city of Babylon; its first inhabitants indicated by the name. The better class of people in the neighborhood soon rid them out. It is now deserted by all but a solitary saloon keeper.

A large amount of development has been had in the neighborhood of Tideoute in the past two years, with excellent average success. During that period the daily production has averaged from 1,200 to 1,500 barrels. The territory developed is but a small proportion, and the producing area is being constantly enlarged by new developments in the surrounding localities. The main producing localities of the present time are the hills and table lands. Very few wells are in operation on the flats.

### SHAMBURG.

The extension of the development toward the headwaters of West Pithole Creek and its numerous branches, during 1866 & 1867, which resulted in the striking of some fair wells, at differents points, finally culminated, late in 1867, in the striking of a number of wells on the lands of the Pittsburgh and Cherry Run Petroleum Company, the Shamburg Petroleum Company, Tallman and Atkinson farms, at the headwaters of Upper Cherry Run, on the first of the table lands extending beyond Pleasantville. This oil field is about one mile and a half from the Miller Farm Station of the Oil Creek and Allegheny River Railroad, on Oil Creek, three miles southwest of Pleasantville, and six miles from the Allegheny River. The largest production of this locality was between two and three thousand barrels per day. In the Spring of 1868, when the development was at its height, Pleasantville bloomed forth in its full glory, and caused many of the Shamburg operators to commence operations in the new and more promising field. As usual, a considerable town grew up with the first development, containing hotels, livery stables, saloons, and stores of all kinds. While many of the first operators left Shamburg, a large number remained, and as a general thing have achieved a greater average success than has been met with at Pleasantville. Some of the wells in this locality have produced from four to five hundred barrels of oil per day, while one hundred barrel wells have been quite numerous. We have in no place met with a better class of operators. The operations so far have been principally confined to a narrow strip of the table lands, where it forms a basin-like depression on the west, and upon the flat and bluff of Upper Cherry Run, on the east. The first embraces the Tallman and Atkinson farms, and the latter the lands of Shamburg and Pittsburgh and Cherry Run Petroleum Companies. The sandrocks here are clear and well defined, being of hard close texture, as compared with those of Pleasantville. The frequency of mud-veins occasion great annoyance to oil miners. But the most serious obstacle met with, is the crumbling, loose nature of the rocks or slates met with at considerable depths. This annoyance can only be avoided by the use of large casing.

On the eastern portion, on the lands of the Shamburg & Pittsburgh and Cherry Run Petroleum Companies, is the dividing line between the green oil and the black. This line is defined sharply, as if by a plummet, and is a seeming idiosyncracy of nature. The rock in which the black oil is found is thin—15 to 20 feet thick—coarse, and extremely porous in its nature. The rock of the green oil is seldom less than 25 feet in thickness, and generally averages 40 to 50 feet on elevated lands, the third, or last, sand, in all cases except Pleasantville and vicinity, increasing in thickness with the altitude. As, for instance, the oil-bearing sandrocks at the mouth of Dennis Run, Tideoute, is only 15 to 20 feet thick; further up, on Triumph Hill, an elevation of 610 feet above the level of the Allegheny river, the same rock has been drilled to the depth of from 50 to 60 feet without getting through it. The same fact is observable in all other localities, both on river and on the creeks. The wells producing green oil are distant but a few rods from those producing black oil. The depth drilled is about the same average, and the general features met with in drilling exhibit great similarity, except a slight difference of the last regular sandrocks met with, these being coarser and darker than those met in the wells of the so-called "green oil belt." The cause of this coloring of the oil is doubtless owing to the rock containing it. So far as we can learn, the gravity and general nature of the black oil is the same in all general respects as the green oil found over the entire oil region. Some well-informed operators ascribe the coloring of the oil to the presence, at great depths, of oxide of iron. Oxide of iron manifests its presence in the third sandrock, in a number of localities where only green oil is found. The most marked instance of this kind is found in the last sandrock of Triumph Hill, Tidionte.

The general machinery used by operators in this vicinity is the best that can be had. In earlier years, the average height of derricks for the shallower wells of Oil Creek were from 20 to 30 feet. Those of Shamburg, as well as all other localities where deep wells are required, are 56 feet in height, and 24 feet square at the base. In earlier years, the weight of the drilling-tools was about 600 pounds. The weight of the drilling apparatus used for the deep wells of the present day is from 1,600 to 1,800 pounds, and the rest of the machinery is in proportion. The force with which this ponderous mass strikes the rock at the depth of several hundred feet, having the weight of the heavy rope in addition, can be imagined. The extrication of the same when it becomes fastened from some unfortunate accident is comparatively no greater than in the case of smaller apparatus of the same kind, owing to the invention of implements corresponding to the strength of tools and difficulties met with.

The present daily production of the Shamburg Oil field is from 1,200 to 1,400 bbls. per day. The oil is conveyed to the Miller Farm Station by Abbott & Harley's Oil Pipe Line, the main terminus of which is at Miller Farm. The oil is here stored in large iron tanks, being generally purchased by the Pipe Company, at the ruling market prices. Shamburg and vicinity is well located, being convenient to both the creek and railroad, thus afford ample facilities for transportation and obtaining supplies.

On the western portion of the Atkinson farm, the third sandrock is not found, having doubtless cropped out in the first elevation above the valley of Oil Creek, the altitude of this farm above the level of Oil Creek being, we should estimate, some three to four hundred feet. We have no record of the third sandrock being generally found in the valley of Oil Creek, from the Noble well, on the Farrel farm, to Titusville. On Church Run, about two miles from Titusville, it again makes its appearance. The absence of the third sandrock caused a cessation of operations in a Western course, yet we are of opinion it will again be found continuous in a more southerly direction. The table lands of Shamburg are high and rolling, covered in places with heavy timber; pine, oak, and cherry prevailing. Several streams emptying into the Allegheny take their rise in this vicinity, affording ample water, while excellent springs are quite numerous. We give below a detailed description of the principal producing farms.

*Pittsburgh and Cherry Run Petroleum Company Tract.*—Located on the headwaters of Cherry Run. Original owner, O. Stowell. Present owners, Pittsburgh and Cherry Run Petroleum Company, of Pittsburgh, Dr. G. Shamburg being the largest owner of the same. Purchased in 1864. Contains seventy-eight acres. Bounded on the north by lands of Huidekoper Petroleum Company, east by Clark and Brown Bros., south by the A. Clark farm, west by J. R. Tallman farm. Two miles from Miller Farm Station, three and a half miles from Pithole, three miles from Pleasantville, six miles from Titusville, and six miles from the Allegheny River. Cherry Run heads three quarters of a mile north, and runs through the western portion of the farm. Bearing of stream south of east and south of west. The road leading from Plumer to Titusville passes through east end of farm. Have twelve producing wells on farm; no non-producing wells. The two first wells pumped and flowed, but it was considered best to pump them. Four new wells in progress. The first well, the Shamburg, commenced producing February 19, 1866. Total production estimated 300,000 barrels. Average depth of wells, 840 feet. Average cost of wells, $4,500. Wells located principally on the flat. Owned in part as follows: First well, Dr. G. Shamburg; second well, Pittsburgh & Cherry Run Petroleum Company; third well, J. B. Fink, fourth and fifth, Dr. D. Bly; fifth, sixth, and seventh, Dr. Shamburg. Commenced the drilling of one well with pole-tools, but failed to finish the same by that means, the pole-tools proving a failure. Depth of first sandrock, 112 feet—30 feet thick; second sandrock, 290 feet—28 feet thick; the third sandrock, which is considered by Dr. Shamburg—a close observer, by the way—to be the *first sandrock* of the valley of Oil Creek, is found at an average depth of about 520 feet, and is about 25 feet thick. On the southwest corner of this farm, and on the west end of the Clark farm below, this sandrock is not found. Fourth sand is found at 640 feet—25 feet thick; fifth sand, or oil-bearing rock, 775 feet—50 to 60 feet thick. The oil is principally pumped at about twenty feet from the bottom of this rock, except in few wells, that pump their oil from the bottom of the rock. Between the sandrocks are layers of slate-rock, with an occasional layer of soapstone. The third sandrock contains considerable oxide of iron, and by some is called the red rock. The quality of the fifth sand is pebble, mixed with fine white sand. Close to Cherry Run, 72 feet of driving-pipe is used. Some five rods back, 45 feet, and in some places, 27 feet only is required.

We give the following detailed record of well No. 12, on the property of the Pittsburgh and Cherry Run Petroleum Company, kindly furnished by Dr. Shamburg:

38 feet, depth of driving-pipe.
38 feet, soft slate rock.
70 feet first sandrock.
71 feet, water crevice.
91 to 112 feet, crevices.
130 feet, through fine white sandrock, 60 feet thick.
132 feet, gray sandrock, bluish cast.
152 feet, passed through same, 20 feet thick.
153 feet, slate-rock; good drilling to 245 feet.
245 to 256 feet, hard, dark slate and sand to 278 feet.
278 feet, hard pebble sand-shell, 18 inches thick.
280 to 289 feet, hard, gray sand and slate.
289 feet, second sandrock, hard pebble, 11 feet thick.
300 feet, sand, bluish cast, white pebbles, 5½ feet thick.
305½ feet, gray and white shells for 29½ feet.
338 to 440 feet, blue sandy rock, mixed with slate.
420 to 480 feet, blue and red rock, alternate.

505 feet, hard, blue rock shell, 15 feet thick.

520 feet, third sand; very hard, white and yellow pebbles, 10 feet thick.

530 feet, mud vein.

545 feet, through third sand, 25 feet thick; two crevices, and gas very strong.

545 to 575 feet, blue sand and slate to 605 feet.

608 feet, hard shell, 2 feet thick.

610 to 636 feet, blue slate.

636 feet, hard, white sand, mixed with pebble; hard shell, 4 to 5 feet thick.

640 feet, top of the fourth sand.

648 to 654 feet, hard pebble.

654 feet, large gas vein and show of oil.

655 feet, bad mud vein.

668 feet, through fourth sand, 28 feet, 6 inches thick.

745 feet, slate; hard shell, 6 inches thick.

745 to 748 feet, hard slate.

748 feet, hard shell, yellow pebble, and good gas vein.

750 feet, slate rock.

768 feet, slate and hard shells.

776 feet, top of fifth sand.

776 to 778 feet, pebble rock, open and porous.

778 feet, crevice, gas vein, and good show of oil.

781 feet, rock becomes darker.

783 feet, dark rock, gassy.

784 feet, porous rock.

792 feet, white and yellow pebble, crevice, oil and gas.

794 feet, white rock, coarse and porous.

806 feet, mud vein.

828 to 830 feet, white and yellow pebble.

830 feet, hard, close white sand.

834 feet, slate and sand mixed, to 835 feet, bottom of the well.

This proved to be an excellent producing well, one of the best on the farm.

*Shamburg Petroleum Company.*—Located on the

headwaters of Upper Cherry Run, in Oil Creek township. Original owner, Marshall Goss. Present owners, Shamburg Petroleum Company. Contains 100 acres. Bounded on the north by lands of J. E. Benninghoff, east by Widow Lytle and Huidekoper Petroleum Company, south by J. R. Tallman, west by Atkinson. Upper Cherry Run heads on east end of farm, bearing north and south. One mile and a half from Miller Farm Station, two miles from Pleasantville. Road leading from Titusville to Plumer passes through the farm. Total of producing wells, sixteen. All the wells drilled on the farm have produced oil. Two new wells drilling. Commenced producing August, 1867. Total production to date of January 1, 1869, 100,000 barrels, estimated. Average depth of wells, 900 feet. Average cost, $4,500. Wells located on high table lands. Owned by company and lessees. About twenty engines on the farm. Sandrocks about the same as on Pittsburgh and Cherry Run Petroleum Company. Depth of driving-pipe, 25 to 40 feet.

*Tallman Farm.*—On headwaters of Upper Cherry Run, in Oil Creek township. Original owner, John R. Tallman. Present owners, Frank W. Andrews, F. L. Backus, Lyman Stewart, Milton Stewart, C. H. & W. C. Andrews, John W. Irvin. Purchased in 1865 by W. C. Andrews. Contains 110 acres. Bounded on the north by Sheridan farm (or Shamburg), east by lands of Pittsburgh and Cherry Run Petroleum Company, south by A. Poor and Atkinson. Upper Cherry Run passes through farm south. Pithole and Miller Farm plank-road passes through the southern boundary of the farm. Have twenty-eight producing wells, and nine non-producing. All the wells, except one, have produced oil. Daily production at date of January 1,

1869, 400 barrels. Commenced producing in September 1867. Total production to date of January 1, 1869, over 200,000 barrels. Well No. 112, struck in April, 1867, flowed at the commencement 400 barrels per day. Average depth of wells, 860 feet. Average cost, $4,800. Wells located on high table lands. Owned by landowners and lessees. Have thirty-two engines on the property. Depth to top of first sandrock, 90 to 100 feet; to top of second sandrock, 315 feet—20 feet thick; top of third sandrock, 545 feet—25 feet thick; to top of fourth sand, 685 feet—40 feet thick; to top of fifth sand, 810 feet—50 feet thick. Depth of driving-pipe, 25 to 40 feet.

The following estimate of the cost of drilling a well on this farm, in detail, was kindly furnished us by Mr. Stewart, the superintendent:

| | |
|---|---|
| Engine, new, set up | $1,150 |
| Drilling-rig, complete | 550 |
| Driving-pipe, belting, &c | 200 |
| Drilling-contract price | 2,000 |
| Tubing, casing, sucker-rods, &c | 900 |
| Total | $4,800 |

Average time required to drill a well, 40 days.

The cost of running wells on this farm, including contingencies, is from $200 to $500 per month. Gas is used exclusively for fuel. Where coal is required, this amount would be increased, say $500 to $750 per month. This estimate is made for wells pumping twenty-five barrels per day and upward.

The following table shows the sales of oil made from the farm, and the value of same, from February 1, 1868, to January 1, 1869:

| MONTHS. | BARRELS. | VALUE. |
|---|---|---|
| February | 9,313 | $17,632.88 |
| March | 4,948 | 11,348.36 |
| April | 4,584 | 11,288.38 |
| May | 19,222 | 60,500.66 |
| June | 22,791 | 89,504.17 |
| July | 25,403 | 121,113.65 |
| August | 12,859 | 55,881.43 |
| September | 54,372 | 211,478.52 |
| October | 15,445 | 58,538.14 |
| November | 17,754 | 67,744.65 |
| December | 11,914 | 50,668.36 |
| Totals | 198,665 | $755,699.20 |

Average price for 1868, per barrel $3.06.

*Atkinson Farm.*—On branch of headwaters of Cherry Run, in Oil Creek township. Former owner, —— Fleming. Present owners, Sherman, Barnsdall & Co., of New York. Contains about 180 acres. Purchased in 1866. Is one and a half mile east of Miller Farm Station, or Oil Creek, four miles from Pleasantville, Pithole four miles, Titusville eight miles. Bounded on the north by F. H. Bennehoff farm, east by lands of Shamburg Petroleum Company and Tallman farm, south by A. Poore, west by H. Fleming, and S. J. Sutley. Roads, Miller Farm Station west, Pithole southeast, Pleasantville northeast, Titusville, north. Have six producing wells on the farm, and thirty-one non-producing wells. Commenced producing in July, 1867. Present daily production, sixty barrels. Total production of farm to date of January, 1, 1869, 210,000 barrels. All but six of the non-producing wells have produced oil. The Jack Brown and Fee wells have been the most productive on the farm.

The Jack Brown well commenced producing December 27, 1867, flowing 431 barrels per day. The following is the production and value of the same of the well for each month of its existence:

| MONTHS. | BARRELS. | VALUE. |
|---|---|---|
| January | 12,500 | $18,800 |
| February | 12,500 | 20,500 |
| March | 12,500 | 24,600 |
| April | 12,500 | 36,000 |
| May | 9,000 | 20,000 |
| June | 9,000 | 3,700 |
| July | 1,800 | 6,000 |
| August | 1,700 | 8,000 |
| September | 1,500 | 4,000 |
| Totals | 73,000 | $141,000 |

The Fee well commenced flowing in January, 1868, 300 barrrels per day, and averaged for five months 8,000 barrels per month. For July the production was 3,500 barrels; August, 1,500 barrels; September about 100 barrels, and ceased. Total value of product, $133,-500.

Average depth of wells, 880 feet. Vary according to elevation. The Atkinson farm is nearly 400 feet above the level of Oil Creek.

The following is a record of the sandrocks: Jack Brown well No. 2, on lease 108—distance from surface to top of first, or "mountain sand," 100 feet—thickness of same, 100 feet; distance from surface to top of second sand, 310 feet—thickness, 25 feet; distance from surface to top of third sand, 557 feet—thickness 13 feet; distance from surface to top of fourth sand, 769 feet—25 feet thick; distance from surface to top of fifth sand, 817 feet—thickness 44½ feet. Total depth of well, 875 feet. Depth of seed-bag, 325 feet; mud veins, 827 feet; working-barrel, 25 feet from bottom of well. Plenty of gas to run two engines, and daily production of well when struck, 75 barrels. Depth of driving-pipe varies considerably according to elevation. At one well on the flat 100 feet of pipe were required, there being, as was supposed, a deep fissure in the rock. In other places on the farm, from 25 to 35 feet of pipe only is required. Fifteen engines on the property. Wells owned by lessees.

A number of wells will be drilled upon this farm during 1869. It is well located, and has proved highly productive. But a small portion of the farm has been developed.

### PLEASANTVILLE AND VICINITY.

The excitement prevailing upon the first developments at Pleasantville and vicinity, in the Winter and Spring of the past year, 1868, ranks next to Pithole in former years, and in general was characterized by similar features.

A few wells were struck in this vicinity as early as 1865 and 1866, producing, in small quantity, for some two or three years. The story related of one of these, that fish without eyes were brought up in the sand pump from a depth of 500 feet, is, we are inclined to believe, highly apocryphal. The story related of the locating of the first well, the Harmonial No. 1, on the Porter farm, in the borough limits of Pleasantville, that led to the extensive development of 1868, by spiritualistic agency, is stoutly affirmed by the party who claims to have been the instrument selected for locating the famed well. We give it for the consideration of the reader:

During the Fall of 1867, a gentleman named James was journeying from Pithole to Titusville. When arriving opposite the site of the Harmonial well, which is

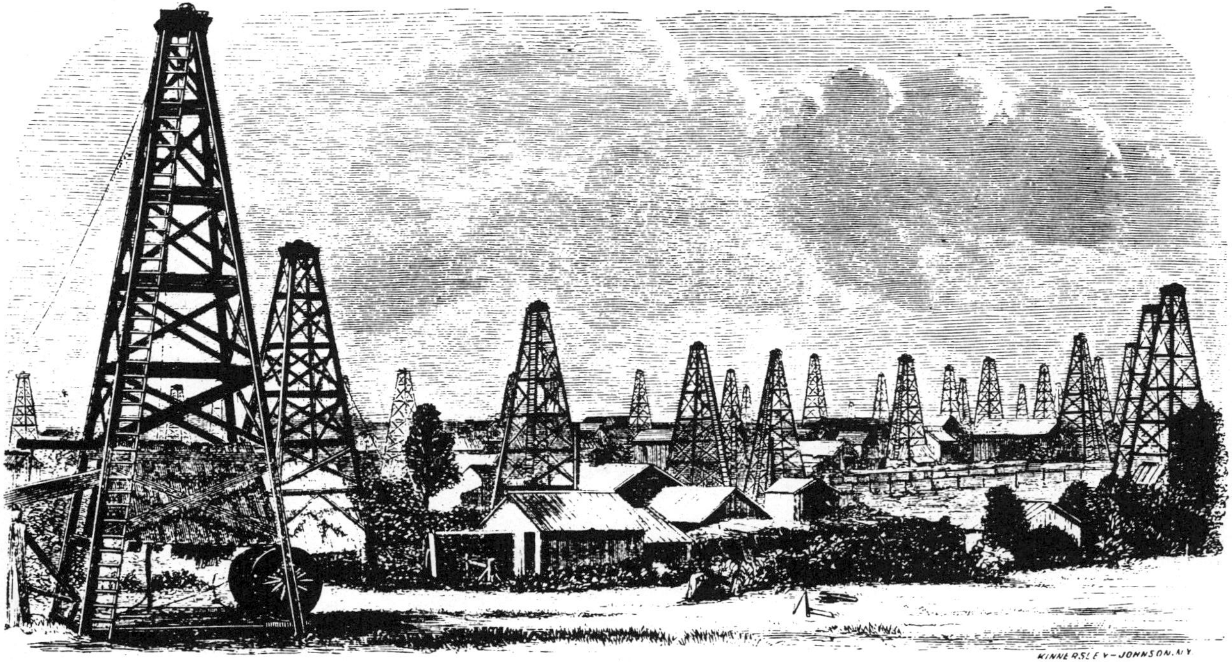


VIEW AT PLEASANTVILLE IN 1868.

convenient to the main road, he became so strongly magnetized by supernatural agency, that he was compelled to alight from his carriage, and was guided to the exact location of the well, by the same mystic power, and the spot for the sinking of the same indicated to him in some unmistakably manner that only those initiated into the mysteries of spiritualism fully comprehend.

We have no evidence that the "spirits" aforesaid supplied the capital necessary to drill the well thus located. But certain it is that Mr. James procured the necessary sum, went to work on the spot thus indicated, and in February, 1868, succeeded in obtaining a well producing at the rate of considerably over 100 barrels per day of black oil, and now ranks among our wealthiest operators.

The striking of the new well created great excitement among oil miners generally, and all who could hastened to secure leases in the new oil field. Of course, no one put any faith in the highly supernatural narrative of Mr. James. But strange to relate, every operator located his well as near to the "Harmonial" as they could get, and were generally rewarded by obtaining excellent wells.

In a short time, scores of new wells in the neighborhood were in process of drilling. Never has there been shown so great a spirit of competition between the operators of the different localities, as was here exhibited. It became an established fact that every well drilled would prove a paying one. Thus secure in making the investment, the operators were each eager to sink the greatest number of wells, and thus secure the largest amount of the liquid treasure. The striking of each new well served to increase the excitement, and by the time Spring had fairly opened, the oil scenes of

Pithole were being reënacted at Pleasantville. The plodding landowners of that thrifty neighborhood sold large amounts of their lands at prices ranging from $500 to $2,000 per acre, and in some instances at double the latter figure. Every road leading to the place was thronged with travellers, while the few hotels afforded scarcely standing-room for the crowds who flocked to them. The prices of engines and well-machinery advanced materially, and all the machine shops of the Oil Region were kept running night and day to fill the orders that were crowded upon them by the eager operators. Property changed hands rapidly. Interests in wells "going down," were freely vended, and but few were willing to be so much out of the fashion as to acknowledge that they owned no interests at Pleasantville. All who ever had any inclination to invest in new oil fields went into the new enterprise with a rush. The long-headed, cautious folks, who in the beginning gave vent freely to unfavorable predictions of the future of the new oil field, waited a brief time, and then went into the speculation too. The only class, within the writer's acquaintance who did not invest at Pleasantville during the memorable season of 1868, were those impecunious mortals who found it impossible to "finance" through a sixteenth interest. During the Summer of 1868, the arrivals at Pleasantville frequently averaged 1,000 persons per day. The excitement culminated late in the summer.

The fourth sandrock at Pleasantville is found all over the entire field at an average depth of about 900 feet. It is from fifteen to twenty feet thick only, and exceedingly coarse and porous in its nature. The successes in this field will doubtless present a greater number of producing wells in proportion to the number drilled, than in any that have preceded it. But the aggregate of production has been less than that of any of the others. The highest daily production of the entire field has scarcely exceeded 2,000 barrels. The present production ranges from 1,500 to 2,000 barrels. So thickly have the wells been clustered together, that the whole supply of oil, or at least a greater portion, in the localities first developed, seems to have been exhausted by the pumps. The wells that have been struck latterly are of but brief duration—generally yielding an average of first production but a few weeks, and then rapidly declining in the yield. Many operators are inclined to the belief that the supply first found "sweated" through the coarse fourth sandrock, being forced up from the depths below, and is now about exhausted. Green oil in any appreciable quantity has not as yet been found. Traces of it have been discovered, it is stated, in the fifth sandrock, and many are preparing to drill their wells still deeper during the present season, hoping to find other and more inexhaustible supplies of oil at a greater depth. The *fifth* sandrock, where found, is considerably thicker, and harder, and closer in its texture, than the fourth. We hope the new venture will prove successful. Operators are still following the extension of the "oil belt" toward Enterprise and to the river, at or about Tideoute, and occasionally get good wells. But by far the greater number are prospecting new oil fields in various portions of the Oil Region.

We have found it about impossible to get an intelligible description of the development on each of the different tracts of oil lands at Pleasantville and vicinity, the same having been cut up and sold in small parcels. We shall, therefore, be compelled to give a detailed description of a number of the principal producing farms.

Considerable additions have been made to the town since the commencement of the oil excitement, both of population, dwellings, and business-houses, and some excellent hotels. Pleasantville is located on the summit of the high table lands, and in point of altitude may be called the summit in this region of country. The land is well adapted to farming purposes, and has been highly cultivated. It has ever been one of our thriftiest inland towns, and its people noted for their wealth, pleasant homes, and high culture. We should estimate its present population at about 2,000 to 2,500. The oil from this neighborhood is transported to Titusville by an oil-pipe line, and an excellent plank-road leads to Titusville, only six miles distant. With a large extent of territory yet undeveloped, Pleasantville bids fair to remain for a long time the centre of a large producing oil field. We give the main producing farms as follows:

*S. Q. Brown, Bates, and others' Tract*—In Oil Creek township. Original owner, A. Dawson. Present owners, S. Q. Brown, Bates, and others. Purchased in 1868. Contains seventy-five acres. Bounded on the north by Brown, Fertig & Hammond, east by Morrison, south by J. H. & H. B. Hebert, west by lands of Vesta and National Oil Companies. One mile from Pleasantville, six from Titusville, and four miles from Miller Farm Station. West branch of West Pithole Creek runs through the property in a southerly direction. Pleasantville and Shamburg road passes through the farm. Eleven producing wells on the property, all pumping. Average depth of wells, 800 feet. Average cost, $5,000. No non-producing wells. Commenced producing in September, 1868. Total production to January 1, 1869, about 15,000 barrels. Wells located on the high table lands. Owned by company and lessees. Have ten portable and stationary engines on the property. First sandrock, of varied thickness, is found at depths varying from 200 to 300 feet; second sand, 490 feet—40 feet thick; third sand, 720 feet—40 feet thick; fourth sand, 800 feet—15 to 20 feet thick. Depth of driving-pipe, 15 to 75 feet. Present daily production, January 1, 1869, 500 barrels.

*Brown Bros. Tract.*—On west branch of West Pithole Creek, in Oil Creek township. Original and present owner, Brown Bros. Contains sixty-four acres. Purchased some thirty years since. Bounded on the north by Byers tract, east by Porter & Wright, south by Armstrong, west by lands of Vesta Petroleum Company and Charles Hyde. Is half a mile from Pleasantville. West branch of West Pithole Creek passes through property in a southerly direction. Road to Pleasantville and Shamburg southeast. Have twenty-nine producing and ten non-producing wells on the property. All the wells have formerly produced oil in paying quantities. Average depth of wells, 800 feet. Average cost, $500. Commenced producing in July, 1868. Present daily average production, 150 barrels. Total production to date of January 1, 1869, 50,000 barrels. Wells located on table lands. Owned by company and lessees. Sandrocks and depth of driving-pipe about same as previously described. In fact, one description of sandrocks will answer for the entire district, the same being generally uniform in general respects. Have twenty-eight engines, portable and stationary, on the property.

*Holeman & Newkirk Tract.*—In Pleasantville borough. Original owner, Thos. Mitchell. Present owners, Holeman & Newkirk. Purchased in September,

1868. Contains twenty-four acres. Bounded on the north by Dunham road, east by Anderson, Carroll & Co., south and west by Pithole and Pleasantville plank-road. Five producing wells on the farm. Present daily average production, 160 barrels. Total production to January 1, 1869, 4,000 barrels. Average depth of wells, 880 feet. Average cost, $5,000. Commenced producing, October, 1868. Wells located on table lands. Owned by company and lessees. Average thickness of sandrocks, 18 feet. Depth of driving-pipe, 12 to 20 feet.

*Brown & House Tract.*—In the borough of Pleasantville. Original owner, Aaron Benedict. Present owners, Brown & House. Purchased in July, 1868. Contains forty acres. Bounded on the north by Watkins, east by Dawson, south by Third street, west by Main street. Five producing wells. Daily production at date of January 1, 1869, 100 barrels. Total production to same date, 2,500 barrels. Commenced producing November, 1868. Average depth of wells, 875 feet. Cost, $5,000. Wells located on table lands. Thickness of sandrock, 18 feet. Have seven engines on the property. Depth of driving-pipe, 20 feet.

*Tyrrell Farm.*—On Dunham Run, in Oil Creek township. Original owner, Benjamin Tyrrell. Present owners, Brown, Watson & Porter. Contains 175 acres. Purchased in 1868. Bounded on the north by Dunham, east by Crossman, and W. D. Byles, south by Lamb, west by Wrigglesworth and Dunham. Is one mile from Pleasantville, southeast. Dunham Run passes through the farm in a southerly direction. Dunham road and Warren and Franklin turnpike bearing east. Have two small wells, with a nominal production. Just commenced active developing. Average depth of wells, 850 feet. Cost, $5,000. Commenced producing in July and August, 1868. Sandrocks are the general average. Depth of driving-pipe, 20 feet. Have four engines on the property.

*Porter Farm.*—In Oil Creek township, borough of Pleasantville. Original and present owner, William Porter. Purchased about thirty years since. Contains forty-four acres. Bounded on the north by lands of Dawson's heirs & Jack, east by Hebbard, south by Wright, and west by Main street. Number of producing wells, twenty, and nine in progress; one well partly drilled and plugged up. All the wells are producing. Commenced producing in February, 1868. Present daily production not ascertained. Total production of farm to date of January 1, 1869, 25,000 barrels. Average depth of wells, 850 feet. Cost, $5,000. Wells located on table lands. Owned principally by lessees. Have twenty-seven engines on the the property. Depth of first sandrock, 350 feet—25 feet thick; second sand, 575 feet—20 feet thick; stray sand, 688 feet; distance from stray to third sandrock, 28 feet; third sand, 715 feet—20 feet thick; mud vein in third sand at 735 feet; fourth sand, 820 feet—21 feet thick.

This was the first producing farm in the Pleasantville oil field, the Harmonial well No. 1 was the first one struck, being located, as alleged, by spiritualistic direction, by A. James. This well commenced producing in February, 1868, at something over 100 barrels per day. Several excellent wells were obtained in the immediate neighborhood of the Harmonial.

*J. H. Jack Farm.*—In Allegheny township, one mile north of Pleasantville. Original owner, Wm. Dawson. Present owners, McGrew Bros. & Co. Contains fifty acres. Purchased in August, 1868. Bounded on the north by Newkirk, east by Dunham, south by Dunham & Morrison, west by Brown and Neillsburg, or township road. The Milltown and Pleasantville road passes through, bearing north. Have eight producing wells on the farm. Production not ascertained. Average depth of wells, 850 feet. Cost, $5,000. Commenced producing in 1867. Eight wells in progress. Wells on table lands. Sixteen portable and stationary engines. Sandrocks, general average. Depth of driving-pipe, 18 to 35 feet.

*Hebert Tract.*—In Pleasantville borough. Original owner, —— Zuver. Present owners, J. H. & H. B. Hebert. Purchased in 1864. Contains sixty acres. Bounded on the north by Jack & Fisher Bros., east by plank-road, south by Anderson & Carroll, west by Wm. Porter. Twelve producing wells, all pumping. Four non-producing wells. Daily production, 150 barrels. Average depth of wells, 860 feet. Cost, $5,000. Commenced producing in August, 1868. Total production to January 1, 1869, 25,000 barrels. Wells owned in part by Fisher Bros., Goss & Curll, Bennehoff Run Company, Dalzell Petroleum Company, J. W. Jenkins, Kernochan & Weed. Have eighteen engines on the property. Second sandrock is found at 490 feet —40 feet thick; third sand, 720 feet—40 feet thick; fourth sandrock, 840 feet—18 feet thick. Depth of driving-pipe, about 50 feet.

*Byles Farm.*—In Oil Creek township. Original owner, Edwin Byles. Present owners, Vesta Petroleum & Refining Company. Purchased in 1864. Contains sixty acres. Bounded on the north by S. Q. Brown, Bates, and others, east by public road, dividing it, and Morrison, south by Turner and Benedict, west by Bates Petroleum Company. Is one and a half miles southwest from Pleasantville. West branch of West Pithole Creek runs through it, bearing southwest. Road from Pleasantville to Plumer on east side. Five producing wells on the farm, all pumping; three non-producing wells. Commenced producing in December, 1867. Total production to date of January 1, 1869, 4,650 barrels. Average depth of wells, 825 feet. Cost, $5,000. Wells as follows: W. & I. N. Porter, Fisher Bros., Woodford and others, Narrington, and others, Samuel Fertig, Bryan, Dillingham & Co., and Galloway. Have eight engines on the property. Second sandrock is found at the depth of 420 feet—40 feet thick; third sand, 700 feet—40 feet thick; fourth sand, 800 feet—20 feet thick. Depth of driving-pipe, 30 feet.

*Beebe Farm.*—In Oil Creek township. Original owner, M. C. Beebe. Present owners, Vesta Petroleum and Refining Company. Purchased in 1864. Contains sixty acres. Bounded on the north by lands of Hyde and others, east by Brown Bros. & Armstrong, south by Geo. K. Anderson and New York and Providence Petroleum Company, west by John F. Carll. Is located about three quarters of a mile southwest of the borough of Pleasantville. West Pithole Creek runs through in a southerly direction. Road from Pleasantville to Shamburg, and Holbrook well. Have four producing wells, and one non-producing. Daily product not ascertained. Commenced producing in July, 1868. Total production to date of January 1, 1869, 2,250 barrels. Average depth of wells, 800 feet. Cost, $5,000. Wells owned by Kernochan, Weed & Co., J. W. Jenkins, A. H. Bronson, and others. Four engines on the property. Depth of second sandrock, 400 feet —40 feet thick; third sand, 680 feet—40 feet thick; fourth sand, 770 feet—12 feet thick; fifth sand, 800 feet—30 feet thick. Depth of driving-pipe, 30 feet.

*Mill Farm.*—On West Pithole Creek, in Oil Creek township. Original owner, E. L. Davis. Present owners, J. H. & H. B. Hebert. Contains thirty-four acres. Purchased in 1864. Bounded on the north by Bates Petroleum Company, east by Vesta Petroleum and Refining Company, south by M. Taylor, west by Syle farm. Is two miles southwest from Pleasantville. West Pithole Creek runs through in a southerly direction. Road from National wells to Shamburg. Have two producing wells, and one old well, non-producing. Commenced producing September, 1868. Total production to date of January 1, 1869, 4,150 barrels. Average depth of wells, 760 feet. Cost, $5,000. Wells owned by Williams, Say & Co, Semkins, Brett & Barr. Have two engines, twelve horse-power. Depth of second sandrock, 360 feet—35 feet thick; third sand, 620 feet—40 feet thick; fourth sand, 740 feet—20 feet thick. Depth of driving-pipe, 40 feet.

*Ensign Farm.*—In Oil Creek township. Original owner —— Ensign. Present owners, Anderson & Carroll. Contains sixty-eight acres. Purchased in 1868. Have seven producing wells. Daily production, 100 barrels, estimate; and six wells in progress. Total production not ascertained. Twelve engines on the property. Depth of wells, 800 feet. Sandrocks, general average.

*Gerow Farm.*—In Allegheny township. Original owner, —— Gerow. Present owners, Anderson, Carroll & Hawes. Contains fifty acres. Has one producing well—the Hawes; estimated daily production, seventy-five barrels. Have nine wells in progress, and ten engines.

*Marcy Farm.*—In Oil Creek township, on the dividing tp. line. Original owner, —— Connelly. Present owners, Anderson, Carroll & Bates. Contains 100 acres. Have just commenced developing.

*Drake Tract.*—In Oil Creek township. Present owner, G. K. Anderson. Contains twelve acres. One producing well, Blue Flag well, estimated as yielding twenty barrels per day, on the tract, and two in progress. Have three engines on the property.

*Armstrong Farm.* — In Oil Creek township, near borough line of Pleasantville. Original owners, Brown Bros. Present owner, John Armstrong. Contains 100 acres, more or less. Thirty-four acres sold to J. Hinkly, of Westfield, N. Y., in 1864. Purchased in 1863. Bounded on the north by Brown Bros., east by Pleasantville road, south by J. C. Burtis and N. C. Bates, west by lands of New York and Providence Petroleum Company, Bronson Tract and Vesta Petroleum and Refining Company. Is three quarters of a mile southwest of Pleasantville. West branch of West Pithole Creek runs through lower part of farm. Thirty-five wells have been drilled on the farm. Number producing at date of January 1, 1869, thirty-one, and twenty-nine wells in progress. Commenced producing in Spring of 1868. Total production to date, 48,881 barrels. Average depth of wells, 800 feet. Cost, $5,000. Sandrocks general average.

*S. M. Dunham Farm.*—In Allegheny township.

Original owner, E. Dunham. Present owner, S. M. Dunham. Purchased by E. Dunham in 1819. Contains seventy-four acres, with allowance, seventy-four acres in all. Bounded on the north by Guild & Wright, east by E. Dunham, south by Anderson & Carroll and E. Dunham, west by lands of Collins Bros., J. L. Connelly, S. Harsh, and A. W. Brown. Adjoins the borough of Pleasantville. Dunham's Run passes through the southern portion of the farm. Twenty-seven acres were sold off the west part of this tract in the Summer of 1868 to Meade & Lowrey. Three wells have been drilled on the property, one of which is on the Meade & Lowrey tract. The Magee well produced when first struck, as estimated, 150 barrels per day. A dozen of wells are reported in progress on the farm. Average depth of wells, 800 to 900 feet. Sandrocks very coarse and pebbly. Depth and thickness about the general average. Total production not fully ascertained, estimated at 5,000 to 6,000, to date of January 1, 1869. Commenced producing in December, 1868. The original Dunham farm, from which this and the other Dunham tracts are taken, contained 250 acres. Two wells are in progress on the E. Dunham tract. No development as yet on the E. C. Dunham tract nor the S. M. Dunham, adjoining.

*National Petroleum Company's Tract.*—On West Pithole Creek, in Oil Creek township. Present owners, National Petroleum Company, of New York. Contains fifty acres. Have six producing wells, all pumping. The old Pitcher well commenced producing February 8, 1866, at the rate of about eighty barrels per day. Is still pumping at this date, January 1, 1869, twenty to thirty barrels per day. Daily production, 100 barrels. Total production not ascertained. Average depth of wells, 780 to 800 feet. Depth of first sandrock, 310 to 330 feet—20 feet thick; second sand, 521 to 545 feet—24 feet thick; third sand, 636 to 660 feet; fourth sand, 739 to 759 feet—15 feet of dark pebble rock. Depth of wells varies according to elevation. The absence of the superintendent prevented the obtaining of full information.

*Lyle Farm.*—Adoining the lands of the National Petroleum Company on the south. Contains 113 acres. Former owner, Captain Lyle. Present owner, Samuel Mitchell, of New York City. Four producing wells on the farm, and five in progress. Present daily production, 100 barrels. Commenced producing in September, 1868. Sandrocks, depth of wells, &c., about the same as on National Petroleum Company's tract.

The foregoing embraces the majority of the principal producing farms of Pleasantville and vicinity, and are as full in detail as our time and opportunities could make them. A large number of the owners of tracts and farms we visited were absent, and we could find no one on the ground to furnish the desired information. And, as before stated, we could not spare the time to communicate with the numerous absentees, scattered as they were throughout the length and breadth of the land, and await their answers. We have given ample to serve as landmarks. We have found no other locality where so many small parcels of main tracts were purchased by operators and speculators. This latter class being numerous, and as a general thing non-resident, the reader can judge the difficulty in obtaining accurate details.

## VALLEY OF OIL CREEK.

Though it has undergone many changes within the last few years, and its chief glories, the huge flowing wells, that spouted forth from 1,000 to 4,000 barrels of oil each per day, vanished into the realms of the shadowy past, Oil Creek and its tributaries is still preeminent as an oil producing locality, and stands peerless among its numerous rivals. Rich in its resources of mineral wealth, and with an historic fame that no future can dim, we gladly hail it as the chief source of the Petroleum supply of the present day. The main supply of Petroleum has been obtained from the locality above named, and, in the aggregate, a larger amount of wealth has been realized from the farms bordering on Oil Creek and its tributaries, by at least one half, than all the other producing localities combined. Though at present a number of the farms bear a dismantled, deserted appearance, its territory has retained its old reputation for excellence. No other locality can present more substantial successes, or cases where the investment has been more often realized with princely profits. The local causes that have operated to render so many of these farms comparatively non-producing at the present day, can be overcome by steady application and skill, when the scarcity of supply and remunerative prices warrant the outlay necessary for their successful recuperation.

Nearly the entire flat portion of the lands on the Creek and tributaries have been literally drowned out with the water that, in its normal state, was located from 300 to 500 feet above the oil measures. The drilling of so large a number of wells in a limited space was, in a great measure, the cause of the flooding of the territory, the usual custom in former years being, when the well proved unproductive, to pull up the tubing and seed-bag, and allow the water to flow into the oil measures below. Hundreds of wells being thus left, will account greatly for the general ruin that resulted in so many localities. The exodus of operators from the old producing localities to the new, as the latter have been successively opened up, has likewise been a prolific source of evil. The landowners, as a class, have regarded the desertion of their lands not as a loss, but as a gain. The leases are made for a term of years, and generally, by the terms of the lease, if the well is allowed to remain idle for any considerable length of time, such cause works a forfeiture of the lease. The landowner having generally received a handsome profit on his investment from the royalty, can usually turn the deserted leases to good account, by plugging up or closing such number as he may see fit, and successfully work the remainder, having his land unincumbered by leases. The Story farm now has only ten lessees, the remainder of the wells having been forfeited by their former owners.

On several of the farms, within the last three years, the bluffs have been partially developed, with excellent success. The most noted localities are Blood and Buchanan farms. The hill farms, the Stevenson, Wood & Benninghoff farms, at Petroleum Centre, were among the first developments on elevated lands. The hills and bluffs are now generally preferred by operators. This latter class of territory is sufficiently ample for the development of the present century at least.

The present daily production of the locality under consideration is over double that of any other producing locality in the Oil Region. The low prices of the previous three years have prevented the working to advantage of scores of small pumping wells. A steady price of four to five dollars per barrel at the wells, would render the working of even a well producing three barrels per *diem* profitable. Preparations have been and are now making to commence pumping a large number of this class of wells during the coming season. This, together with the aid of a number of new wells in same localities, will materially increase the daily production of the famed valley.

We have previously given the main history of these farms. Of the entire Petroleum product, from the commencement to the present date, the farms bordering on Oil Creek and its tributaries have supplied, we estimate, three fourths of the amount. During some seasons, in the first year of its development, the locality we speak of has yielded a daily production estimated at from 12,000 to 15,000 barrels. The waste that has occurred in all the years of production, all over the Oil Region, has in no place presented an aggregate equal to that of Oil Creek. The seventh of the entire production for the successive years, we consider a moderate estimate for the loss.

We still regard the future of Oil Creek valley as promising to maintain an average production of several thousand barrels per day. The operations on many of the farms since 1867 has been limited, and the production merely nominal. On others, the production has been kept up to a full average. We shall essay to give, in brief, the operations had upon the principal farms.

### PIONEER RUN.

The production in the neighborhood of Pioneer Run was quite large in the Summer and Fall of 1866, culminating during the Winter of 1866 and 1867, and the Spring of same year. The low prices ruling during 1867 rendered the working of the smaller wells unprofitable. The developments of the new field at Tideoute and Shamburg took away a large number of the operators. By the Fall of 1867, operations had become limited in this neighborhood, and at present the numerous derricks are deserted, and silence reigns where a few years since all was activity and bustle. The average daily product during its best season, of the district known as Pioneer Run, was estimated at about 2,500 barrels. The wells averaged all the way from 25 to 200 barrels per day in production. We have no means of definitely ascertaining the total production, which is variously estimated at 150,000 to 175,000 barrels. Arrangements are making for the operating of a number of the old wells during this spring. The number of wells we are unable to ascertain fully. The sandrocks, &c., of the locality are spoken of elsewhere.

*Upper and Lower M'Ilhenny Farms.*—The upper M'Ilhenny farm embraces a considerable portion of Pioneer Run, at its mouth. The extent of operations for the last year has been extremely limited, and the production slight. Several good wells were in operation on the upper M'Ilhenny during 1867. The lower M'Ilhenny farm adjoins the Boyd farm, and fronts Oil Creek for its entire length. Both farms contain some 180 acres. Total production of both farms, from December 31, 1864, to December 31, 1869, 349,785 barrels. Total number of wells drilled to same date, 175. Total number of wells drilled from the first development to date of December 31, 1866, 139. Of these, sixty-three were known to have produced more or less oil, ten were worthless, thirteen never completed, and fifty-three so old that the superintendent making the report, could give no account of them. Very little doing on these farms at present, and production merely nominal.

The bluff portion of the farm, which is well located for oil development, has not as yet been operated on to any material extent.

*John Benninghoff Farm.*—This farm lies on the opposite side of Oil Creek from the M'Ilhenny, fronting on the Creek and running back to a considerable distance. This farm produced largely during portions of 1867 and early part of 1868. Total production for 1867 and 1868 estimated 150,000 barrels. Commenced to fall off in production in Spring of 1868. Has not been doing much for a year past. Has lately commenced producing again, several of the old wells having been cleaned out and pumped, with good success. One old well lately started up is yielding eighty-one barrels per day. Present daily production of the farm, about 300 barrels.

The Benninghoff Run tract, which adjoins this farm, and the Warner farm, both famed producing localities in 1866 are doing nothing now, having been deserted by the majority of its operators since 1867. The production on each of the tracts exceeded at one time in 1866, 1,000 barrels per day. Total production of these localities not ascertained.

The Patterson farm, on Bull Run, yielded largely in 1866, as also did the Farrel and Skinner farms, in same locality. All these farms began to fail in production during 1867. The production for 1868 of the entire locality embraced by these farms was light. Considerable development is contemplated on Bull Run and vicinity during the present season. The entire production of the locality, we should estimate at 100,000 barrels.

*Stevenson Farm.*—This farm which is located on the hill back of and adjoining Petroleum Centre was a famous producing locality in 1866 and part of 1867. Nothing of consequence is doing on the farm at present, most of the wells having been rendered unproductive by being flooded with water from each other. This is a hill farm, located at an altitude of from 300 to 400 feet above the level of Oil Creek. Estimated total of entire production, about 200,000 barrels.

*Woods Farm.*—This farm is located on the hill just back of Petroleum Centre, about one mile from Oil Creek, at an altitude of about 500 feet above the level of Oil Creek. Was purchased in 1865 by Thomas Woods, of Pittsburgh, from John Sherman, and contains seventy acres. Bounded on the north by Stevenson farm, east by lands of Central Petroleum Company, south and west by lands of Bartlett and Canusett Oil Company. Have forty-six producing wells on the farm, at date of January 1, 1869. Average of these from ten to fifty barrels each. Twenty-four non-producing wells. All but four of these have been productive. Had one flowing well on the farm. Struck July 5, 1867, and commenced with a daily production of 150 barrels, and contined to yield for over one year. Is still being pumped. Commenced producing in 1866. Total production of farm to date of January 1, 1869, 525,000 barrels, as estimated. Average depth of wells, 875 to 900 feet. Cost of wells, about $4,000. Wells located on the hill plain. Owned by Woods and lessees. Have sixty-two stationary and portable engines on the property. Average depth of sandrocks as follows: first sandrock, 570 feet—40 feet thick; second sand, 720 feet—2 to 20 feet thick, irregular and shelly; third sand, 850 feet—40 to 50 feet thick. Present daily production, January 1, 1869, from 350 to 400 barrels. Depth of driving-pipe, 18 to 25 feet.

The farm is well located for oil purposes, and has been among the best producing farms on Oil Creek. Though what is called well developed, so far as the number of its wells are concerned, there are quite a number of acres on the back part of the farm that have not as yet been developed. The water used by the engines is pumped from Oil Creek, and conveyed to the wells by iron pipes. The oil is conveyed to the receiving tanks on the railroad by same means.

The Canusett Oil Company's lands, consisting of about thirty acres, adjoin the Woods farm on the west. A number of wells have been drilled on this tract, but the success so far has been any thing but flattering. The production we were unable to ascertain, but is very small. This tract is about on a level with the Woods farm, if any thing portions of it are somewhat higher. The whole surface of the hill is extremely rough and stony, and a few years since was covered with a heavy growth of timber.

*John Pearson Farm.*—Located in Cornplanter township, one mile from Petroleum Centre, and one and a half miles from Cherry Tree. Original owner, John Pearson. Present owners, Western Pennsylvania Oil Company. Purchased in 1865. Contains eighty-three acres. Bounded on the north by lands of Claremont Oil Company, and east by Woods and Canusett Oil Company. Cherry Tree Run passes through a portion of the farm, bearing south. Roads to Petroleum Centre east, to Story farm south, Cherry Tree north. Have fifteen producing wells on the farm, and one non-producing. Commenced producing about September 1, 1868. Total production to date of February 1, 1869, 37,391 barrels. Shipped to same date, 33,381 barrels. Average depth of wells, 940 feet on the hill, and 780 feet in the valley. Average cost of wells, $4,500 to $5,000. Wells located on the flat and on hill. Owned in part as follows: Bronson & Harrington, five wells; Vera & Blake, one; Joseph Overy, one; Blood & Pennington, Panton & Newton, J. Adams, Thomas Frothingham, R. W. Evans, —— Klinordlinger. Daily production in January 1869, 269 barrels. Average depth of sandrocks: first sand on the hill, 625 feet; in the valley, 470 feet; third sand on the hill, 890 feet; in the valley, 780 feet. Average thickness of third sandrock, 40 to 50 feet. Depth of driving-pipe, from 5 to 20 feet. Twenty engines on property. The wells averaged daily as follows for January, 1869: Lease No. 1, Bronson & Harrington, struck January 8, 1869, 21$\frac{1}{2}$ barrels; lease No. 2, December 3, 1868, Bronson & Harrington, 30$\frac{1}{2}$ barrels; lease No. 3, Vera & Blake, September 17, 1868, 77 barrels; lease No. 4, Joseph Overy, December 14, 1868, 53 barrels; lease No. 5, same, October, 1868, 3$\frac{1}{2}$ barrels; lease No. 9, Ward & Wait, November 1, 1868, 6 barrels; lease No. 14, Blood & Pennington, 3$\frac{1}{4}$ barrels; lease No. 15, Bronson & Harrington, 4 barrels; lease No. 17, Panton & Newton, 3$\frac{1}{4}$ barrels; lease No. 29, J. Adams, 6$\frac{1}{3}$ barrels; lease No. 44, Bronson & Harrington, 25$\frac{1}{3}$ barrels; lease No. 45, Thos. Frothingham, November 8, 1868, 28$\frac{1}{3}$ barrels; lease No. 46, R. W. Evans, 4 barrels; lease No. 34, Klinordlinger, 3$\frac{1}{4}$ barrels. The average of lease No. 35 not given in superintendent's report. The daily average for January, 1869, 269 barrels. This farm gives promise of a large production, and is well located. Developments are in progress on a number of the adjoining farms, a majority of which have been previously described, the course of operations being, as a general thing, in the direction of Cherry Tree Run. A number of small wells have been in operation at various points between the Pearson farm and the mouth of Cherry Tree Run, where it empties into Oil Creek at the Rynd farm. The general average of these is from one to five barrels per day each. Many operators are of opinion that the so-called "oil-belt," extends from the Woods hill along the hills back of Oil Creek, to Charley and Shaffer Runs, at Oil City. With this view, about all the farms located along and back of the range of hills along Oil Creek, have been purchased, and are now being held by operators awaiting events. Some development has taken place on the Lichtenthaler farm, adjoining the Story, which is supposed to be in the "belt," with only tolerable success. The largest well yet obtained is reported at twenty-five barrels per day. A series of developments, at various points between Pearson farm and Charley Run, has been commenced, which by the beginning of summer will fully test the territory.

*Hyde and Egbert Farm.*—Operations have ceased on this farm, in fact there have been none of consequence for two years past. All the wells have been affected to such a degree by water as to be worthless, except at considerable expenditure. The desire of the landowners to regain full possession may have considerable to do with the present state of inactivity. In times past this farm has been among the most noted for a large production, and we can see no valid reason why it may not prove measurably so again.

Nothing of consequence is being done on the Macrae farm, adjoining the Hyde & Egbert. The Dalzell & Hays farm, just below and adjoining the Hyde & Egbert are practically deserted.

For the operations of the Story farm, or more properly the Columbia Oil Company, of Pittsburgh, to whom it belongs, we refer our readers to the admirable report of the company for 1868, which will be found in the chapter on Statistics and Results.

*Tarr Farm.*—From date of our report in 1866 to January 1, 1869, the production of this farm has varied all the way from 150 to 1,200 barrels per day. The flat and a portion of the bluff has become flooded, and the production of the present is from 150 to 200 barrels per day. When means are devised to rid the numerous wells of the surplus water, the farm will again take rank among the first-class producing ones. For aggregate production, see *Herald* report, in Chapter on Statistics and Results. The Phillips and Woodford wells, noted in the history of Petroleum, are located on this farm. The Phillips was struck in October, 1861, and commenced flowing at the rate of over 4,000 barrels per day. A stop-cock was applied shortly afterward, litigation having ensued between the working and landed interests, regarding a clause in the lease compelling the lessees to furnish the royalty, or land-owner's share, in barrels. Oil at this time was worth twenty cents per barrel, and barrels from $3.50 to $4.00 each. In December, over two months after it was struck, the Phillips well flowed by actual measurement, in twenty-four hours, three thousand nine hundred and forty barrels of oil. The total production of this well to date of Jan. 1, 1869, is estimated at 500,000 barrels, and the Woodford at 100,000. These two wells, it will be remembered, flooded each other at intervals for several years. The total production of the Noble well, as estimated, we have given at 1,000,000 barrels. The Noble, when first struck, was thought by competent judges to be about as large in point of production as the Phillips. It was not interffered with by other wells, and flowed without interruption from January, 1863, to Feb. 28, 1865. From this we are inclined to the belief that its production was fully one third, and may have been one half, greater than the Phillips.

*Blood Farm.*—For two years previous to December 24, 1867, the date of the striking of the first well on the bluff, operations on the farm had been merely nominal. The flat portion adjoining the Tarr farm, in 1861 and 1862 boasted of *thirteen* flowing wells, and a daily production almost equalling that of the entire Oil Creek valley, of the present date. One by one the wells became unproductive from being flooded by water and by other causes, and were deserted. The company owning the farm, or at least the one owning the present producing portion, with an energy highly creditable, concluded to test the bluff portion of the farm. Operations were commenced in the Fall of 1867, and as before stated, the first well was struck December 24, 1867. All the wells are on the hill or bluff, which is elevated from 300 to 400 feet above the level of Oil Creek. Total number of producing wells at date of January 1, 1869, twenty-nine. Non-producing wells, two. Total production from December 24, 1867 to January 1, 1869, 90,422 barrels. Daily production at date of January 1, 1869, 540 barrels. Highest daily average previous, 620 barrels. Average depth of wells, 800 feet. Number of wells in progress, seventeen. Number of engines on the farm, forty-eight. Average depth of sandrocks: top of first sand, 563 feet—thickness, 35 feet; top of second sand, 615 feet—24 feet thick; top of grey rock, 716 feet—15 feet thick; top of third sand, 762 feet—38 feet thick. With a large extent of territory yet to develop, the owners of this property can reasonably hope for abundant prosperity for years to come.

The Rynd farm, next below, has never been classed among the large producing farms of Oil Creek, though the development has been large. Development since 1866, considerable. Particulars, not definitely ascertained.

Widow M'Clintock (or J. W. Steele), farm, next farm below the Rynd, has been comparatively idle since latter part of 1866, until past season. The farm is now held by trustees of creditors of former owner. Several of the old wells have been cleaned out and started up again, and a few new wells drilled. The average production of the farm during 1868 has been from forty to seventy-five barrels. Gives promise of yet proving a good producing farm.

*John M'Clintock Farm.*—Located on same side of Creek, and adjoining Rynd farm. Formerly had many good wells, the Excelsior, struck in 1863, flowed several hundred barrels per day. The development has been extensive. Total production to date not fully ascertained. Commenced to develop largely several years ago upon the portion of the farm bordering on Cherry Run, and were highly successful. Wells became watered during 1867 and 1868. Production previous to this, from 150 to 200 barrels per day. Have commenced to develop on hill part of farm, with good prospects.

*A. & J. Buchanan Farms.*—Within last two years active operations have been transferred from the flats to the hill-portions of these farms, the entire flats having become so watered as to be nominally worthless for production. Production on the hills commenced in 1867. For 1867 and 1868, the daily production has averaged from 400 to 500 barrels. Total number of wells on both farms, new development, to date of January 1, 1869, sixty-five, and quite a number in progress. The total production from first commencement we have been unable to fully ascertain. (See *Herald* Report.)

*H. M'Clintock Farm.*—Operations on this farm since 1866 have been very limited. In commencement of 1868, a number of the old wells were cleaned out, and have since yielded a considerable amount by pumping. The Hebbard well, No. 1, struck several years since, and which has yielded largely, is now pumping twenty to thirty barrels per day. The M'Kinley and others of the old wells are producing from three to five barrels each per day. A number of wells have been drilled during last year on the upper or hill part of the farm, known as Hamilton Run, adjoining the line of the A. Buchanan farm, with excellent success. Present daily production of farm, 80 to 100 barrels. The prospect is good for a large and successful development on the farm during the coming season.

*Cornplanter Tract.*—But little has been done for three years past on this tract. It is a narrow strip lying between the M'Clintock and Clapp farms, and has been excellent producing territory. The desire of the company to whom it belongs to secure the leases granted in former years, has effectually checked active operations. One or two of the wells on the Creek bluff are being pumped. Production merely nominal.

*Clapp Farm.*—This farm adjoins the Graff, Hasson, or Petroleum Farms Association, on which a portion of Oil City, including Cottage Hill, is built. Very little doing on the farm at present. The owners have secured a majority of the old leases and plugged up the wells with water lime, cement, gravel, and sand. The Williams and Stanton wells, on the Creek bluff, were good, yielding a large production. But little change since report of 1866, which will be found in former description of the farm.

### OIL CITY AND VICINITY.

*Hickman Farm, Shaffer Run.*—On Shaffer Run, in Cornplanter and Sugar Creek townships. Original owner, Simeon Hickman, by settlement. Present owners, Hasson, Cornwall & Co., of Oil City. Contains 200 acres. Purchased in 1867. Bounded on the north by M'Fate, east by Fee, south and west by lands of Reno Land and Oil Company. Is one and a half miles nearly west of Oil City, and about the same distance from Reno. Shaffer Run passes through the farm bearing southwest. No roads. Reno railroad passes through property. Number of producing wells, five. Daily production, 100 barrels. Non-producing wells, two. The Hasson well, struck in November, 1867, commenced to flow about 300 barrels per day. Is still producing. Total production not ascertained. Average depth of wells, 780 to 820 feet. Cost, $5,000. Depth of first sandrock, 468 feet—48 feet thick; second sand, 622 feet—25 feet thick; third sand, 760 feet—20 feet thick. The above record was of the shallowest well, 780 feet in depth. The depth of the wells varies according to altitude. Wells located on Run and bluffs of the same. The flat along the Run is very narrow. The farm consists chiefly of hill land. Promises to be an excellent producing farm. The wells are owned by lessees and the landowners. Convey the oil from wells to large iron tanks at mouth of Charley Run, on the line of the Atlantic and Great Western Railway by an oil pipe running over the hill, a distance of nearly a mile. No extensive development prevailing at present.

*Lackawanna Oil Company's Tract.*—On Shaffer Run. Adjoins the Hickman farm. Original owner, not ascertained. Present owners, Lackawanna Oil Company. Purchased in 1864. Have two producing wells, located a few rods from the Hasson well. The first of these was struck in ——— 186 . Largest average production, 150 barrels per day. Both wells are still producing. Present daily production ——— barrels. Depth of wells and sandrocks similar to those on Hickman farm. This tract extends across the valley of the Run.

*Oil City Land Company's Farm*, formerly Nevins, or Michigan Rock Oil Company.—On upper part of Charley Run. Have eight producing wells. Present daily product, 250 barrels. No non-producing wells. Twenty-five wells in progress. Average depth of wells, 750 to 800 feet. Estimate of total production to date, 25,000 barrels. Commenced producing July, 1868. The Blakeley, or Gas well, was the first of any size struck. The gas was forced out with such violence as to be heard for a considerable distance, after it had been tested for a time. A majority of the wells struck at first flowed. The largest well commenced at 100 to 150 barrels. The show of gas in this locality equals any portion of the Oil Region in former years. Operations are active on the farm. A wagon-road from Oil City to the wells, up Haliday Run, has been constructed recently. The farm consists chiefly of hill lands. The hill tops are gently rolling, giving ample room for operating. From fifteen to twenty engines on the property. Average depth of first sandrock, 460 feet, irregular—15 to 20 feet thick; second sand, 610 feet—15 to 20 feet thick; third sand, 730 feet—25 to 30 feet thick. The altitude above level of Allegheny River, about 300 feet. A number of old wells on Haliday Run are on this property, and will be found in our former description.

The Moran farm adjoins this property on Charley Run. Have just commenced to drill a number of wells on the same. This farm extends along the Run to its mouth at the Allegheny River.

The Cornen farm on the hill adjoining the lands of Oil City Land Company, lies between Charley and Haliday Run, or rather the headwaters of these streams. A few leases have been let, and a number of wells will be commenced as soon as the season opens.

On the Foster farm, on headwaters of Haliday Run, a few wells have been drilled with indifferent success so far. The farm will be thoroughly tested during the coming season.

A few small producing wells are found at the head of and along Haliday Run, the largest scarcely exceeding five barrels per day.

Two new wells have been drilled during the last year on the lands of the Latonia Town and Oil Company, in Venango City, at the edge of the hill at the rear of the town, and one near the old Clarion road, back of Imperial City. In the latter a large amount of gas was obtained, with very little oil, and in the two former not enough of any thing to justify operations. Sage Run, which empties into the Allegheny just above the borough line of Imperial, now included in Venango City, was tested slightly several years since near its mouth, and only a few small wells obtained. Preparations are making for development on an extensive scale in the spring. Several oil stock companies own large tracts of land on different portions of the Run.

Upon the Lee, Moran, Farran, Hayes, and Huff farms, occupying the river front from Venango City to a point opposite Reno, but slight operations have been had since 1866. During that time a number of wells were drilled on the Lee and Farran farms, a number of which were productive. Very little doing now, except pumping a number of the wells by "heads," that is one or more times during the twenty-four hours, for an hour or more. The wells yield from one to five barrels each. Upon the Huff farm was obtained in former years an excellent quality of lubricating oil of about 35° gravity, which sold at the wells for twenty-five dollars per barrel.

On the Siverly and Alcorn farms, on the Allegheny, above Oil City, a number of wells have been drilled during last two years with some success, several of the wells commencing with a daily production of fifty to seventy-five barrels. A few small wells are being operated at present. Upon the Reno property a number of wells are being pumped, and arrangements are making for the drilling of some new wells when the season opens. All the above-mentioned farms are considered good for oil production. In the coming season the development on these will be considerable. The total of wells drilled in and around Oil City, upon the various farms named as being in the vicinity, since 1866, will not exceed one hundred in number.

ALLEGHENY RIVER FARMS.

Upon none of the farms bordering on the Alle gheny River above Franklin, except at Oil City, have three been any material operations since the years 1865 and 1866. Only here and there are any of the numerous wells drilled in former years being operated. On several of the farms below Franklin, between that point and Scrubgrass, at Emlenton and Parker's Landing, active operations have prevailed for several years. A number of new wells have been drilled during the last year on the Hoover and Cochran farms, and many of the old wells are still producing. The present production of the Cochran farm is estimated at forty barrels per day, and that of the Hoover at ninety barrels. From the Cochran and Hoover to the Foster farm, six miles below Franklin, nothing of consequence is being done at present. A legion of oil companies are interested in the fee of these intervening farms, both of those bordering upon and lying back of the river for some distance, on each side, all the way to Parker's Landing, forty-five miles below Franklin, and even beyond, as well as on the numerous tributary streams emptying into the Allegheny. As before mentioned, at several of these farms operations have been active, especially so for the year 1868. The increase of development along the entire lower Allegheny, during 1869, promises to be large. In fact, the tendency of general development is from Pleasantville to the Allegheny River, at Tideoute, at the northeastern portion of the Oil Region, and from Oil City to Parker's Landing, the present limit, fifty miles below, at the southwestern limit. Every locality for the latter entire distance, that has been thoroughly tested within the last year, has been found of undoubted excellence for oil production. The sandrocks present great uniformity, the wells are of fair size, and are above the general average, even on Oil Creek, in longevity. The depth of wells on the flats and bluffs is about the same average as in the Oil Region generally. These favorable features are exceedingly attractive to oil operators, as is also the large extent of the territory affording such signs of abundant production. We give below the prinicipal producing points:

### SCRUBGRASS.

The farms adjoining and opposite the mouth of Scrubgrass creek, where it empties into the Allegheny, have been profitably operated for several years. This point is generally known as Scrubgrass, although the post-office title is Witherup. It is eighteen miles below Franklin, eight miles from Fosters. The Allegheny Valley Railroad passes through these farms, on the east side of the river. The scope of territory here is ample, and so far as tested, of undoubted excellence. The principal production commenced here in 1865. The farms bordering on the river between Foster's and Scrubgrass, are regarded as good oil territory, and preparations are being made to fully test them. Witherup —or Scrubgrass, as it is generally called—is a thriving little village. Several stores and shops appear to be doing a thriving business. There are two hotels, one of which, the Gregory House, is new, and is a first class establishment. A ferry crosses the river here, connecting with the road leading to Kennerdell's Mills. We give the following description of the principal producing farms:

*M'Millin Farm.*—On Allegheny river, east side, Rockland township. Original owner, Daniel M'Millin. Present owners, M'Millin Farm Oil Company, of Philadelphia. Contains 190 acres. Has two miles of river front. Purchased in 1864. Commenced producing in 1865, and ever since. Total number of wells, thirty-three. All but three of these have produced oil. One well, No. 10, produced, when struck, over 300 barrels per day, by flowing. Daily production from commencement has averaged from 60 to 300 barrels. Average depth of wells on river, 600 feet; on the hill, 700 to 750 feet. Cost, $4,000. Depth of first sandrock on the river, 330 feet, and 400 feet on the hill—white, and 90 feet thick. 2d sandrock, 440 feet on the river, and 500 feet on the hill—30 feet thick. 3d sand, 585 to 590 feet on the river, and 710 to 720 feet on the hill—18 to 20 feet thick. Depth of driving-pipe, 8 feet on the river, and 40 feet on the hill. Wells located on flat and hill. Owned by lessees. This farm is bounded by the Allegheny river and the property of the Philadelphia and Boston Petroleum Company. Have 16 engines on the property. Bearing of Allegheny river, all points of the compass as it passes through farm. Some of the principal wells are owned as follows: Nos. 10, 3, and 7 by Rittock & Squiers; No. 4, A. M. Hoover; No. 99, P. T. Ridgway. The village of Scrubgrass, or Witherup, is located on this farm. The flat here is of moderate extent, hills high and abrupt to river in most places. A. V. R. R. station is located on this farm. Some development is being made at Kennerdell's Mills, on Scrubgrass creek, two miles west of south direction, and on opposite side of River from M'Millin Farm.

*Belle Island.*—In Allegheny river, just below Scrubgrass station, on A. V. R. R., and opposite M'Millin Farm. Patented by Dr. McCoy. Present owners, Belle Island Petroleum Company. Contains about eight acres, bounded by river. Is four and a half miles in a direct line from Foster's, and nine miles by river. Twenty-three miles by river from Reno, and eighteen from Emlenton. Commenced producing in 1867. Non-producing wells, 4; producing, 3. All the wells have produced oil by pumping and flowing. Average depth of wells, 598 to 612 feet. Average cost, $3,000. Daily production not ascertained. Total production to date of January 1, 1869, 26,788 barrels. Sandrocks similar to M'Millin Farm on river. Wells owned principally by company and C. D. Angell. This has been highly productive territory, paying from ten to forty per cent. a month on the first investment.

*Philadelphia and Boston Petroleum Company's Lands.*—On east side of Allegheny river, in Rockland Township. Formerly known as Tract No. 4, of the Bingham Land Estate. Present owners, Philadelphia and Boston Petroleum Company. Contains 262 acres. Purchased in 1864. Total number of wells, 6, all producing in July, 1868, and on to date. Bounded by M'Millin Farm and river. Average depth of wells 900 feet. Cost, $5,000. All the wells located on the hill at an elevation of about 400 feet above the river. Daily production not ascertained. Depth of first sandrock, 575 feet—70 feet thick; second sandrock, 720 feet —40 to 50 feet thick; third sandrock, 875 feet—20 feet thick—a coarse white pebble. Producing wells owned as follows: Nos. 12 and 13, Bartlett, Wilson & Co.; Nos. 10 and 11, Thomas McDonough Oil Co.; No. 15, Bartlett & Co.; No. 14, Bartlett Oil Co. Main portion of the farm lies on top of the hill, is level, and favorable to operate upon. So far as developed, this territory has proved highly productive. Depth of driving-pipe varies from 12 to 100 feet. Sandrocks also vary according to elevation. Total production about 13,000 barrels.

*Tract No.* 1, Bingham Land Estate, at Foster's, on Allegheny river, is also owned by this company. Original owner, Bingham Land Estate. Located on east side of river, in Rockland township. Contains 75 acres.

Purchased in 1864. Bounded on north by river, east by Coal City property, west by P. McCalmont, and south by Geo. Snider. Total of wells, 12, all producing. Daily production, 60 barrels. Aggregate of production to date of January 1, 1869, 50,000 barrels. Average depth of wells, 600 feet. Cost, $3,000. Depth of first sandrock, 300 feet—20 feet thick; second sandrock, 440 feet—25 feet thick; third sandrock, 595 feet—16 feet thick. Dig to the rock for driving-pipe. No flat. A. V. Railroad passes along river front of property. Commenced producing in 1865. Is but slightly developed, and promises well for a large future production.

*Foster Farm.*—On north bank of Allegheny River, in Sandy Creek township. Original owner, —— Foster Present owners, Foster Farm Oil Company, of Philadelphia. Purchased in 1864. Contains 230 acres. Bounded on the east by lands of Miller Farm Oil Company, north by J. Foster and others, west by lands of Excelsior Oil Company. Six miles south of Franklin. Allegheny River bears west. No regular roads. Have twelve producing wells on the farm. The Rogers well, lease No. 4, was struck in March, 1865, and commenced to flow at the rate of 300 barrels per day. Is still flowing twenty barrels per day. Well on lease No. 5, after flowing for two years, commencing at twenty barrels per day, was torpedoed ten months since, and has yielded a daily average of fifty barrels ever since. Have nine non-producing wells; all have produced some oil. Two of them have only been drilled to the second sandrock. The total production of the farm is estimated at 175,000 to 200,000 barrels. Average depth of wells, 610 to 725 feet, according to altitude. Average cost, $3,500 to $4,000. Commenced producing in 1865. Depth of driving-pipe, 40 feet on river bank. On the bluff none. Have thirteen engines on the property. Average depth of first sandrock, 160 feet—16 to 18 feet thick; second sand, 240 feet—20 feet thick; third sand, 598 feet—12 to 16 feet thick. There is but little flat land on the farm. The bluffs come abruptly to the water's edge. The wells are located chiefly on the bluff, and are owned by lessees.

*Foster's Island.*—Is in the Allegheny River, opposite the Foster farm. Owned by Foster Island Oil Company. Has two producing wells, and five non-producing. Daily production, fourteen barrels. Commenced producing in 1865. Total production estimated at about 50,000 barrels. The Island contains from four to six acres.

*Miller Farm.*—Adjoins Foster farm, on same side of river. Is owned by Miller Farm Oil Company, of Philadelphia, and a New York Oil Company. Contains about 150 acres. Some fifteen wells have been drilled on the farm; most of them in 1865 and 1866. One of these, the Buckholtz well, struck in 1865, flowed at the commencement 150 barrels per day. Has never been considered very productive. The daily production at present is from ten to fifteen barrels. Cost, depth of wells, and sandrocks, about the same as on the Foster farm above.

*Lewis, Bonsall & Co's. Tract*—Is part of Bingham Land Estate, on south side of Allegheny, in Rockland township. Former owners, Bingham Land Estate. Present owners, Lewis, Bonsall & Co., of Philadelphia. Purchased in 1864. Contains about 120 acres. Bounded on the north by Allegheny River, east by lands of Granite Oil Company, south by Highfield and Hamby, west by lands of Snider and Venango Central & Duck Creek Oil Company. Is six miles south of Franklin. Allegheny River bears west. Have three producing wells. Two non-producing. One of these was struck two years since, and commenced to flow 300 barrels per day. Was burned soon after being struck. The rope to which the tools were attached burned off, and the tools dropped into the well. These were taken out after several months' labor, and the well produced 20,000 barrels of oil. In sinking the well deeper, a short time since, the tools became fast again. Are now engaged in taking them out by means of Luther's Patent Well Rods. These rods are made of two and a quarter inch iron, and are very heavy. They are fastened to the top of the refractory tools, when practicable, and the same pulled up, after being jarred loose. When this cannot be done, the fastened tools are unscrewed, piece by piece, the rods being arranged to act as a left-handed screw. By means of suitable gearing attached to the top of the rods, either the bull-wheel or the full power of the engine can be attached. This is the most effective apparatus yet devised for extraction of tools from wells. The owner of the patent rods contracts with owners of wells to extract tools for a specified sum, and the apparatus seldom fails. Have one well drilling on the property. Allegheny Valley Railroad passes through property on river bank west. Public road to Freedom passes through portion of tract bearing southeast. Total production, 30,000 barrels. River front of farm, abrupt bluff, railroad forming only flat portion. Average depth of wells, 620 to 720 feet, according to altitude. Average cost, $4,000. Commenced producing in April, 1866. Wells on the bluff. Owned by Lewis, Bonsall

& Co. Have five engines on the property. Depth of first sandrock, 310 to 370 feet—20 feet thick; second sand, 455 to 540 feet—28 feet thick; third sand, 610 to 700 feet—12 feet thick. Two feet of the third sand, at the top is a beautiful transparent and variegated pebble, extremely hard, being in appearance pure quartz. Depth of driving-pipe on river bank 8 feet. On bluff, 60 feet.

The hills are gently sloping, and the property has proved excellent so far as developed.

*Snider Farm.*—On south bank of Allegheny River, in Rockland township. Former owner, John Snider. Present owners, Venango Central & Duck Creek Petroleum Company. Purchased in 1861. Contains about fifty acres. Bounded on the north by the river, east by Lewis, Bonsall & Co., south by John Hamby, west by Ogden Mining Company. Six miles south of Franklin. Allegheny River runs through, bearing west. Railroad west. Road to Freedom southeast. Have seven producing wells on the farm. Daily production, eighty barrels. Thirteen non-producing wells. Most of them have produced some oil; one not tested. Commenced producing in 1861. Total production estimated at over 50,000 barrels. Wells owned by company and lessees. Depth of wells, cost, sandrocks, &c., about same as on Lewis, Bonsall & Co.'s tract. Bluff on river front, abrupt to water.

*Excelsior Oil Company's Tract.* (Smith farm.)—On north bank of Allegheny River, Sandy Creek township. Former owner, David Smith. Present owner, Excelsior Oil Company, of Philadelphia. Purchased in 1864. Contains from 150 to 200 acres. Bounded on the north by Highfield and J. Foster, east and south by river, west by lands of Foster Farm Oil Company. Five and a half miles south of Franklin. Bearing of Allegheny River south and west. Road to Franklin over the hill northwest. No producing wells at present. Seven or eight of their wells formerly produced. Commenced producing in 1863 or 1864. Total production, estimated, 4,000 to 5,000 barrels. Wells only drilled to second sandrock. Average depth of wells, 475 feet. Located on the flat, and owned by company. Average cost, $3,000. Have two engines on the property. Depth of first sandrock, 300 feet—16 to 18 feet thick; second sand, 460 feet—20 feet thick. Nothing doing on the farm.

*Ogden Mining Company's Tract.* (Snider farm), —Former owner, Snider. Present owner, Ogden Mining Company, of New York. Purchased in 1864. Contains sixty acres. Bounded on the north by river, east by lands of Venango Central & Duck Creek Oil Company, south by Hamby & Snider, west by Philadelphia & Boston Petroleum Company. Six miles south of Franklin. Bearing of Allegheny River, west. Roads, railroad and public road to Freedom passing through. Have one producing well, two barrels per day. Eleven non-producing wells. Wells only drilled to second sandrock, so we were informed. Total production estimated at 500 barrels. The company, we were told, was originally formed to mine for coal on this tract. Could not learn of any success on their part. The one well pumping is all sign of life visible on the tract at present.

## PARKER'S LANDING.

Parker's Landing, the present centre of an extensive oil field, judging from present appearances, is one of the principal steamboat landings on the west bank of Allegheny River, fifty-two miles below Oil City. The development is confined at present chiefly to the Parker farms, three in number, and containing about 300 acres, and the Pittsburgh and Tom's Run Petroleum Company's lands, containing about 200 acres. The general appearance of the country at this point is similar in general respects to the other portions of the river valley of the tortuous Allegheny. High hills, the altitude of which decrease slightly with the ascent, with occasional valleys or flats, as they are called, the hills being cut through or divided by numerous ravines. The wells are located on the flats, ravines, and bluffs, or sides of the hills. The first well in this locality was struck in October, 1868. In the Summer and Fall of 1868, a number of good wells were struck. Since then the excitement has been considerable, and a large number of operators are now engaged in development. In addition to the wells already drilled and producing, at the time of our visit, thirteen new wells were in different stages of completion and more are being commenced daily. The following is a list of the present producing wells in the locality during the month of February, 1869: Colfax well, 25 barrels per day; Grant well, 6 barrels; Red Bank Oil Company, 8 barrels; J. & S. D. Carnes, 12 barrels; West Sunbury Oil Company, two wells, 18 barrels; Clarion and Allegheny River Oil Company, two wells, 25 barrels; Sherland Oil Company, two wells, 28 barrels; all the above wells on Parker farms. On Pittsburgh and Tom's Run Oil Company's lands are the following: Company's two wells, 15 barrels; Polecat, 4 barrels; Hovey well, 8 barrels.

Average depth of wells, 830 to 850 feet. Average cost of same, $3,500 to $4,000. Depth of first sandrock, 212 feet—15 to 20 feet thick; second sand, 540 feet—15 to 20 feet thick; third sand, 800 to 820 feet—30 to 35 feet thick. The third sand found here is a beautiful white pebble, mixed with quartz. Depth of driving-pipe, 16 to 20 feet. About forty engines are on these properties.

At Foxburgh, or Foxton, on east side of Allegheny River, just above Parker's Landing, are two producing wells, with a daily production of about ten barrels. The Colden well, on opposite side of river, is producing a small amount. Foxburgh is in Clarion County.

On the Crawford farm, at Emlenton, in Venango County, is one producing well, small, and two more drilling.

At Brady's Bend, on the Allegheny, below Parker's Landing, are several small producing wells. Average depth of wells deeper than those above.

A number of farms, upon the Clarion River, were purchased by oil stock companies during 1864 and 1865. The development on these has not been sufficient to test them thoroughly. The same is the case with the numerous tributaries of the Allegheny, from Parker's Landing, and even below, to its headwaters. It is reasonable to assume that all these farms will be thoroughly tested during coming years.

### CHURCH RUN,

Comes into the northern portion of the city limits of Titusville, and passes through a portion of the same. The wells are located about one mile up the Run. The principal developments are on the Alcorn, Weed, and M'Knight farms, embracing in the aggregate from 1,000 to 1,200 acres. Development was commenced in the first years of Petroleum discovery, but never became extensive until within the last few years. The number of wells producing and non-producing we estimate at about 100. Total of producing wells and production not ascertained fully. Producing wells at present average from ten to thirty barrels each. Daily production of entire locality, about 450 barrels. The longevity of wells on Church Run, is greatly above the average. The Eureka well, owned by Atlantic & Great Western Oil Company, commenced producing over three years ago, at the rate of seventy-five barrels per day. By exploding a torpedo in the well last spring, the production was increased to 280 barrels. Is still producing about thirty barrels per day. Depth of wells, from 575 to 650 feet, according to elevation. Cost, $4,000. Depth of first sandrock, 225, feet—60 feet thick; second sand, 495 feet—15 feet thick; third sand, 500 feet—60 feet thick, hard and pebbly. Some twenty-five to thirty new wells have been commenced, and a large extent of the territory will be developed during 1869. The reappearance of the *third sandrock* of Oil Creek at this locality is one of the marvels which only patient investigation can satisfactorily settle. Of its extent we can as yet form no correct idea. It possibly may have some connection with that found in New York State, at Cuba, and in portions of Chautauque and Cattaraugus counties. It is remarkable that so far as traced the oil-bearing sandrocks have one general direction, northeast and southwest. The streams tributary to the Allegheny River have this same general bearing throughout the oil-producing region. The source or fountain has not been clearly defined, but will be in a few years of development. We shall not be surprised if Church Run and vicinity proves as abundantly productive as any of the oil fields that have preceded it. At any rate, its capacity for oil production will be fully tested during the coming season.

This ends the chapter on the oil farms of the Petroleum Region. The impossibility of procuring the exact details of operations upon each farm, must be apparent to all who have any knowledge of the reckless manner in which the business was transacted during the first several years following its commencement. Enough has been ascertained, however, to give the reader not only a correct idea of the development and results of same in each locality treated of, but also of the general development, in the entire Oil Region, its characteristics, and the general facts that an operator or even the stranger would be desirous to learn.

The labor of obtaining the facts, descriptive of the various farms, has been great. We have only been actuated by a desire to present such details as would prove a guide to the oil operator, or as information to parties wishing to make investments. The general description given of each farm and locality will enable both classes to commence understandingly. The tabular statement which follows, gives a general summary of the entire development from the commencement:

| Original Owner of Farm. | Present Owner of Farm. | No. of Acres. | No. of Producing Wells. | No. of Non-Producing Wells. | No. in Progress. | No. Flowing Wells. | No. Pumping Wells. | Total No. of Wells. | Average Daily Production. | Average Depth of Wells. | Average Cost of Wells. | No. of Engines. | Depth of First Sand Rock. | Depth of Second Sand Rock. | Depth of Third Sand Rock. | Depth of Fourth Sand Rock. | Thickness of First Sand Rock. | Thickness of Second Sand Rock. | Thickness of Third Sand Rock. | Thickness of Fourth Sand Rock. | Depth of Driving Pipe. | No. of Refineries. | Weekly Capacity Bbls. Crude. |
|---|---|---|---|---|---|---|---|---|---|---|---|---|---|---|---|---|---|---|---|---|---|---|---|
| F. Holliday........ | F. Holliday and Midas Pet. Co.... | 500 | 3 | 5 | .. | .. | 1 | 8 | .. | 650 | $5000 | 8 | 120 | 430 | 570 | .. | 25 | 20 | 40 | .. | 15 | | |
| J. Nevins........ | ................................ | ... | 6 | 4 | .. | .. | 6 | 10 | 35 | 650 | 5000 | 6 | 120 | 380 | 495 | .. | 25 | 20 | 40 | .. | 12 | 2 | 500 |
| T. Moran........ | T. Moran, north side river........ | 400 | 3 | 1 | .. | .. | 3 | 4 | 2 | 560 | 5000 | 2 | 202 | 352 | 469 | .. | 29 | 25 | 40 | .. | 7 | | |
| L. H. Piaget........ | Oil City Pet. Co.................. | 147 | 5 | 3 | 1 | .. | 5 | 9 | 20 | 500 | 6000 | 6 | 215 | 350 | 475 | .. | 25 | 50 | 22 | .. | 20 | | |
| Jas. Wilson........ | Reno Oil and Land Co............. | | | | | | | | | | | | | | | | | | | | | | |
| J. Shafer........ | " " " .......... | 290 | | | | | | | | | | | | | | | | | | | | | |
| Ormsby........ | " " " .......... | 916 | 10 | .. | 50 | .. | 10 | 60 | 50 | 600 | 5000 | 10 | 230 | 400 | .. | .. | 20 | 25 | .. | .. | 12 | 2 | 500 |
| Brannon........ | Pembrook Oil Co................ | 40 | 4 | 1 | 2 | .. | 4 | 7 | 12 | 400 | 4000 | 6 | 220 | 400 | 583 | .. | 20 | 30 | 15 | .. | 30 | | |
| C. Shirk........ | Powell Oil Co., Eureka Oil Co., Shirk Farm Oil Co., Great Western Consolidated Oil Co. | 217 | 1 | 6 | 12 | .. | .. | 19 | 3 | 500 | 4000 | 4 | 110 | 245 | 483 | .. | 25 | 30 | 30 | .. | 20 | | |
| W. Hiiands........ | Hoffman Pet. Co., N. Y........... | 400 | .. | 13 | .. | .. | .. | 13 | .. | 400 one 1041 | 5000 | 10 | 248 | 490 | 1000 | .. | 30 | 30 | 20 | .. | 37 | 1 | 200 |
| A. Plummer........ | Plumer Oil and Refining Co., Mutual Interest Pet. Co., and Enterprise Co. | 400 | 5 | 3 | 2 | .. | 5 | 10 | 41 | 400 to 600 one 900 | 3000 | 7 | 252 | 540 | 440 | .. | 27 | 20 | .. | .. | 34 | 1 | 250 |
| Martin & Epley.... | Lycoming and Clinton County Oil Co., Fay and Great Northern Oil Co. | 22 | 8 | 8 | .. | .. | 8 | 16 | 23 | 300 to 542 | 6000 | 9 | 280 | 510 | .. | .. | 40 | 30 | .. | .. | 40 | 1 | 1200 Lubricating. |
| Franklin Borough.. | ................................ | ... | 3 | 30 | 17 | .. | 3 | 50 | 10 | 120 to 700 one 1090 | 2000 to 5000 | 20 | 205 to 465 | 502 to 715 | .. | .. | 28 | 30 | .. | .. | 35 | 3 | 620 |
| Blakeley........ | N. B. Mosley & Co., Phila., Murphy & Irwin. | 230 | .. | 8 | .. | .. | 1 | 8 | .. | 400 | 4000 | 1 | 75 | 209 | .. | .. | 8 | 25 | .. | .. | 39 | 1 | 250 |
| Hastings........ | Pacific Oil Co., N. Y., A. W. Raymond. | 20 & more. | 2 | .. | 1 | .. | 2 | 3 | 2 | 500 to 600 | 4000 | 3 | 320 | 442 | 600 | .. | 25 | 20 | .. | .. | 28 | | |
| Hannon........ | Hoover Oil Co., Phila., and Eureka Oil Co. | 14 | 6 | .. | 7 | .. | 6 | 13 | 6 | 525 | 5000 | 13 | 275 | 438 | .. | .. | 45 | 18 | .. | .. | 24 | | |
| Hoover........ | Geo. H. Bissell & Co., J. P. & C. M. Hoover and Cameron Petroleum Co. | 272 | 15 | 15 | .. | .. | 15 | 30 | 150 | 460 | 5000 | 15 | 300 | 440 | .. | .. | 40 | 25 | .. | .. | 45 | | |

| | | | | | | | | | | | | | | | | | | | | | | | |
|---|---|---|---|---|---|---|---|---|---|---|---|---|---|---|---|---|---|---|---|---|---|---|---|
| Snake Island | Phila. Co. | ... | 1 | 1 | .. | .. | 1 | 2 | .. | 440 | 5000 | 1 | | | | | | | | | | | |
| Wm. Porter | Porter Farm Oil Co., Pittsburgh | 100 | 1 | 4 | .. | .. | 1 | 5 | 5 | 500 | 5000 | 3 | 160 | 355 | .. | .. | 35 | 25 | .. | .. | 45 | | |
| Porter Island | Porter Island Oil Co. | ... | 1 | .. | .. | .. | 1 | 1 | 20 | 365 | 5000 | 1 | | | | | | | | | | | |
| Widow Pope | Pope Farm Oil Co., Phila. | 90 | 5 | 5 | .. | .. | 5 | 10 | 50 | 400 | 5000 | 5 | 200 | 400 | .. | .. | 20 | 15 | .. | .. | 36 | | |
| Smith farm | | ... | .. | 6 | .. | .. | .. | 6 | .. | 550 | 5000 | 2 | | | | | | | | | | | |
| Miller | | ... | 8 | 7 | .. | .. | 8 | 15 | 50 | 600 | 5000 | 8 | 150 | 350 | 600 | .. | 20 | 15 | 15 | .. | 4 | | |
| Amos Martin | Tolpehocan Oil Co., Honey Comb. Oil Co., and Suffolk & Venango Oil Co. | 160 | 6 | 4 | .. | .. | 6 | 10 | 44 | 450 | 4000 | 6 | 180 | 420 | 640 | .. | 40 | 40 | .. | .. | 33 | | |
| Rice | Butler County Co. | 130 | 2 | 7 | .. | .. | 2 | 9 | 4 | 600 | 5000 | 3 | 160 | 355 | .. | .. | 35 | 25 | .. | .. | 48 | | |
| Nickland | Mammoth Mining Co. | ... | 6 | 2 | .. | .. | .. | 8 | 15 | 500 | 4000 | 6 | 300 | 450 | .. | .. | 30 | 35 | .. | .. | 32 | | |
| Irvine | Penn. Oil & Coal Co., Balt. | 5–101 below on river | 1 | 2 | .. | .. | 1 | 3 | 10 | 480 | 3000 | 2 | 300 | 435 | .. | .. | 40 | 15 | .. | .. | 25 | | |
| Cochrane | Cochrane | 1011 | 14 | 5 | .. | .. | 14 | 19 | 90 | 450 | 4500 | 3 | 285 | 440 | 575 | .. | 50 | 20 | 5 | .. | 54 | | |
| Hoover | River Oil Co. of Phila. | 100 | 1 | 4 | 1 | .. | .. | 6 | 4 | 500 | 3500 | 2 | 330 | 440 | .. | .. | 20 | 20 | .. | .. | 40 | | |
| Bissell & Stewart | Cameron Pet. Co. of Phila. | ... | .. | 2 | 10 | .. | .. | 12 | .. | 500 | 4000 | 10 | 300 | 445 | .. | .. | 40 | 25 | .. | .. | 30 | | |
| Fuller | Fuller Farm Oil Co. | ... | .. | 7 | .. | .. | .. | 7 | .. | 550 | 3500 | 3 | 60 | 470 | .. | .. | 4 | 9 | .. | .. | 25 | | |
| Boothe | Simmons, Jacobs & Booth's heirs. | 60 | 1 | 2 | .. | .. | 1 | 3 | 12 | 500 | 3500 | 2 | 330 | 440 | .. | .. | 20 | 20 | .. | .. | 40 | | |
| J. Chambers | | ... | .. | 2 | .. | .. | .. | 2 | .. | one 800 | | | | | | | | | | | | | |
| P. Bowser | | ... | .. | 4 | .. | .. | .. | 4 | .. | 300 | | | | | | | | | | | | | |
| J. Milton | Higate Petroleum Co. | ... | .. | 2 | .. | .. | .. | 2 | | 400 | | | | | | | | | | | | | |
| Hogue Island | Hogue Island Lubricating Oil Co. | 45 | 2 | 2 | .. | .. | 2 | 6 | 8 | 350 | 3000 | 2 | 240 | 620 | .. | .. | 32 | 40 | .. | .. | 37 | | |
| P. Meely | Penn. Pet. Co. and Great Northern Oil Co. | 150 | .. | 3 | .. | .. | .. | 3 | | | | | | | | | | | | | | | |
| J. Plowman | Middletown Lubricating Oil Co. | ... | .. | 3 | .. | .. | .. | 3 | .. | ... | ... | 2 | | | | | | | | | | | |
| J. & S. Huff | F. Prentice (leased to Ballard & Filly). | 40 | 6 | .. | 5 | 1 | 5 | 11 | 10 | 550 | 5000 | 10 | 233 | 370 | 495 | .. | 32 | 32 | 30 | .. | 19 | | |
| J. P. & A. Hays | Buchanan Oil Co. and Artesian Oil Co. | 120 | 6 | 7 | .. | 1 | 5 | 13 | 60 | 525 | 5000 | 6 | 240 | 360 | 485 | .. | 25 | 25 | 25 | .. | 30 | | |
| J. Farran | Buchanan Farm Oil Co. and Passaic Oil Co. | 175 | 6 | .. | 9 | .. | 6 | 15 | 120 | 525 | 4000 | 15 | 225 | 360 | 475 | .. | 30 | 35 | 30 | .. | 45 | | |
| Hickory Island | Hickory Island Oil Co, | 7 | 1 | 3 | .. | 1 | .. | 4 | 3 | 535 | 6000 | 2 | 240 | 360 | 515 | .. | 30 | 30 | 15 | .. | 40 | | |
| T. Moran, south side | T. Moran and Orange Oil Co. | 200 | 2 | 1 | .. | .. | 2 | 3 | 3 | 520 | 4000 | 3 | 225 | 360 | 475 | .. | 35 | 30 | 30 | .. | 50 | | |
| Jas. H. Lee | Heirs of Lee and Wm. Reed | 180 | 3 | 7 | 1 | .. | 3 | 11 | 9 | 550 | 4000 | 5 | 230 | 365 | 485 | .. | 30 | 35 | 30 | .. | 50 | | |
| H. Bastian | Latonia Town and Oil Co. | 80 | 3 | 3 | .. | .. | 3 | 6 | .. | 520 | 4000 | 3 | 230 | 365 | 485 | .. | 35 | 30 | 30 | .. | 30 | | |
| T. G. Downing | Imperial Oil Co. | | | | | | | | | | | | | | | | | | | | | | |
| Graff, Hasson & Co. | United Pet. Farms Association | 400 | 30 | 15 | .. | 8 | 22 | 45 | .. | 400 | 4000 | 20 | 200 | 340 | 465 | .. | 15 | 15 | 20 | .. | 15 | 7 | 1380 |
| Clapp Farm, south. | G. H. Bissell and A. Plumer | 249 | 19 | 87 | .. | .. | 19 | 106 | 50 | 530 | 4500 | 30 | 190 | 330 | 460 | .. | 33 | 30 | 25 | .. | 25 | 1 | 200 |
| " north. | Cornplanter Oil Co. | 249 | 10 | 60 | .. | .. | 10 | 70 | 150 | 515 | 5000 | 20 | 200 | 340 | 460 | .. | 40 | 20 | 20 | .. | 30 | | |
| Rural Island | Haldeman, Heydrick & Co., Patentees. | 4 | 1 | 3 | .. | .. | 1 | 4 | .. | 500 | 5000 | 3 | 200 | 340 | 460 | .. | 40 | 20 | 20 | .. | 30 | | |

| Original Owner of Farm. | Present Owner of Farm. | No. of Acres. | No. of Producing Wells. | No. of Non-Producing Wells. | No. in Progress. | No. Flowing Wells. | No. Pumping Wells. | Total No. of Wells. | Average Daily Production. | Average Depth of Wells. | Average Cost of Wells. | No. of Engines. | Depth of First Sand Rock. | Depth of Second Sand Rock. | Depth of Third Sand Rock. | Depth of Fourth Sand Rock. | Thickness of First Sand Rock. | Thickness of Second Sand Rock. | Thickness of Third Sand Rock. | Thickness of Fourth Sand Rock. | Depth of Driving Pipe. | No. of Refineries. | Weekly Capacity Bbls. Crude. |
|---|---|---|---|---|---|---|---|---|---|---|---|---|---|---|---|---|---|---|---|---|---|---|---|
| H. McClintock..... | McClintockville Pet. Co.......... | 350 | 20 | 90 | 15 | 3 | 17 | 125 | 400 | 520 | $5000 | 40 | 190 | 330 | 490 | .. | 20 | 25 | 15 | .. | 30 | 2 | 825 |
| J. & A. Buchanan.. | Buchanan Farm Oil Co., Oil Basin Pet. Co., Haldeman & Co., J. Alexander (working interest). | 426 | 21 | 138 | .. | .. | 21 | 159 | 280 | 510 | 5000 | 75 | 196 | 340 | 480 | .. | 30 | 25 | 18 | .. | 25 | 2 | 215 |
| John McClintock... | Wallace Oil Co., Blivin Oil Co., Allegheny & Pittsburgh Oil Co., and others (leased). | 200 | 32 | 23 | 14 | 10 | 22 | 99 | 200 | 550 | 5000 | 50 | Flat 200<br>Hill 260 | 340<br>400 | 520<br>560 | .. | 25 | 25 | 12 | .. | 30 | 1 | 200 |
| Widow McClintock | J. W. Steel.......................... | 100 | 20 | 25 | 5 | 2 | 18 | 50 | 300 | 520 | 6000 | 30 | 160 | 330 | 473 | .. | 40 | 21 | 30 | .. | 25 | | |
| J. Rynd........... | Rynd Farm Oil Co............... | 196 | 12 | 49 | 4 | 1 | 11 | 65 | 56 | 550 | 5500 | 25 | 169 | 345 | 470 | .. | 30 | 30 | 25 | .. | 20 | 5 | 970 |
| Rynd Island....... | Island Oil Co.................... | .. | .. | 6 | .. | .. | .. | 6 | .. | .... | ... | 2 | | | | | | | | | | | |
| J. Blood........... | Home Pet. Co. and Blood Farm Pet. Co. | 440 | 15 | 60 | .. | .. | 15 | 75 | 200 | 500 | 5000 | 40 | 190 | 315 | 465 | .. | 25 | 25 | 25 | .. | 35 | 1 | 200 |
| J. Tarr............ | J. Tarr, Tarr Farm Pet. Co., and Clark & Sumner Association. | 190 | 26 | 52 | 11 | 2 | 24 | 89 | 1000 | 500 to 580 | 5000 | 50 | Flat 190<br>Hill 233 | 325<br>367 | 475<br>507 | .. | 20 | 25 | 35 | .. | 40 | | |
| Wm. Story......... | Columbia Oil Co.................. | 500 | 60 | 62 | 10 | 4 | 56 | 132 | 650 | 515 | 4000 | 75 | 233 | 367 | 480 | .. | 20 | 25 | 25 | .. | 27 | 3 | 1080 |
| Hayes.......... | Dalzell Pet. Co.................. | 108 | 2 | 14 | 5 | .. | 1 | 21 | 30 | 520 | 4000 | 10 | 180 | 360 | 480 | .. | 25 | 30 | 20 | .. | 25 | | |
| A. Davidson....... | Egbert & Hyde.............. ..... | 38 | 18 | 21 | 21 | 1 | 17 | 60 | .. | 550 | 5000 | 40 | 170 | 320 | 456 | .. | 30 | 30 | 40 | .. | 40 | | |
| G. W. McClintock.. | Central Pet. Co................. | 207 | 33 | 5 | 40 | 10 | 23 | 78 | 900 | 550 | 6000 | 65 | 180 | 340 | 460 | .. | 20 | 25 | 25 | .. | 50 | 5 | 935 |
| John Stevenson.... | Ocean Oil Co..................... | 153 | 9 | .. | 34 | 6 | 3 | 43 | 1000 | 690 | 6500 | 43 | 350 | 330 | 650 | .. | 8 | 9 | 40 | .. | 15 | 1 | 200 |
| James McCray..... | James McCray.................... | .. | .. | 6 | .. | .. | 1 | 6 | .. | 600 to 700 | | | | | | | | | | | | | |
| James Boyd........ | James Boyd....................... | .. | 1 | 21 | .. | .. | 1 | 22 | 10 | 480 | 4000 | 5 | 165 | 325 | 455 | .. | 20 | 25 | 25 | .. | 50 | 3 | 550 |
| McElheny ..... | Hussey & McBride and Haldeman & Co. | 100 | 14 | 65 | .. | .. | 14 | 79 | 217 | 480 | 5000 | 35 | 155 | 300 | 446 | .. | 30 | 32 | 27 | .. | 30 | | |
| John Benninghoff.. | John Benninghoff................ | 150 | 8 | 10 | 7 | 3 | 5 | 25 | 445 | 460 | 5000 | 12 | 152 | 308 | 440 | .. | 20 | 25 | 27 | .. | 30 | 1 | 320 |
| " " .... | Benninghoff Reserve Co.......... | 24 | 1 | 2 | 1 | .. | 1 | 4 | 15 | 460 | 5000 | 3 | 165 | 316 | 440 | .. | 20 | 25 | 27 | .. | 30 | | |
| Geo. P. Espy....... | Geo. P. Espy...................... | 154 | 2 | 5 | 1 | 1 | 1 | 8 | 22 | 460 | 4500 | 6 | 165 | 316 | 440 | .. | 15 | 15 | 30 | .. | 7 | | |
| McElheny...... | Hussey & McBride and Haldeman & Co. | 80 | 11 | 47 | .. | .. | 11 | 58 | 310 | 480 | 5000 | 25 | 151 | 305 | 440 | .. | 30 | 32 | 27 | .. | 35 | 2 | 320 |

| | | | | | | | | | | | | | | | | | | | | | | | | |
|---|---|---|---|---|---|---|---|---|---|---|---|---|---|---|---|---|---|---|---|---|---|---|---|---|
| McElheny | Reserve | 20 | 1 | 4 | .. | .. | .. | 5 | 4 | 480 | 5000 | 3 | | 151 | 305 | 440 | .. | 50 | 32 | 27 | .. | 55 | | |
| Caldwell | Caldwell Oil Co. | 220 | 4 | 5 | 5 | .. | 4 | 14 | 105 | 480 | 6000 | 10 | | 165 | 306 | 445 | .. | 40 | 35 | 30 | .. | 15 | | |
| J. Foster | ........ | .. | .. | 50 | .. | .. | 1 | 50 | .. | 470 | 6000 | 5 | | 140 | 290 | 435 | .. | 15 | 20 | 30 | .. | 50 | | |
| J. Farrell | J. Farrell and Commonwealth Oil Co. | 36 | 1 | 12 | .. | 1 | 2 | 13 | 40 | 515 | 4000 | 6 | | 185 | 345 | 478 | .. | 10 | 10 | 36 | .. | 31 | | |
| R. P. Beaty | Clinton Oil Co. | 60 | 2 | 5 | 3 | .. | 2 | 10 | 80 | 500 to 650 | 5000 | 5 | Flat | 170 | 210 | 480 | .. | 20 | 25 | 40 | .. | 5 | | |
| | | | | | | | | | | | | | Hill | 280 | 420 | 550 | | | | | | | | |
| D. Gregg | D. Gregg | 300 | .. | 40 | .. | .. | .. | 40 | .. | 500 | ... | 10 | | 170 | 210 | .. | .. | 10 | 15 | .. | .. | 45 | 2 | 300 |
| Sanny | Potter & Nurse, and others | 125 | .. | 11 | 3 | .. | 2 | 14 | .. | 500 | 4000 | 6 | | 170 | 210 | .. | .. | 15 | 15 | .. | .. | 40 | 2 | 300 |
| G. Shaffer | (Not ascertained.) | 54 | .. | 24 | .. | .. | 1 | 24 | .. | 300 to 650 | 5000 | 6 | | 180 | 230 | .. | .. | 19 | 10 | .. | .. | 40 | | |
| R. Miller | Indian Rock Oil Co. of N. Y. | 303 | .. | 20 | .. | .. | .. | 20 | .. | 600 | 4000 | 10 | | 143 | 290 | .. | .. | 20 | 25 | .. | .. | 40 | 2 | 240 |
| Fleming | (Not ascertained.) | .. | .. | 10 | .. | .. | .. | 10 | .. | 500 | ... | 5 | | 143 | 290 | .. | .. | 20 | 25 | .. | .. | 40 | | |
| Widow Jones | " " | .. | .. | 15 | .. | .. | .. | 15 | .. | 500 | ... | 2 | | | | | | | | | | | | |
| J. Stanford | Henderson and Brewer, Watson & Co. | .. | .. | 10 | .. | .. | .. | 10 | .. | 500 | ... | 1 | | | | | | | | | | | | |
| Stackpole & Fletcher | (Not ascertained.) | .. | .. | 10 | .. | .. | .. | 10 | .. | 500 | ... | 3 | | ... | .. | .. | .. | .. | .. | .. | .. | .. | 2 | 150 |
| O. Stackpole | Northern Light Oil Co. | .. | .. | 2 | .. | .. | .. | 2 | .. | 500 | ... | 1 | | | | | | | | | | | | |
| Geo. H. Bissell | Geo. H. Bissell | .. | .. | 2 | .. | .. | .. | 2 | .. | 400 | ... | .. | | | | | | | | | | | | |
| Conley | (Not ascertained.) | .. | .. | 6 | .. | .. | .. | 6 | | | | | | | | | | | | | | | | |
| B. Griffin | N. Y. & Penn. Pet. Co. | 11 | 3 | 3 | .. | .. | 3 | 6 | 20 | 600 | 5000 | 3 | | 145 | 315 | 471 | .. | 20 | 59 | 14 | .. | 45 | | |
| Geo. H. Bissell | Original Pet. Co. | 81 | 6 | 18 | .. | .. | 6 | 24 | 80 | 500 | 5500 | 10 | | 145 | 315 | 471 | .. | 20 | 59 | 14 | .. | 30 | | |
| Hibbert | Watson Pet. Co. | 36 | 2 | 4 | 1 | .. | .. | 7 | 5 | 400 | 5000 | 3 | | 150 | 370 | .. | .. | 10 | 55 | .. | .. | 45 | | |
| Kingsland | Original Pet. Co. | .. | 1 | 6 | 4 | .. | 1 | 11 | 12 | 400 | 5000 | 6 | | 150 | 370 | .. | .. | 10 | 55 | .. | .. | 45 | | |
| Jonathan Watson | Oil Creek Pet. Co. and Penn. & Oil Creek Co. | 90 | 6 | 54 | .. | .. | 6 | 60 | 30 | 475 | 5000 | 20 | | 150 | 365 | .. | .. | 10 | 55 | .. | .. | 60 | | |
| Parker | Parker Farm Pet. Co. | .. | 12 | 26 | .. | .. | 12 | 38 | 50 | 500 | 4000 | 20 | | 150 | 365 | .. | .. | 15 | 50 | .. | .. | 60 | 2 | 800 |
| Titusville Borough | ........ | 800 | .. | 1 | .. | .. | .. | 1 | .. | 1600 | ... | 1 | | 150 | 360 | .. | .. | 50 | 25 | .. | . | 75 | 10 | 1500 |
| Above Titusville | Various and different parties | .. | .. | 100 | .. | .. | .. | 100 | .. | 500 | 5000 | 25 | | | | | | | | | | | | |
| John Watson | John Watson | 200 | 2 | 18 | .. | .. | 2 | 20 | 5 | 500 | 5000 | 5 | | 150 | 350 | .. | .. | 50 | 25 | .. | .. | 80 | | |
| Dr. Brewer | Hopeman & Co. | .. | . | 1 | .. | .. | .. | 1 | .. | 709 | 5000 | 1 | | | | | | | | | | | | |
| Benedict | Mutual Benefit Pet. Co. | .. | 3 | .. | 2 | .. | 3 | 5 | .. | 800 | 6000 | 5 | | | | | | | | | | | | |
| Patterson | Patterson | .. | 3 | .. | 3 | 1 | 2 | 6 | .. | 625 | 6000 | 5 | | | | | | | | | | | | |
| J. B. Skinner | M. C. Egbert | 216 | 3 | .. | 4 | .. | 3 | 7 | 75 | 760 to 860 | 8000 | 7 | | 280 | 440 | 730 | 850 | 25 | 30 | 30 | 40 | 30 | | |
| Jas. O. Guild | Watson & Howland and Morrill Pet. Co. | 200 | .. | 16 | 2 | .. | .. | 18 | .. | 500 | 4000 | 8 | | 160 | 440 | .. | .. | 13 | 30 | .. | .. | 50 | | |
| Part of G. W. McClintock. | In small tracts on Benninghoff Run. | 30 | .. | 18 | 2 | .. | .. | 20 | .. | 550 | 5000 | 6 | | 225 | 360 | 485 | .. | 20 | 20 | 50 | .. | 15 | 4 | 640 |
| Part of John Stevenson. | Titus & Drysdel Tract | 13 | .. | 1 | 1 | .. | .. | 2 | .. | 600 | 5000 | 1 | | 240 | 360 | 490 | .. | 20 | 25 | 50 | .. | 15 | 1 | 160 |
| ........ | " " | 3 | 2 | 4 | .. | 1 | 1 | 6 | 75 | 600 | 5000 | 3 | | 260 | 375 | 500 | .. | 20 | 25 | 54 | .. | 15 | | |
| George Warner | Benninghoff Pet. Co. | 37 | 15 | 3 | 10 | 12 | 3 | 28 | 1000 | 600 | 4500 | 20 | | 270 | 380 | 525 | .. | 20 | 25 | 54 | .. | 15 | | |

| Original Owner of Farm. | Present Owner of Farm. | No. of Acres. | No. of Producing Wells. | No. of Non-Producing Wells. | No. in Progress. | No. Flowing Wells. | No. Pumping Wells. | Total No. of Wells. | Average Daily Production. | Average Depth of Wells. | Average Cost of Wells. | No. of Engines. | Depth of First Sand Rock. | Depth of Second Sand Rock. | Depth of Third Sand Rock. | Depth of Fourth Sand Rock. | Thickness of First Sand Rock. | Thickness of Second Sand Rock. | Thickness of Third Sand Rock. | Thickness of Fourth Sand Rock. | Depth of Driving Pipe. | No. of Refineries. | Weekly Capacity Bbls. Crude. |
|---|---|---|---|---|---|---|---|---|---|---|---|---|---|---|---|---|---|---|---|---|---|---|---|
| Weiss........ | Cornwall & Co. ................ | 80 | .. | 6 | .. | .. | .. | 6 | .. | 600 | $5000 | 3 | | | | | | | | | | | |
| Widow Foster...... | McCorns......................... | 12 | .. | 3 | .. | .. | .. | 3 | .. | 600 | 4500 | | | | | | | | | | | | |
| Martin Benninghoff | Benninghoff Pet. Co. and Martin Benninghoff Pet. Co. | .. | 1 | .. | 1 | 1 | .. | 2 | 4 | 600 | 4500 | 2 | 290 | 400 | 550 | .. | 25 | 30 | 40 | .. | 20 | | |
| On Church Run.... | Atlantic & Great Western Pet. Co. | .. | 1 | .. | .. | .. | 1 | 1 | 50 | 550 | 5000 | 1 | | | | | | | | | | | |
| Widow Stevenson.. | Petroleum Centre Co ............ | 13 | 5 | 4 | .. | .. | 5 | .. | .. | 550 | 4000 | 6 | 200 | 380 | 480 | .. | 20 | 30 | 25 | .. | 70 | | |
| .................... | Cold Water Oil Co............... | .. | .. | 6 | .. | .. | .. | .. | 25 | 550 | 5000 | 5 | 200 | 280 | 480 | .. | 20 | 30 | 25 | .. | 70 | | |
| Judge McCalmont.. | Knox & Co..................... | 36 | 4 | 10 | 2 | 1 | 3 | 16 | 30 | 550 | 5000 | 10 | 195 | 245 | 480 | .. | 20 | 30 | 50 | .. | 40 | | |
| Hayes & Gibson.... | Hays & Gibson.................. | .. | .. | 8 | .. | .. | .. | 8 | .. | 500 | 5000 | 4 | 170 | 350 | 475 | .. | 30 | 30 | 25 | .. | 30 | | |
| Geo. Tarr.......... | (Not ascertained.)............... | 25 | .. | 2 | .. | .. | .. | 2 | .. | 600 | 5000 | 1 | 240 | 372 | 500 | .. | 20 | 12 | 22 | .. | 14 | | |
| Robert Ward...... | (Not ascertained.)............... | 22 | .. | 3 | .. | .. | .. | 3 | .. | 600 | 5000 | 2 | 243 | 385 | 525 | .. | 40 | 35 | 25 | .. | 12 | | |
| J. R. Newman...... | Kingston & Randolph Pet. Co., and Plymouth Rock Oil Co. | 30 | .. | 3 | .. | .. | .. | 3 | .. | 600 | 6000 | 2 | 390 | 465 | 528 | .. | 40 | 35 | 25 | .. | 7 | | |
| Brown......... | Metropolitan Oil Co. of N. Y..... | 22 | 1 | 1 | .. | .. | 1 | 2 | 5 | 600 | 6000 | 1 | 248 | 405 | 537 | .. | 43 | 22 | 20 | .. | 12 | | |
| Wm. Sutley........ | Wm. Sutley...................... | 75 | 1 | 1 | .. | .. | 1 | 2 | 12 | 600 | 6000 | 1 | 265 | 407 | 573 | .. | 40 | 22 | 20 | .. | 11 | | |
| Black.......... | Big Tank Oil Co.................. | .. | .. | 2 | 12 | .. | .. | 14 | .. | 600 | 6000 | 5 | 265 | 407 | 575 | 700 | 40 | 22 | 20 | 10 | 10 | | |
| T. Dempsey........ | T. Dempsey...................... | 160 | 2 | 6 | 20 | .. | 2 | 28 | 7 | 600 | 6000 | 10 | 280 | 450 | 573 | .. | 20 | 22 | 32 | .. | 10 | | |
| Mallory........ | (Not ascertained.)............... | .. | .. | 4 | 20 | .. | .. | 24 | .. | 600 | 6000 | 5 | | | | | | | | | | | |
| ................... | Chichester................ | .. | .. | 2 | 5 | .. | .. | 7 | .. | 800 to 1000 | 6000 | 5 | | | | | | | | | | | |
| Baney.......... | Day & Co., of N. Y., and others.. | 125 | 1 | 4 | .. | .. | 1 | 5 | 5 | 600 | 5000 | 3 | 285 | 450 | 575 | .. | 20 | 22 | 32 | .. | 12 | | |
| J. Pierson.......... | McClintock Farm & Cherry Tree Pet Co. | .. | .. | 2 | 4 | .. | 1 | 6 | .. | 780 | 5000 | 2 | 290 | 455 | 595 | .. | 35 | 35 | 25 | .. | 15 | | |
| J. B. Stewart....... | Claremont Oil Co. of Balt. and Potomac Oil Co. of Washington. | 230 | .. | 4 | .. | .. | 2 | 4 | .. | 828 | 6000 | 4 | 330 | 435 | 725 | .. | 30 | 40 | 17 | .. | 54 | | |
| S. W. Irvine........ | Dervon Oil Co. of Chicago....... | 119 | .. | 2 | .. | 4 | .. | 2 | .. | 660 | 7000 | 2 | 340 | 596 | 750 | .. | 45 | 25 | 30 | .. | 45 | | |
| ..................... | Village of Cherry Tree and above. | .. | .. | 10 | .. | .. | .. | 10 | .. | 650 | 6000 | 5 | Hill Flat 240 | 370 | 535 | | | | | | | | |
| John McClintock... | Union Pet Co. and Brevoort Pet. Co. | 138 | 27 | 5 | 12 | 4 | 23 | 44 | 310 | 610 | 6000 | 35 | 311 | 453 | 600 | .. | 30 | 20 | 25 | .. | 20 | | |

| | | | | | | | | | | | | | | | | | | | | | | | |
|---|---|---|---|---|---|---|---|---|---|---|---|---|---|---|---|---|---|---|---|---|---|---|---|
| Knapp | St. Nicholas Oil Co. | 50 | 3 | .. | 12 | .. | 3 | 15 | 59 | 760 | 6000 | 12 | 430 | 565 | 730 | .. | 39 | 41 | 25 | .. | 20 | | |
| Duff Tract | Mingo Oil Co. of Phila. and Wadsworth & Wynkoop. | 5 | 9 | 5 | 6 | 1 | 9 | 20 | 200 | 600 | 6000 | 15 | 240 | 365 | 550 | .. | 35 | 25 | 25 | .. | 20 | | |
| Tolls | Murray Oil Co. of Pittsburg | 87 | 1 | .. | 12 | .. | 1 | 13 | .. | 800 | 7000 | 8 | 450 | 600 | 775 | .. | 39 | 41 | 25 | .. | 20 | | |
| J. Rynd | Curtin Oil Co. and North American Pet. Co. | 36 | 21 | 6 | 7 | 6 | 14 | 34 | 175 | 600 | 6000 | 25 | Hill Flat 245<br>430 | 390<br>565 | 530<br>710 | .. | 39 | 41 | 36 | .. | 20 | | |
| Wm. Smith | Cherry Valley Oil Co. of Conn. | 57 | 20 | 7 | 13 | 7 | 13 | 40 | 350 | 600 | 7000 | 35 | 270 | 401 | 550 | .. | 40 | 40 | 40 | .. | 15 | | |
| McFate & Knapp | Cherry Run Pet. Co. of Phila. | 401 | .. | 25 | .. | .. | .. | 25 | .. | 625 | 6000 | 10 | 285 | 425 | 570 | .. | 15 | 15 | 14 | .. | 15 | | |
| Joseph McFate | Oil Creek & Cherry Run Pet. Co. | 160 | .. | 21 | .. | .. | .. | 21 | .. | 625 | 6000 | 12 | 285 | 425 | 570 | | | | | | | | |
| Widow McFate | Balliet & Co | .. | .. | .. | 4 | .. | .. | 4 | .. | 700 | 7000 | 4 | 335 | 543 | 670 | | | | | | | | |
| J. Lamb | Springfield Pet. Co., Cherry Run Central Oil Co., and Hogg Farm Co. | 230 | .. | 38 | .. | .. | .. | .. | .. | one 1000 | .... | .. | 163 | 400 | 552 | 908 | 40 | 36 | 35 | .. | 30 | 1 | 1500 |
| J. R. Slents | Various parties. | | | | | | | | | | | | | | | | | | | | | | |
| J. A. Turner | (Not ascertained.) | .. | .. | .. | 40 | .. | 3 | 78 | .. | 600 | 7000 | 20 | 100 | 336 | 500 | 650 | 42 | 37 | 34 | 14 | 18 | | |
| Robert McFate | (Not ascertained.) | 80 | .. | 10 | .. | .. | .. | 10 | .. | .... | .... | 3 | 163 | 400 | 550 | .. | 42 | 36 | 30 | .. | 15 | | |
| McCalmont tract | Village of Plumer | .. | .. | 7 | .. | .. | .. | 7 | .. | .... | .... | 2 | 165 | 400 | 555 | .. | 40 | 23 | 40 | .. | 15 | 2 | 1000 |
| T. W. McCalmont | Rochester Pet. Co. | 45 | | | | | | | | | | | | | | | | | | | | | |
| Balliett | Lehigh Co. | 40 | .. | 5 | 3 | .. | 2 | 8 | .. | 740 | 8000 | 5 | 283 | 455 | 548 | 710<br>5th 613 | 54 | 20 | 35 | 25 | 25 | | |
| A. C. Prather | Prather heirs & Liberty Oil Co., McAboy & Cherry Run Oil Co. | 237 | .. | 12 | 12 | .. | .. | 24 | .. | 500 | 5000 | 10 | 150 | 275 | 500 | 740 | 35 | 20 | 55 | 8 | 50 | | |
| Prather Bros | Prather Bros | 112 | .. | 1 | .. | .. | .. | 1 | .. | 650 | .... | .. | 150 | 275 | 500 | .. | 35 | 20 | 55 | .. | 50 | | |
| J. Rickets | J. Rickets' heirs and Plumerville Oil Co. | .. | .. | 7 | 2 | .. | 1 | 9 | .. | 650 & 837 | 6000 | 4 | 150 | 275 | 500 | 740 | 35 | 20 | 55 | 8 | 50 | | |
| Morrison and tracts above. | (Not ascertained.) | .. | .. | 6 | .. | .. | .. | 6 | .. | 700 | 6000 | 3 | 150 | 275 | 500 | 740 | 40 | 20 | 60 | 8 | 53 | | |
| Stowel | Titus Oil Co. in part | .. | 1 | .. | 1 | .. | 1 | 2 | 25 | 740 | 7000 | 2 | 150 | 275 | 500 | 740 | 40 | 20 | 60 | 8 | 53 | | |
| Above Stowel | Great Republic Oil Co. | .. | 2 | .. | 1 | .. | 2 | 3 | 20 | 750 | 7000 | 3 | 150 | 275 | 500 | 740 | 40 | 20 | 60 | 8 | 53 | | |
| Huidekoper | Crescent City Oil Co. | 76 | .. | .. | 5 | .. | .. | 5 | .. | 750 | 5500 | 5 | 125 | 275 | 512 | .. | 25 | 60 | 15 | .. | 36 | | |
| .......... | Green Well | .. | .. | 1 | .. | .. | 1 | 1 | 25 | 750 | 6000 | 1 | 125 | 275 | 512 | .. | 25 | 60 | 15 | .. | 36 | | |
| J. Nevins | Woods, McAboy & Co. and Chamberlin. | 185 | 1 | 8 | .. | .. | 1 | 9 | 3 | 600 | 4000 | 3 | 190 | 330 | 460 | 560 | 20 | 40 | 15 | .. | 12 | | |
| Two-Mile Run | Various parties | .. | .. | 10 | 6 | .. | .. | 16 | .. | 600 | 5000 | 10 | | | | | | | | | | | |
| W. C. & R. S. McCormick. | McCormick heirs, Eclectic Oil Co., Pierson Pet. Co., N. Y. & Franklin Oil Co., Ashland Oil Co., Cattaraugus & French Creek Pet. Co., Beacon Oil Co. | 421 | 3 | 18 | 2 | .. | .. | 23 | 8 | one 1000<br>350 to 800 | 3000 to 6000 | 7 | 265 | 500 | .. | .. | 30 | 28 | .. | .. | 25 | | |

| Original Owner of Farm. | Present Owner of Farm. | No. of Acres. | No. of Producing Wells. | No. of Non-Producing Wells. | No. in Progress. | No. Flowing Wells. | No. Pumping Wells. | Total No. of Wells. | Average Daily Production. | Average Depth of Wells. | Average Cost of Wells. | No. of Engines. | Depth of First Sand Rock. | Depth of Second Sand Rock. | Depth of Third Sand Rock. | Depth of Fourth Sand Rock. | Thickness of First Sand Rock. | Thickness of Second Sand Rock. | Thickness of Third Sand Rock. | Thickness of Fourth Sand Rock. | Depth of Driving Pipe. | No. of Refineries. | Weekly Capacity Bbls. Crude. |
|---|---|---|---|---|---|---|---|---|---|---|---|---|---|---|---|---|---|---|---|---|---|---|---|
| S. F. Dale | Dale Farm Oil Co. | 200 | 1 | 1 | .. | .. | 1 | 2 | 6 | 466 | $3000 to 6000 | 1 | 217 | 248 | .. | .. | Shells | 12 | .. | .. | 10 | | |
| Samuel Hays | Ohio Carbon Oil & Mining Co. | 272 | .. | 4 | .. | .. | .. | 4 | .. | 650 | 5000 | 2 | | | | | | | | | | | |
| Bowman | Mammoth Oil Co. | 200 | 2 | .. | .. | .. | .. | 2 | .. | 450 | 5000 | 2 | 80 | 330 | .. | .. | 60 | 14 | .. | .. | 10 | | |
| R. C. Longwell | Centralia Oil Co. of Phila. | 106 | 1 | 3 | 1 | .. | 1 | 5 | 3 | 250 | 3800 | 3 | 80 | 330 | .. | .. | 70 | 14 | .. | .. | 42 | | |
| J. W. Sutley | Sutley Lubricating Oil Co., Eureka Oil Co., McCormick & McKissick Oil Co. | 100 | 4 | 13 | 2 | 1 | 3 | 19 | 15 | 400 | 3000 | 6 | 250 | 600 | .. | .. | 25 | 25 | .. | .. | 10 & 60 | | |
| J. Dubbs | French Creek Lubricating Oil Co. | 125 | 1 | 5 | 1 | .. | 1 | 7 | 5 | 300 to 650 | 5000 to 7000 | 4 | 90 | 240 | .. | .. | 20 | 25 | .. | .. | 11 | | |
| Newel | December Lubricating Oil Co. | 160 | 1 | 10 | .. | .. | 1 | 11 | .. | 400 | 6000 | 6 | 50 | 290 | .. | .. | 25 | .. | .. | .. | 30 | | |
| W. Brown, | Venango Lubricating Oil Co. | 220 | .. | 3 | .. | .. | .. | 3 | .. | 350 | 5000 | 2 | 40 | 290 | .. | .. | 21 | .. | .. | .. | 35 | | |
| D. McElrath | McElrath Oil Co., leased to Sugar Valley Oil Co. | 174 | 2 | 3 | 2 | .. | 2 | 7 | 10 | 325 | 5000 | 7 | 40 | 285 | .. | .. | 20 | 25 | .. | .. | 9 | | |
| H. W. Brown | Junction Oil Co. and Brown Farm Oil Co. | 130 | 4 | 5 | .. | .. | 4 | 9 | 30 | 350 | 4000 | 4 | 75 | 264 | .. | .. | 10 | 30 | .. | .. | 10 to 35 | | |
| ......... | South side of creek—no account of. | .. | .. | 3 | .. | .. | .. | 3 | | | | | | | | | | | | | | | |
| R. Roberts | Indian Spring Oil Co., and Phila. Pet. Co. | 275 | 1 | 7 | .. | .. | 1 | 8 | 5 | 400 | 5000 | 5 | 120 | 320 | .. | .. | 12 | 45 | .. | .. | 12 to 96 | | |
| J. Hayes | Hayes' Farm Oil Co. | 150 | .. | 2 | .. | .. | 1 | 2 | .. | 415 | 5000 | 2 | 120 | 320 | .. | .. | 12 | 45 | .. | .. | 12 to 90 | | |
| W. Hannah and others. | Hannah Gap Oil Co. and others. | .. | 3 | 11 | .. | .. | 1 | 14 | .. | .... | .... | 5 | | | | | | | | | | | |
| ......... | In progress above and abandoned. | .. | .. | 6 | 4 | .. | .. | 10 | | | | | | | | | | | | | | | |

| | | | | | | | | | | | | | | | | | | | | | |
|---|---|---|---|---|---|---|---|---|---|---|---|---|---|---|---|---|---|---|---|---|---|
| Blakely & Clark.... | Franklin Lubricating Oil Co...... | 313 | 2 | 3 | .. | .. | 2 | 5 | .. | 325 | 5000 | 3 | 130 | 280 | .. | .. | 20 | 25 | .. | .. | 40 |
| .................... | Tract above, Blakely & Clark..... | .. | .. | 3 | .. | .. | .. | 3 | .. | 650 | 6000 | .. | 130 | 280 | .. | .. | 20 | 25 | .. | .. | 10 |
| J. M. Russel........ | Russel Farm Oil Co., or Sugar Dale. | 112 | 1 | 2 | .. | .. | 1 | 3 | 1 | 400 | 3000 | 1 | | | | | | | | | |
| H. H. Homan and others. | Sugar Creek Oil Co. and Development Co. | 363 | 4 | 1 | .. | .. | 4 | 5 | 60 | 325 to 700 | 6000 | 3 | 100 | 275 | .. | .. | 20 | 60 | .. | .. | 60 |
| J. W. McCalmont.. | Rochester Pet. Co................ | .. | 1 | 1 | .. | .. | 1 | 2 | 5 | .... | .... | 2 | | | | | | | | | |
| Frazier and tract above........................ | | .. | .. | 4 | .. | .. | .. | 4 | .. | .... | .... | 4 | | | | | | | | | |
| Between the last tract and Coopertown...... | | .. | .. | 4 | .. | .. | .. | 4 | .. | .... | .... | 4 | | | | | | | | | |
| Above Coopertown ........................ | | .. | .. | 4 | .. | .. | .. | 4 | .. | .... | .... | 4 | | | | | | | | | |
| .................... | Phila. & Sheridan Oil Co. ........ | .. | .. | 6 | .. | .. | .. | 6 | .. | .... | .... | 4 | | | | | | | | | |
| Graff, Hasson & Co. | Graff, Hasson & Co................ | 600 | 4 | 1 | .. | .. | 3 | 5 | 12 | 442 | 2500 | 3 | 180 | 240 | 475 | .. | 30 | 30 | 25 | .. | 15 |
| P. H. Siverly....... | John W. Knox.................... | 118 | 4 | 1 | 20 | .. | 4 | 25 | 70 | 530 | 5000 | 10 | 250 | 370 | 490 | .. | 30 | 40 | 30 | .. | 60 |
| Cornplanter Island. | Riverside Pet. Co................ | .. | .. | 2 | .. | .. | .. | 2 | .. | 520 | 5000 | 1 | 250 | 370 | 490 | .. | 30 | 40 | 30 | .. | 60 |
| Ellen Clapp........ | Ocean Oil Co. and Sage Run & Oil Creek Pet. Co. | 164 | .. | 1 | .. | .. | .. | 1 | .. | 660 | 6000 | 1 | 240 | 350 | 500 | .. | 30 | 25 | 15 | .. | 40 |
| R. Alcorn.......... | Alcorn Oil Co. and Congress Oil Co. | 175 | 2 | 2 | .. | .. | 2 | 4 | .. | 576 | 5000 | 1 | 130 | 250 | 540 | .. | 30 | 25 | 10 | .. | 28 |
| W. P. Alcorn...... | Seneca Oil Co.................... | 300 | 1 | 1 | .. | .. | 1 | 2 | .. | 556 | 5000 | 1 | 150 | 300 | 540 | .. | 30 | 25 | 10 | .. | 30 |
| N. Downing........ | Balt. Pet. Co. and W. B. Wier.... | .. | 2 | 5 | .. | .. | 2 | 7 | 4 | 550 | 5000 | 2 | 212 | 350 | 475 | .. | 28 | 28 | 10 | .. | 28 |
| E. R. Eaker........ | Riverside Pet. Co................ | 120 | 1 | 2 | .. | .. | 1 | 3 | .. | 550 | 6000 | 2 | 160 | 380 | 485 | .. | 20 | 26 | 30 | .. | 28 |
| .................... | Ogden Oil Co..................... | 48 | 2 | 2 | 2 | .. | 2 | 6 | 2 | 600 | 6000 | 4 | 190 | 420 | 552 | .. | 12 | 33 | 16 | .. | 10 |
| .................... | Sage Run Oil Co.................. | .. | .. | 2 | 4 | .. | .. | 6 | .. | 650 | 6000 | 4 | 195 | 430 | 560 | .. | 12 | 33 | 16 | .. | 10 |
| .................... | Forward Oil Co................... | .. | .. | .. | 4 | .. | 1 | 4 | .. | 800 | 7000 | 3 | 200 | 435 | 570 | .. | 12 | 30 | 20 | .. | 10 |
| E. Evans........... | Horse Creek Oil & Manufact'g Co. | 2190 | 1 | 2 | 1 | .. | 1 | 4 | 3 | 500 | 5000 | 3 | 200 | 300 | .. | .. | 30 | 25 | .. | .. | 20 |
| H. Carey........... | Rockwood Oil Co. and American Oil Co. | .. | 3 | 10 | .. | .. | 3 | 13 | 30 | 312 to 500 | 4000 | 6 | 220 | 350 | 485 | .. | 30 | 25 | 10 | .. | 10 |
| S. Lamb............ | Rogers & Macy; N. Y. & Balt. Pet. Co.; Phila., Lancaster & Cherry Run Co., and Tarr Homestead Co. | .. | 4 | 6 | .. | .. | 4 | 10 | 56 | 380 | 5000 | 4 | 190 | 298 | .. | .. | 7 | 15 | .. | .. | 12 |
| Rennoff........ | McClintock & Cornwall Pet. & Mining Co. and Rose Oil Co. | .. | 2 | 18 | .. | .. | 2 | 20 | 3 | 500 | 5000 | 6 | 270 | 330 | .. | .. | 28 | 28 | .. | .. | 50 |
| J. Blakeley......... | Harrington Oil Co. and Star Oil Co. | .. | 1 | 6 | .. | .. | 1 | 7 | 15 | 500 | 6000 | 4 | 210 | 350 | .. | .. | 15 | 25 | .. | .. | 50 |
| J. Nellis........... | Panther Run & Allegheny Oil Co. Clapp & Linderman. | .. | .. | 3 | 1 | .. | .. | 4 | 60 | 600 | 6000 | 3 | 210 | 350 | .. | .. | 15 | 25 | .. | .. | 40 |
| Philip Krotzer..... | Continental Oil Co. and others.... | 95 | 7 | 6 | .. | .. | 7 | 13 | .. | 450 | 5000 | 7 | 170 | 260 | .. | .. | 30 | 20 | .. | .. | 45 |
| Henry Dille........ | U. S. Pet. Co.................... | 95 | .. | 3 | .. | .. | .. | 3 | .. | 200 to 500 | 3000 | .. | 165 | 265 | 315 | .. | 40 | 16 | 12 | .. | 40 |
| Alex. Dotson ...... | New World Pet. Co................ | 25 | .. | 3 | .. | .. | .. | 3 | .. | 483 | 5000 | 2 | 170 | 270 | 320 | .. | 40 | 16 | 12 | .. | 45 |
| Henry Krotzer..... | Krotzer Farm Oil Co. and Gibson Oil Co. | 50 | 2 | 8 | .. | .. | 2 | 10 | 57 | 450 | 5000 | 6 | 185 | 285 | 335 | .. | 40 | 16 | 12 | .. | 60 |
| H Marsh.......... | Walnut Bend Oil Co............... | 6 | .. | 3 | .. | .. | 1 | 3 | .. | 530 | 5000 | 2 | 170 | 270 | 320 | .. | 40 | 16 | 12 | .. | 50 |

| Original Owner of Farm. | Present Owner of Farm. | No. of Acres. | No. of Producing Wells. | No. of Non-Producing Wells. | No. in Progress. | No. Flowing Wells. | No. Pumping Wells. | Total No. of Wells. | Average Daily Production. | Average Depth of Wells. | Average Cost of Wells. | No. of Engines. | Depth of First Sand Rock. | Depth of Second Sand Rock. | Depth of Third Sand Rock. | Depth of Fourth Sand Rock. | Thickness of First Sand Rock. | Thickness of Second Sand Rock. | Thickness of Third Sand Rock. | Thickness of Fourth Sand Rock. | Depth of Driving Pipe. | No. of Refineries. | Weekly Capacity Bbls. Crude. |
|---|---|---|---|---|---|---|---|---|---|---|---|---|---|---|---|---|---|---|---|---|---|---|---|
| Jacob Barr | Walnut Bend Shoe & Leather Oil Co. and Star Oil Co. | 30 | 1 | 10 | 1 | .. | 1 | 12 | 5 | 500 | $5000 | 4 | 175 | 275 | 325 | .. | 40 | 16 | 12 | .. | 58 | | |
| Tolls | Brooklyn Pet. Co. and others | .. | 2 | 9 | .. | .. | 2 | 11 | 10 | 500 | 5000 | 3 | 150 | 360 | 480 | .. | 40 | 25 | 20 | .. | 30 | | |
| S. Conver | Plainfield Pet. Co. | .. | .. | 12 | .. | .. | .. | 12 | .. | 500 | 5000 | 4 | 150 | 300 | 400 | .. | 25 | 30 | 20 | .. | 20 | | |
| J. Kintzler | Humboldt Pet. & Refining Co. | 116 | 1 | 6 | .. | .. | 1 | 7 | 25 | 400 | 4000 | 3 | 160 | 280 | 340 | .. | 10 | 25 | 20 | .. | 12 | | |
| J. McMahan | McMahan Farm Oil Co. | 300 | 1 | 10 | | .. | 1 | 11 | 15 | 400 | 3000 | 3 | 180 | 280 | .. | .. | 30 | 20 | .. | .. | 12 | | |
| A. Russel | Lamb Bros. & McClintock | 100 | .. | 3 | .. | .. | .. | 3 | .. | 200 | 2000 | 1 | 160 | 280 | .. | .. | 10 | 25 | .. | .. | 20 | | |
| Samuel McFate | Lamb Bros. | 25 | 4 | 8 | .. | .. | 4 | 12 | 24 | 350 | 4000 | 4 | 160 | 280 | 340 | .. | 10 | 25 | 15 | .. | 12 | | |
| Hogg | (Not ascertained.) | .. | .. | 5 | .. | .. | .. | 5 | .. | 400 | 4000 | 5 | | | | | | | | | | | |
| W. Bruner | Bruner Oil Co. and Almeda Oil Co. and others. | 160 | .. | 17 | .. | .. | 1 | 17 | .. | 300 to 800 | 5000 | 5 | 180 | 320 | .. | .. | 30 | 30 | .. | .. | 40 | 1 | 42 |
| Lower Walnut Isl'd. | Farley, Hodgkiss, and others | 3 | 1 | .. | .. | .. | 1 | 1 | 5 | 500 | 5000 | 1 | 165 | 265 | 315 | .. | 40 | 15 | 12 | .. | 40 | | |
| Middle Walnut Isl'd | Whitmore & Hall | 7 | .. | 3 | .. | .. | .. | 3 | .. | 550 | .... | .. | 174 | 256 | .. | .. | 20 | 20 | .. | .. | 40 | | |
| Upper Walnut Isl'd. | Upper Walnut Island Oil Co. | 20 | 4 | 6 | 3 | .. | 4 | 13 | 35 | 400 | 5000 | 9 | 174 | 256 | .. | .. | 20 | 20 | .. | .. | 40 | | |
| J. Anderson | Great Western Consolidated Pet. Co., Anderson Pet. Co. (leased to First National Pet. Co.) | 200 | 4 | 14 | .. | .. | 4 | 18 | 200 | 400 | 3000 | 8 | 180 | 230 | .. | .. | 30 | 30 | .. | .. | 40 | | |
| Foster | S. D. Ulrick | .. | .. | 2 | .. | .. | .. | 2 | | | | | | | | | | | | | | | |
| McCalmont | N. Y., Phila. & Balt. Pet. Co. | 7 | .. | 2 | .. | .. | .. | 2 | .. | 400 | 4000 | 2 | 170 | 290 | .. | .. | 20 | 15 | .. | .. | 40 | | |
| C. W. Heydrick | .................. | .. | .. | 2 | 1 | .. | .. | 3 | | | | | | | | | | | | | | | |
| Howe | Hazelton Pet. Co. | 133 | 2 | 5 | .. | .. | 2 | 7 | 32 | 350 | 5000 | 4 | 165 | 210 | .. | .. | 15 | 30 | .. | .. | 36 | | |
| Alexander | Alexander | 44 | 2 | 2 | 1 | .. | 2 | 5 | 2 | 400 | 5000 | 3 | 165 | 210 | .. | .. | 15 | 30 | .. | .. | 25 | | |
| C. W. Heydrick | Heydrick Oil Co. and Heydrick Bros. | 194 | 1 | 8 | 6 | .. | 1 | 15 | 15 | 375 | 4000 | 5 | 175 | 350 | .. | .. | 20 | 30 | .. | .. | 40 to 70 | | |
| McKissick | Barrows, Hazelton & Co. and Pithole and Allegheny Pet. Co. | 400 | .. | 15 | 5 | .. | .. | 20 | .. | 400 to 700 | 4000 | 6 | 170 | 290 | .. | .. | 20 | 15 | .. | .. | 50 | | |
| Judge McCalmont. | Farrar Oil Co. of Boston | 291 | .. | 1 | .. | .. | .. | 1 | .. | 650 | 6000 | 1 | 125 | 245 | .. | .. | 12 | 25 | .. | .. | 35 | | |
| S. & J. Armstrong. | Winona Oil Co. | 175 | .. | 8 | .. | .. | .. | 8 | .. | 300 | 4000 | 4 | 125 | 250 | .. | .. | 10 | 30 | .. | .. | 45 | | |

| | | | | | | | | | | | | | | | | | | | | | |
|---|---|---|---|---|---|---|---|---|---|---|---|---|---|---|---|---|---|---|---|---|---|
| J. Henry | Hussey & McBride | 150 | .. | 6 | 1 | .. | .. | 7 | .. | 300 | 4000 | 4 | 125 | 250 | .. | .. | 10 | 30 | .. | .. | 45 |
| H. Culbertson | Beekman Oil Co. | 150 | .. | 4 | 1 | .. | .. | 5 | .. | 350 | 5000 | 4 | 125 | 250 | .. | .. | 10 | 30 | .. | .. | 45 |
| Ludlow & McCabe | Michigan Co. and Boston Co. | .. | .. | 8 | .. | .. | .. | 8 | | | | | | | | | | | | | |
| A. Culbertson | Commercial Oil Co. | 190 | .. | 7 | .. | .. | .. | 7 | .. | 300 | 4000 | 4 | 130 | 260 | .. | .. | 12 | 25 | .. | .. | 50 |
| H. & J. McCrea | Eagle Rock Oil Co. | 404 | .. | 7 | 3 | .. | .. | 10 | .. | 550 | 6000 | 8 | 178 | 275 | .. | .. | 14 | 30 | .. | .. | 50 |
| R. Clapp | President Pet. Co. | 8400 | .. | 15 | .. | .. | .. | 15 | .. | 500 to 900 | 4000 | 5 | 124 | 230 | 724 | .. | 11 | 10 | 24 | .. | 36 |
| R. E. Harper | McCrea & Cherry Run | .. | .. | 1 | .. | .. | .. | 1 | .. | 600 | 6000 | 1 | | | | | | | | | |
| " " | Great Western Oil Co. | 100 | .. | 2 | .. | .. | 2 | 2 | 6 | 450 | 5000 | 3 | 124 | 230 | .. | .. | 11 | 10 | .. | .. | 25 |
| R. P. Elliott | Farrar Oil Co. | 200 | .. | 3 | .. | .. | .. | 3 | .. | 483 | 4000 | 2 | | | | | | | | | |
| Dr. Smiley | Richardson & Brognard | 4 | .. | 3 | 4 | .. | .. | 7 | .. | 300 | 3000 | 2 | 124 | 300 | .. | .. | 11 | 10 | .. | .. | 19 |
| H. Caldwell | H. Caldwell | 17 | .. | 3 | .. | .. | .. | 3 | .. | 350 | .. | 2 | 230 | 320 | .. | .. | 22 | 24 | .. | .. | 10 |
| John Shaw | Syracuse Oil Co. | 100 | .. | 1 | .. | .. | .. | 1 | .. | 700 | 6000 | 1 | 280 | 390 | .. | .. | 20 | 25 | .. | .. | 12 |
| E. Shaw | Dill, Hope, Penn Yan, International, and Bailey Oil Cos. | 10 | .. | 6 | 1 | .. | .. | 7 | .. | 650 | 6000 | 5 | 280 | 390 | .. | .. | 20 | 25 | .. | .. | 12 |
| Keyser | Baltimore Pet. Co. | 100 | .. | 1 | 1 | .. | .. | 2 | .. | 650 | 6000 | 1 | | | | | | | | | |
| Luther Wood | Venango Region Pet. Co., Excelsior Co., and others. | 139 | .. | 10 | 4 | .. | .. | 14 | .. | 600 | 6000 | 6 | 280 | 390 | .. | .. | 20 | 25 | .. | .. | 12 |
| J. A. Rickets | Faunce & Auburn and Rickets Bros. | 160 | .. | 7 | .. | .. | .. | 7 | .. | 600 | 6000 | 6 | 290 | 300 | .. | .. | 20 | 25 | .. | .. | 12 |
| Luther Wood | Phillips Pet. Co. and New World Pet. Co. | 150 | .. | 9 | .. | .. | .. | 9 | .. | 700 to 1000 | 6000 | 7 | 290 | 300 | .. | .. | 20 | 25 | .. | .. | 12 |
| Prather | Prather Bros. | 50 | .. | 3 | .. | .. | .. | 3 | .. | .... | .. | 2 | | | | | | | | | |
| Lake | N. Y. Star Pet. & Mining Co. | 60 | .. | 2 | .. | .. | .. | 2 | .. | 750 | 7500 | 2 | 36 | 235 | 575 | 640 | 84 | 18 | 20 | 15 | 30 |
| Reynolds | Lehigh Co. | 40 | .. | 2 | 3 | .. | 2 | 5 | .. | 740 | 8000 | 5 | 25 | 233 | 455 | 548 613-710 | 54 | 20 | 35 | 25 | 25 |
| .... | Tract above Lehigh | .. | .. | 4 | .. | .. | .. | 4 | .. | .... | .. | 2 | | | | | | | | | |
| Reynolds | Central Basin | .. | .. | 1 | 1 | .. | .. | 2 | .. | 883 | 7000 | 2 | | | | | | | | | |
| Wood & Copeland | Phillips Bros. and American Illuminating Co. | 160 | .. | 9 | 5 | .. | .. | 14 | .. | 630 | 5000 | 6 | 60 | 309 | 342 | 436 5th-595 | 80 | 5 | 40 | 40 | 16 |
| Reynolds | Philpot, Sherman & Co. and Garden City Co. | .. | .. | 11 | .. | .. | .. | 11 | .. | 650 | 6000 | 5 | 60 | 309 | 342 | 436 5th-595 | 80 | 5 | 40 | 40 | 16 |
| Blackmer | Prentice, Clark & Seeley | 104 | .. | 2 | 18 | .. | .. | 20 | .. | 650 | 6000 | 10 | 180 | 350 | 460 | 605 | 40 | 30 | 30 | 13 | 12 |
| Mrs. S. Rooker | Bonta & Bates, and Drew and others. | 100 | 7 | 2 | 45 | 3 | 4 | 58 | 2000 | 615 | 6000 | 50 | 95 | 340 | 440 | 595 | 35 | 30 | 18 | 14 | 24 |
| Thos. Holmden | Wright & Chittendon, and others. | 160 | 20 | .. | 115 | 4 | 16 | 135 | 3000 | 612 | 6000 | 75 | 95 | 340 | 440 | 595 | 35 | 30 | 18 | 14 | 24 |
| .... | Miller Farm & Pithole Pipe Co. | .. | .. | 2 | .. | .. | .. | .. | .. | .... | .. | 4 | Surface | 120 | 500 | 600 | | | | | |
| R. Hyner | Botolph Oil Well Co. and Hyner Oil Co. | 150 | 7 | 2 | 17 | 4 | 3 | 26 | 250 | 615 to 670 | 6000 | 20 | 120 | 395 | 500 | 600 | 50 | 80 | 30 | 10 to 68 | 10 |
| P. T. Copeland | Albert G. Mowry & D. H. Burtiss and Copeland Reserve. | 106 | 8 | .. | 62 | 4 | 4 | 70 | 1500 | 625 | 6000 | 60 | 120 | 395 | 480 | 687 | 65 | 65 | 35 | 18 | 25 |
| W. McKinney | Second National Pet. Co. and U. S. Pet. Co. | 75 | 5 | .. | 50 | 4 | 1 | 55 | 300 | 640 | 7000 | 30 | 150 | 400 | 472 | 633 | 52 | 24 | 27 | 12 | 54 |

| Original Owner of Farm. | Present Owner of Farm. | No. of Acres. | No. of Producing Wells. | No. of Non-Producing Wells. | No. in Progress. | No. Flowing Wells. | No. Pumping Wells. | Total No. of Wells. | Average Daily Production. | Average Depth of Wells. | Average Cost of Wells. | No. of Engines. | Depth of First Sand Rock. | Depth of Second Sand Rock. | Depth of Third Sand Rock. | Depth of Fourth Sand Rock. | Thickness of First Sand Rock. | Thickness of Second Sand Rock. | Thickness of Third Sand Rock. | Thickness of Fourth Sand Rock. | Depth of Driving Pipe. | No. of Refineries. | Weekly Capacity Bbls. Crude. |
|---|---|---|---|---|---|---|---|---|---|---|---|---|---|---|---|---|---|---|---|---|---|---|---|
| C. M. Ball | S. Zentz, J. G. Dale, and M. Henry | 106 | 1 | .. | 17 | .. | 1 | 18 | 30 | 650 | $6000 | 15 | 124 | 357 | 457 | 595 | 24 | 40 | 30 | 20 | 30 | | |
| J. Dawson | W. B. Tuell & Co., J. T. Sawyer & Co., and others. | 175 | 1 | 10 | 21 | .. | 1 | 32 | 30 | 630 | 5500 | 20 | 124 | 357 | 457 | 595 | 30 | 40 | 30 | 20 | 35 | | |
| Samuel Blank | Irving Pet. Co. of N. Y. | 106 | .. | 3 | 2 | .. | 1 | 5 | .. | 688 | 7500 | 3 | 130 | 400 | 448 | 645 | 12 | 8 | 14 | 6 | 18 | | |
| J. Blank | Pithole Creek Pet. Co. | 80 | .. | .. | 4 | .. | .. | 4 | .. | 620 | 7000 | 4 | 123 | 380 | 443 | 500 | 12 | 10 | 22 | 15 | 30 | | |
| Haworth | G. L. & J. W. Haworth | 35 | .. | 2 | 1 | .. | 2 | 3 | .. | 650 | 7000 | 2 | 125 | 365 | 455 | 5th 598 510 | 25 | 45 | 25 | 25 | 9 to 27 | | |
| Upper Haworth | " " " | .. | .. | 8 | .. | .. | .. | 8 | .. | 650 | 7000 | 8 | 130 | 370 | 460 | 515 | 25 | 45 | 25 | 25 | 27 | | |
| G. L. Haworth | Genesee & Venango Oil Co. | 126 | .. | 2 | 6 | .. | .. | 8 | .. | 625 | 5500 | 7 | 125 | 365 | 455 | 510 | 25 | 45 | 25 | 25 | 9 | | |
| Stewart | (Not ascertained.) | .. | .. | 2 | 2 | .. | .. | 4 | .. | 625 | 6000 | 2 | | | | | | | | | | | |
| Conley | Rochester Pet. Co. | .. | .. | 5 | 5 | .. | 5 | 10 | .. | 600 | 6000 | 7 | | | | | | | | | | | |
| J. Siggins | Geneva Oil Co. | 53 | 1 | 2 | .. | .. | 2 | 3 | .. | 631 | 8000 | 2 | 332 | 398 | 488 | 602 | 45 | 45 | 25 | 25 | 18 | | |
| Pratt | Chichester, Mich. Co., and Rochester Co. | .. | .. | 3 | .. | .. | 1 | 3 | .. | 600 | 6000 | 2 | 232 | 398 | 488 | 602 | 45 | 45 | 25 | 25 | 18 | | |
| Vanwykle | (Not ascertained.) | .. | .. | 2 | .. | .. | .. | 2 | .. | 525 | 6000 | 1 | 203 | 375 | 450 | 600 | 15 | 30 | 12 | .. | 20 | | |
| Conley | Brooklyn Pet. Co. | .. | .. | 1 | .. | .. | .. | 1 | .. | 600 | 6000 | 1 | 268 | 410 | 570 | .. | 15 | 30 | 12 | .. | 20 | | |
| Magarrah | (Not ascertained.) | .. | .. | 1 | .. | .. | .. | 1 | .. | 625 | 6000 | 1 | 285 | 425 | 573 | .. | 15 | 20 | 14 | .. | 20 | | |
| Tyrell, Carson & Griffin | | .. | 1 | 3 | .. | .. | 1 | 4 | .. | 600 | 6000 | 3 | 285 | 425 | 573 | .. | 15 | 30 | 14 | .. | 20 | | |
| Walter Holmden | Walter Holmden's heirs | 120 | .. | 2 | .. | .. | .. | 2 | .. | 615 | 6000 | 2 | 95 | 340 | 450 | 500 | 35 | 30 | 18 | 13 | 23 | | |
| Widow Lyons | Webster Pet. Co. | 110 | .. | 3 | .. | .. | 2 | 3 | .. | 700 | 8000 | 3 | 120 | 250 | 450 | 560 | 75 | 3 | 12 | 13 | 12 | | |
| Clark | Pittsb. & Phila. Pet. Co. | 200 | .. | .. | 4 | .. | .. | 4 | .. | 715 | 8000 | 4 | | | | | | | | | | | |
| Steen | Paxton Pet. Co. | 107 | 5 | .. | 14 | .. | 5 | 19 | 55 | 700 | 6000 | 19 | 245 | 550 | 608 | 713 | 35 | 22 | 25 | 13 | 10 | | |
| A. Turner | Geo. Cathrall, Dr. Egbert, S. Fell, Beacon Oil Co., Benton Oil Co. | 106 | .. | .. | 7 | .. | .. | 7 | .. | 735 | 6000 | 7 | 101 | 583 | 653 | 745 | 26 | 22 | 20 | 15 | 10 to 40 | | |
| Wm. Vose | West Pithole Pet. Co. | 106 | 1 | .. | 6 | .. | 1 | 7 | 25 | 750 | 7000 | 7 | 145 | 456 | 610 | 710 | 40 | 40 | 25 | 12 | 45 | | |
| R. Brumagin | Brumagin Farm Pet. Co. | 134 | .. | .. | 4 | .. | .. | 4 | .. | .... | ... | 4 | | | | | | | | | | | |

| | | | | | | | | | | | | | | | | | | | | | |
|---|---|---|---|---|---|---|---|---|---|---|---|---|---|---|---|---|---|---|---|---|---|
| S. Austin | Davis & Parker | 107 | .. | .. | 3 | .. | .. | 3 | .. | 700 | 6000 | 3 | 54 | 160 | 598 | 648 | 28 | 20 | 23 | .. | 52 |
| Mill | (Not ascertained.) | .. | 1 | .. | 4 | .. | .. | 5 | .. | 800 | 7000 | 1 | 330 | 436 | 735 | 760 | | | | | |
| | Second National Oil Co. | .. | 1 | .. | .. | .. | .. | 1 | 75 | 806 | 7000 | 1 | 7-Sand rocks | | | | | | | | |
| | At Pleasantville | .. | 1 | .. | 1 | .. | .. | 2 | 75 | 891 | 7000 | 1 | | | | | | | | | |
| J. S. Elliott | Walter Scott Pet. Co., Boston, Hemlock Pet. Co., Pittsburgh. | 150 | 1 | 2 | .. | .. | 1 | 3 | .. | 600 | 5000 | 2 | 300 | 340 to 560 | .. | .. | 10 | 5 to 14 | .. | .. | 40 |
| M. McCrea | Culbertson Run Oil Co. | .. | .. | 1 | .. | .. | .. | 1 | .. | 630 | ... | 1 | | | | | | | | | |
| McCalmont | Fryburg Co. of Phila. | 100 | .. | 2 | .. | .. | .. | 2 | .. | 500 | 3000 | 2 | 220 | 340 | Doubtful 440 | .. | 15 | 40 | 80 | | 40 |
| D. Smith | Oceanic Oil Co., Mercantile Co., Farrar Oil Co. | 238 | .. | 4 | 2 | .. | 1 | 6 | .. | 575 | 5000 | 6 | 172 | 287 | .. | .. | 45 | 8 | .. | .. | 31 |
| H. P. & | H. C. McCalmont | 200 | .. | 1 | .. | .. | .. | 1 | .. | 360 | | | | | | | | | | | |
| Thomas McCalmont | Farrar Oil Co., Boston, and Samuel Duff, Pittsburgh. | 212 | 1 | 1 | .. | .. | 1 | 2 | .. | 475 | 5000 | 2 | 140 | 230 | 430 | .. | 12 | 15 | 40 | .. | 8 |
| J. Brewer | Germania Oil Co., Progress Oil & Mining Co. | 74 | .. | 2 | .. | .. | .. | 2 | .. | 600 to 1000 | 8000 | 2 | 215 | 295 | .. | .. | 15 | 32 | .. | .. | 49 |
| James Myres | Wa-Bun Oil Co. | 14 | .. | 1 | .. | .. | .. | 1 | | | | | | | | | | | | | |
| A. Sayer | Caledonia Oil Co. | 120 | .. | 3 | .. | .. | 1 | 4 | .. | 600 | 6000 | 3 | | | | | | | | | |
| Vangiesen | Vangiesen heirs | .. | .. | 3 | .. | .. | .. | .. | .. | 500 | 5000 | 3 | | | | | | | | | |
| J. S. McCalmont | Onondaga Oil Co., Baum tract, Cicero Oil Co. | 400 | .. | 5 | .. | .. | .. | 5 | .. | 478 | 6000 | 3 | | | | | | | | | |
| Alexander Holeman | { C. Curtiss<br>{ Also Holeman Island | 60<br>475 | .. | 2 | .. | .. | .. | 2 | .. | 650 | 5000 | 1 | 16 | 260 | 564 | .. | 200 | 20 | 86 | .. | 14 |
| Maple Islands | Drs. Harrison & Mockeridge, of Phila. | 4 | | | | | | | | | | | | | | | | | | | |
| Alex. Holeman, Jr. | Phœnix Oil Co. of Phila. | 154 | .. | 2 | .. | .. | .. | 2 | .. | 450 to 750 | 6000 | 1 | | | | | | | | | |
| J. M. Clapp | J. M. Clapp | .. | .. | 2 | .. | .. | .. | 2 | .. | 300 | | | | | | | | | | | |
| Wm. & Eli Pierson, and M. Elder. | C. Curtiss | 316 | .. | 1 | .. | .. | .. | 1 | .. | 930 | 5000 | 1 | 183 | None regular | .. | .. | 20 | .. | .. | .. | 47 |
| D. Hunter | Illinois Petroleum & Mining Co. of Chicago. | 100 | .. | 4 | .. | .. | .. | 4 | .. | 1000 | 6000 | 3 | 125 | 260 | 310 | .. | 4 | 30 | 30 | .. | 50 |
| A. Dustin | Dustin Run & Chicago Oil Co., Phila. | .. | .. | 1 | .. | .. | .. | 1 | .. | 450 | 7000 | 1 | | | | | | | | | |
| Wm. N. Watson | Security Pet. Co., N. Y. | 50 | .. | 3 | .. | .. | .. | 3 | .. | 740 | 7000 | 3 | 135 | .. | .. | .. | 40 | .. | .. | .. | 40 |
| Chas. Holeman | " " " | 125 | .. | 4 | .. | .. | .. | 4 | | | | | | | | | | | | | |
| W. McClatchey | " " " | 116 | .. | 1 | .. | .. | .. | 1 | .. | 750 | 7000 | 1 | | | | | | | | | |
| S. McCray | Hamlin & Moore | .. | .. | 1 | .. | .. | .. | 1 | | | | | | | | | | | | | |
| A. Towner | Lanier, N. Y. | .. | .. | 1 | .. | .. | .. | 1 | | | | | | | | | | | | | |
| Ananias Dustin | Perry Oil Co., Pittsburgh | 25 | .. | 2 | .. | .. | .. | 2 | .. | 200 to 600 | | | | | | | | | | | |
| Wm. Halings | Phila., Lancaster & Cherry Run Oil Co. | 100 | .. | 1 | .. | .. | .. | 1 | .. | 600 | 5000 | 1 | .. | .. | .. | .. | .. | .. | .. | .. | 15 |
| J. Shreves | Shreve Farm Oil Co., N. Y. | 84 | .. | 3 | .. | .. | .. | 3 | .. | 525 to 910 | 8000 | 3 | 245 | 560 | 890 | .. | 23 | 11 | 40 | .. | 30 |

| Original Owner of Farm. | Present Owner of Farm. | No. of Acres. | No. of Producing Wells. | No. of Non-Producing Wells. | No. in Progress. | No. Flowing Wells. | No. Pumping Wells. | Total No. of Wells. | Average Daily Production. | Average Depth of Wells. | Average Cost of Wells. | No. of Engines. | Depth of First Sand Rock. | Depth of Second Sand Rock. | Depth of Third Sand Rock. | Depth of Fourth Sand Rock. | Thickness of First Sand Rock. | Thickness of Second Sand Rock. | Thickness of Third Sand Rock. | Thickness of Fourth Sand Rock. | Depth of Driving Pipe. | No. of Refineries. | Weekly Capacity Bbls. Crude. |
|---|---|---|---|---|---|---|---|---|---|---|---|---|---|---|---|---|---|---|---|---|---|---|---|
| James Watson | Hamlin & Moore | 100 | .. | 1 | .. | .. | .. | 1 | .. | 370 | .... | 1 | 240 | .. | .. | .. | 14 | .. | .. | .. | 50 | | |
| Wm. Hunter | Pennock, Ball & Co., Pittsburgh | 200 | .. | 2 | .. | .. | .. | 2 | .. | 600 | .... | 1 | 10 | 280 | 530 | .. | 40 | 15 | .. | .. | 10 | | |
| G. S. Hunter | G. S. Hunter | 15 | .. | 1 | .. | .. | .. | 1 | .. | 530 | $7000 | 1 | .. | 280 | 530 | .. | 15 | 15 | .. | .. | 10 | | |
| C. Sowers | E. & J. Fawcett | 100 | .. | 10 | .. | .. | .. | 10 | .. | 150 to 500 | .... | 2 | | | | | | | | | | | |
| Ensign & Jamison | Denver Pet. Co., N. Y. | 162 | .. | 13 | .. | .. | .. | 13 | .. | 250 to 730 | 5000 | 2 | 101 | 149 | 206 | .. | 35 | 14 | 14 | .. | 13 | | |
| R. & G. Jamison | Jamison Oil Co., Phila. | 69 | .. | 1 | .. | .. | 1 | 1 | .. | 408 | 5000 | 1 | 101 | 149 | 206 | .. | 35 | 14 | 14 | .. | 13 | | |
| Gilbert Jamison | Gilbert Jamison | 80 | .. | 1 | .. | .. | .. | 1 | .. | 557 | | | | | | | | | | | | | |
| John Jamison | Sinclair & Co., Pittsburgh | 250 | .. | 5 | .. | .. | .. | 5 | | | | | | | | | | | | | | | |
| J. Dale's heirs | (Not ascertained.) | 250 | .. | 2 | .. | .. | .. | 2 | | | | | | | | | | | | | | | |
| J. G. Dawson's heirs | J. G. Dawson's heirs | 300 | .. | 1 | .. | .. | .. | 1 | .. | 160 | .... | 1 | | | | | | | | | | | |
| Dawson & Gordon | Dawson Run Oil Co. | 200 | .. | 5 | .. | .. | .. | 5 | .. | 575 | 6000 | 4 | 220 | 489 | 780 | .. | 16 | 30 | 30 | .. | 6 | | |
| J. Gordon | C. Curtis | 437 | .. | 3 | .. | .. | 1 | 3 | .. | 315 to 812 | 4000 | 4 | 180 | 387 | 780 | .. | 30 | 48 | 30 | .. | 44 | | |
| J. Green | Hickory Farm Oil Co., West Hickory Oil & Mining Co. | 800 | 4 | 4 | 50 | .. | 3 | 56 | .. | 111 to 240 & 420 | .... | 7 | 13 to 31 | 75 to 83 | 128 to 154 | 409 to 425 | 15 | .. | .. | .. | 16 | | |
| Luther Bennett | (Not ascertained.) | 50 | .. | 1 | .. | .. | .. | 1 | .. | 250 | | | | | | | | | | | | | |
| M. K. Riddle | Forrest Landing, Pittsburgh and Phila. | 100 | .. | 4 | .. | .. | .. | 4 | .. | 350 | .... | 2 | 74 | 171 | .. | .. | 36 | 36 | .. | .. | 18 | | |
| C. Richardson | C. Richardson | .. | .. | 1 | .. | .. | .. | 1 | .. | 300 | .... | 1 | | | | | | | | | | | |
| S. Hunter | (Not ascertained.) | .. | .. | 1 | .. | .. | .. | 1 | .. | 300 | .... | 1 | | | | | | | | | | | |
| T. H. Prather | A Philadelphia company | .. | .. | 3 | .. | .. | .. | 3 | .. | 700 | .... | 2 | | | | | | | | | | | |
| Ball. | | .. | .. | 1 | .. | .. | .. | 1 | | | | | | | | | | | | | | | |
| Various Farms | Between this point and Tidioute. | .. | .. | 32 | .. | .. | .. | 32 | .. | 200 to 500 | .... | 8 | | | | | | | | | | | |
| J. Green | | .. | .. | 2 | .. | .. | .. | 2 | .. | 500 | .... | 2 | | | | | | | | | | | |
| G. S. Siggins | | .. | .. | 1 | .. | .. | .. | 1 | .. | 250 | .... | 1 | | | | | | | | | | | |
| J. C. Siggins | N. Y. & Phila. Pet. Co. | 100 | .. | 3 | .. | .. | .. | 3 | .. | 300 | .... | .. | 100 | 200 | .. | .. | 25 | 40 | .. | .. | 14 | | |
| S. Hawthorn | S. Hawthorn | .. | .. | 1 | .. | .. | .. | 1 | .. | 300 | | | | | | | | | | | | | |
| D. Scott | D. Scott | .. | .. | 2 | 2 | .. | .. | 4 | .. | 400 | | | | | | | | | | | | | |
| E. & E. L. Jones | E. Ferrero and others | .. | .. | 3 | .. | .. | .. | 3 | .. | ... | .... | 1 | | | | | | | | | | | |

| | | | | | | | | | | | | | | | | | | | | | | | |
|---|---|---|---|---|---|---|---|---|---|---|---|---|---|---|---|---|---|---|---|---|---|---|---|
| S. Smith | G. S. Long | .. | .. | 4 | .. | .. | .. | 4 | .. | .... | .... | 1 | | | | | | | | | | | |
| | T. H. Prather heirs, or Woodford | .. | .. | 1 | .. | .. | .. | 1 | .. | 250 | .... | 1 | | | | | | | | | | | |
| W. Hunter | Hopewell & Jennings | .. | .. | 1 | .. | .. | .. | 1 | .. | 280 | .... | .. | 96 | 130 | .. | .. | .. | .. | .. | .. | 14 | | |
| P. H. & J. Roortman | Grove Farm Oil Co. of Pittsburgh. | 500 | .. | Several | 1 | .. | .. | 1 | -- | | | | | | | | | | | | | | |
| J. Dunn | Hopewell & Jennings | .. | .. | 3 | 1 | .. | .. | 4 | .. | 300 | .... | 1 | | | | | | | | | | | |
| L. D. Richardson | R. A. Wilder and others | .. | .. | 6 | .. | .. | .. | 6 | .. | 60 to 500 | .... | 1 | .. | .. | .. | .. | .. | .. | .. | .. | 10 | | |
| D. & J. Ralston | Ralston and others, Philadelphia Co. | .. | .. | 18 | .. | .. | .. | 18 | .. | 300 | .... | 3 | | | | | | | | | | | |
| Wm. Davidson | Economy Oil Co. | 6000 | 22 | 32 | .. | .. | 22 | 54 | 600 | 500 | 3500 | 40 | 165 | 312 | 408 | .. | 10 | 18 | 23 | .. | 15 | | |
| Robert Henry | New York & Allegheny Oil Co. of N. Y. | 300 | 5 | 51 | 25<br>30 } To be drilled | .. | 5 | 65 | 365 | 325 | 3000 | 30 | 70 | 215 | 325 | .. | 18 | 20 | 40 | .. | 10 | | |
| Samuel Parshall | Tidioute & Warren Oil Co. | 400 | 21 | 26 | .. | .. | 21 | 56 | 300 | 400 | 4000 | 34 | On river 70<br>125 | 215 | 325 | .. | 18 | 20 | 40 | .. | 10 | | |
| E. Jones | New York & Dennis Run Oil Co. of N. Y. | .. | .. | .. | 12 | .. | .. | 12 | .. | .... | .... | 8 | | | | | | | | | | | |
| S. Tipton | Tipton Farm Oil Co. of Phila. | 100 | . | A large number | .. | .. | | Large number | | | | | | | | | | | | | | | |
| McGuyr | McGuyr Run Pet. Co. | .. | .. | 1 | .. | .. | .. | 1 | .. | .... | .... | 1 | | | | | | | | | | | |
| Parshall tract | Oil City Pet. Co. | 40 | .. | 1 | .. | .. | .. | 1 | | | | | | | | | | | | | | | |
| Henry farm | Midas Oil Co. of N. Y. & Pittsb. | 300 | .. | .. | 4 | .. | .. | 4 | .. | .... | .... | | .. | .. | .. | .. | .. | .. | .. | .. | .. | 6 | 1200 |
| Brown tract | Minturn & Bro. | 600 | .. | .. | 2 | .. | .. | 2 | .. | .... | .... | 2 | 120 | 210 | 320 | .. | 42 | 30 | 13 | .. | 10 | | |
| Kinnier tract | Tidioute Association & Empire Oil Co. of N. Y. | .. | .. | 6 | .. | .. | .. | 6 | .. | .... | .... | 3 | | | | | | | | | | | |
| Arter estate | .... | .. | .. | 4 | .. | .. | .. | 4 | .. | .... | .... | 1 | .. | .. | .. | .. | .. | .. | .. | .. | .. | 1 | |
| Jones' property | Demming mill property | 15 | .. | 1 | .. | .. | .. | .. | .. | .... | .... | 1 | | | | | | | | | | | |
| Bristol & Preston | Tidioute Oil & Lumbering Co., of N. Y. | 1000 | .. | .. | 2 | .. | .. | 2 | .. | .... | .... | 2 | | | | | | | | | | | |
| Thompson | Oil City Petroleum Co. | 300 | .. | .. | 1 | .. | .. | 1 | .. | .... | .... | 1 | | | | | | | | | | | |
| .... | Arcade Oil Co. of Rochester, N. Y. | .. | .. | 1 | 1 | .. | .. | 2 | .. | .... | .... | 1 | | | | | | | | | | | |
| C. Brown | Tremont Co. | .. | .. | 3 | .. | .. | .. | 3 | | | | | | | | | | | | | | | |
| H. Stowe | H. & H. H. Stowe | 5000 | .. | 1 | .. | .. | .. | 1 | | | | | | | | | | | | | | | |
| Bennett | Beaver Lumber & Mining Co. of Phila. | .. | .. | 1 | .. | .. | .. | 1 | | | | | | | | | | | | | | | |
| Tyndall property | .... | .. | .. | 1 | .. | .. | .. | 1 | .. | .... | .... | 1 | | | | | | | | | | | |
| Manross | West Hickory Mining Association of Phila. | 1000 | 7 | .. | 15 | .. | 7 | 22 | 225 | 240 | 3000 | 22 | 60<br>to | 147 | 190 | .. | 25 | 18 | 14 | .. | 10 | | |
| J. White | Hickory Creek Pet. Co. | 87 | 4 | .. | 25 | .. | 4 | 29 | 65 | 400 | 4000 | 29 | 115 | 200 | .. | .. | 25 | 18 | .. | .. | 10 | | |
| J. Fleming | Cherry & Trout Run Oil Co. of Phila. | 133 | .. | .. | 4 | .. | .. | 4 | .. | .... | .... | 4 | | | | | | | | | | | |

| Original Owner of Farm. | Present Owner of Farm. | No. of Acres. | No. of Producing Wells. | No. of Non-Producing Wells. | No. in Progress. | No. Flowing Wells. | No. Pumping Wells. | Total No. of Wells. | Average Daily Production. | Average Depth of Wells. | Average Cost of Wells. | No. of Engines. | Depth of First Sand Rock. | Depth of Second Sand Rock. | Depth of Third Sand Rock. | Depth of Fourth Sand Rock. | Thickness of First Sand Rock. | Thickness of Second Sand Rock. | Thickness of Third Sand Rock. | Thickness of Fourth Sand Rock. | Depth of Driving Pipe. | No. of Refineries. | Weekly Capacity Bbls. Crude. |
|---|---|---|---|---|---|---|---|---|---|---|---|---|---|---|---|---|---|---|---|---|---|---|---|
| J. Fleming, subdivision 2. | Pennock, Bagaley & Co., of Pittsburgh. | 160 | .. | .. | 2 | .. | .. | 2 | .. | .... | .... | 2 | | | | | | | | | | | |
| J. Fleming, subdivision 3. | Pennock, Bagaley & Co., of Pittsburgh. | 219 | .. | .. | .. | .. | .. | .. | .. | .... | .... | .. | 85 | 155 | ... | ... | 12 | 7 | .. | .. | 10 | | |
| J. Fleming, subdivision 4. | Pennock, Bagaley & Co., of Pittsburgh. | 219 | .. | .. | 11 | .. | .. | 11 | .. | .... | .... | 5 | 85 | 155 | ... | ... | 12 | 7 | .. | .. | 10 | | |
| J. Fleming, subdivision 5. | Pennock, Bagaley & Co., of Pittsburgh. | 269 | .. | .. | 5 | .. | .. | 5 | .. | .... | .... | 5 | 125 | 250 | 390 | ... | 10 | 15 | 15 | .. | 12 | | |
| R. H. Giles.......... | Shaw Giles, Howe & Eastman | 100 | .. | No developments. | | | | | | | | | | | | | | | | | | | |
| W. Gorman......... | W. Gorman.................. | 350 | .. | 2 | .. | .. | .. | 2 | .. | .... | .... | .. | 85 | 20 | ... | ... | 12 | 15 | .. | .. | 18 | | |
| H. Miles............ | Warren & Venango Oil Co. .. | 600 | 1 | .. | .. | .. | .. | 1 | 10 | 370 | .... | .. | ... | ... | ... | ... | .. | .. | .. | .. | 18 | | |
| Michael Gorman... | Adnah Neyhart & Salem Oil & Lumber Co. | 185 | .. | 2 | .. | .. | .. | 2 | .. | .... | .... | 2 | | | | | | | | | | | |
| " " ... | Reserve...................... | .. | .. | .. | 1 | .. | .. | 1 | .. | .... | .... | 1 | | | | | | | | | | | |
| D. Beaty............ | D. Beaty...................... | 421 | .. | .. | 2 | .. | .. | 2 | .. | .... | .... | 2 | | | | | | | | | | | |
| .................... | West Hickory Oil & Lumber Co. | 200 | .. | .. | 1 | .. | .. | 1 | .. | .... | .... | 1 | | | | | | | | | | | |
| Michael Gordon.... | Tremont Oil Co. of Phila..... | 94 | .. | .. | 1 | .. | .. | 1 | .. | .... | .... | 1 | | | | | | | | | | | |
| .................... | Drew property............... | 20 | .. | .. | 1 | .. | .. | 1 | .. | .... | .... | 1 | | | | | | | | | | | |
| Wm. Wallace....... | J. Pierce and Cherry Run & West Hickory Pet. Co. | 285 | .. | .. | 2 | .. | .. | 2 | .. | .... | .... | 2 | | | | | | | | | | | |
| R. C. Scott......... | Chicago Co.................... | .. | .. | .. | 4 | .. | .. | 4 | .. | .... | .... | 4 | | | | | | | | | | | |
| .................... | J. R. Smith & Co............. | 14 | .. | .. | 2 | .. | .. | 2 | .. | .... | .... | 2 | | | | | | | | | | | |
| .................... | Grove Farm Oil Co. of Pittsburgh. | 950 | .. | .. | 2 | .. | .. | 2 | .. | .... | .... | 2 | | | | | | | | | | | |
| Wm. Wallace....... | Wm. Wallace................. | 78 | .. | 1 | .. | .. | .. | 1 | .. | | | | | | | | | | | | | | |
| " " ...... | Triumph Oil Co.............. | 225 | 31 | 13 | 2 | .. | 31 | 44 | 600 | 700 | $4000 | 35 | 450 | 500 | 650 | ... | 20 | 25 | 50 In rock. | | | | |
| M. Walter.......... | Frazer & Co., of Rochester, N. Y. | 600 | .. | 1 | 1 | .. | .. | 2 | .. | .... | .... | 1 | | | | | | | | | | | |

| | | | | | | | | | | | | | | | | | | | | | |
|---|---|---|---|---|---|---|---|---|---|---|---|---|---|---|---|---|---|---|---|---|---|
| Walter Barnes..... | A Pittsburgh company....... | 300 | .. | 1 | .. | .. | .. | 1 | .. | .... | .... | 1 | | | | | | | | | |
| Barnes & Church... | Little Hickory Oil Co. of Phila. & N. Y. | 460 | .. | 1 | .. | .. | .. | 1 | .. | .... | .... | 1 | | | | | | | | | |
| Hatch.......... | Cincinnati & Pittsburgh Oil Co. | .. | .. | 2 | .. | .. | .. | 2 | .. | 760 to 1160 | 8000 | 2 | 225 | · 310 | ... | 1004 | 75 | 6 | Grey | .. | 18 |
| Jamison........ | Fahnestock & Bros., of Pittsburgh. | 40 | .. | 1 | .. | .. | .. | 1 | .. | 715 | .... | 1 | | | | | | | | | |
| Henry Sutley....... | Buffalo Co................... | .. | .. | 1 | .. | .. | .. | 1 | .. | 440 | .... | 1 | | | | | | | | | |
| G. G. Sickles....... | G. G. Sickles................. | .. | 1 | 1 | .. | .. | .. | 2 | .. | 600 | .... | 2 | 245 | 560 | 890 | ... | 23 | 11 | 40 | .. | 30 |
| D. Stowe........... | Pittsburgh, N. Y., and Tionesta Pet. Co. | 75 | .. | 3 | .. | .. | .. | 3 | .. | 600 to 1000 | .... | 2 | 300 | 900 | ... | ... | 40 | 60 | .. | .. | 36 |
| Conroy......... | Conroy................... | 50 | .. | 1 | .. | .. | .. | 1 | .. | 600 | .... | 1 | | | | | | | | | |
| May Bros........... | S. Q. Brown................. | 120 | .. | 1 | .. | .. | .. | 1 | .. | 713 | .... | 1 | | | | | | | | | |
| Hall & Leavy ...... | Tionesta Lumbering and Mining Co. | 5000 | .. | 3 | .. | .. | .. | 3 | .. | 700 | .... | 3 | | | | | | | | | |
| Wm. White........ | Wolf & Co. and Eastman & Co. | 1000 | .. | 2 | .. | .. | .. | 2 | .. | 400 | .... | 2 | | | | | | | | | |
| Johnson........ | Tionesta Creek Oil Co., of Buffalo, N. Y. | 960 | .. | 4 | .. | .. | .. | 4 | .. | 650 to 1074 | .... | 4 | | | | | | | | | |
| .................... | Marcy Oil, Lumbering and Mining Co. | 500 | .. | 2 | .. | .. | .. | 2 | .. | 650 | .... | 2 | 103 | 248 | ... | ... | 6 | 3 | | | |
| Wm. & Jno. Hunter | Banner Oil Co., of Pittsburgh | 200 | .. | 4 | .. | .. | .. | 4 | .. | 355 to 800 | .... | 4 | 200 | 310 | 370 | 420 | 10 | 20 | 10 | 25 | 8 |
| Richard Irvine..... | John Anderson Pet. Co., of N. Y. | 1500 | .. | 4 | .. | .. | .. | 4 | .. | 700 | 7000 | 4 | 100 | 250 | ... | ... | 8 | 15 | .. | .. | 17 |
| Edmonds & Sligo .. | Little Tionesta Pet. Co....... | 213 | .. | 2 | .. | .. | .. | 2 | .. | 800 | 6000 | 2 | 35 | 103 | ... | ... | 100 | .. | .. | .. | 35 |
| Taylor.......... | S. Q. Brown.................. | .. | .. | 2 | .. | .. | .. | 2 | .. | 200 | .... | 2 | | | | | | | | | |
| McKinley...... | Church & Hurbert........... | .. | .. | 4 | .. | .. | .. | .. | .. | 650 | .... | 3 | | | | | | | | | |
| Jonathan Hickard.. | Hutchins................. | 200 | 1 | .. | 1 | .. | .. | 2 | 3 | 650 | 8000 | 2 | 94 | 5–635 344 | 6–659 455 | 7–700 602 | 5–8 40 | 6–23 30 | 7–3 20 | 19 | 40 |
| Fair............ | Fair Farm Oil Co............. | 120 | .. | 1 | .. | .. | 1 | 2 | .. | 704 | 8000 | 2 | 94 2–344 | 3–455 4–602 | 5–635 6–659 | 7–700 | 40 30 | 20 19 | 8 23 | 3 | 40 |
| James Dawson..... | Pittsb. & Stewart's Run Oil Co. | 300 | 1 | 2 | .. | .. | 1 | 3 | .. | 655 | 8000 | 3 | 119 2–231 | 3–374 4–393 | 5–431 6–455 | 7–473 | 18 7 | 16 22 | 19 11 | 12 | 22 |
| M. Abbott.......... | Stewart's Pet. Co., of N. Y... | 180 | .. | 3 | 1 | .. | .. | 4 | .. | 700 | 8000 | 4 | " | " | " | " | " | " | " | " | 22 |
| John McKinley .... | Chicago Oil Co.............. | 87 | .. | .. | 1 | .. | .. | 1 | .. | .... | .... | 1 | | | | | | | | | |
| J. H. Dawson...... | Queens County Association... | 100 | .. | 1 | .. | .. | .. | 1 | .. | 606 | 5500 | 1 | 102 | 241 | 553 | ... | 16 | 26 | 70 | .. | 58 |
| Robert Green...... | Jocelyn Oil Land Association, N. Y. | 350 | .. | 2 | 1 | .. | 2 | 2 | .. | 730 | 6000 | 3 | 112 | 145 | 309 580 | 636 676 | 10 | 16 | 22 40 | 22 34 | 45 |
| John Jonnson...... | Boston Oil Co. & Capt. Pray, agent. | 30 | .. | 1 | .. | .. | .. | 1 | .. | 600 | 7000 | 1 | 100 | 244 | 584 | ... | 22 | 22 | 40 | .. | 18 |
| Leedham....... | Northern N. Y. Pet. Co....... | 67 | .. | 2 | 1 | .. | .. | 3 | .. | 708 | 7500 | 2 | 146 | 290 | 630 to 684 | | 22 | 22 | 17 | 15 | 18 |
| Vandine........ | Oak Ball Co. of Phila......... | 146 | .. | | | | | | | | | | | | | | | | | | |

| Original Owner of Farm. | Present Owner of Farm. | No. of Acres. | No. of Producing Wells. | No. of Non-Producing Wells. | No. in Progress. | No. Flowing Wells. | No. Pumping Wells. | Total No. of Wells. | Average Daily Production. | Average Depth of Wells. | Average Cost of Wells. | No. of Engines. | Depth of First Sand Rock. | Depth of Second Sand Rock. | Depth of Third Sand Rock. | Depth of Fourth Sand Rock. | Thickness of First Sand Rock. | Thickness of Second Sand Rock. | Thickness of Third Sand Rock. | Thickness of Fourth Sand Rock. | Depth of Driving Pipe. | No. of Refineries. | Weekly Capacity Bbls. Crude. |
|---|---|---|---|---|---|---|---|---|---|---|---|---|---|---|---|---|---|---|---|---|---|---|---|
| Wm. Willings | Oak Ball Co. of Phila. | 66 | .. | 1 | .. | .. | .. | 1 | .. | 630 | $5000 | 1 | 130 | 390 | 520 | ... | 18 | 30 | 40 | .. | 14 | | |
| J. Hoffman | " " | 100 | .. | 4 | .. | .. | .. | 4 | .. | 563 | 5000 | 4 | 170 | 460 | 520 | ... | 18 | 30 | 40 | .. | 14 | | |
| J. S. McCalmont | Onondaga Oil Co. | .. | .. | 1 | .. | .. | .. | 1 | .. | 750 | | | | | | | | | | | | | |
| Thos. McCalmont | Farrar Oil Co. | .. | .. | 1 | .. | .. | 1 | 1 | .. | 450 | | | | | | | | | | | | | |
| " | Pithole & Oil City Co. | .. | .. | 2 | .. | .. | .. | 2 | .. | 550 | | | | | | | | | | | | | |
| Youngs | Oak Ball Co., of Philadelphia | 100 | .. | 1 | .. | .. | .. | 1 | .. | 500 | 5000 | 1 | 170 | 460 | 520 | ... | 18 | 30 | 40 | .. | 14 | | |
| Switzer | ... | .. | .. | 4 | .. | .. | .. | 4 | .. | | | | | | | | | | | | | | |
| Stowell | Pittsburgh & Cherry Run Co. | 78 | 12 | .. | .. | .. | 12 | 12 | .. | 840 | 4500 | 15 | 112 | 290 | 520 | 5th 775<br>640 | 30 | 28 | 25 | 5th 50 | 14 & 40 | | |
| M. Goss | Shamburg Pet. Co. | 100 | 16 | .. | 2 | .. | 16 | 16 | .. | 900 | 5000 | 20 | " | " | " | " | " | " | " | " | | | |
| Tallman | Andrews, Backus, Stewart, and others. | 110 | 28 | 9 | .. | .. | 28 | 37 | 400 | 860 | 5000 | 32 | 100 | 315 | 545 | 5th 810 | 20 | 25 | 25 | 50 | | | |
| A. Dawson | S. Q. Brown, Bates, and others | 75 | 10 | .. | .. | .. | 10 | 10 | 500 | 850 | 5000 | 10 | 250 | 490 | 720 | 800 | .. | 40 | 40 | 20 | | | |
| Brown Bros. | Brown Bros. | 64 | 29 | 10 | .. | .. | 29 | 39 | .. | " | " | 28 | " | " | " | " | .. | " | " | " | | | |
| Thos. Mitchell | Holeman & Newkirk | 24 | 5 | .. | .. | .. | .. | 5 | 160 | " | " | 6 | " | " | " | " | .. | " | " | " | | | |
| A. Benedict | Brown & House | 40 | .. | .. | .. | .. | .. | .. | 100 | " | " | 7 | " | " | " | " | .. | " | " | " | | | |
| Benj. Tyrrell | Brown, Watson & Porter | 175 | 2 | .. | .. | .. | .. | 2 | .. | " | " | 2 | " | " | " | " | 20 | " | " | " | | | |
| Wm. Porter | Wm. Porter | 44 | .. | .. | .. | .. | .. | .. | 100 | " | " | 27 | " | " | " | " | " | " | " | " | | | |
| Wm. Dawson | McGrew Bros. & Co. | 50 | 8 | .. | 8 | .. | .. | 8 | .. | " | " | 16 | " | " | " | " | " | " | " | " | | | |
| Zuver | J. H. & H. B. Hebert | 60 | 12 | 4 | .. | .. | .. | 16 | 150 | .... | .... | 18 | " | " | " | " | " | " | " | " | | | |
| E. Byles | Vesta Pet. & Ref. Co. | 60 | 5 | 3 | .. | .. | .. | 8 | .. | .... | .... | 8 | | | | | | | | | | | |
| M. C. Beebe | Vesta Refining & Pet. Co. | 60 | 4 | 1 | .. | .. | 4 | 5 | .. | 800 | 5000 | 4 | .. | 400 | 680 | 770 | .. | 40 | 40 | 12 | 30 | | |
| E. L. Davis | J. H. & H. B. Hebert | 34 | 2 | 1 | .. | .. | 2 | 3 | .. | 760 | " | 2 | .. | 360 | 620 | 740 | .. | 35 | 40 | 20 | 40 | | |
| Ensign | Anderson & Carroll | 68 | 7 | .. | .. | .. | 7 | 7 | 100 | 800 | " | 12 | .. | " | " | " | .. | " | " | " | " | | |
| Gerow | Anderson, Carroll & Hawes | 50 | 1 | .. | 9 | .. | .. | .. | 75 | .... | " | 10 | | | | | | | | | | | |
| Connelly | Anderson, Carroll & Bates | 100 | | | | | | | | | | | | | | | | | | | | | |
| Drake tract | G. K. Anderson | 12 | 1 | .. | 2 | .. | .. | 3 | 20 | .... | " | 3 | | | | | | | | | | | |

| | | | | | | | | | | | | | | | | | | | | | |
|---|---|---|---|---|---|---|---|---|---|---|---|---|---|---|---|---|---|---|---|---|---|
| Brown Bros. | John Armstrong | 100 | 31 | 4 | 29 | .. | .. | 64 | .. | .... | " | 40 | | | | | | | | | |
| Dunham tracts | Dunhams | 350 | 3 | .. | 12 | | | | | | | | | | | | | | | | |
| .................. | National Petroleum Co. tract | 50 | 6 | .. | .. | .. | .. | .. | 100 | 800 | " | 9 | 330 | 545 | 660 | 759 | 20 | 24 | 20 | 15 | 25 |
| Capt. Lyle | Samuel Mitchell, of N. Y. | 113 | 4 | .. | 5 | .. | .. | 9 | 100 | .... | .... | 9 | | | | | | | | | |
| John Sherman | Thomas Woods | 70 | 46 | 24 | .. | .. | 46 | 70 | 350 | 875 | 4000 | 62 | 570 | 720 | 850 | .. | 40 | 20 | 45 | .. | 25 |
| John Pearson | Western Pennsylvania Oil Co. | 83 | 15 | 1 | .. | .. | 15 | 16 | 269 | 900 | " | 20 | 625 | .. | 890 | .. | .. | .. | .. | .. | 50 |
| Foster | Foster Farm Oil Co., of Phila. | 230 | 12 | 9 | .. | .. | 12 | 21 | 100 | 725 | " | 13 | 160 | 240 | 598 | .. | 18 | 20 | 16 | | |
| Foster | Foster Island Oil Co. | 6 | 2 | 5 | .. | .. | 2 | 7 | 14 | " | " | 2 | " | " | " | .. | " | " | " | .. | 25 |
| Miller | Miller Farm Oil Co. | 150 | .. | 15 | .. | .. | .. | 15 | | | | | | | | | | | | | |
| Bingham Land Estate. | Lewis, Bonsall & Co. | 120 | 3 | 2 | 1 | .. | .. | 6 | .. | 720 | " | 5 | 370 | 540 | 700 | .. | 20 | 28 | 12 | | |
| John Snider | Ven. Cen. & Duck Creek Pet. Co. | 50 | 7 | 13 | .. | .. | 7 | 20 | 80 | | | | | | | | | | | | |
| David Smith | Excelsior Oil Co. | 200 | .. | .. | .. | .. | .. | 15 | .. | 475 | .... | .. | 300 | 460 | .. | .. | 18 | 20 | | | |
| John Snider | Ogden Mining Co., of N. Y. | 60 | 1 | 11 | .. | .. | 1 | 12 | | | | | | | | | | | | | |
| Simeon Hickman | Hasson, Cornwall & Co. | 200 | 5 | 2 | .. | .. | 5 | 7 | 100 | 820 | 4000 | .. | 468 | 622 | 760 | .. | 48 | 25 | 20 | | |
| .................. | Lackawanna Oil Co. | | | | | | | | | | | | | | | | | | | | |
| Atkinson farm | Sherman, Barnsdall & Co. | 180 | 6 | 31 | 9 | .. | 6 | 37 | 60 | 880 | " | 15 | 100 | 310 | 557 | 769 | 100 | 25 | 15 | 25 | 35 |
| .................. | Church Run & Titusville | .. | 46 | 54 | 35 | .. | 46 | 100 | 450 | 600 | " | 60 | 225 | 410 | 500 | .. | 60 | 15 | 75 | 44½ | |

# CHAPTER XV.

## FLOOD, FIRE, AND FINANCIAL DISASTER.

No locality of similar extent has suffered more severely from the causes enumerated at the head of the chapter, than has the Oil Region. Though depressing to the most buoyant spirit, and well nigh ruinous at times, the energy of the people has always proved equal to the emergency. Scarce had the waters receded from the site of buildings which the flood had carried away or wrecked, ere a new foundation for a structure was being laid. After a destructive conflagration, new buildings were commenced often before the embers had ceased to smoulder. In like spirit has every disaster been met. We shall essay to give a brief sketch of these in the order they occurred:

The crush of oil boats at Oil City by an ice-gorge, December 7, 1862, was the first disaster of any magnitude. Owing to the great scarcity of oil in the Pittsburgh market, the low water, and extremely high rates of freight, an unusually large number of oil boats had accumulated at the wharves of Oil City. A large per centage of these were laden with bulk and barrel oil waiting anxiously to take advantage of the first flood or rise of water, and thus be enabled to reap the rich harvest that awaited them at Pittsburgh. A large amount of this oil had been purchased at the wells during the early portion of the season, at a cost of seventy-five cents per barrel. At the time of the disaster,

Pittsburgh quotations for crude were thirty-one and thirty-two cents per gallon.

On Friday snow fell, and the weather became cold, slush-ice forming in the river and Creek, with thirteen inches of water in the river channel. In addition to the previous large accumulation, 30,000 to 40,000 barrels of oil had come out of the Creek on pond freshets, a few days previous, and the boats containing it were moored at the different landings. The cold became more intense, and ice formed rapidly in the river on Saturday, when another heavy fall of snow occurred, slush-ice forming fast in the Creek. On Saturday night the river was effectually closed, ice having formed several inches in thickness, the cold increasing in intensity. The Creek also became blocked up with ice, but not to so great an extent as the river. Extra lines were put on the boats, and every preparation made to secure them from danger of wreck. Late in the forenoon of Sunday, the ice, which had formed a "gorge" in the Creek, broke loose and came out.

An "ice-gorge," as it is called, is occasioned by the blocks of ice, when the same is broken, grounding in shallow places, where it is often frozen to the bottom and is called "ground-ice." These form a dam in a short time, forcing the water back. This soon forces its way under the sheet-ice, causing it to break and pile up, frequently to the height of ten to fifteen feet in narrow streams. The water forced back by this mass of ice, in a short time bores its way through like an immense auger, causing it to give way, and sweeps it along, a mighty torrent of ice and water, that carries every thing before it.

The ice coming out of the Creek, and the great volume of water thus forced, went under the river ice, and lifted it up, causing a gorge. In a short time this began to move, the river being clear of ice for several miles below the lower landings. This torrent of ice and water received in addition the gorge of the Creek. The mass coming against the boats swept them from their moorings, breaking the strongest lines like yarn threads, crushing boats and barges into splinters, emptying their greasy cargoes into the swollen stream, making a general wreck. The whole affair only lasted a few minutes. Hemp, wood, and iron, all gave way before the mighty avalanche of ice and water. As estimated at the time, from 50,000 to 60,000 barrels of oil, and 200 boats were lost. The total loss, real and contingent, was over $500,000. In a few days after this disaster the river was clear of ice, and in good navigable order. But in this short time, the price of oil in Pittsburgh had fallen to nine cents per gallon, and receded during the rest of the season. Crude oil was selling at the time of this disaster, for five and six dollars at the wells, and in February, 1863, was any thing but brisk sale at forty cents per barrel.

The scene presented at the commencement and during this disaster, was one that is vividly impressed upon the memory of the writer. The oil boats for the entire length of the wharf, were thickly clustered together, in places half a dozen deep out into the stream. At the first pressure of the ice and water, the lines fastening the boats to the shore, became taut, then stretched to their utmost tension, inch and a half hawser-laid cable seeming scarcely thicker than one's finger, and singing like fiddle-strings. A few surges more of the flood, and these broke asunder with sounds like pistol-shots in quick succession, sweeping the boats into a maelstrom of destruction.

Similar recurrences took place in succeeding seasons, but no damage of moment ensued.

During 1863 and 1864 several fires occurred among the oil boats, one or two of them proving disastrous. The first of these was occasioned by one of the boatmen holding a lantern in the compartment of a bulk boat, in order to ascertain if the boat was leaking. The gas generated by the loose oil caught fire from the blaze of the lantern, and exploded with a shock that was felt all over the neighborhood, sadly damaging both man and lantern, and throwing jets of blazing oil over the adjoining boats, which ignited in a moment. In a very short time some forty boats, laden with oil, were blazing. Fortunately the fire occurred at the upper end of the wharf. By means of poles the burning boats were pushed out into the stream as they took fire, and floated down past the fleet with the current. The river was soon covered with the blazing fleet. The fire occurred at night. A dense black cloud, formed by the smoke of the burning oil, hung over the entire locality like a pall. The lower part of the cloud from the reflection of the flames was rose-colored, and frequently a beautiful pink. Every few moments a flash of light, resembling lightning, illumined the cloud, making a combination of colors exceedingly beautiful. The blazing boats floating on, the crowds of men stationed along the outside of the boats that remained fastened keeping off those that were on fire, their forms frequently obscured by the smoke, boats sinking, and the general confusion that reigned, furnished a fair *fac simile* of a naval battle, the explosion of the oil barrels from the intense heat, furnishing the artillery accompaniment. The oil from the burning boats ran out upon the surface of the water, and at times the whole river was ablaze.

One of the blazing boats passed under the suspension bridge across the Allegheny at Franklin, setting on fire and destroying it. Beyond the boatman, who was seriously burned by the explosion, no one was injured. The loss, including the bridge, was about $150,000.

Another fire occurred among a cluster of boats aground at the head of the island opposite the mouth of Oil Creek, at a subsequent period. It was produced from the same cause as the one described, and was similar in its general results. The adjoining boats became enveloped in flames almost instantaneously. One of these, not being grounded firmly, floated off in a direct line for the ferry landing, on the Venango City side of the river. When near the shore, as if impelled by some invisible force, it came nearly in a straight line against the current, back again to the Oil City side, and drifted alongside of the first of the boats of the oil fleet. For a few moments the destruction of the entire wharves, oil, boats, and even a large portion of the town seemed inevitable. Two row-boats were hastily manned, chains fastened to the bow of the burning boat, the oarsmen bent lustily to their work, and the blazing mass was towed past the entire fleet, and grounded on the island at the foot of the eddy. Anxious eyes followed the progress of the burning boat towed along by the gallant crews, and hearts throbbed audibly. The other boats being firmly grounded, burned as long as a drop of oil remained, and were entirely consumed. For a few hours the oil burned in the water, cracking and snapping with a din truly infernal, the water only increasing its intensity. Small streams of oil flowed over the water in different directions. These momentarily ignited, and the blaze, many yards in length, leaped upward to a height of fifteen to twenty feet, scattering sparkling spangles of fire in showers.

It is difficult to conceive the terrible aspect presented by a blazing mass of Petroleum. A thick black smoke is evolved by the burning oil, dense and suffocating in its nature, beyond any from the combustion of other substances. The heat is intense, beyond description; all combustible substances melt before it like frost work in the rays of the sun. To deluge it with water, only adds to its combustion and volume. All that can be done is to confine it to as small a space as possible, and wait until it ceases for want of fuel.

In some cases it is smothered out by the introduction of a powerful jet of steam, but such instances are rare. At burning wells earth has been used to cover up the mouth of the well, with invariable success. The Burning well, on the Allegheny River, is partially an exception, and has been burning from the date of its striking, over three years ago. When the flame was extinguished at the mouth of the well, it broke out from the surface of the earth surrounding in different directions, and even from the surface of the river. To prevent damage to the adjacent property, it was allowed to burn. This wonderful phenomenon is some fifteen miles below Franklin, on the Allegheny River, and a few rods distant from the line of the Allegheny Valley Railroad. This well has recently ceased to burn.

A number of frame buildings, on Main street, Oil City, among which was Michigan Block, burned during the period we have mentioned.

An extensive conflagration occurred in October, 1862, at the Blood farm, on Oil Creek. The gas from a flowing well taking fire from an engine near by was the cause. The space burned over embraced an extent of from twenty to thirty acres, on which were 100 wooden tanks, containing between 20,000 and 30,000 barrels of oil. Seven flowing and three pumping wells, with all their machinery were burned. The burning oil ran over the flats and the Creek. The scene was both grand and terrible. The loss was estimated at fully $1,000,000, some of the best wells being ruined.

The losses by pond freshets was very great, during the first and subsequent years. The most noted disaster of this kind transpired in May, 1864. It was occasioned by some boats becoming lodged against the pier of the bridge at Oil City, thus causing a "jam." From 25,000 to 30,000 barrels of oil and a large number of boats were lost. Oil was dipped from the surface of the water for days afterward, some of those engaged in the work procuring from 50 to 100 barrels per day. The business of dipping oil from the surface of the Creek and river after pond freshets or other disasters, was quite profitable, and furnished employment to a large number of persons. Operators with a limited amount of capital used boards or an oar stem, which was thrown out into the stream, acting as a boom, and forming an eddy into which the floating oil was collected. It was dipped or skimmed off the surface of the water with long-handled tin or iron dippers. Affluent dippers used an oil boat, swung broadside to the current. The oil collected against the side, and was dipped into the boat as described. After being subjected to a certain degree of heat to rid it of the water, an old boiler being used as a still, this oil became merchantable. Certain enterprising parties in the business, not content to await accidents for a harvest, bored holes with a gimlet in the bottoms of bulk boats. This plan was attended with such success, that augers were finally brought into requisition. The business soon became in bad odor with shippers, and gradually unprofitable.

The great flood of March, 1865, was the most rüinous in extent of any disaster that has ever transpired in the Oil Region. The total loss, from the Oil Region alone, was estimated at $5,000,000. We witnessed the flood, and can safely assert that but few have occurred that exceeded it in destructive force. During the winter, snow had accumulated to the depth of several feet on the hills of the Upper Allegheny, its tributaries and the headwaters of Oil Creek. Heavy rains fell for several days preceding the flood, causing the snow to melt rapidly, and swelling the streams.

On Wednesday the river commenced to rise, the rain still continuing. On Thursday both river and Creek came up rapidly, and during the day occurred the heaviest and most continuous showers of rain we have ever witnessed. It seemed, indeed, for a short time, that the "flood-gates of Heaven" were opened, and the Deluge, on a small scale, was about to be reënacted.

Haliday Run, a small stream ordinarily, passing through the centre of the west side of Oil City, swollen into a torrent, came pouring through Main street like a cataract, flooding the cellars of the buildings in its course, and setting afloat the oil barrels in Fisher Bros., and the adjoining oil yards. The rain continued during the entire day and night, and the water rose rapidly. At four o'clock, P. M., the water of the Creek was within four feet of the bridge that divides the town. The whole upper portion of the Hasson flats on east side of Creek, were covered with water, that stream having run over its banks, as was also Centre street, in places to a depth of three to four feet. A large number of oil boats, and great quantities of drift of various kinds came down the Creek. Two large oil boats wedged fast under the west end of the Creek bridge, directly against the abutment, lifting it from the foundation, and a large boat laden with oil was lying athwart the pier. The current of the Creek came down with terrific force, bearing upon its surface drift-wood, logs, whole boats, and parts of boats, and oil tanks; these latter being partly filled with water, were sufficiently low to pass under the bridge.

On the west side of the Creek, the water of the river came up to the doors of the upper warehouses on the landings, Shirk & Co.'s, and Fisher Bros.; while the warehouse of Parker & Castle, just below, was submerged to the first floor. Booms were constructed along the front of the upper wharves to keep the barrels from floating off.

The night was one of pitchy darkness, and the rain poured down unceasingly. At six o'clock, P. M., the river was rising at the rate of six inches an hour, and at twelve, M., it exceeded a foot. At seven, P. M., one half of the bridge gave way, and was carried into the stream. At ten, P. M., the other half of the bridge was swept away, severing the only mode of communication between the east and west portions of the city.

From this time the work of destruction went on. The river current was diverted from its usual channel ten to twelve rods, and poured in resistless force through the western part of the Oil City, directly through about the centre of the flat, between Main street and the river, on which were located the oil yards. Into this current the waters of the Creek discharged. Trees, drift-wood, and wreck of every description were borne along on this current, which swept every thing before it. Heavy boilers and engines were torn from their foundations, and carried considerable distances, as though they had been feathers, derricks thrown down, and heavy foundation timbers crushed like reeds. Houses, lifted from their foundations, became mingled with the general wreck. Nearly 100 buildings, it is estimated, were afloat on both sides of the Creek. In many of these were the occupants, not thinking for an instant that their houses would be carried away, until it was too late. They were rescued by the earnest efforts of the people, but lost most of their effects. But few of the participants will soon forget the horrors of that memorable Thursday night.

Never was the dawning of day more gratefully hailed than by these sufferers. The full realization of the destruction of the few previous hours came with it. The entire east side of the Creek was a complete wreck. Houses were floating over the flat, both above and below Centre street, a majority of the buildings on the upper portion of the flat having been carried away. Along Centre street, from the Gibson House, over midway to the bridge, most of the buildings were submerged to the second story. Several buildings on the upper side of the street had been lifted from their foundations, and carried into the street. The street itself was filled with floating fragments of boats, houses, wagons, furniture, and horses, and, in fact, débris of every description.

But the saddest sight of all was the groups of families assembled upon the higher ground, who had been compelled to leave their houses during the night and earlier part of the morning. Some had saved a portion of their effects, while others had scarcely saved wearing apparel sufficient to shield them from the chilly air. Suitable provision was made for them at the earliest practicable moment, and in a short time they were all suitably cared for and comfortably housed.

The space between Centre street and the river, was full of timber, drift, and other wreck, including a large number of oil barrels, full and empty. One of the large iron tanks of Graff, Hasson & Co., 8,000 barrels in capacity, was lifted from its foundation and partly turned over. A boom, made of empty oil boats, securely lashea together, prevented the wrecked matter from being carried away. Fortunately, too, the current of the Creek kept in its accustomed channel. The depth of water in Centre street was, in places, ten feet.

The appearance presented by the western portion of Oil City was equally as disastrous. For some distance below the bridge, Main street was flooded, the buildings, nearly to the Linden well, being submerged to a considerable depth. The whole distance to a point below was in a similar condition. Nearly opposite the Petroleum House, between Main street and the river, the current of the water from the Creek had set in toward the shore, and was materially assisted in its volume by the rushing flood of the river, now at its height. This current struck the river bank a short distance above Ferry street. The high elevation from the river to the street kept the current from going into Main street itself. Below this the water had struck with great force, clearing away every thing before it.

The warehouse of Shirk & Co., the first on the wharf, was about in the centre of this current. A large amount of iron tubing and heavy castings were stored in this warehouse, enabling it to withstand the current. The warehouses of Fisher Bros. and Parker & Castle, just below, held out well also, the three warehouses acting as a breakwater. At seven o'clock, A. M., the warehouse of Parker & Castle was carried away, and that of Fisher Bros. at nine o'clock, A. M. At ten o'clock, A. M., Shirk & Co.'s warehouse, after struggling long and well against the combined forces of the mighty element, succumbed, and went sailing out into the river. The force of the current being now unobstructed, swept with fearful violence through the oil yards below, carrying away oil barrels and every thing else that came in its course. During this time the surface of the river was covered with oil barrels, thousands having been carried out during the night and day. Huge giants of the forest, that had been torn from the banks by the force of the water, came sailing majestically along. Sixteen large wooden oil tanks went by in the course of an hour. Boats, derricks, and engine-houses, added to the variety of the wreck.

The water continued to rise slowly during Friday. On Friday night it commenced to fall so gradually as to be almost imperceptible, and during the night there was a severe storm of wind and rain. The wind blew a gale, and the rain fell in torrents. For a time it seemed that every thing would be leveled with the earth. Fortunately no serious damage ensued from the storm. During the latter part of the night the weather became several degrees colder, and several inches of snow fell. This checked the melting of the snow above, causing the river to fall rapidly, and in a few days it had receded to its accustomed channel. Strange to relate, but one life was lost during the recurrence of the flood. A young man lost his life while endeavoring to save a horse at one of the lower landings, being swept into the rapid current where all human aid was powerless to save him. The total height of the water, at the highest point, was estimated at twenty-eight feet.

Sixty thousand barrels of oil and 40,000 empty barrels were lost; also, boats and barges in great number. Many of the wells on the bank of the river and Creek, and especially those on the islands were ruined by the water and sand, besides having their derricks and engine-houses swept away. It required months to clear away the wreck, and many persons were irretrievably ruined. Business of all kinds was practically suspended for weeks. After a time the wreck was cleared away, the buildings destroyed replaced, and business in general went on as usual. A slight repetition of this flood occurred during subsequent seasons, but no serious damage ensued.

A number of destructive fires occurred during 1866 and 1867, at the different points along Oil Creek, at Pithole, and Tidioute, consuming oil, varying in amounts from 5,000 to 20,000 barrels, and a number of buildings, &c. The most disastrous of these was the extensive conflagration that occurred at Oil City, May 26, 1866, destroying the entire business portion of the eastern side of the place.

The fire occurred in a building occupied as a laundry, on Low street, at about eleven o'clock, A. M. Having no fire apparatus, it was impossible to check the headway of the flames. In a few hours some 200 buildings were totally consumed, and 150 familes rendered homeless.

Coming as it did so closely upon the heavy losses of the preceding year, this was the most disheartening blow yet sustained by the community. The delay, however, was brief. Before the smoke had fairly cleared away from the burning buildings, new structures could be seen rising phœnix-like from the ashes, and in a few months the greater portion of the burned district was covered with a better style of buildings than before. The total loss was estimated at $1,000,000. An efficient fire apparatus was also purchased, which has on several occasions since saved the town from a general conflagration.

The effect was made worse by the subsequent crash of the oil-bubble that had been inflated to such an extent by the formation of over 1,000 oil companies, in different portions of the land. The total losses incurred by the people of the Oil Region during the successive years from the causes enumerated we estimate at not less than $10,000,000. The indirect or prospective loss from suspension of business, &c., amounted to a far greater sum.

Each disaster seemed to inspire the people with fresh energy. Firm believers in the doctrine of manifest destiny, so far as their own immediate business future was concerned, no obstacle daunted or impeded their progress. The successful results of each subsequent year amply prove that their faith was well founded.

# CHAPTER XVI.

COAL—BITUMINOUS CLAYS AND SHALES—BITUMEN—PETROLEUM—TABLE OF VOLATILE MATTERS, COKE AND CRUDE OIL FROM COALS, &C.

The best general description of the substances enumerated as above, and their properties, we find in Gesner's valuable work on Coal Oils. From this eminent authority we compile the following:

The varieties of coal have heretofore been classed under the heads of

Anthracite, or Hard Coal,
Caking Coal,
Cherry Coal,
Splint Coal, and
Cannel Coal.

These five varieties have the following composition:

| | | Richardson. | | | Thompson. |
|---|---|---|---|---|---|
| | Anthracite. | Caking Coal. | Cherry Coal. | Splint Coal. | Cannel Coal. |
| Carbon | 92.56 | 87.952 | 83.052 | 82.924 | 76.25 |
| Hydrogen | 3.33 | 5.239 | 5.250 | 5.491 | 5.50 |
| Nitrogen | " | " | " | " | 1.61 |
| Oxygen | 2.53 | 3.806 | 8.566 | 8.847 | 13.83 |
| Ashes | 1.58 | 1.393 | 1.549 | 1.128 | 2.81 |

Other varieties of combustibles have been arranged by Berthier in the following manner:

| Composition in 100 parts. | Peat or Turf. | Lignite or Brown Coal. | Bituminous Coal. | Anthracite, Pa. | Plumbago or Graphite. |
|---|---|---|---|---|---|
| Carbon | 38 | 54 | 73 | 94 | 95 |
| Hydrogen | 38 | 05 | 05 | 2.55 | 95 |
| Oxygen | 38 | 26 | 20 | 2.56 | 95 |
| Ashes | 17.4 | 14 | 02 | 2.56 | 95 |
| Volatile matter | 28 | 14 | 02 | 2.56 | 95 |
| Iron | 28 | 14 | 02 | 2.56 | 5 |

The names given to combustible substances have been applied with reference only to their common characters and uses. Frequently coals bear the names of the places where they are mined. Few of their appellations have any reference whatever to their chemical composition, and therefore in seeking for *oil coals* (if the term may be used), the manufacturer must be directed by an actual test of the material itself.

In the same coal field, the same series of strata, and in the same stratum there are important differences of composition. It is as providential as wonderful that the carbonaceous material of the same deposit is adapted to different uses.

The varieties of coal may have been produced from the different kinds of plants from which coal has been derived, and the peculiar conditions of the districts where those plants flourished before their downfall and inhumation, or submersion. The changes that have taken place in the original plants during their passage from woody fibre into coal are ascribed to the evolution of a part of their carbon, hydrogen, and oxygen, as there is less of those elements in the coal than in wood. This will be observed by viewing the following table:

| | Carbon. | Hydrogen. | Oxygen and Nitrogen. |
|---|---|---|---|
| Composition of the anthracites of the *Transition Rocks* | 90. | 2.50 | 3.69 |
| Bituminous coal of the *Secondary Rocks* | 86. | 4.86 | 7.11 |
| Lignites of the *Tertiary Rocks* | 60.36 | 5.00 | 25.62 |
| Wood (*recent*) | 49.60 | 5.80 | 42.56 |

It will also be observed, that the older the formation the greater the amount of carbon contained in its coal, the amount of hydrogen being diminished. This fact may be ascribed, chiefly or in part, to the greater degree of heat and pressure to which the lower and older coal strata have been, and still are subjected.

The gases of deep coal mines are very similar to those of gas manufactories, and such as are produced by a high temperature. The deeper the mine, the greater is the discharge of carburetted hydrogen. It is to the internal heat of the earth, and other chemical agencies combined with causes of less force, that we must chiefly ascribe the transmutation of wood into coal. The similarity of the distilled products of wood and coals, and of charcoal and coke, should not be overlooked in seeking for proofs of the vegetable origin of coal. In mines of lignite and cannel coal, carbonic acid or *choke damp*, is almost the only gas present. In lower coal mines, or those that have been longer under the influence of heating and other chemical agents, carburetted hydrogen, or *fire damp*, predominates.

Liebig has shown the great loss of oxygen and increase of hydrogen and carbon in lignite and brown coal, during their transition from a vegetable to a fossil state ; still there is much that remains unexplained regarding other kinds of coal

### BOGHEAD COAL, OR BITUMINOUS CLAY.

This peculiar mineral occurs at Torbane Hill, in the carboniferous limestone of the Frith of Forth, Scotland. It is the material from which Mr. Young obtains paraffin oil and paraffin, and his manufactory is in the immediate vicinity. It has been extensively shipped to the United States, and employed in the manufacture of kerosene oil at New York and Boston. During the year 1859, the North American Kerosene Gas Light Company imported upwards of 20,000 tons of this material for the supply of their works at Newtown Creek, Long Island, and at an average cost of eighteen dollars per ton. * * *

Although this mineral possesses but few of the characteristics of a true coal, the term coal has been applied to it for convenience. It has been the source of long continued and expensive lawsuits. The point in dispute affected the ownership, whether it was coal, or not coal. Numbers of the most scientific men in Europe were arraigned before courts and juries to decide whether the so-called Boghead coal is coal, or bituminous clay. There was a decided preponderance against the term "*coal*," and in favor of "bituminous clay." Finally the contending parties compromised, and the term coal was permitted to be applied, although the bitumen of the Great Pitch Lake of Trinidad has an equal right to the appellation.

Boghead coal is among the most valuable minerals for the manufacture of oils. It has an uneven fracture, is of an earthy color, and burns with a long smoky flame. It yields 13,000 cubic feet of gas, of specific gravity, 0.775 per ton. As it contains only traces of nitrogen, the quantity of ammonia given off is small. The following is the medium result of four trials in testing its qualities:

| | |
|---|---|
| Volatile matter | 70.10 |
| Carbon in coke | 10.30 |
| Ash | 19.60 |
| | 100.00 |

The ton of coal run in common retorts gives 120

gallons of crude oil of which sixty-five gallons may be made into lamp oil, seven gallons of paraffin oil, and twelve pounds of pure paraffin. The coke is worthless, and the ash consists chiefly of silica and alumina. At the price of eleven dollars per ton for the coals, the cost of the oil is estimated at sixty-three cents per gallon.

### ALBERT COAL.

This bituminous mineral occurs at Hillsboro', Albert County, in the province of New Brunswick, and within four miles of the Peticodiac River. It is an injected vein, situated almost vertically in the earth, and from one to sixteen feet in thickness. It is associated with rocks highly charged with bitumen, and has neither roof, floor, under-clay, nor stratum of *stigmaria*, nor other accompaniments which distinguish coal deposits from all others.

The Albert coal, so-called, is extremely brilliant, breaks with a conchoidal fracture, does not soil the fingers, and is strongly electric. It melts, and drops in the flame of a candle, and dissolves in naphtha and other solvents, forming a varnish. It has all the essential properties of asphaltum, while it is void of those which constitute true coal. Like the mineral of Torbane Hill, it has been the subject of disputes and lawsuits, the total cost of which has exceeded $30,000. If the substance were coal, the coal was the property of one party; if asphaltum, the asphaltum belonged to another. Coal had been reserved by the Crown of Great Britain; but asphaltum was not mentioned in the grants of the land. In April, 1852, an intelligent jury, who analysed the mineral at Halifax, decided that it was asphaltum, and not coal. Another trial was held in the county where the so-called Albert coal is mined in July of the same year. It lasted eleven days. Chemists were summoned from every quarter; and under the most conflicting testimony, and with a jury of farmers, the advocates for coal obtained a verdict, and the asphaltum has since been called Albert coal. The composition of the Albert coal is as follows:

| | | |
|---|---|---|
| Carbon | 86.207 | 85.400 |
| Hydrogen | 8.962 | 9.200 |
| Nitrogen | 2.930 | 3.060 |
| Sulphur | traces. | a trace. |
| Oxygen | 1.971 | 2.220 |
| Ash | 0.100 | 0.120 |
| | 100.000 | 100.000 |
| | G. M. Wetherell. | Gesner. |

The average yield of crude oils by four trials in large retorts was 110 gallons per ton, and

| | |
|---|---|
| Volatile matters | 61.050 |
| Coke | 30.650 |
| Hygroscopic moisture | 0.860 |
| Ash | 7.440 |
| | 100.000 |

Of the crude oil seventy per cent. may be made into lamp oil, ten per cent. is heavy oil and paraffin. The coke is exceedingly bright and cellular; it burns rapidly, and gives a strong heat.

### BRECKINRIDGE COAL.

The Allegheny or Appalachian coal field of the United States has been estimated to embrace 63,000 square miles. Interstratified with the common bituminous coals, in this vast region, there are very numerous strata of cannel coals, adapted to the manufacture of oils. In the numerous surveys and valuable reports made on the coal districts, cannel coals are seldom described as a distinct variety.

A peculiarity of the great Western coal field is, that it does not appear to be separated into basins, or lake-like depressions in the earth, as it is in Europe, and in the anthracite coal districts. The bituminous coal is found in the tops of hills, and even in the Allegheny Mountains, in beds nearly horizontal, and it displays the same peculiarity as it stretches away towards the Gulf of Mexico, the Canadian Lakes, and the Rocky Mountains.

Among cannels that have been discovered, Breckinridge coal holds an important place. This coal occurs in Breckinridge County, Ky. It is a rich variety of cannel, three feet in thickness, and has already supplied a large amount of oil and paraffin. The lamp oil, when properly purified, is of good quality. At a red heat this coal yields:

| | |
|---|---|
| Volatile matters | 61.300 |
| Fixed Carbon | 30.000 |
| Ash | 8.055 |
| Hygroscopic moisture | .645 |
| Sulphur | a trace. |
| | 100.000 |

By the ordinary methods of working, this coal yields 130 gallons of crude oil per ton, of which fifty-eight per cent. was manufactured into lamp oil and twelve gallons into paraffin oil and paraffin. The quality of the coal is variable, and the products are very much influenced by the degree of heat applied to the retorts in the distillation.

### CANDLE TAR.

The tar and pitch resulting from the manufacture

of stearine have been employed for the production of oils. Large supplies have been imported from England to the United States, and sold under the names of "grease," and candle tar. The ordinary yield of crude oil from this material is 200 gallons per ton, of which seventy per cent. may be made into lamp oil, and ten per cent. into lubricating oil. The oils are excellent in quality, but heretofore the first distillation of the tar has been attended with inconvenience, as it "foams up" in the retorts, and the coke adheres very firmly to their sides. The price has varied from twenty-five to forty dollars per ton. * * * The distillate runs from 65° to 30° Beaumé. There is a hard coke or pitch left in the still.

### SOUTH BOGHEAD COAL.

Near Poole (England), there occurs a peculiar kind of shale, which has been sold as "South Boghead Coal." It abounds in the remains of fishes and *crustacea*. It gives out forty-two per cent. of volatile matter, and, therefore, has offered an object of trial to oil makers; but the oils made from this rock contain a greater number of the equivalents of carbon than those derived from coals, or bitumens, and with the ordinary density they smoke in the common lamp. It seems quite evident that the elements of the oil—carbon, hydrogen, nitrogen, and oxygen, now composing the shale, in part, have been derived from fishes and other marine animals, and not from the vegetable kingdom as in the case of coal.

### BROWN COAL.

Singular beds of brown coal have been discovered on the Ouachita River, Arkansas, and at other places in that quarter. They contain the remains of *sphagneous* plants and woody fibre. It appears that peat bogs have been overflown, or otherwise saturated with Petroleum, and hardened by time and oxydation. The oils distilled from this material abound in paraffin. It has the following composition in 100 parts:

| | |
|---|---|
| Volatile matters condensed into oils and gas uncondensed........................ | 60.10 |
| Fixed Carbon........................... | 32.85 |
| Ash..................................... | 7.05 |
| | 100.00 |

Crude oil at the rate of sixty-eight gallons per ton was obtained from it. It is semi-solid when the thermometer is at 80° Fahrenheit, and, besides lamp and lubricating oils, it produces 143 pounds of paraffin per ton.

Nova Scotia and New Brunswick produce a variety of shales, which at one time could be profitably used in the manufacture of oil. The "Pictou shale" of Nova Scotia, and the "asphalte rock," of Dorchester, New Brunswick, yield from twenty to thirty gallons of refined oil to the ton. A sample of cannel coal from Hunter's River, Australia, gave a yield of sixty gallons of crude and forty gallons refined to the ton. Australia evidently abounds in bituminous substances. Mr. H. H. Hall, of Sidney, N. S. W., handed the author several fine samples of cannel coal, from the vicinity of Sidney, together with a small specimen of shale, which, in appearance and burning properties, resembled the Boghead coal.

### BITUMEN AND BITUMINOUS SANDS AND CLAYS.

At the various localities mentioned in the preceding chapter as Petroleum deposits, bitumen in different stages of solidity is generally found.

The bitumen of Trinidad was the article from which the author first obtained "kerosene," which differs in some degree from "coal oil." The bitumen is of a grey color, somewhat brittle, but still yielding to the heat of the sun. A cargo of broken masses will consolidate in a ship's hold in such a manner as to require mining before it can be discharged. The following is the result of several trials made with reference to its application for the manufacture of oils:

| | | | | |
|---|---|---|---|---|
| Specific gravity | 0.882 | gals. | per | ton. |
| Crude Oil | 70 | " | " | " |
| Refined Oil | 42 | " | " | " |
| Lubricating Oil | 11 | " | " | " |

This bitumen varies in quality, owing to the sand and debris over which it flows. Taken from the lake itself, it would probably yield twice the above quantity of oil.

A vein of bitumen has been discovered near Cairo, thirty miles east of Parkersburg, West Va. It is represented as a perpendicular mass jutting out from the side of a hill 290 feet. The strata of the hill are nearly horizontal, and they are cut at right angles by the continuous vein of the bituminous mineral, which is four feet eight inches in thickness. The position of the vein has been ascertained by the proprietors, who have sunk a shaft upon the line of the outcrop. A sensible description represents that it appears the hill has been split, a perpendicular chasm opened, and afterwards filled with asphaltum in a liquid state, and which has since hardened into a compact material. Coal never occurs in this manner; but is always interstratified with its associate sandstones, shales, and fine clays. In all its geological relations and characters, the Cairo deposit is like the asphaltum of Albert County, New Brunswick.

Some of the Cuba bitumens yield 120 gallons of crude oil to the ton. The finest varieties are used in making varnish. On the borders of the river Acarahy, forty miles south of Bahia, there are extensive beds of lime and clay saturated with bitumen, and capable of yielding oil in large quantities. The oils resemble those obtained from Trinidad or Cuba bitumen.

The annexed table will afford some guide to the manufacturer regarding the proportion that crude oil bears to the volatile matters of the material, and also regarding the localities of coals, shales, bitumen, etc. The refined products of crude coal oil depend so much upon their treatment that it is difficult to express in figures their actual amount:

Breckinridge Coal, as has been shown, gives 13 gallons crude oil per ton

| | |
|---|---|
| From which we obtain | 80 galls. illuminating oil. |
| And | 12 galls. paraffin oil. |
| | — |
| Making............. | 92 in all of marketable oils. |

Boghead Coal yields 120 galls. crude oil per ton.

| | |
|---|---|
| From which we obtain | 65 galls. illuminating oil. |
| | 7 galls. paraffin oil. |
| And | 12 barrels paraffine. |
| | — |
| Equal to about....... | 84 galls. of marketable oil. |

Yet by experimental distillation, Boghead coal yields more volatile matter and leaves less coke than the Breckinridge:

| Locality. | Volatile Matters. | Coke. | Yield of Crude Oil per Ton. |
|---|---|---|---|
| England. | | | |
| Derbyshire | 48.36 | 53 | 82 gallons. |
| Wigan Cannel | 44 | 56 | 74 " |
| Liverpool | 39 | 61 | 50 " |
| Poole (shale) | 42 | 58 | 50 " |
| Newcastle | 35 | 65 | 48 " |
| Scotland. | | | |
| Boghead | 70.10 | 29.90 | 120 " |
| Scotch Cannel | 38 | 62 | 40 " |
| Leshmahago | 51 | 49 | 96 " |
| Provincial. | | | |
| Albert Coal, New Brunswick | 61.050 | 30.65 | 110 " |
| Asphalte Rock, New Brunswick | 43 | 57 | 64 " |
| Pictou Shale, Nova Scotia | 27 | 73 | 47 " |
| American. | | | |
| Breckinridge | 61.30 | 38.55 | 130 " |
| Erie Railroad | 35 | 65 | 47 " |
| Newburg | 38 | 52 | 72 " |
| Falling Rock | 50 | 50 | 80 " |
| Pittsburgh | 36 | 64 | 49 " |
| Kanawha | 46 | 54 | 71 " |
| Elk River | 41 | 59 | 60 " |
| Cannelton | 34 | 66 | 86 " |
| Coshocton, Ohio | 45 | 64 | 74 " |
| Darlington, Pa | 42 | 58 | 56 " |
| Ouachita River, Arkansas | 60 | 40 | 54 " |
| Bitumen, etc., United States. | | | |
| Ritchie County, Virginia; Pennsylvania | .... | .... | 170 galls. per ton. |
| Petroleum Springs, Alabama, Georgia, Tennessee, Kentucky, Virginia, Maryland, Ohio, Pennsylvania, Canada | .... | .... | From 75 to 85 per cent. of lamp oil. |
| Cuba. | | | |
| Bitumen | 71 | 29 | 120 gallons. |
| Trinidad. | | | |
| Bitumen | 38 | 52 | 70 " |
| Canada, etc. | | | |
| Bitumen | 70 | 30 | 118 " |
| Illinois "Gas Stone" | 26 | Limestone. | 18 " |
| California | 70 | 30 | 116 " |
| Brazil | 78 | 22 | |
| Peat | 71 | 25 | 6 to 8 " |

Peat has been distilled for oils in Ireland, and in Kildare the extensive works of the Irish Peat Company have been in operation for that purpose for some time. One ton of peat yields on an average :

| | |
|---|---|
| Liquids (not oily) | 65 gallons |
| Tar | 6 " |
| From which are produced: | |
| Lamp oil | 2 " |
| Lubricating oil | 1 gallon |
| Paraffin | 3 barrels |
| Ammonia | 3 " |
| Acetic acid | 5¼ " |
| Naphtha | 8 " |
| And 25 per cent. of charcoal. | |

Thus far, however, peat has not been a successful competitor of coal, bitumen, or other more compact carbonaceous materials.

### VARIETIES AND PRODUCTS OF PETROLEUM.

The oil wells of Pennsylvania yield generally greenish oils of rather unpleasant odors. Their specific gravity ranges from .820 to .782 or from proof 40° to proof 48° Beaumé. The oil by distillation yields from seventy-five to eighty-five per cent. of a lamp oil which should not vaporize and inflame under a temperature of from 110° to 116° Fahrenheit. A very heavy lubricating oil is obtained from many of the wells in Pennsylvania and Ohio, of specific gravities .880 to .860, or from proof 28° to 32° Beaumé.

An oil was obtained from the wells of the Tarentum Oil and Coal Company, of the specific gravity of .795, or 45° Beaumé. It is of a dark amber color, and will burn in lamps for a time without being refined. It yields five per cent. of naptha by distillation, and ninety per cent. of oil. It is used by woollen manufacturers instead of lard oil.

The Canada Petroleum is of specific gravity from .880 to .860, or proof 28° to 32° Beaumé. It is a dark-colored and offensive oil. The odor can only be removed by extra distillation. It yields more lamp oil than the Pennsylvania, as it will burn in a lamp at proof 36° Beaumé, or specific gravity .838.

We take the following account of the manufacture of illuminating oils from various mineral substances from the report of the United States Revenue Commissioners, made in 1866:

"The most recent observations, in July and August, 1865, showed the following facts: In North Wales and Staffordshire, the material used was the cannel coal. Some fifteen or twenty mines and oil works were visited there. The amount of capital employed was quite large. The wages paid miners were two dollars per day, equivalent to about fifty cents per ton of coal. The best cannel coal was worth three to four dollars per ton; common coal, on the ground, one dollar. The coal was usually broken fine and retorted; the crude oil was treated with chemicals, and subjected to several distillations. The product was an oil, nearly, but not quite, equal to our best refined, worth from two to five cents per gallon less. The supply of material was not limited; the total cost of production of the refined oil was 1*s.* 3*d.* (thirty cents), per imperial gallon, or less.* The selling price, at wholesale, was 1*s.* 6*d.* to 1*s.* 10*d.* (thirty-six to forty-four cents), package included, delivered in London or Liverpool.

"In Scotland, the manufacture was very extensive, both from Boghead coal and the shales. Mr. Young had recently bought a large tract of shale lands, at an outlay of from $500,000 to $1,000,000. The shales, though yielding less oil, were said to be equally profitable with the cannel coals, owing to the greater ease with which they were reduced, and the greater purity

* The imperial gallon is $1\frac{20}{100}$ American gallons.

of the product. The shales were supposed to be inexhaustible. Manufacturers were in high spirits, and undoubtedly making money fast. The total cost of the purified oil there was about the same as in Wales—thirty cents per gallon or under. The increase in production since 1861 was very remarkable, and it was evident that the business had not been injured by the large shipments from this country, but had assumed a permanent and lucrative character. On the continent the same increase in the business, and the same evidences of prosperity, were visible. The chief centres of oil production were about Autun, in France, from the shales and schists; about Bonn, in Switzerland, in the departments of Halle and Mersberg, in Saxony, and in Bohemia, from the shales and brown coal. The supply of these materials was very great.

"At one mine and factory, at Weissenfels, 500 laborers were employed. The highest wages paid for skilled labor, mechanics and artisans was seventy-five cents per day; for miners and common laborers, thirty to forty cents. Much of the work was done by women at still lower rates. The bed of brown coal at this mine was sixty feet thick. The average cost of production of the several grades of refined oil did not exceed twenty-five cents per gallon; the selling prices at the neighboring railway stations were as follows:

"First quality, ten Prussian dollars* per 100 German pounds; second quality, nine and a half Prussian dollars per 100 German pounds; third quality, thirty-three cents (gold) per gallon.

"Or reduced to American money and gallons:

"First quality, forty-one and a half cents (gold) per gallon; second quality, thirty-nine and a half cents

"* The rix dollar or thaler of Prussia and the northern States of Germany is equal to sixty-nine cents in American coin.

(gold) per gallon; third quality, thirty-three cents (gold) per gallon.

"The production of coal in 1863 and 1864, in the two small districts of Mersberg and Halle, were as follows:

"1863, Mersberg, 12,255,365 German tons; 1864, Mersberg, 14,421,551 German tons; 1863, Halle (No. of mines, 371), 24,149,214 German tons; 1864, Halle, (No. of mines, 371), 26,260,856 German tons.

"The number of miners employed in the 371 mines of Halle, in 1864, was 11,219. A German ton is 331$\frac{2}{10}$ pounds avoirdupois. The total product of Mersberg and Halle, in 1864, was 6,015,184 American tons of 2,240 pounds each, worth, at three dollars per ton, $18,-045,552.

"Twelve oil factories near the town of Weissenfels in Halle, produced in 1864:

| | |
|---|---|
| Crude tar | 55,000,000 gals. |
| Burning Oil | 15,000,000 |
| Refined paraffin | 3,500,000 |
| Lubricating oil | 2,000,000 |

"Reduced to our avoirdupois weight—100 German pounds being equal to 110 $\frac{4}{1000}$ pounds avoirdupois—the figures would be:

| | POUNDS AVOIRDUPOIS. |
|---|---|
| Crude tar | 60,500,000 |
| Burning oil | 16,500,000 |
| Refined paraffin | 3,850,000 |
| Lubricating oil | 2,200,000 |

"In regard to the extent of the brown coal deposits, they are pronounced to be inexhaustible. A single company, 'The Brown Coal Works,' at Weischen, in their report, dated July 1, 1865, estimate the amount of coal in the ground owned by them at 95,900,000 German tons, and their lands comprised but a minute portion of the territory in which this coal is found in paying quantities.

"Taking into account, also, the product and supply of material in other parts of Saxony, in Bohemia, in Switzerland, in France, and in the north of Italy, it will be seen the witnesses have good grounds for the opinion they express, that the foreign production of oil from minerals will be a permanent and paying business under any contingency, even supposing that our oil should be produced at the wells at a nominal price, and sold at a very low figure to our dealers for shipment.

In France the production from the shales is being very much extended, and large investments of capital are being made in that direction. There are also extensive refineries near Paris and at Marseilles, where they refine our crude oil, when it is sufficiently low in price. When prices advance beyond their views, they use the crude oil from the shales and coals. The French tariff duty of three francs per 100 kilogrammes on refined, crude being admitted duty free, lessens the importation of refined oil.* The best refined oil was selling at these refineries at from sixty-five to seventy-two francs per 100 kilogrammes. It advanced subsequently to about eighty. These prices would be equivalent to from thirty-two to thirty-eight cents, gold, per American gallon.

"The quality of oil produced on the continent is

"* Both crude and refined are admitted free of duty in all other European countries, so far as the Commission is informed. The duty in France on oils of Petrole and schiste, rectified and refined, is as follows: Per 100 kilogrammes (221 pounds avoirdupois), from the country of production, by land, five francs (ninety-three cents); by sea, in French vessels, three francs (55.8 cents); in foreign vessels, five francs; from any other than the country of production, five francs, whether in French or foreign vessels.

good, but somewhat inferior to ours. The odor is stronger, and, containing more carbon, it burns with more smoke; yet it can be produced so cheaply that our merchants must always expect a sharp competition, even if they should not be driven out of the markets of Europe.

"The conditions there are most favorable to cheap production, very low wages, very low interest, and abundant capital, as well as abundant raw material. Add to these the advantage of being in the midst of the dense oil-consuming population, where we seek our customers, and it becomes evident that the prospects of our foreign trade are not such as to warrant the continuance of any heavy burdens upon it. *It is also evident that not only the American Petroleum, but the foreign Petroleums also might cease to be obtained, and yet the markets of the world could be amply supplied by the products of the foreign coals, shales, and schists.*"

Petroleum, similar in all general respects to that of the United States, has been found in Peru, and American enterprise is already engaged in its development. In 1864, Messrs. Geo. H. Bissell and James Bishop, of the city of New York, leased of Don Diego Llama his estate of Mancora, consisting of about 4,500,000 acres. A company called the Peruvian Petroleum Company was formed, with a capital stock of $5,000,000. Operations were soon after commenced by the company on the northern portion of the tract, at Zorritos, twenty miles south of the Tumbez river, near the southern terminus of the equatorial rain belt. This portion of the property is easily accessible to Lumbez, which is connected with Paita to the south, and Panama to the north, by steamer, three times a month. The works are immediately on the coast. The company have a full force of experienced Oil Creek miners employed.

Several wells have been drilled with good success, and several cargoes of oil shipped by the company. In one well, 531 feet deep, the following is the order in which the rocks were found :

Soapstone and slates,
Sandrock and slates,
Conglomerate limestone,
Conglomerate rocks,
Hydrate of Iron, or Reddle,
Cretaceous sandstone,
Carboniferous slate,
Gravel, in which the oil was found.

The oil where struck seemed to be in veins, similar to those of water under the earth. Oil was met with at a depth of eighteen feet from the surface.

The wide region of Petroleum, or Rock Oil, so far as traced on the North American continent, extends from the southeast portion of the Ohio Valley to the Georgian Bay of Lake Huron, in Upper Canada, and from the Alleghenies to the western limits of the bituminous coal fields of the Missouri river. It has been found in Virginia, Maryland, Pennsylvania, New York, Ohio, Michigan, Kentucky, Tennessee, Kansas, Illinois, Texas, Louisiana, and California. This general statement does not, of course, imply that Petroleum is to be met with everywhere throughout the vast territory thus vaguely defined. On the contrary, it has been found only in scattered localities within the above described limits.

Of the productive oil fields the best known and hitherto most abounding one is a broad area embracing a part of Canada West, from Lake Huron to Lake Ontario, portions of western New York, western Pennsylvania, the southeastern half of Ohio, all of northwestern

Virginia, and the eastern portion of Kentucky. It is stated that the geographical centre of this wide region, so profusely abounding in Petroleum, is at or near Marietta, in Ohio, and the superficial extent of it cannot be less than 100,000 square miles.

Geologists familiar with the great Petroleum tract, entertain no doubt that the rocky strata within its limits, are, in almost every square mile of it, impregnated with the greasy fluid and its gassy adjuncts, but not everywhere in the same degree. A merely superficial exploration of the country shows that the subterranean oil and gases are distributed very unequally.

The general surface indications whereby Petroleum manifests itself, have been previously described. They are so peculiar as so be unmistakable. The quantity to be obtained can only be ascertained by drilling to a depth sufficient to reach the strata of rock in which it is found.

# CHAPTER XVII.

## THE PRINCIPAL TOWNS OF THE PENNSYLVANIA OIL REGION.

Oil City, from the natural advantages presented by its location, has been from the commencement of the business the principal shipping point of the Petroleum Region. It is located on both sides of Oil Creek, at its junction with the Allegheny river. A small settlement was made on the present site of Oil City about a quarter of a century previous to the oil discovery. It had then a grist-mill, an iron furnace, store, warehouse, hotel, and a steamboat landing. This settlement flourished for a short time only, and then fell into rapid decay. A store was established here, in 1852, by Mr. John Hopewell.

The first settler on that portion of territory on which the western part of our city is located, was Mr. Francis Haliday, who purchased the land from the Government. The amount of this purchase embraced several hundred acres. Subsequently a greater portion of this tract was sold to Dr. John Nevins, who in turn sold it to Plummer & Drum, of Franklin. These gentlemen afterward sold it to the Michigan Rock Oil Company, who, in 1860 laid the first foundation of the present city.

The land on the east side of the Creek, which is now owned by the United Petroleum Farms Association, of New York, having been purchased by them in 1865, of Graff, Hasson & Co., originally belonged to Cornplanter, the renowned chief of the Seneca tribe of Indians, one of the famous Six Nations. Congress presented him with a grant of some three hundred acres of this property, for distinguished services rendered by him during the wars of the Revolution. Like many of his prototypes of the present day, Cornplanter was addicted to a free indulgence in "fire water." In one of his drunken frolics, as stated, he sold this land for a mere trifle.

The first commencement of the present town, as before stated, was in 1860, soon after the striking of Drake's well. The town increased rapidly with the growth of the oil business, and in 1865 its population was estimated at eight to ten thousand inhabitants. At the present date we should estimate the number at about five to seven thousand. There are four churches, Presbyterian, Methodist, Episcopal, United Presbyterian and Catholic, also a Baptist and Episcopal Congregation, who have erected no church buildings as yet; a public library, two large school-houses, affording ample room for one thousand pupils, both graded, many elegant private residences, hotels, stores, machine shops, &c., ample for its own and surrounding population. A substantial iron bridge over Oil Creek unites the eastern and western portions of the City; the Railroad bridge of the A. & G. W. Railway spans the Creek at its mouth, and a fine wooden bridge has been built over the Allegheny river, connecting Oil City and Venango City. The cost of these bridges in the aggregate was about $225,000. The Meadville and Franklin Branch of the A. & G. W. Railway, has its terminus at Oil City, and was the first railroad extended through the borough limits, commencing to run to Oil City in March, 1866. The Oil City & Pithole Railroad was completed about the same time, also having its terminus here. The Farmers' Railroad, extending from Oil City to Petroleum Centre, was projected and finished early in 1867. The Oil City and Pithole Railroad connected with the Warren & Franklin Railroad, which had its terminus at Irvington, connecting at that place with the Philadelphia and Erie Railroad. The Farmers' Railroad connected at Petroleum Centre with the Oil Creek Railroad. Subsequently, the Warren and Franklin Railroad Company, by purchase and otherwise, obtained control of the Farmers' and Oil City and Pithole roads, as well as the right to run over the Oil Creek road, the latter having its terminus at Corry, connecting at that place with the A. & G. W. and Philadelphia and Erie roads, the whole now being consolidated under the name of the Oil Creek and Allegheny River Railroad. In January, 1868, the extension of the Allegheny Valley Railroad from the mouth of Mahoning to Oil City, was completed and has its terminus at Venango City. The Cranberry Coal Company also built a road from their coal mines, some seven miles distant, to Oil City, in 1867. The Jamestown and Franklin Railroad are extending their line to Oil City. In addition to the above railroads centering and terminating at Oil City, the Allegheny River and Oil Creek, at different seasons of the year, furnish means of transportation, cheap and ample. All these roads have direct communications with the principal railroads of the country, in all directions, and with the principal cities, east, west, north, and south. Oil City is the headquarters of those engaged in the Petroleum business.

Oil City was erected into a borough in the spring of 1862, and after the Charter of incorporation was obtained, the first borough election held. The following were the officers chosen: Burgess Capt. Wm. Hasson; Councilmen, Chas. Robson, F. M. Bishop, Chas. Haines, T. B. Hoover; Constable, C. E. Hueston. W. R. Johns was appointed Clerk to the Council, and Hugh M'Clintock, Street Commissioner. Notwithstanding the many serious checks from fire and flood, the growth of Oil

City has been steady, and it gives promise of being one of the most permanent, as it is already one of the most important, commercial towns of the Oil Region. The amount of iron tankage at this point is half a million barrels, being about half the entire amount of iron tankage of the Oil Region. In addition to its shipping facilities, there are a number of small producing wells within the borough limits, and the City is surrounded on all sides with territory that has proved highly productive, even with the slight development had. The rapidly increasing business is sufficient to build up in a few years a considerable city; with the development of its rich surrounding oil field, this result will be greatly hastened.

VENANGO CITY,—Just opposite, the present terminus of the Allegheny Valley Railroad, is a thrifty town, located on portions of the Bastian, Downing and Lee farms. In the spring of 1863, Mr. Wm. L. Lay, formerly of Cincinnati, Ohio, purchased the Bastian Farm, consisting of eighty acres. Subsequently it was incorporated into a stock company, styled the Laytonia Town & Oil Company, laid off into town lots, and a considerable town built up, called Laytonia. In the spring of 1864, Mr. Jas. Bleakley and others, of Franklin, purchased the Downing farm adjoining the Bastian. This property was afterward formed into a stock company, called the Imperial Oil Company, of Philadelphia. A portion of this property was also laid off into town lots, and a thriving town called Imperial built up. Subsequently both towns were consolidated under the name of Venango City. The consolidated towns contain a population of from two to three thousand inhabitants. It has one church building, a public school-house, and some elegant private residences, several stores and hotels, and in former years was regarded as excellent oil territory.

Good wells were obtained on this farm in 1860, by Phillips, Frew & Co.

M'Clintockville, Rouseville, and Petroleum Centre, are small towns along the valley of Oil Creek, between Oil City and Titusville, projected during the different years of the petroleum development, and are located on the line of the Oil Creek and Allegheny river Railroad. All of them are surrounded by productive territory, and have but little business outside of oil shipping, and the supplying of such articles as are needed by the oil operators. Petroleum Centre is the largest of these, located in the midst of productive oil farms, and is an important shipping point. It has two church buildings, a banking office, a number of stores, hotels, and machine shops, and contains about two thousand inhabitants. It has also an excellent daily paper, "The *Petroleum Centre Record.*

Oleopolis, at the mouth of Pithole Creek, on the Allegheny river, was laid out in 1865, and for the first few months increased in number of buildings and population rapidly. It is a place of no importance now, having gone into decay.

Tionesta, just above, on the opposite side of the Allegheny, is a thriving village of about five hundred inhabitants, many fine private residences, several stores, hotels, and churches. It is the county seat of Forest county. A new court-house and jail is about completed. Two newspapers, the *Journal* and the *Press*, representing each of the political parties, are published here. There is also a banking office. Tionesta Creek here debouches into the Allegheny.

Tideoute, in Warren county, just above Tionesta, on the Allegheny river, is the centre of an excellent oil producing region. It is one among the oldest towns in this part of Pennsylvania, is well built, has several churches, school-houses, hotels, stores, banking office, machine shops, and a weekly newspaper, "*The Journal.*" It contains about 2,500 inhabitants.

FRANKLIN, the County seat of Venango county, is pleasantly located on French Creek, at its confluence with the Allegheny river, on the south side of the former and the west side of the latter. The French fort of Venango was built here in 1754, but was abandoned by them in 1859. The fort was situated on the bank of the Allegheny river, some thirty rods below the present bridge. In 1765, a large party of Senecas by treachery gained admission into Fort Venango, then in possession of the English, massacred the garrison, tortured Lieutenant Gordon, the commandant, over a slow fire, and burned the Fort. In 1787, a company of United States soldiers arrived at what is now Franklin, from Fort Pitt, for the purpose of erecting a fort to protect the country in its early settlement. The Fort was built, and a garrison consisting of one hundred men kept in it until 1796, when a new fort was erected at the mouth of French Creek. The first of these forts was called Fort Franklin, and the other the "Old Garrison." This second fort was simply a strong building, surrounded by pickets, and was garrisoned by soldiers until 1803, when they were withdrawn.

The town was laid out in the year 1795, by Gen. William Irvine and Andrew Ellicott, the commissioners appointed by the State under the act of assembly, on a tract reserved by the State for a County-town; was surveyed and laid out into inlots, out-lots, and out-tracts. It was organized and constituted a Borough in the year 1829, with Judge Wm. Conelly, Burgess, to whom we are indebted for the items of its early history. During subsequent years, it became a thriving town, and at one time had a rolling-mill that manufactured eight tons of bar-iron and nails per day, from ore obtained in Venango County. It has assumed its chief importance since the Petroleum development, and has largely increased in general prosperity and population. It is not only admirably located, but comprises many excellent buildings. Among these are several brick blocks for business houses. The private residences, of which there is a great number, are exceedingly tasteful in style, and great attention is paid to the ornamentation of the grounds surrounding them. There is a sufficient number of business houses, machine-shops, foundries, one large flour-mill, &c., to accommodate the wants of the community and surrounding country. A new Court-house is in process of erection at the present date, the old one being totally inadequate to meet the public requirements. Several church buildings have been erected during the last few years. Among these are the Episcopal, Presbyterian, and Methodist churches. These are built of brick and stone, and rare good judgment has been displayed in their design and construction. The three buildings alluded to will compare favorably with any of the kind in the State. A three story brick school-house has been built during the last two years, at a cost of $45,000, and has ample accommodations for one thousand to twelve hundred pupils. Two weekly newspapers, the *Spectator* and *Citizen*, representing the two great political parties of the country, are published here. Franklin is the terminus of the Jamestown & Franklin Railroad. The Meadville branch of the Atlantic & Great Western passes through the borough limits, and the Allegheny Valley Railroad along the bank of the river on the opposite side. The town has lately been incorporated with a city charter, has already a large trade, and bids fair to become a healthy inland city. The population is estimated at from 7,000 to 8,000.

TITUSVILLE, one of the principal cities of the Oil Region, is in Crawford County, at the headwaters of Oil Creek, one mile from the boundary line of the Venango and Crawford Counties. It is at the extreme northeast corner of Crawford County, twenty-eight miles from the junction of the Philadelphia & Erie, and the Atlantic & Great Western Railroads at Corry. It owes all its importance to the Petroleum development, and has been built up to its present proportions by the business. The prosperity of the place dates from the completion of the Oil Creek Railroad, the only one that runs through it. Previous to the oil discovery, it was a village of one hundred and fifty inhabitants. At the present time it has an estimated population of about 10,000. The City is built mainly upon the flats, the suburbs reaching to and up the hill-side on the north. Crossing the creek on the south side are also considerable improvements. The general bearing of Oil Creek is south by east. Titusville was first incorporated as a borough in 1847, by a special act of Assembly, with Joseph L. Chase as Burgess, Wm. Barnsdall, S. S. Bates, J. K. Kerr and G. C. Pettis, council; three school directors, one constable, &c. This organization continued until 1861, at which time it was reorganized under the general act of 1834, with a burgess, six councilmen, six school directors, and all other officers rendered requisite by the charter. It was incorporated as a city in 1867.

The business portion of the city is built of brick, and the streets paved. The residences of the wealthy class of citizens are elegant in style and finish. There are seven churches, one large public school, graded, and an ample fire department. There are also several private schools, one literary society and reading-room, and a board of trade, six or eight first-class hotels, and a score or so of smaller ones, four or five public halls, two national banks, several machine shops and foundries, a paper mill; fifteen or twenty practising physicians, nine or ten lawyers, a Mayor, police, and all the necessary adjuncts of a modern city. It has two daily newspapers, the *Daily Morning Herald*, and *Titusville Morning Star*, and one weekly, the *Long Roll*, published by the pupils of the Soldier's Orphan Asylum located here. The *Morning Herald* was established several years since, is a standard authority upon oil matters, and is conducted with decided ability.

Its citizens are noted as being the most enterprising in the Oil Region, and among them are many of the most successful operators. In 1865 and 1866, there were twelve refineries in the borough limits, with a capacity of one hundred and forty to six hundred and forty barrels each of refined oil per week. Many of these have since been discontinued or are idle. The stores and business houses are well filled with goods, and are generally first-class establishments. The development of the new oil field at Pleasantville, six miles from Titusville, with which it is connected by an excellent plank road, has given a new impetus to the former place. Titusville is the headquarters of the Oil Pipe Companies, and bids fair to retain its position as one of the most thriving inland cities of our State.

MEADVILLE is a well built, thriving city of eight to ten thousand inhabitants; is the County seat of Crawford County, and one of the principal inlets for travel to the Oil Region, being, like Titusville, on the border of the present developed portion of the Oil Region. The city is beautifully located on a tributary of French Creek, thirty miles from Franklin. It is noted as a seat of learning, having an excellent system of public schools and two colleges—one, the Allegheny, belonging to the Methodist, the other, a theological school, to the Unitarian denomination. The Atlantic & Great Western Railway have their headquarters here, also their principal work-shops, which have added materially to the prosperity of the place. A large portion of the trade comes from the Oil Region, and the city is surrounded by a good farming country. It is well supplied with all the business branches requisite, banks and banking houses, three newspapers, viz.: the *Republican* (daily and weekly), *Journal* and *Democrat* (weekly), churches and schools. Its hotels are numerous, and the M'Henry House is one of the best hotels in the country.

CORRY, in Erie County, situated at the point where the Atlantic & Great Western Railway intersects the Philadelphia & Erie, and the Oil Creek Railroad terminates, is a flourishing city of about the same size as Titusville, and has been built up since the oil discovery. It has two banks, several churches, numerous hotels, public schools, one large and several small oil refineries, quite a number of manufacturing establishments, and two newspapers, both weekly. The inhabitants are enterprising, and the town is one of the most thriving in the surrounding country.

WARREN, the County seat of Warren County, like Corry and Meadville, is on the outskirts of the Oil Region, fifty miles from Oil City by river. The Philadelphia & Erie Railroad passes through it. It is a handsome town of about four thousand inhabitants. It is well located, has the usual public buildings, churches, school-houses, a bank, and two newspapers.

PITHOLE, one of the creations of the oil developments, located on Pithole Creek, has been previously described. But a small portion of the original city remains.

Such is a brief description of the principal cities and towns located in, and bordering on the Pennsylvania Oil Region. Before the oil discovery by Col. Drake, in 1868, the older towns were generally small villages, and owe their growth to the Petroleum development. The number of cities projected by sanguine speculators in the flush times of 1864 and 1865, were legion. Beyond a few board shanties, in hopeless decay, nothing now remains of them.

# CHAPTER XVIII.

## METHOD OF REFINING, AND THE OIL REFINING INTEREST.

THE refining of Petroleum, so as to prepare it for illuminating purposes, is a new branch of industry created during the last eight years, employing at the present time an investment of probably not less than thirty to forty millions of dollars, and requiring several thousand operatives in its manufacture. The crude oil is refined for the purpose of ridding it of the smoke, its offensive smell, and to render it non-explosive. The average percentage of the refined article obtained is about 60 to 70 per cent., or two barrels of refined to three of crude. Great success has crowned the efforts of the refiner in all these respects, the refined article being rendered colorless as water, free from offensive odor or smoke, and as safe from explosion as any illuminator known. The business has been greatly system atized within the last few years.

At the commencement of the Petroleum business, refineries were erected in all the principal cities and towns of the country, and a large number in the Oil Region. They have proved unprofitable for the most part, save in a few commercial centres like Pittsburgh and Cleveland, where the refining interest is now mainly centred. At these two points is now manufactured the great bulk of the article, the foreign supply being obtained from the refineries of these two cities. The crude oil is purchased at the place of production by agents of the refiners, or more frequently by one of the firm, resident here for the purpose. A number of these refineries are vast in their proportions and capacity, costing a quarter of a million of dollars, and manufacturing from five hundred to three thousand barrels of refined oil per week, besides the other products derived from the article.

In the opinion of the writer, the place of production is the proper location for the refineries, and we believe it eventually will be. The reasons for locating them at Pittsburgh and Cleveland, in the first instance, is accounted for by the lack of proper facilities of transportation in the Oil Region. This great want has been amply supplied, and with the systematized facilities now offered by the Oil Region, its advantages as a manufacturing point for the product are certainly apparent.

Several parties claim to have been the first to introduce refined Petroleum for public use. From the best evidence before us, to Mr. Samuel M. Kier, of Pittsburgh, is due to a great extent the honor of its first utilization and introduction for public use, as an illuminator. This gentleman gave his attention to the matter as early as 1849. Being extensively engaged in boring for salt at or near Tarentum in Allegheny County, a few miles above Pittsburgh, he obtained a considerable quantity of Petroleum from some of his wells, it coming out of the same with the salt water. A quantity of it was sent to Philadelphia for analyzation. The report of the chemist was favorable for its use as an illuminator, provided a suitable lamp could be devised for burning it. This was soon supplied. Mr. Kier erected a small refinery and commenced its manufacture. From 1850 to 1855, he disposed of all the Petroleum he could obtain from his own and the neighboring salt wells, selling the oil and the lamps for burning it. This oil was sold at from one dollar to a dollar and a half per gallon. Other parties, in the mean time, had been experimenting upon Petroleum as an illuminator, and their experiments had proved successful. Hence, at the time of the striking of the first oil well by Col. Drake, in 1859, the value of the article, as an illuminator at least, had been satisfactorily demonstrated by the practical and scientific men of the country.

Strange to relate, the best lamps furnished at the commencement of its use as an illuminator, were manufactured at Vienna, in Austria. American inventors, however, soon greatly improved upon the models thus furnished.

The manufacture of coal oil from Cannel Coal, had been commenced some time prior to this date, and large refineries were erected at various points. The most important of these were the Kerosene Works, in New York; the works of Samuel Downer, of Boston; and a large establishment at Maysville, Kentucky. The supplies of the two first establishments were obtained from Nova-Scotia, and consisted of the substance known as Albert Coal. Breckinridge County furnished ample supply for the establishment at Maysville. Several other refineries for distillation of illuminating oil from coal were in operation in different parts of the country, and about the time of the Petroleum discovery this bid fair to become an immense business. A double distillation is required for coal oil, rendering the cost of this process far greater than in the case of Petroleum. The crude article could also be obtained much cheaper than coal. The cheap rate at which refined Petroleum could be furnished, the seemingly inexhaustible supply of the crude product, soon overshadowed coal oil. In a very short time its manufacture became unprofitable, and Petroleum reigned supreme as an illuminator.

The process of refining Petroleum we will essay to

describe in general terms, many minor improvements being constantly making in the different details, and ev n the general plan is varied or modified to suit the quality of oil under treatment. We are indebted for much of the following to Messrs. Hill & Thumm, proprietors of the Economy Oil works, Oil City, Pa., who are among our best practical refiners :

The first process in distillation is to pump the oil from the crude oil tanks into the stills. The stills range in capacity all the way from ten barrels to one thousand barrels each, in the different refineries. These stills are made of heavy boiler-plate iron, capable of withstanding a high degree of heat. The liquid, or oil, is then subjected to a degree of heat necessary to bring it up to the distillation point, which commences at 180 degrees, increasing as the distillation progresses, to 1,000 degrees Fahrenheit. From twenty-four to thirty-six hours are required to "run off" the charge or contents of a one-hundred-barrel still, after the fire is started in the furnaces under the still or retort. As the oil evaporates under such powerful heat, it passes from the retort into condensers attached to same,— these consisting generally of long iron tubes or pipes. The condensing pipes are immersed in a current of cold water, which causes the vapor to return to a liquid condition. On emerging from the tube, the separation of its products is made. The first product of distillation is gasoline, of seventy to seventy-two degrees, Beaumé. The second product is Benzine, which is 62° to 65°, Beaumé. At this stage, the oil has a bluish white color, instead of its natural dark green.

From the condenser it next passes into a receiving tank, the oil in this stage being called " distillate. " Out of this it is transferred, without any further change, into the "treating tank," or "agitator." This is of sufficient size to allow the mixing of the oil thoroughly with sulphuric acid (oil of vitriol). The quantity of acid used is usually from one to two per cent. This is poured in, and the whole mass stirred or agitated by means of iron paddles, or a blast of air in the bottom of the vessel, for about ten or fifteen minutes. The object of this process is to separate from the oil all foreign ingredients, such as tar, dirt, or other impurities that may have become mingled with it. These settle down on the concave bottom of the vessel by their own greater specific gravity, and are removed separately.

The oil is next washed with clean water, and agitated afresh for a period of fifteen or twenty minutes, this process being repeated several times, so as to remove from it every particle of the acid. After washing, it is treated with a solution of caustic soda, standing at 12 to 15 degrees Beaumé, in the proportion of one half to one per cent. The alkali gives the oil brilliancy, and removes every particle of acid that may have remained in it. Next, it is drawn off into the "bleachers" or "settling tubs," which are large shallow wooden tubs metallic lined. The liquid has now a whitish or bluish white color. Weighed by the hydrometer, this instrument is found to rest at the point marked forty-five to forty-seven degrees on the scale, sinking to the level of a higher figure in proportion to the lightness of the oil. It remains in the "bleacher" until the gaseous and lighter portions of the oil have evaporated to an extent that the oil will stand a fire test of 110 degrees and upward without igniting, when a lighted match is applied to it. From the settling tanks or bleachers, it is drawn off and barrelled. The barrels are made of the best white oak, the inside of the barrels being carefully coated with glue, or other suitable compounds, previous to being filled. This process is repeated each time the barrels are refilled, the old coat being removed by means of a jet of steam.

After distillation, as before stated, the second liquid that comes off is naphtha, or benzine, a very light, volatile, and inflammable substance, its hydrometrical test varying from sixty-five to seventy-five degrees. When the discharge coming from the condensers descends to sixty or sixty-two degrees, the benzine is cut off, and let run to oil. If cut off at sixty-five or seventy-five degrees, the oil will be of inferior quality, inflammable, and insufficient to stand the fire test, being liable to explode. The standard now adopted for refined oil requires it to stand a fire-test of one hundred and ten degrees and upward ; this being specified in all contracts.

The mode of determining the " fire tests " is by placing a small quantity of the oil taken from the " bleacher " in a metal cup. Heat is applied by means of the flame of a lamp beneath it. A thermometer is immersed in the oil, and from time to time a lighted match is applied to the oil. The point or degree marked upon the scale of the thermometer at which ignition takes place, is the " fire test." When the oil is too inflammable to come up to the standard stated, it is allowed to remain in the " bleacher " until further evaporation takes place, sufficient to bring it up to the requisite point. Oil that ignites below the specified standard is in danger of igniting when placed in a lamp, causing it to explode, since a certain portion of heat is communicated to it all through from the flame above. The point of ignition is between one hundred and ten and one hundred and twenty degrees Fahrenheit, in the best qualities of refined Petroleum. It may happen, however, that in consequence of its more rapid heating by some than by others, this standard will vary, the difference between the figures in testing the same sample running from one to five degrees. There are three grades of refined oil—"Prime White," "Standard White," and "Straw Color," such being their commercial classification. All these are required to stand a fire test of 110° and upward. The residuum from the condenser is tar, the average ratio of which to all others is estimated to be about one per cent. This substance is used for fuel, as a substitute for pitch, and a lubricating oil for heavy machinery is also made from it. The average per centage of refined obtained from crude Petroleum is about sixty-six to seventy per cent.

As before stated, the refining of Petroleum is a much cheaper process than the procuring of oil from coal. Distillation over carbonate of soda has been tried with success. Distillation by superheated steam is recommended by eminent authorities, and has been extensively practised. But the large majority of refiners use the process we have described, varying the same according to circumstances.

J. Geisner, the eminent chemist, says: "In refining Petroleum or coal oils, care should be taken that the acid used be wholly removed by the alkali or water washing; many samples of Petroleum are found to contain sulphuric acid in sufficient quantity to produce a most disagreeable and dangerous sulphuric acid gas in burning. This has been noticed by physicians upon visiting patients in the country, where the oil is most used. The atmosphere of the sick room is very soon made poisonous by the gas evolved by the night-lamp. Sulphurous acid gas is very irritating to the lungs and mucous membrane. The presence of sulphuric acid in the oil may be detected by adding a solution of chloride of barium to the oil, when a white precipitate will fall if any acid be present."

From Dr. Gesner's valuable work on "Coal Oils," we make the following extracts relative to acids, alkalies, and other oxidating agents used in the purification of Petroleum and their action:

"Acids, alkalies, peroxide of manganese, permanganate of potash, bichromate of potash, etc., have been unsparingly used in the purification of hydrocarbon oils, on account of their oxidating qualities. The object of chemists has been to impart oxygen to the impurities, by which they separate themselves from the oils, and generally fall to the bottom of the vessel that contains them.

"The oxidation of organic compounds takes place in several different ways. In combustion, atmospheric oxygen is aided by a high temperature. If the supply of air be deficient, as in the case of a burning lamp, the hydrogen, from a greater attraction for oxygen, is oxidated, and the carbon of the oil appears in smoke or soot. The decay of wood is produced by oxidation, and ulmine is the result. So also in some of the impurities in hydro-carbon oils; their combination with oxygen gives them new characters, by which they no longer remain with their native liquids. Reagents may be applied to oils that will not separate from them until exposed to the heat of distillation. By its oxidating properties, permanganate of potash converts sugar into oxalic acid. Bichromate of potash diluted with sulphuric acid converts salicine into the hydruret of salicile or oil of spirea. Organic substances are oxidated by the atmosphere, and its action promoted by a higher temperature. Hot air has therefore been forced through hydro-carbon oils during the process of purification, and, in some instances, with advantage.

"*Action of Sulphuric Acid.*—In general, when sulphuric acid is applied to organic compounds, (and such are the oils under consideration,) it decomposes or chars them. By the aid of heat, its effects are more powerful, and it transmutes starch and lignine into grape sugar. Paraffine is not sensibly affected, when boiled with sulphuric acid. For that reason it is employed in the purification of that substance, as it absolutely burns out all its impurities. Sulphuric acid, or oil of vitriol, is now universally used in the purification of coal oils, by which some of their impurities are converted into tar, or rendered soluble in water. The acid may be separated from the tar by distillation. This acid always decomposes a part of the oils in proportion to its strength and the quantity employed. It is a powerful purifier. It removes one kind of odor and substitutes another less disagreeable. How far it changes the characters of the oils has not been determined; but in some instances, when it is used in large quantities, there can be no doubt it produces what may be called *sulpho-oils*, which are unchangeable by the use of alkalies. Certain it is that these sulpho-oils are quite dissimilar to the natural oils obtained by the fractional distillation of coal oils, and inferior to them for the purposes of illumination. The powerful effects of the before-mentioned acid in removing impurities from the distillates of coal and petroleum, and its cheapness, have brought it into general use.

"*Action of Nitric Acid.*—The operations of nitric acid upon organic substances are very numerous. It usually, if not always, produces one or more acids. From gum there comes mucic acid; from indigo, indigotic and nitro-picric acids; from stearic acid, margaric acid, etc., Laurent has clearly described the action of nitric acid upon naphthaline, &c.

"Benzole admits of having its hydrogen replaced by one, two, or three equivalents of nitric acids. This remark applies equally to eupion and all the lighter products distilled from coals, petroleum, etc. All these compounds have an aromatic odor. As an instance, when benzole is saturated with fuming nitric acid, and water is added to the hot solution, nitro-benzole subsides as a yellow oil with the odor of cinnamon. It is sold as the oil of bitter almonds. Other light hydro-carbons give similar results, and a great number of oils, useful for perfumery and cookery, may be produced from them.

"As an oxidator, nitric acid is more powerful than sulphuric acid, but it exerts a greater action on the oils themselves, changing them into *nitro*-oils, and removing them farther away from the natural products of the material first employed.

"Permanganate of potash must be included among the materials used for oxidating the impurities contained in distilled oils. Its effects are feeble when compared with those of sulphuric acid, and its price is too great a drawback on the profits of the manufacture.

"As the oils here treated consist of carbon and hydrogen, some notice may be taken of these two elements. Carbon occurs abundantly in the animal, vegetable, and mineral kingdoms. In its pure and crystallized state it constitutes the diamond. It is the chief substance of plumbago, and frequently forms more than ninety per cent. of anthracite coal. It is essential to the organization of animals, and enters extensively into the composition of minerals, especially the varieties of coal, bitumen, Petroleum, &c., and all substances of vegetable origin. Carbon appears also in the gases of coal mines as carburetted hydrogen, or fire-damp, or carbonic acid, or 'choke-damp.' When organic matter is heated in close vessels, volatile substances are expelled; these consist of carbon, hydrogen, nitrogen, and oxygen; the residue is carbon mixed with the ashes—the minerals that enter into the composition of the wood. Carbon is without taste or smell, and insoluble. It resists decomposition, and, when buried in the earth, is imperishable. Combined with oxygen, carbon forms two gaseous compounds, carbonic acid and carbonic oxide. Carbonic oxide may be considered a compound radical. It combines with chlorine, oxygen, and the metals. It is a transparent, colorless gas, without taste or smell, and, when inhaled, is fatal to animal life. This gas takes fire, and burns with a fine blue flame, which is often seen on the surface of coals burning in a grate.

"Carbonic acid is formed by the respiration of animals, and by vinous fermentation. It is a product of combustion, and is produced artificially by the action of acids upon carbonate of lime. It is a colorless gas, and so much heavier than air, that it may be contained in open vessels. The effervescing properties of wine, beer, soda-water, and some mineral waters, arise from the presence of this acid. It forms the food of growing plants, a part of which they retain in their structures. Another part is expelled, and is found in the atmosphere.

"*Hydrogen* forms one ninth part, by weight, of water, and exists in vegetable and animal substances. It has neither taste, color, nor smell, and is the lightest substance discovered in nature. It is nearly sixteen times lighter than oxygen, and fourteen and a half times lighter than air. It was, therefore, first employed in floating air balloons. A pressure of a thousand atmospheres has no sensible effect in the condensation of hydrogen gas. Sound moves with three times the velocity in hydrogen that it does in common air, and it refracts light with more power than any other gas. The greater the quantity of hydrogen present in any body, the less will be its weight, or specific gravity. It is thus with the hydro-carbon oils. Hydrogen is also the most inflammable substance in nature; it burns with an almost colorless flame and great heat. The opinion is entertained by some, that hydrogen is a gaseous metal, as mercury is a liquid metal.

" *Carbon* and *Hydrogen*, *hydro-carbons*.—Carbon and hydrogen combine in a great number of proportions, and consequently produce numerous compounds; and as both elements are combustible, their compounds are also combustible and inflammable. By some, these compounds are called carbo-hydrogens. At the ordinary temperatures, some of these are solid, such as paraffin, naphthalin, etc.; others are liquid, as the oils of lemons, naphtha, etc. Two of them are gaseous, namely, light carburetted hydrogen gas, and olefiant gas, which are the roots of two, if not more, series of compounds. All these compounds are the products of vegetables, or they are produced from the decay or destructive distillation of organic matter.

" *Carburetted Hydrogen* ($C_1$ $H_2$), mixed with atmospheric air, is the explosive fire-damp of the coal mines, and it frequently rises from the earth through fissures connected with beds of coal, or collections of Petroleum. When mixed with twice its volume of oxygen, it explodes with great violence. If mixed with six times its volume of atmospheric air, it also explodes. By this mixture gasometers have been blown up with terrible effect.

*Bi-carburetted hydrogen, or olefiant gas* ($C_2$ $H_2$), mixed with the above and other gases, occurs in coal mines. It is also transparent and colorless. It takes fire readily, and burns with a white flame, giving out much light. It is also the root of an extensive series of hydro-carbons. This gas and the preceding carburetted hydrogen, when pure, form what is known as coal gas, now extensively employed to light cities. Its value depends much upon the quantity of olefiant gas contained in the mixture.

"The light produced by the combustion of the hydro-carbon oils is like that of coal gas. It is from gas in both instances. The oils are put in lamps, and inflamed; the gas is produced at the top of the wick, and decomposed instantaneously. In the other instance, the gas is made by heating the coals in retorts, and storing it in gasometers ready for use, and its distribution through pipes and burners. In the benzole, or atmospheric light, the vapor of the hydro-carbon is conveyed in the air to the burner, and there burned as coal gas. The fluctuations in the condensation of this vapor by changes of temperature are impediments to this mode of supplying artificial light."

## CHAPTER XIX.

### GENERAL RESULTS—STATISTICAL, &C.

THE cost of the entire Petroleum development, from the date of the striking of Drake's well, in August, 1859, to the date of January 1, 1869, can be approximated from the tabular statement preceding this chapter. The prices paid for the land we were unable to fully ascertain, save in a few cases that were exceptional, and would not serve as a reliable basis. To the date of the speculative era of 1864 and 1865, the prices were generally an average of those paid for improved farming land. The prices of 1864 and 1865 paid to original owners, scarcely averaged more than from $300 to $500 per acre. The prices at which they were put in as the basis of the oil stock companies was far greater. The average price paid for lands from the beginning to the present date, will not exceed, we think, 100 per cent. that of first-rate improved farming land in any desirable locality, say from $100 to $300 per acre. The farms of this and the adjoining counties are small, averaging from 100 to 150 acres each. The total number of farms purchased by oil companies in the Pennsylvania Oil Region we have been unable to ascertain definitely.

From the best information we can get, the following report, made early in 1865, will show the approximate number of oil stock companies formed, and the aggregate amount of the par capital of the same. The capital stock issued by each company varied in amount from $5,000 to $10,000,000. At the close of March, 1865, the amount of par capital of the companies in the several cities where the trade had centred, was estimated as follows:

| | No. of Companies | Amount of par Capital. |
|---|---|---|
| In Philadelphia | 436 | $ 230,885,000 |
| In New York | 220 | 214,400,000 |
| In Pittsburgh | 91 | 22,930,000 |
| In Boston | ... | 3,000,000 |
| In Baltimore | ... | 2,000,000 |
| In Chicago | ... | 5,250,000 |
| In Detroit, Mich | ... | 500,000 |
| In Erie, Pa | 4 | *1,500,000 |
| In Titusville, Pa | 3 | *4,750,000 |
| In Washington, D. C | 2 | 700,000 |
| In Indiana | 7 | 1,780,000 |
| In Kentucky | 81 | †24,000,000 |
| In Ohio | 103 | 25,373,500 |
| In places not enumerated | ... | 7,500,000 |
| Ninety-two companies in New York, Philadelphia, and Pittsburgh, whose capital is not published, estimated | | 36,000,000 |
| Total | | $580,568,500 |

* About. † Estimated.

From this it would appear that, at the close of March, the combined par capital of the companies organized in the United States amounted to over $580,000,000. The total number of companies in existence at the beginning of June, 1865, was 1,085, with an aggregate capital of $600,000,000. Estimating that the paid-up capital averaged one sixth of the whole amount, it would appear that $116,000,000 had at that time been actually invested in oil lands, machinery, and boring operations. The enormous growth, as here stated, was attained in four years.

The return from this large investment of capital is stated in the same report as follows:

"PRODUCTION AND VALUE OF PETROLEUM.

"From a careful comparison of all the available

statistics, relative to the yield of Petroleum, we conclude the production since 1860 to have been about as follows:

| | | |
|---|---|---|
| In 1861 | galls. | 24,000,000 |
| 1862 | " | 40,000,000 |
| 1863 | " | 70,000,000 |
| 1864 | " | 87,000,000 |
| Total, four years | " | 221,000,000 |

"During the current year, the yield of the wells has, thus far, fallen short of that of 1864; it is, however, anticipated that the large number of new wells now 'going down' will largely increase the daily production over that of last year, before the close of the year.

"The following comparison, showing the production and average price of oil at the wells during each year, will give the aggregate value of the product during each of the four years:

| | Product. Galls. | Av. Price. | Total Value. |
|---|---|---|---|
| 1861 | 24,000,000 | 6½ cents. | $1,560,000 |
| 1862 | 40,000,000 | 4 cents. | 1,400,000 |
| 1863 | 70,000,000 | 9½ cents. | 6,650,000 |
| 1864 | 87,000,000 | 23 cents. | 20,010,000 |
| Total for four years | | | $29,820,000 |

"The product of 1864 yielded about 62,000,000 gallons of refined oil, the average price of which, in bond, at New York, was sixty-two cents per gallon; giving as the value of the refined article, in 1864, $40,300,000. Considering that the value of the raw product at the wells was $20,010,000, it appears that the refined product is worth just double the value of the raw oil from which it is derived. This fact illustrates the largeness of the annual gains to labor and capital resulting from the manufacture of the Petroleum products.

"The yield of oil in 1864, as valued at the wells, was worth $20,000,000. As the cost of producing the oil is comparatively light, it would perhaps be safe to estimate $15,000,000 of this amount as representing the net profit on this return. This gain would be equivalent to a profit of twenty per cent. upon $75,000,000 of paid up capital; which amount may be considered as about the sum actually invested in the Petroleum enterprises in 1864, upon an average for the year. During the closing months of 1864, and the first three months of the current year, 1865, the creation of companies has progressed at a rate that would make up the difference between the $75,000,000 of paid-up capital here estimated, and the $116,000,000 above stated, as the probable paid-up capital at the close of March, 1865."

The currency price of crude and refined oil in New York for each month of 1861, 1862, 1863, and 1864 is given as follows:

| Months. | Crude. | | | Refined in Bond. | | | |
|---|---|---|---|---|---|---|---|
| | 1862. Cents. | 1863. Cents. | 1864. Cents. | 1861. Cents. | 1862. Cents. | 1863. Cents. | 1864. Cents. |
| January | 22½ | 23⅞ | 31⅛ | 78½ | 40 | 40 | 46⅛ |
| February | 17½ | 22⅝ | 30⅜ | 79½ | 32½ | 38¼ | 49½ |
| March | 19 | 21⅝ | 31⅜ | 72½ | 30 | 34¼ | 49⅛ |
| April | 14¾ | 21 | 37⅛ | 66¾ | 27¼ | 33¾ | 54⅛ |
| May | 13 | 26⅛ | 38¼ | 62¾ | 25 | 39½ | 59½ |
| June | 14 | 27¼ | 44⅞ | 50 | 26¾ | 44½ | 68⅞ |
| July | 15 | 30⅞ | 52⅜ | 50¾ | 30 | 49 | 86 |
| August | 19½ | 35⅛ | 52⅝ | 59¾ | 34 | 53½ | 84⅞ |
| September | 18¾ | 36 | 45½ | 58 | 34¾ | 58 | 75 |
| October | 22 | 32⅞ | 41¼ | 55 | 36¼ | 52½ | 63⅜ |
| November | 36 | 27⅝ | 45⅝ | 48½ | 60¾ | 41½ | 64¾ |
| December | 38 | 30⅞ | 51⅞ | 48½ | 59½ | 46½ | 72¾ |
| Average | 20⅝ | 28 | 41⅞ | 61½ | 36⅜ | 44¼ | 64¾ |

The lack of proper management, the large number of corrupt companies formed, and absorption of the inadequate working capital of the companies, the closing of the war, and consequent reduction of the values prevailing, caused a large proportion of the oil companies to collapse. Owing to the small amount of capital devoted to actual development, but few of the companies had paid any dividends, and their stock, after the first issue, only represented a merely nominal value, even where the company was organized on a legitimate basis. With the bursting of the bubble at the close of the war, the shares of stock were, in a large number of companies, scarcely worth the paper they were printed upon. Thousands of shareholders were ruined, and the large investment rendered valueless for all practical benefit. A number of companies still retain the lands purchased in 1864 and 1865. A large number have been sold out by the sheriff for taxes and claims, real or otherwise, brought against them.

The failure of the oil companies proved highly detrimental to the Petroleum interests, and but for the indomitable energy of those engaged legitimately in the trade, would have been ruinous. But few oil companies were organized in the Oil Region proper, and a majority of these proved successful. Though a temporary injury, and one from which the Petroleum business has not as yet fully recovered, the good results of that exciting, though brief era, have become apparent in the last two years. The large and general development inaugurated by the numerous oil companies during the two years of their reign, could not have been effected successfully by individual effort in a quarter of a century. The scores of wells drilled by them in all portions of the Oil Region, now serve as landmarks or beacons to guide the oil operator in his search for the oily treasure. From the effort made in previous days or years, he can form his own plans with more reasonable hope of success. The unfortunate stockholders of a majority of the oil companies, deserve our sympathy, though their investments were too often made without reflection and contrary to the established laws which govern legitimate business. Yet from the development of 1868 and '69, a hope, far from faint, can be held out to stockholders whose lands here are still retained. The area of the general oil-producing field is being extended greatly from year to year, and many of the lands of these companies are being included in it, and are proving highly productive.

We are no apologists for the system of oil companies as inaugurated in 1864 and 1865. On the contrary, we consider them as having been of great detriment, and as causing the ruin of thousands. People became excited over the fanciful reports of those who wished to obtain their money, and invested their means recklessly, losing it, of course. But there are many companies organized on a sound basis who have been enabled to outride the storm, whose chances for success are better to-day than ever before, and are constantly improving. Excitements in oil stocks may possibly occur again, but we are sure the speculation will be on a better basis. By reducing their capital stock to a figure approximating somewhat to the real value of their possessions, and economical management, with the advantage of superior knowledge of the practical details of the business, the chief drawback to their success in former years, the chances for realizing their investments in whole or in part, is certainly flattering. We have preferred to treat the subject in general terms, as a detailed statement would take up too great an amount of space, and could be productive of no beneficial results to the parties concerned.

The following, copied from the exceedingly valuable Annual Petroleum Report for 1868, of the Titusville

*Morning Herald*, furnishes, in brief form, as reliable statistical information of the business for 1868, as well as the preceding years, as can be obtained. It is so interesting that we give it nearly entire :

## SECOND ANNUAL PETROLEUM REPORT.

### REVIEW OF THE PETROLEUM MARKET ON THE CREEK.

*January.*—There was a quiet feeling during the first week, and prices were quoted at \$2.70@\$2.85, but about the 15th the feeling became stronger under an increase in the demand for home consumption, and prices appreciated 15 to 30 cents. Before the 20th there was another advance of 15 cents, and the market closed firm at \$2.00@\$2.15.

*February.*—In the first half of this month prices were unusually steady, having varied only about five cents from those at the close of January. Between the 14th and 17th there was an advance of 10 cents, and from the latter date there was a steady appreciation until the 26th, when the rise culminated at \$2.40@\$2.50, and subsequently prices declined, and were quoted on the 29th at \$2.30@2.40.

*March.*—The market was firm at the beginning and continued with an upward tendency. Sales were made on the 9th as high as \$2.65, on the 18th at \$2.85, and on the 19th at \$2.90@\$3.00, after which the feeling broke and prices declined slowly until the close when quotations were given at \$2.75.

*April.*—There was a steady decline from the opening, and on the 8th there were sales at \$2.40 which was the lowest point reached, prices having subsequently advanced, and on the 20th they were quoted at \$2.70 @\$2.90. The market after the 20th was very strong and at the close sales were made as high as \$3.50.

*May.*—The upward movement of the previous month was continued until about the 6th, when prices reached \$3.80@\$4.10 at all points along the Creek. There was then a decline, but it was slight, and only lasted a few days, the feeling having become strong about the 13th, and on the succeeding days sales were made at from \$3.50 to \$4.00. The advance, however, lasted but a day or two, and was followed by a weak feeling which carried prices slowly downward, and on the 30th there were sales as low as \$3.00 and \$3.50. The wide difference between the two last prices was caused by the freight charges on the local railroad.

*June.*—Early in this month there was a sharp advance, and prices reached \$4.50 on the 8th. There was then considerable excitement and the upward movement continued, and at the close quotations were given at \$4.90@\$5.00. The highest figures of the month along the Creek were \$5.25@5.50, and at Oil City \$5.25@\$5.35. These prices were maintained but a few hours.

*July.*—There were but few fluctations in the price, and it was more steady than during any other month in the year. At the opening there was a moderate activity at \$5.25@\$5.50 along the Creek, and \$5.85 at Oil City. Toward the end of the second week there was a decline, but prices rallied during the last week, and closed at \$5.10.

*August.*—The closing quotation of the previous month was sustained until about the 6th, when the market broke and prices declined, and on the 13th sales were made at \$4.75@\$4.90. There was then a steady decline, and on the 28th the quotations were \$3.70@\$4.05. During the last two days the market recovered, and at the close it was firm.

*September.*—During the first three days the feeling

was quiet,and prices were steady, but about the 5th there was an advance, but it was lost early in the second week under a strong "bear" movement, and prices receded until the 18th. At that date sales were effected at $3.65@$3.75. Subsequently prices were carried up by the efforts of a combination of dealers to $4.00@$4.25, at which large quantities were thrown on the market, and quotations fell off 30@35 cents.

*October.*—At the beginning there was a firm feeling, and prices advanced slightly. On the 5th sales were made at $4.00@$4.25, after which the market broke; but the decline was not great, and about the middle of the month the feeling rallied, and on the 21st round lots were sold at $4.25. During the next six days the tone was firm, but after the 26th the demand decreased, and on the 30th the feeling was weak.

*November.*—From the opening quotation—say $4.10—there was a gradual decline, and on the 13th sales were made at $3.00@$3.15, and there were some small lots changed hands at 5@10 cents under those prices. The lowest prices were reached during the 13th and 14th. The decline had the effect of bringing speculative buyers into the market, and on the 16th there was a stronger feeling, and prices advanced until about the 25th, when quotations were given at $4.30@$4.50. Toward the close there was a weaker feeling, and prices receded 25@35 cents.

*December.*—The market was firm during the most of this month, and advanced materially. On the 12th sales were made as high as $4.75. There was then until the 20th a more quiet feeling, but after that date prices began to decline, and on the 20th there were sales at $4.20@$4.30. During the last three days there was a better feeling, and at the close the tendency was upward.

## RANGE OF PRICES IN 1868.

The following table shows the range of prices of crude per barrel on the Creek; of crude per gallon at Pittsburgh and New York, and of refined per gallon at New York:

| | | Crude per barrel of 43 gallons, on the Creek. | Crude per gallon in Pittsburgh. | Crude per gallon in New York. | Refined per gallon in New York. |
|---|---|---|---|---|---|
| Jan. | 2 | 1 70@1 90 | 6¾ | 10½ | 24½ |
| " | 5 | 1 70@1 80 | 7 | 10½ | 23½@24 |
| " | 8 | 1 70@1 85 | 7 | 10½ | 24½ |
| " | 11 | 1 70@1 85 | 7 | 10½ | 24½ |
| " | 14 | 1 80@2 00 | 7 | 10½@11 | 25½ |
| " | 17 | 1 90@2 10 | 7 @7¼ | 10½ | 25 |
| " | 20 | 1 95@2 10 | 7 @7¼ | 10¾ | 24¾ |
| " | 23 | 1 95@2 10 | 7¼@7½ | 9½ | 24½ |
| " | 26 | 1 90@2 10 | 7 | 10½ | 24 |
| " | 29 | 2 00@2 15 | 7¼ | 10 | 22 @22¾ |
| Feb. | 2 | 2 00@2 15 | 7 @7¼ | 10½ | 24½ |
| " | 5 | 2 00@2 15 | 7 @7¼ | 10½ | 24 |
| " | 8 | 2 15 | 7 | 9½@10 | 22 |
| " | 11 | 2 10@2 15 | 7½ | 11½ | 24¾ |
| " | 14 | 2 10@2 25 | 7¼ | 11 | 24¾ |
| " | 17 | 2 10@2 25 | 7¼@7½ | 11½ | 24½ |
| " | 20 | 2 15@2 30 | 7½ | 12 | 25 |
| " | 23 | 2 35@2 50 | 8 | 11 | 24½ |
| " | 26 | 2 40@2 50 | 8 | 11 | 24½@25 |
| " | 29 | 2 30@2 45 | 7½@7¼ | 12 | 24½@26 |
| March | 3 | 2 20@2 40 | 7½@7¾ | 12 | 25½@26 |
| " | 6 | 2 40@2 45 | 7½ | 13 | 25½@26 |
| " | 9 | 2 50@2 65 | 8 | 11 | 26 |
| " | 12 | 2 55@2 70 | 8 @8¼ | 14 | 26 |
| " | 15 | 2 60@2 70 | 8¼ | 14 | 26 |
| " | 18 | 2 60@2 70 | 8¼ | 14 | 26 |
| " | 21 | 2 70@2 85 | 8 @8¼ | 14 | 25½ |
| " | 24 | 2 85@2 90 | 8½ | 12½ | 25½ |
| " | 27 | 2 75@2 80 | 8¾ | 12¼ | 25 @25½ |
| " | 30 | 2 65@2 70 | 8¾ | 12½ | 25¾@26 |
| April | 2 | 2 65@2 75 | 8¾@9 | 12 | 25½@26 |
| " | 5 | 2 65@2 75 | 8¾ | 12 | 26 |
| " | 8 | 2 40@2 50 | 8¾ | 11 | 26 |
| " | 11 | 2 40@2 65 | 8½ | 11 | 25¾ |
| " | 14 | 2 50@2 75 | 8½ | 10½ | 25½ |
| " | 17 | 2 60@2 85 | 8¾ | 11 | 25½ |
| " | 20 | 2 70@2 90 | 8¾ | 11 | 26 |
| " | 23 | 2 75@3 00 | 9¼ | 12½ | 26½ |
| " | 26 | 3 00@3 10 | 9½ | 12½ | 26½ |
| " | 29 | 3 00@3 30 | 10 | 13 | 27½ |
| May | 2 | 3 60@3 75 | 10½ | 13½ | 28 |
| " | 5 | 3 80@4 10 | 12½ | 15 | 31 |
| " | 8 | 3 55@3 80 | 11½ | 14 | 30 |
| " | 11 | 3 25@3 50 | 12 | 14 | 30 |
| " | 14 | 3 50@4 00 | 10½@11 | 14 | 30½ |
| " | 17 | 3 40@3 50 | 11 | 14 | 29 |

| | | Crude per barrel of 43 gallons, on the Creek. | Crude per gallon in Pittsburgh. | Crude per gallon in New York. | Refined per gallon in New York. |
|---|---|---|---|---|---|
| May | 20 | 3 40@3 60 | 10½ | 13 @13½ | 29 |
| " | 23 | 3 40@3 75 | 10½@11 | 13 @13½ | 30 |
| " | 26 | 3 30@3 50 | 11 | 13½ | 29½ |
| " | 29 | 3 30@3 50 | 10½ | 13½ | 28½ |
| June | 2 | 3 30@3 60 | 10½ | 13½ | 27½ |
| " | 5 | 4 00@4 50 | 11 @11½ | 13½ | 30 |
| " | 8 | 4 25@4 50 | 11¾@12 | 14 @14½ | 31 |
| " | 11 | 4 35@4 50 | 11½ | 15½ | 31½ |
| " | 14 | 5 00@5 25 | 12½ | 15½@16 | 32 |
| " | 17 | 5 15@5 35 | 12½@13 | 16¾ | 32½@33 |
| " | 20 | 4 65@4 85 | 12 | 16 | 31½@32 |
| " | 23 | 4 75@4 85 | 12½ | 15½@16 | 31 |
| " | 26 | 5 00@5 25 | 12½ | 16 @16½ | 31 |
| " | 29 | 4 90@5 00 | 13 | 16¼@16½ | 32 |
| July | 2 | 5 00@5 25 | 13½@14 | 16 @16½ | 32 @32½ |
| " | 6 | 5 00@5 50 | 15 | 17 | 32½@33 |
| " | 8 | 5 50@5 75 | 15 | 17½@18 | 34 @34½ |
| " | 11 | 5 00@5 25 | 14½ | 17½ | 34 |
| " | 14 | 4 40@4 60 | 12½ | 16½ | 33 |
| " | 17 | 5 00@5 25 | 14½@15 | 17¼@17½ | 34 @34½ |
| " | 20 | 4 85@5 00 | 14 | 17 @17½ | 34 @34½ |
| " | 23 | 4 90@5 00 | 13½@14 | 16½@17 | 33½@34 |
| " | 26 | 4 80@5 00 | 14 | 17 | 34 @34½ |
| " | 29 | 5 00@5 10 | 14 | 17½ | 34½ |
| August | 1 | 5 00@5 10 | 14½@14¾ | 18 @18½ | 34 |
| " | 4 | 5 00@5 19 | 14 @14½ | 17 @17¼ | 34½ |
| " | 7 | 5 00 | 14 @14½ | 17½@17¾ | 34¼@34½ |
| " | 10 | 4 95 | 14 | 17½ | 34¼@34½ |
| " | 13 | 4 75@4 90 | 13½@14 | 16½@16¾ | 34 |
| " | 16 | 4 50@4 80 | 13 @13½ | 16 | 33 @33½ |
| " | 19 | 4 65@4 75 | 13¾ | 16 | 34 |
| " | 22 | 4 50@4 60 | 13 | 15 @15½ | 32½@32¾ |
| " | 25 | 4 00@4 25 | 12½ | 15 @15½ | 31½@32 |
| " | 28 | 3 55@3 90 | 11¾@12 | 14 @14½ | 30½@30¾ |
| Sept. | 1 | 3 70@4 05 | 11¾@12 | 14½ | 31¼@31½ |
| " | 4 | 4 00@4 25 | 12 | 15 @15½ | 31½@32 |
| " | 7 | 4 25@ | 11½@12 | 16½@17 | 32 |
| " | 10 | 3 85@4 00 | 11½ | 15½@16 | 32 |
| " | 13 | 3 70@3 90 | 11 | 15 | 30 |
| " | 16 | 3 50@3 75 | 11 | 14¾ | 30 |
| " | 19 | 3 75@3 80 | 10¾ | 16 | 30 @30½ |
| " | 22 | 3 80@4 00 | 11¼@11½ | 15⅞ @16 | 31¾@32 |
| " | 26 | 4 00@4 30 | 12 @12½ | 16½@17 | 32½ |
| " | 29 | 4 00 | 11¾ | 15½@16 | 30 |
| October | 2 | 4 00@4 10 | 12 | 16 | 30½ |
| " | 5 | 4 10@4 25 | 13 | 16 | 30½@31 |
| " | 8 | 4 00 | 12½ | 16 | 30 |
| " | 11 | 4 00@4 25 | 12½ | 15¼@15½ | 29½@30 |
| " | 14 | 4 00@4 15 | 12½ | 16 | 29 @30 |
| " | 17 | 4 00@4 15 | 11½@11¾ | 15½@16 | 29½ |
| " | 20 | 4 00@4 20 | 11¾ | 15½ | 29 @29½ |
| " | 23 | 4 15@4 20 | 11½ | 15¾@16 | 30 |
| " | 26 | 4 00@4 20 | 12 | 16½ | 31 |
| " | 29 | 4 00@4 15 | 11¾@12 | 17 | 30½@31 |
| Nov | 1 | 4 00@4 10 | 11½@11¾ | 17 | 29½ |
| " | 4 | 3 60@3 90 | 11½ | 16¼ | 29½ |
| " | 7 | 3 30@3 75 | 10⅞ @11¼ | 17 | 27 |

| | | Crude per barrel of 43 gallons on the Creek. | Crude per gallon in Pittsburgh. | Crude pe. gallon in New York. | Refined per gallon in New York. |
|---|---|---|---|---|---|
| Nov. | i1 | 3 10@3 40 | 11 | 15 | 27 |
| " | 13 | 3 00@3 15 | 10 @10½ | 12½@13 | 27½ |
| " | 16 | 3 00@3 35 | 10 | 14 | 27½ |
| " | 19 | 3 50@3 60 | 10 @10½ | 15 | 28¾ |
| " | 22 | 4 20@4 35 | 11¾ | 16¾ | 33½@34 |
| " | 25 | 4 30@4 50 | 12¾ | 18 | 35 |
| " | 28 | 3 90@4 25 | 11½ | 18 | 35 |
| Dec. | 1 | 3 80@4 20 | 12 | 17 @18 | 31 @32 |
| " | 3 | 3 60@3 90 | 11½ | 16¼ | 29½ |
| " | 6 | 4 30@4 40 | 11½ | 17½ | 31½ |
| " | 9 | 4 40@4 50 | 12 @12½ | 18 | 33½@34 |
| " | 12 | 4 75@4 90 | 12½ | 18 | 33½@34 |
| " | 15 | 4 55@4 70 | 12½@ 13 | 19 | 32 |
| " | 18 | 4 40@4 50 | 12½ | 18¼@18½ | 32¼ |
| " | 21 | 4 25@4 40 | 12½ | 18 @18½ | 31½@32 |
| " | 24 | 4 25@4 30 | 12 | 18 | 31½ |
| " | 27 | 4 10@4 25 | 12 @12½ | 17¾ | 31 @31½ |
| " | 30 | 4 40@4 50 | 12 @12½ | 18 | 31¼ |

EXPORTS DURING THE LAST EIGHT YEARS.

The exports of crude, refined, naphtha, &c., from the United States were as follows:

| | | | | | |
|---|---|---|---|---|---|
| In 1861.... | galls. | 1,112,476 | equal to | bbls. | 27,812 |
| 1862.... | " | 10,887,701 | " | " | 272,192 |
| 1863.... | " | 28,250,721 | " | " | 706,268 |
| 1864.... | " | 31,872,972 | " | " | 796,824 |
| 1865.... | " | 29,805,523 | " | " | 745,138 |
| 1866.... | " | 67,430,451 | " | " | 1,685,761 |
| 1867.... | " | 67,051,020 | " | " | 1,676,300 |
| 1868.... | " | 91,281,750 | " | " | 2,482,044 |

Messrs. Eagle & Blakeslee, Petroleum Freight Brokers, of New York, make the amounts of crude, refined, &c., exported from New York, Philadelphia, Boston, and Baltimore, in 1868, as under:

| | | |
|---|---|---|
| Crude.... .................... | galls. | 9,473,625 |
| Refined......................... | " | 79,345,165 |
| Lubricating oil.................. | " | 1,701,561 |
| Naphtha......................... | " | 9,128,972 |
| Residuum...................... | " | 125,685 |

At this late date it is extremely difficult to obtain the product of the earlier years, and we have only ventured on an approximation of it. In 1859 there were small amounts produced on the flats near Titusville on the W. M'Clintock and A. and J. Buchanan farms. The total product for that year was about 82,000 barrels, and in 1860 it is set down at 90,000 barrels. The subjoined tables show the production of the farms and districts as near as could be ascertained during the years named:

| Names of Farms and Districts. | 1861 Barrels. | 1862 Barrels. | 186 Barrels. |
|---|---|---|---|
| Titusville and vicinity | 40,000 | 36,000 | 35,000 |
| Bull Run and vicinity | ... | .... | 150,000 |
| Sherman flats | .... | .... | 255,000 |
| Benninghoff farm | 3,500 | 4,000 | 60,000 |
| McElheny farm | 380,000 | 400,000 | 200,000 |
| Hyde & Egbert farm | .... | .... | 110,500 |
| Story farm | 41,600 | 179,106 | 159,859 |
| Tarr farm | 40,000 | 100,000 | 190,000 |
| Blood farm | 13,000 | 300,000 | 160,000 |
| Steele farm | 29,000 | 25,000 | 200,000 |
| A. Buchanan farm | 912,500 | 1,095,000 | 180,000 |
| J. Buchanan farm | 365,000 | 547,500 | 72,000 |
| J. McClintock farm | 170,000 | 180,000 | 300,000 |
| Allegheny River district | 84,000 | 90,000 | 44,000 |
| Other farms and districts | 35,000 | 100,000 | 495,000 |
| Total | 2,113,600 | 3,056,606 | 2,611,359 |

| Names of Farms and Districts. | 1864 Barrels. | 1865 Barrels. | 1866 Barrels. |
|---|---|---|---|
| Titusville and vicinity | 25,000 | 40,000 | 55,000 |
| Bull Run and vicinity | 212,000 | 199,197 | 153,268 |
| Sherman flats | 109,500 | 28,000 | .... |
| Foster farm, Pioneer Run | .... | .... | 64,713 |
| Benninghoff farm | 4,000 | 8,000 | 92,953 |
| McElheny farms | 119,803 | 77,550 | 93,482 |
| Bennehoff Run | .... | 100,500 | 132,000 |
| Stevenson farm | .... | 140,000 | 300,000 |
| W. McClintock farm | 58,403 | 89,192 | 289,906 |
| Hyde & Egbert farm | 360,000 | 240,000 | 120,000 |
| Dalzell farm | .... | 8,277 | 10,688 |
| Story farm | 241,976 | 210,996 | 177,567 |
| Tarr farm | 75,000 | 25,000 | 56,000 |
| Blood farm | 50,000 | 60,000 | 34,000 |
| Steele farm | 250,000 | 136,000 | 17,000 |
| A. Buchanan farm | 200,000 | 21,000 | 40,000 |
| J. Buchanan farm | 159,000 | 37,000 | 12,000 |
| J. McClintock farm | 41,000 | 27,000 | 72,000 |
| Pithole | .... | 350,000 | 800,000 |
| Tidioute | 60,000 | 300,000 | 803,000 |
| Allegheny River district | 75,000 | 150,000 | 134,000 |
| Other farms and districts | 75,000 | 250,000 | 140,000 |
| Total | 2,116,182 | 2,497,712 | 3,597,527 |

PRODUCTION—1868.

At the commencement of 1868 the then recent discovery of two important producing districts, and an improved prospect for an advance in the price of oil, made it look probable that there would be a very large production during the summer, but owing to the wells in both districts becoming exhausted much sooner than it was expected they would, the anticipations of a large production were not realized. There were wide fluctuations, the extreme daily averages having been 8,600 and 12,500 barrels. The former average was in May, and the latter during the last two days of September.

The most important feature during the year was that in the last three months the product continued to decrease, although the number of wells being completed was larger than previously. This unusual occurrence in oil producing was owing to more than one half of the wells that were completed having been located in the Pleasantville district, where the oil-producing sand-rock is thin and the supply of oil in it proportionally light.

The annexed table shows the average daily production from January 1st to April 8th, from April 8th to May 1st, and during each succeeding month in the year 1868:

| | | |
|---|---|---|
| From January 1st to April 8th | bbls. | 9,280 |
| From April 8th to May 1st | " | 8,650 |
| During May | " | 9,740 |
| " June | " | 10,102 |
| " July | " | 10,698 |
| " August | " | 11,981 |
| " September | " | 11,486 |
| " October | " | 11,556 |
| " November | " | 10,275 |
| " December | " | 9,737 |

The following is a recapitulation of the production since 1859. The production during 1867 is taken from our last Annual Report, and that for 1868 from our monthly reports:

| | | | |
|---|---|---|---|
| Produced in | 1859 | bbls. | 82,000 |
| " | 1860 | " | 500,000 |
| " | 1861 | " | 2,113,600 |
| " | 1862 | " | 3,056,606 |
| " | 1863 | " | 2,611,359 |
| " | 1864 | " | 2,116,182 |
| " | 1865 | " | 3,497,712 |
| " | 1866 | " | 3,597,527 |
| " | 1867 | " | 3,347,306 |
| " | 1868 | " | 3,715,741 |
| | Total | bbls. | 23,638,033 |

NUMBER OF PRODUCING WELLS—AVERAGE CAPACITY—NUMBER OF WELLS COMPLETED IN 1868.

As the producing capacities of the wells have grown smaller, the number of wells has been enlarged in order to meet the requirements of an increasing demand, and on January 1, 1869, the total number in the region was 1,186—an increase of probably 200 from the same date last year. The wells are located over an extent of territory about sixty miles in length, and from one to sixteen miles in width. Most of them, however, are in the Oil Creek, Pleasantville, Tidioute, and Upper Cherry Run districts.

The total daily capacity for producing of all the wells, at the date above named, was equal to 13,819 barrels. This is what the wells would produce when worked continuously under favorable crircumstances. But as it is only in rare instances that wells can be, or are worked without stopping for any length of time, it necessarily follows that the capacity of wells is not a criterion of the production; nor can any entirely reliable rule for ascertaining the product from the capacity of the wells be made, because the length of time which wells are worked steadily depends much on the weather, but more on the skill with which they are managed. Assuming, however, that the capacity of the wells during January was the same as on the 1st of that month, and as the average daily product for January was 10,192 barrels, it is seen that the capacity was 3,669 barrels, or about one third greater each day than the production.

The subjoined table shows the number of wells that were producing on each farm, or in each district, on January 1, 1869; the total daily average capacity for production and the number of wells that were completed in 1868:

| NAMES OF FARMS AND DISTRICTS. | No. of Producing Wells. | Total Capacity for Production. | Average Capacity for Production. | No. of Wells completed in '68 about |
|---|---|---|---|---|
| Titusville and Church Run | 61 | 715 | 11⅔ | 46 |
| Pleasantville District | 214 | 3,164 | 14¾ | 300 |
| Atkinson Farm, Upper Cherry Run | 6 | 64 | 10⅔ | 35 |
| Tallman Farm, " " " | 27 | 405 | 15 | 31 |
| S. P. Co.'s Tract, " " " | 17 | 427 | 25 | 26 |
| Pittsb'gh & Cherry Run Tract, " | 11 | 165 | 15 | 3 |
| A. Clark Farm, " | 8 | 97 | 12 | 4 |
| Great Republic Oil Co.'s Tract, " | 6 | 30 | 5 | 4 |
| Fink Lease, " | 3 | 33 | 11 | 2 |
| T. Tarr Farm, " | 3 | 36 | 12 | 3 |
| J. E. Bennehoff Farm, " | 3 | 21 | 7 | 2 |
| Keech Farm, " | 3 | 24 | 8 | — |
| Other Tracts and Farms on " | 9 | 65 | 7 | 10 |
| Cow Run, Oil Creek | 9 | 145 | 16 | 6 |
| Caldwell Farm, Bull Run | 3 | 36 | 12 | 5 |
| Patterson Farm, " " | 9 | 54 | 6 | 5 |
| Skinner Farm, " " | 7 | 35 | 5 | 3 |
| Farel Farm " " | 4 | 40 | 10 | 9 |
| Foster Farms, Pioneer Run | 15 | 255 | 17 | 17 |
| Benninghoff Farm, Oil Creek | 13 | 246 | 19 | 20 |
| Benninghoff Run, " " | 4 | 28 | 7 | 5 |

| NAMES OF FARMS AND DISTRICTS. | No. of Producing Wells. | Total Capacity for Production. | Average Capacity for Production. | No. of Wells completed in '68 about |
|---|---|---|---|---|
| Stevenson Farm, Oil Creek | 15 | 158 | 15½ | 22 |
| A. Boyd & Coxton Farm, near Oil Creek | 5 | 45 | 9 | — |
| W. M'Clintock Reserve " " " | 2 | 14 | 7 | — |
| W. M'Clintock Farm, Oil Creek | 47 | 376 | 8 | 13 |
| McElheny Farms, " " | 9 | 45 | 5 | 5 |
| McCray Farm, " " | 2 | 14 | 7 | — |
| Woods Farm, " " | 42 | 714 | 17 | 40 |
| Keneusett Petroleum Company's Tract, Oil Creek | 1 | 12 | 12 | 13 |
| Dalzell Farm, " " | 4 | 12 | 3 | 1 |
| Pierson Farm and Cherry Tree Run, near Oil Creek | 42 | 538 | 12⅔ | 33 |
| Story Farm, Oil Creek | 23 | 322 | 14 | 11 |
| Tarr Farm, " " | 12 | 144 | 12 | 7 |
| Blood Farm, " " | 25 | 575 | 23 | 20 |
| Rynd Farm, " " | 18 | 138 | 7⅔ | 13 |
| Steel Farm, " " | 15 | 60 | 4 | 1 |
| M'Clintockville Petroleum Company's Tract | 17 | 102 | 6 | 8 |
| Cornplanter Farm, Oil Creek | 3 | 38 | 13 | — |
| Clapp Farm, " " | 10 | 76 | 7 | 2 |
| A. Buchanan Farm, Lower Cherry Run | 18 | 180 | 10 | 7 |
| J. Buchanan Farm, " " | 21 | 363 | 17 | 22 |
| John M'Clintock Farm, " " | 11 | 113 | 10 | 2 |
| Union Petroleum Co.'s Tract, " " | 10 | 90 | 9 | 2 |
| Brevoort Petroleum Co.'s Tract, " | 8 | 64 | 8 | 1 |
| Moody Run, " " | 5 | 30 | 6 | 1 |
| Smith Farm, " " | 15 | 120 | 8 | 2 |
| Curtin Oil Company's Tract " " | 3 | 60 | 20 | 2 |
| United Pet. Farms Association Tract, Oil Creek | 32 | 64 | 2 | 8 |
| Hoffman Petroleum Company's Tract, " " | 4 | 7 | 1¾ | 3 |
| Oil City, West side, and Holliday Run | 20 | 36 | 1 | 1 |
| Charley Run, near Oil City | 5 | 140 | 28 | 4 |
| Shaffer Run, near Reno | 6 | 102 | 17 | 5 |
| Reno Company's Property | 15 | 30 | 2 | 1 |
| Cochran Farm, Allegheny River | 12 | 24 | 2 | 1 |
| Honeycomb Oil Company's Tract, Allegheny River | 1 | 7 | 7 | — |
| Hoover Farm, " " | 10 | 55 | 5½ | 1 |
| Snyder Farm, " " | 6 | 108 | 18 | 4 |
| Foster Island, " " | 2 | 18 | 9 | — |
| Foster Farm, " " | 8 | 160 | 20 | 3 |
| Excelsior Petroleum Company's Tract, " " | 2 | 2 | 1 | — |
| Philadelphia and Boston Pet. Co.'s Tract, " | 10 | 100 | 10 | 5 |
| McMillan Oil Company's Tract, " " | 13 | 130 | 10 | 7 |
| Belle Island, " " | 9 | 82 | 9 | 6 |
| Witherup Farm, " " | 10 | 60 | 6 | 9 |
| Parker's Landing, " " | 11 | 121 | 11 | 5 |
| Fox's Landing, " " | 2 | 7 | 3½ | 1 |
| Allegheny River District from Reno to Tidioute | 47 | 141 | 3 | 18 |
| Economy Oil Company's Tract, " | 21 | 444 | 21 | 18 |
| Tidioute and Warren Oil Company Tract, " | 25 | 627 | 5 | 10 |
| Triumph Oil Company's Tract, " | 13 | 130 | 10 | 4 |
| N. Y. & D. Run Co.'s Tract, " | 6 | 42 | 7 | — |
| Pierce & Neyhart Tract, " | 3 | 18 | 6 | — |
| N. Y. & A. O. Co.'s Tract, " | 9 | 54 | 6 | 3 |
| Other Farms and Tracts in vicinity of Tidioute | 7 | 84 | 12 | 5 |
| Pithole and vicinity | 28 | 28 | 10 | 28 |
| Bean Farm, near Pithole, about | 12 | 96 | 8 | 4 |
| Enterprise, Warren Co | 9 | 81 | 9 | 8 |
| Total | 1,186 | 13,8[illegible]9 | 11⅔ | 929 |

About 150 of the wells produced less than two barrels per day, and there were 200 the production of which averaged less than five barrels per day. Of the wells producing on January 1, one, situated on the Buchanan farm on Oil Creek, was struck in 1859, and there were twenty-seven others struck during 1860 and 1861, the production of which is yet considerable. These wells are located along Oil Creek and on the Hoover and other farms on the Allegheny River. The wells producing on January 1, 1869, were completed in the years and proportions as under:

| | |
|---|---|
| In 1859 | 1 |
| 1860 | 10 |
| 1861 | 17 |
| 1862 | 20 |
| 1863 | 29 |
| 1864 | 46 |
| 1865 | 99 |
| 1866 | 92 |
| 1867 | 148 |
| 1868 | 724 |
| Total | 1,186 |

Of the 929 wells completed during 1868, about 100 did not produce in paying quantities. A large proportion of the unsuccessful wells were struck in the first half of the year.

### SHIPMENTS FROM THE OIL REGION.

The shipment of oil from the region was carried on during 1868 almost wholly in bulk, and of 3,669,000 barrels of crude and refined only about 375,000 barrels were moved in packages. The amount moved by the Allegheny River reached about 700,000 barrels.

The following table gives the shipments of crude and refined from the region by all routes, the total shipment of crude and also the shipment of crude in 1867. The barrels are rated at forty-five gallons:

| | | |
|---|---|---|
| To New York | bbls. | 965,862 |
| Cleveland | " | 918,651 |
| Boston | " | 120,981 |
| Philadelphia | " | 236,922 |
| Portland | " | 35,878 |
| Pittsburgh | " | 1,144,956 |
| Other points | " | 245,883 |
| Total | bbls. | 3,669,133 |
| Difference between the Crude and the Refined included in the shipments | " | 117,937 |
| Total | bbls. | 3,787,070 |
| Total in 1867 | " | 2,968,366 |

Of the shipments to other points about 50,000 barrels were moved to Baltimore, 50,000 to Buffalo, 40,000 to Erie, 40,000 to Syracuse, 45,000 to refineries in the interior of New York and Pennsylvania, and about 21,000, mostly refined, to points west of Cleveland.

The annexed table shows the total amount of refined included in the above shipment, and about to what points it was moved, and also the total shipment of refined in 1867:

| | | |
|---|---|---|
| To New York | bbls. | 175,203 |
| Philadelphia | " | 93,660 |
| Boston | " | 11,110 |
| Other points | " | 72,560 |
| Total in 1868 | bbls. | 352,533 |
| Total in 1867 | " | 223,865 |

STOCKS ON HAND IN THE UNITED STATES AND EUROPE JANUARY 1, 1868 AND 1869.

The stock of Petroleum in the United States on

January 1, 1869, according to the closest estimates we have been able to obtain, was 780,000 barrels of crude or its equivalent, against about 1,000,000 barrels at the same date in 1868, showing a reduction during the year of a little over 200,000 barrels. The foregoing applies to oil produced in Pennsylvania, and does not include a stock of about 50,000 barrels on hand in West Virginia and Ohio on January 1, 1868, and one of about 10,000 or 15,000 barrels in the same States on the 1st inst.

The annexed table shows the stocks on hand on the first day of January, 1868 and 1869:

| | | 1869. | 1868. |
|---|---|---|---|
| In New York | bbls. | 75,233 | 91,000 |
| Pittsburgh | " | 225,000 | 100,000 |
| Philadelphia | " | 60,793 | 111,304 |
| Cleveland | " | 100,000 | 46,000 |
| Boston and Baltimore | " | 20,000 | 20,000 |
| Erie, Corry, and Buffalo | " | 5,000 | 25,000 |
| Oil Regions | " | 264,805 | 554,600 |
| On Railroads and Allegheny River | " | 30,000 | 50,000 |
| Total | bbls. | 780,831 | 997,904 |

In making out our estimate for last year, the stock at Pittsburgh was set down at 75,000 barrels, but by later advices we learned that it was fully 100,000 barrels. In the same estimate we did not include the amounts en route by river and railroads from the producing region to the principal home markets, which, then, was 50,000 barrels.

We compile the following account of stocks in, and amounts afloat for, all European ports, from the circulars of John H. Schulken, Petroleum Broker, Bremen; Mordaunt Brothers, London, and others. The figures relating to the seven ports first named and to Genoa and Trieste are considered reliable, but to all others they are but approximations:

| | | |
|---|---|---|
| In Bremen | bbls. about | 88,000 |
| Afloat | " | —— |
| Antwerp | " | 53,000 |
| Afloat | " | 28,000 |
| Hamburg | " | 35,000 |
| Afloat | " | 5,000 |
| Rotterdam | " | 13,555 |
| Afloat | " | 14,570 |
| Stettin | " | 9,523 |
| Afloat | " | —— |
| London | " | 12,816 |
| Afloat | " | 3,380 |
| Liverpool | " | 12,700 |
| Afloat | " | 10,169 |
| Outports of the United Kingdom | " | 5,000 |
| Afloat | " | 3,437 |
| Havre | " | 10,000 |
| Afloat | " | 1,563 |
| Marseilles | " | 26,750 |
| Afloat | " | 7,027 |
| Genoa | " | 20,168 |
| Afloat | " | 10,000 |
| Trieste | " | 5,000 |
| Afloat | " | —— |
| Other European ports | " | 40,090 |
| Afloat | " | 25,000 |
| Total stock and amounts afloat January 1, 1869 | bbls. about | 439,668 |
| Same January 1, 1868 | " | 607,911 |

Nearly all of the above amounts are refined, and after reducing the stock on hand in the United States to that article, the stocks of refined January 1, 1868, and January 1, 1869, compare as follows:

| | | |
|---|---|---|
| In the United States, January 1, 1868..................bbls. | 665,270 | |
| In, and afloat for Europe, January 1, 1868............. " | 607,911 | |
| Total........bbls. | | 1,273,181 |
| In the United States, January uary 1, 1869............. " | 520,588 | |
| In, and afloat for Europe, January 1, 1869.............. " | 439,668 | |
| Total........bbls. | | 960,256 |
| Decrease on January 1, 1869 " | | 312,925 |

Allowing 27,000 barrels of refined for the decrease in the stock in Western Virginia and Ohio, the total decrease in the United States and Europe was about 340,000 barrels, or a crude equivalent of about 450,000 barrels.

### CONSUMPTION IN 1868.

The consumption in the United States in 1868 of oil produced in Pennsylvania, as ascertained by deducting the exports of crude and refined from the amount produced, and allowing for the decrease in stocks during the year, was about 1,250,000 barrels of crude. But these figures do not represent the entire consumption, as there was produced in Western Virginia and Ohio 300 barrels of light oil per day, which was consumed together with about 40,000 barrels of the stock held in those States at the beginning of the year. With the addition of these amounts the total consumption of crude produced in all parts, is made about 1,420,000 barrels. The annexed table will better illustrate the manner in which the above result has been obtained. The export has been brought to a crude equivalent, and the barrels are rated at forty-three gallons.

| | | |
|---|---|---|
| Production in 1868.................bbls. | | 3,715,741 |
| Exported in 1868.................. " | | 2,680,630 |
| Leaving for home use..............bbls. | | 1,035,111 |
| Stock January 1, 1868....bbls. | 997,004 | |
| " " " 1869.... " | 780,831 | |
| Add decrease on January 1, 1869, bbls. | | 217,073 |
| Total consumption of Pennsylvania Oil during 1868................ " | | 1,252,184 |
| Production of West Virginia and Ohio, about.................... " | | 125,000 |
| Decrease in stock to same States, about.......................... " | | 40,000 |
| Total consumption in United States crude................bbls. | | 1,417,184 |

We are without reliable data as to the consumption in 1867, but from what we know of the increase during last year in two or three States, from the shipments to them from this region, we judge that it has been in the aggregate at least twenty per cent.

From Europe we have statements of the amounts of refined taken for consumption from seven of the most important ports during 1866, 1867, and 1868. These statements are from the circulars of Joh. H. Schulken, Bremen; Mordaunt Brothers, London, and Laird & Adamson, Liverpool, and are as follows:

| Deliveries in | | 1866. | 1867. | 1868. |
|---|---|---|---|---|
| From Bremen...... | bbls. | 108,336 | 180,840 | 280,756 |
| Antwerp...... | " | 278,927 | 340,898 | 339,790 |
| Rotterdam... | " | 37,500 | 80,000 | 145,245 |
| Hamburg..... | " | 67,655 | 83,041 | 139,679 |
| Stettin....... | " | 19,000 | 49,500 | 84,827 |
| London...... | " | 99,906 | 94,036 | 88,270 |
| Liverpool.... | " | 62,118 | 63,211 | 5[illegible],236 |
| Total.... | bbls. | 674,472 | 892,126 | 1,118,863 |

The above shows an increase of fully thirty-two per cent. in the consumption of 1867 over that of 1866, and of twenty-five per cent. in 1868. In London and Liverpool the consumption of refined fell off during 1867 and 1868 on account of the introduction of naphtha, and of an increased production of shale oil. But in the other European ports the proportion of naphtha imported was smaller, and shale oil did not enter so largely into competition in them with Petroleum as it did in London and Liverpool, and, therefore, the increase in consumption in the five ports first named above, may be considered a criterion of the increase in Northern Europe. The increase in the deliveries from the ports referred to was about forty-three per cent. greater in 1867 than in the preceding year, and a little over thirty-three per cent. greater in 1868 than in 1867. The total amounts of refined delivered from these ports were:

| | | |
|---|---|---|
| In 1866 | bbls. | 511,848 |
| 1867 | " | 734,879 |
| 1868 | " | 979,357 |

In Southern Europe, the increase was large, and from Genoa the deliveries reached over 100 per cent. more than in 1867. We have not been informed as to the amounts taken for consumption from the remaining European ports, but from the greatly enlarged exports to them we infer that the increase over 1867 must have been from thirty to thirty-five per cent., making the increase in the whole of Europe about thirty-five per cent. As will be seen by the above statement of deliveries from the ports of Antwerp, Bremen, Rotterdam, Hamburg, and Stettin. those of 1868 show an increase of about ninety-five per cent. over 1866."

The foregoing details include the leading points from which the reader can get a clear comprehension of the extent of all branches of the business.

### EXPORTS OF PETROLEUM.

The following table will show the exports for the last seven calendar years from the United States, with the several places of destination of the shipments from New York:

EXPORTS OF CRUDE AND REFINED (INCLUDING NAPHTHA &C.), FROM NEW YORK FOR THE YEARS 1868, 1867, 1866, 1865, 1864, 1863, 1862.

| DESTINATION. | 1868. | 1867. | 1866. | 1865. | 1864. | 1863. | 1862. |
|---|---|---|---|---|---|---|---|
| | Gallons. | Gallons. | Gallons. | Gallons. | Gallons. | Gallons. | Gallons. |
| Liverpool | 1,291,200 | 1,263,042 | 2,003,440 | 1,605,302 | 734,755 | 2,156,851 | 1,781,377 |
| London | 947,310 | 1,599,146 | 2,835,747 | 376,283 | 1,430,700 | 2,576,381 | 1,191,399 |
| Glasgow, &c | .... | .... | 4,189 | 156,167 | 868,402 | 414,943 | 24,151 |
| Bristol | 184,070 | 192,470 | 155,389 | 110,412 | 29,124 | 71,912 | |
| Falmouth, Et | 98,210 | 123,933 | 754,313 | 509,815 | 316,402 | 623,176 | |
| Grangemouth, Et | .... | .... | 247,753 | 102,292 | .... | 425,384 | |
| Cork, &c | 2,272,534 | 4,333,150 | 5,879,868 | 1,157,486 | 3,310,362 | 1,532,257 | 299,356 |
| Bowling, Et | .... | .... | .... | .... | 87,164 | .... | 195 |
| Havre | 2,925,413 | 936,274 | 2,467,482 | 604,330 | 2,324,017 | 1,774,890 | 794,221 |
| Marseilles | 8,269,600 | 1,930,181 | 1,429,753 | 1,333,752 | 1,982,075 | 1,167,893 | 135,765 |
| St. Nazaire and Rouen | 149,460 | 85,267 | 100,135 | 97,841 | .... | 143,646 | |
| Cette | .... | .... | .... | .... | 4,800 | .... | 200 |
| Dunkirk | 369,501 | 223,277 | 63,447 | 110,099 | 232,803 | .... | 2,700 |
| Dieppe | .... | .... | .... | .... | 79,582 | 46,600 | 61,692 |
| Antwerp | 6,886,077 | 4,721,142 | 4,220,860 | 1,749,062 | 4,149,821 | 2,692,974 | 823,090 |
| Bremen | 8,578,076 | 3,818,671 | 3,127,562 | 231,983 | 971,905 | 903,004 | 452,522 |
| Amsterdam | .... | .... | .... | .... | 77,041 | 436 | |
| Hamburg | 2,458,557 | 984,689 | 1,603,484 | 1,049,300 | 1,186,080 | 1,466,855 | 229,384 |
| Rotterdam | 1,695,235 | 2,239,612 | 537,814 | 292,569 | 532,926 | 757,249 | 16,938 |
| Gottenburg | .... | .... | 49,730 | .... | 33,813 | .... | 81,960 |
| Cronstadt | 1,523,387 | 724,121 | 944,240 | 891,389 | 400,376 | 88,060 | |
| Stettin | .... | .... | .... | 53,317 | .... | .... | |
| Cadiz and Malaga | 380,681 | 162,251 | 198,538 | 92,782 | 58,474 | 33,281 | |
| Tarragona and Alicante | 518,260 | 37,014 | .... | 44,988 | 16,823 | 33,000 | |
| Barcelona | 470,929 | 86,808 | 201,531 | 5,128 | 25,500 | .... | |
| Gibraltar and Malta | 4,289,017 | 1,460,032 | 835,301 | 73,751 | 69,188 | 308,450 | 157 |
| Oporto | 251,704 | 68,229 | 89,144 | 28,205 | 17,474 | 2,339 | |

EXPORTS OF PETROLEUM—Continued.

| DESTINATION. | 1868. | 1867. | 1866. | 1865. | 1864. | 1863. | 1862. |
|---|---|---|---|---|---|---|---|
| | Gallons. | Gallons. | Gallons. | Gallons. | Gallons. | Gallons. | Gallons. |
| Naples and Palermo | 1,032,209 | 210,006 | 65,780 | 22,615 | 7,983 | 57,115 | 3,990 |
| Genoa and Leghorn | 2,229,928 | 915,253 | 1,305,974 | 686,611 | 679,606 | 399,674 | 21,000 |
| Trieste | 900,161 | 246,309 | 49,825 | 66,371 | 165,175 | 3,000 | |
| Alexandria, Egypt | 223,000 | 35,776 | .... | .... | 4,000 | | |
| Lisbon | 43,194 | 77,091 | 165,983 | 93,713 | 167,195 | 64,662 | |
| Canary Islands | 16,481 | 21,000 | 10,252 | 5,244 | 3,358 | 5,125 | 1,296 |
| Madeira | .... | .... | .... | .... | .... | 400 | 481 |
| Bilboa | 318,818 | 108,077 | 216,551 | 153,818 | 2,500 | | |
| China and East Indies | 120,300 | 143,850 | 96,031 | 44,630 | 34,338 | 36,942 | 3,970 |
| Africa | 24,560 | 158,655 | 56,670 | 17,090 | 25,195 | 12,230 | 655 |
| Australia | 858,959 | 1,795,542 | 805,219 | 735,891 | 377,884 | 304,166 | 283,699 |
| Otago, N. Z. | 37,500 | 96,000 | 12,000 | 14,880 | 10,810 | 5,500 | 7,850 |
| Sydney, N. S. W. | 224,520 | 314,015 | 213,329 | 162,923 | 97,880 | 48,013 | 113,750 |
| Brazil | 796,516 | 410,166 | 528,865 | 291,752 | 149,678 | 160,152 | 54,967 |
| Mexico | 155,573 | 144,479 | 310,645 | 194,936 | 112,985 | 69,481 | 18,616 |
| Cuba | 972,455 | 942,005 | 1,078,716 | 716,738 | 418,134 | 356,436 | 213,686 |
| Argentine Republic | 169,200 | 228,327 | 184,790 | 68,856 | 20,260 | 24,470 | 7,390 |
| Cisplatine Republic | 94,000 | 248,100 | 212,550 | 72,852 | 78,552 | 117,626 | 13,227 |
| Chili | 188,000 | 98,052 | 90,670 | 53,226 | 92,550 | 66,550 | 17,800 |
| Peru | 233,956 | 240,706 | 351,090 | 110,840 | 169,661 | 256,107 | 56,011 |
| British Honduras | 4,220 | 3,817 | 5,561 | 2,052 | 6,072 | 440 | |
| British Guiana | 39,700 | 10,500 | 22,324 | 5,860 | 7,881 | 15,104 | 9,396 |
| British West Indies | 230,142 | 157,291 | 230,602 | 116,941 | 70,976 | 60,031 | 18,888 |
| British North American Colonies | 42,145 | 114,029 | 242,182 | 104,080 | 28,902 | 16,995 | 2,948 |
| Danish West Indies | 12,255 | 14,168 | 14,691 | 10,947 | 8,463 | 31,503 | 4,102 |
| Dutch West Indies | 17,466 | 24,882 | 22,181 | 18,369 | 26,638 | 12,143 | 7,117 |
| French West Indies | 77,080 | 46,225 | 57,731 | 32,618 | 16,020 | 9,104 | 2,332 |
| Hayti | 8,066 | 7,238 | 14,698 | 13,856 | 7,088 | 12,064 | 4,856 |
| Central America | 2,846 | 5,419 | 2,566 | 5,494 | 993 | 453 | 1,764 |

| | | | | | | | |
|---|---|---|---|---|---|---|---|
| Venezuela | 57,911 | 76,576 | 58,423 | 39,794 | 28,583 | 15,445 | 1,094 |
| New Granada | 64,219 | 83,300 | 90,718 | 58,570 | 57,490 | 107,837 | 37,058 |
| Porto Rico | 34,228 | 21,899 | 25,208 | 43,355 | 20,026 | 59,437 | 2,244 |
| Bordeaux | 184,600 | 104,473 | 84,929 | | | | |
| Nantes | 78,539 | 120,453 | | | | | |
| Konigsberg and Stettin | 2,537,086 | 1,001,413 | | | | | |
| Arendal | .... | 31,324 | | | | | |
| Danzig | 374,671 | 129,253 | | | | | |
| Syria | .... | 50,000 | | | | | |
| Smyrna | .... | 202,930 | 13,500 | | | | |
| Constantinople | 603,013 | 60,060 | 4,200 | | | | |
| Copenhagen | 118,492 | 59,936 | | | | | |
| Palma | 199,163 | 114,540 | | | | | |
| Japan | .... | 8,000 | 2,000 | | | | |
| Ancona | 150,038 | | | | | | |
| | 98,392 | | | | | | |
| Total | 52,599,483 | 33,834,133 | 34,506,385 | 14,626,090 | 21,335,784 | 19,547,604 | 6,720,273 |

## TOTAL EXPORTS FROM THE UNITED STATES.

| From. | 1868. | 1867. | 1866. | 1865. | 1864. | 1863. | 1862. |
|---|---|---|---|---|---|---|---|
| | Gallons. | Gallons. | Gallons. | Gallons. | Gallons. | Gallons. | Gallons. |
| New York | 52,599,483 | 33,834,133 | 34,501,385 | 14,626,090 | 21,335,784 | 19,547,604 | 6,720,273 |
| Boston | 2,367,865 | 2,264,113 | 1,591,694 | 1,511,173 | 1,696,307 | 2,049,431 | 1,071,100 |
| Philadelphia | 38,484,157 | 29,437,429 | 28,811,853 | 12,552,882 | 7,760,148 | 5,395,738 | 2,800,978 |
| Baltimore | 2,587,207 | 1,515,454 | 2,483,419 | 973,117 | 929,971 | 915,866 | 175,100 |
| Portland | 705,107 | 900 | 12,100 | 11,088 | 70,762 | 342,082 | 120,250 |
| New Bedford | .... | .... | 30,000 | 50,000 | | | |
| Cleveland | 270,000 | .... | .... | 81,173 | 80,000 | | |
| Total Gallons | 97,179,919 | 67,052,020 | 67,430,451 | 29,805,523 | 31,872,972 | 28,250,721 | 10,887,701 |
| Equal to Barrels | 2,429,498 | 1,676,300 | 1,685,761 | 745,138 | 796,824 | 706,268 | 272,192 |

The amount of capital invested in the various branches of the business can only be approximated. From the best estimates the total amount is not less than $75,000,000 to $100,000,000. This estimate includes the oil lands, but the latter are averaged at a reduction from the values of 1864 and 1865. The foreign investment, of tanks, warehouses, &c., we have no means of ascertaining.

The two centres for refining Petroleum are Pittsburgh and Cleveland. In point of capacity and amount of refined produced and shipped, Pittsburgh is but slightly in excess. The value of refineries in each place are nearly equal in amount, and the capital required varies but little. No reliable detailed record has been kept in Cleveland of the general business. We take the following able report on the Pittsburgh Petroleum trade for four years, from the Pittsburgh *Commercial* of Jan. 1st, 1869.

### OIL TRADE OF PITTSBURGH.

At present there are forty-eight refineries in operation more or less of the time. These refineries have a capacity to turn out 37,000 barrels per week of refined oil, requiring about 55,000 barrels of crude. This, at forty-two gallons to the barrel, would amount to 80,-808,000 gallons refined per year, or, as will be seen from the annexed figures of exports, from all ports of the United States, within 1,600,000 gallons of the entire amount of refined sent abroad the past year. As it is, our refineries have the capacity to turn out 1,924,000 barrels per year.

The capital invested in these refineries, independent of the immense capital all the time locked up in oil, is estimated at $7,500,000. The amount invested in the works and in barrels, tanks, boats, and other properties necessary to the prosecution of the business, would swell the aggregate capital to over $20,000,000.

The amount on hand January 1, 1869, of all kinds is estimated at 200,000 barrels in crude.

The following official table exhibits the amount of oil in barrels, forwarded over the Pennsylvania Central Railroad, Eastward, for the periods named:

"REPORT OF THE NUMBER OF BARRELS OF OIL FORWARDED OVER THE PENNSYLVANIA CENTRAL RAILROAD DURING THE YEARS 1868, 1867, 1866, 1865, AND 1864.

| Months. | 1868. From Pittsburgh Proper. Refined. | Crude. | Total. | From Line of A. V. Railroad. Refined. | Crude. | Total. | Grand Total. |
|---|---|---|---|---|---|---|---|
| January | 2,710 | 347 | 3,057 | 18,093 | 717 | 18,809 | 21,866 |
| February | 1,008 | 454 | 1,462 | 19,076 | 50 | 19,126 | 20,588 |
| March | 3,974 | 1,321 | 5,295 | 22,272 | .... | 22,272 | 27,567 |
| April | 5,949 | 1,226 | 7,175 | 32,038 | 1,160 | 33,198 | 40,373 |
| May | 20,148 | 230 | 20,378 | 50,311 | 818 | 51,129 | 71,507 |
| June | 21,130 | 971 | 22,101 | 59,902 | 1,426 | 61,328 | 83,429 |
| July | 21,723 | 896 | 22,619 | 55,283 | 6,531 | 61,814 | 84,433 |
| August | 18,623 | 402 | 19,025 | 66,177 | 666 | 66,843 | 85,868 |
| September | 21,485 | 10 | 21,495 | 65,718 | 763 | 66,481 | 87,976 |
| October | 15,318 | 403 | 15,721 | 70,741 | 693 | 71,434 | 87,155 |
| November | 11,919 | 711 | 12,630 | 56,524 | 5,900 | 62,424 | 75,054 |
| December | 14,003 | 730 | 14,733 | 51,309 | 4,311 | 55,620 | 70,353 |
| Total barrels | 157,990 | 7,701 | 165,691 | 567,443 | 23,035 | 590,478 | 756,169 |

| Months. | 1867. From Pittsburgh Proper. Refined. | Crude. | Total. | From Line of A. V. Railroad. Refined. | Crude. | Total. | Grand Total. |
|---|---|---|---|---|---|---|---|
| January | 6,841 | 2,300 | 9,141 | 18,390 | 790 | 19,180 | 28,321 |
| February | 2,467 | 1,616 | 4,083 | 11,647 | 295 | 11,942 | 16,025 |
| March | 5,072 | 2,824 | 7,896 | 23,786 | 303 | 24,088 | 31,984 |
| April | 15,172 | 3,220 | 18,392 | 42,430 | 305 | 42,735 | 61,127 |
| May | 20,216 | 1,324 | 21,540 | 48,280 | 272 | 48,552 | 70,092 |
| June | 11,423 | 420 | 11,843 | 34,539 | 220 | 34,759 | 46,602 |
| July | 20,300 | 203 | 20,503 | 69,292 | 383 | 69,675 | 90,173 |
| August | 9,269 | 864 | 10,133 | 34,886 | 1,373 | 36,259 | 46,392 |
| September | 19,385 | 335 | 19,720 | 38,514 | 455 | 38,969 | 58,689 |
| October | 12,953 | 852 | 13,805 | 46,435 | 231 | 46,666 | 60,471 |
| November | 15,778 | 680 | 13,458 | 35,717 | 423 | 36,140 | 49,598 |
| December | 9,402 | 252 | 9,654 | 10,854 | 539 | 11,423 | 21,077 |
| Total barrels | 145,278 | 14,890 | 160,168 | 414,800 | 5,588 | 420,388 | 580,556 |

| Months. | 1866. From Pittsburgh Proper. Refined. | Crude. | Total. | From Line of A. V. Railroad. Refined. | Crude. | Total. | Grand Total. |
|---|---|---|---|---|---|---|---|
| January | 9,756 | 9,815 | 19,571 | 32,396 | 2,780 | 35,176 | 54,747 |
| February | 4,245 | 5,073 | 9,318 | 8,567 | 881 | 9,448 | 18,766 |
| March | 4,565 | 15,914 | 20,479 | 8,483 | 613 | 9,096 | 29,575 |
| April | 3,175 | 30,559 | 43,734 | 22,287 | 958 | 23,245 | 56,979 |
| May | 14,010 | 26,275 | 40,285 | 40,016 | 863 | 40,879 | 81,164 |
| June | 18,852 | 22,533 | 41,385 | 42,235 | 2,882 | 45,117 | 86,502 |
| July | 17,713 | 5,322 | 23,035 | 28,630 | 1,301 | 39,931 | 62,966 |
| August | 20,330 | 7,052 | 27,382 | 52,540 | 1,487 | 54,027 | 81,409 |
| September | 19,358 | 7,741 | 27,099 | 47,540 | 2,987 | 50,527 | 77,626 |
| October | 20,337 | 12,039 | 32,376 | 49,555 | 2,141 | 51,696 | 84,072 |
| November | 15,591 | 11,539 | 27,130 | 50,948 | 1,283 | 52,181 | 79,311 |
| December | 12,087 | 5,976 | 18,063 | 42,264 | 181 | 42,445 | 60,508 |
| Total barrels | 160,019 | 159,838 | 319,857 | 435,461 | 18,307 | 453,768 | 773,625 |

| Months | 1865. | 1864. |
|---|---|---|
| | Total. | Total. |
| January | 19,276 | 45,581 |
| February | 9,369 | 40,370 |
| March | 10,724 | 27,541 |
| April | 23,350 | 52,311 |
| May | 31,961 | 94,730 |
| June | 48,787 | 58,727 |
| July | 57,566 | 41,526 |
| August | 74,660 | 30,076 |
| September | 75,939 | 28,672 |
| October | 78,372 | 45,122 |
| November | 90,209 | 42,556 |
| December | 90,955 | 38,967 |
| Total barrels | 601,162 | 545,781 |

Increase of 1868 over 1867..........175,613 barrels.
Decrease of 1868 from 1866.......... 17,456 barrels.
Increase of 1868 over 1865..........155,007 barrels.
Increase of 1868 over 1864..........210,388 barrels.

The entire exports from all ports in the United States in 1867 and 1868, were as follows:

The exports of 1867 was:

| | | |
|---|---|---|
| Of refined | galls. | 60,046,334 |
| Of crude | " | 5,262,562 |
| Of benzine | " | 1,771,300 |
| Total | galls. | 67,080,196 |

The export of 1868 was:

| | | |
|---|---|---|
| Of refined | galls. | 82,391,524 |
| Of crude | " | 7,770,833 |
| Of benzine | " | 8,771,599 |
| Total | galls. | 98,833,956 |

Increase of total export...........47 4–10 per cent.
Increase of export of refined alone..37 1–10 per cent.

The total production, as given by the Titusville *Herald*, which is accepted as authority, was:

| | | |
|---|---|---|
| In the year 1868.................... | bbls. | 3,715,741 |
| Total production in the year 1867... | " | 3,320,345 |
| Excess in production....... | bbls. | 395,396 |
| Equal to........................11 | | 9–10 per cent. |
| We exported 22,345,190 gallons refined oil more in 1868 than 1867, at 42 gallons to the barrel, 532,-028 barrels, or reduced to crude. | bbls. | 798,042 |
| We exported 2,508,271 gallons crude, at 42 gallons to the barrel....... | " | 59,721 |
| Excess of export in crude.. | bbls. | 857,763 |
| Deducting excess of production..... | " | 395,366 |
| Leaves a deficiency in home stocks of (crude)...................... | bbls. | 462,367 |

Caused by export alone. In this estimate we have not taken into account the oil consumed at home, which, from the best information we can obtain, we are led to believe, has considerably increased since 1867."

The receipts by river for seven years, ending December 31, 1865, according to a report made by the Pittsburgh *Mining and Manufacturing Journal*, were as follows:

| Years. | Barrels. | Av. price of Crude. | Values. |
|---|---|---|---|
| 1859... | 7,037 | $—— | $—— |
| 1860... | 17,161 | —— | —— |
| 1861... | 94,102 | 2 75 | 258,780 50 |
| 1862... | 171,774 | 1 75 | 300,604 50 |
| 1863... | 175,181 | 3 87 | 677,950 37 |
| 1864... | 476,939 | 10 25 | 4,888,624 75 |
| 1865... | 1,200,000 | 9 50 | 11,400,000 00 |
| T. 7 y.. | 1,142,194 | | $17,525.960 12 |

From the returns of the collector of internal revenue for the district embracing Venango County, it appears that the Government received the following revenues from crude oil for the months indicated:

| | | |
|---|---|---|
| April | 1865 | $111,991 50 |
| May | " | 88,394 50 |
| June | " | 412,258 44 |
| July | " | 184,268 55 |
| August | " | 200,000 00 |
| September | " | 246,734 00 |
| | | $1,243,646 99 |

The tax on refined during the same time in the district amounted to $700,000. The total tax on crude and refined, for the first six months of 1865, was $1,738,095. For the whole year of 1865, was $2,897,032.03. In 1863 the revenue derived was $649,062. In 1864, $2,255,238.80.

The statistics in the foregoing pages need no comment nor elucidation. The reader can speculate from them to the full extent of his inclination. We have endeavored in this work to confine ourself to a brief statement of facts, leaving the reader to make his own deductions and inferences. The home consumption for 1868 is estimated at forty per cent. of the entire production. From the best authorities had upon the subject this percentage will apply to each of the years since 1864, the increase keeping about even pace with the increase of the production.

The workings of an average oil farm, under practical management, is clearly illustrated in case of the Story farm, on Oil Creek. The farm consists of 500 acres, and was purchased by the Columbia Oil Company, of Pittsburgh, in 1861. It adjoins the Tarr, Blood, Central Petroleum. and Hyde & Egbert farms.

The territory of the Story farm is no better in general respects for oil production than are those named. The success achieved by them is due in a great measure to the excellent practical management of the company. We disclaim any intention of partiality in the premises, and insert their report as the best we can find to illustrate the details of oil production, its cost, and the accruing profits. The annual report to the company for 1868 is replete with invaluable details to all parties concerned in the business, and cannot fail to prove interesting to the general reader. We give the report as follows:

## SEVENTH ANNUAL REPORT OF THE COLUMBIA OIL COMPANY, OF PITTSBURGH, PA.

OFFICE OF THE COLUMBIA OIL COMPANY,
PITTSBURGH, January 9, 1869.

The Board of Directors are gratified in being able to present to the Stockholders so favorable an exhibit of the affairs of the Company for the past year.

By reference to tables of Production and Prices, it will be seen that both were better than the previous year, the average daily production being 375 barrels, a gain of 75 barrels per day, and the average price per barrel $3.75, while the average price for the year ending 1867, was only $2.54.

* * * * * *

A large portion of the land of the Company is yet undeveloped, and we have every reason to believe that this part of the farm will be as productive as that already developed. Two or three wells are constantly kept in process of drilling, and by that means the production, we confidently believe, can be kept up to at least what it has been for the last five years.

The extreme range of prices for oil sold was:

| | | | | |
|---|---|---|---|---|
| In 1864 | from | $3.75 | to | $13.00 |
| " 1865 | " | 4.00 | " | 10.00 |
| " 1866 | " | 1.65 | " | 5.00 |
| " 1867 | " | 1.50 | " | 4.00 |
| " 1868 | " | 1.80 | " | 5.50 |

The average price of oil, delivered, for semi-annual and yearly periods, as determined from the books of the company, is as follows:

| | | | |
|---|---|---|---|
| In 1864 | for first six months, | $6.42 | $7.62 |
| " | last " " | 8.70 | |
| In 1865 | for first six months, | 6.64 | 6.18 |
| " | last " " | 5.87 | |
| In 1866 | for first six months, | 4.20 | 3.78 |
| " | last " " | 3.40 | |
| In 1867 | for first six months, | 2.02 | 2.54 |
| " | last " " | 2.86 | |
| In 1868 | for first six months, | 2.96 | 3.95 |
| " | last " " | 4.83 | |

It will be seen, on examination, that the increase in the expenses of the Company for the year 1868 is more apparent than real, as the following must be taken into consideration,—the depreciation for three years on wooden tankage and movables, amounting to $12,349.-20, was carried to profit and loss this year. Also, that as the Supreme Court decided against the Company in the appeal taken from the decision of the Superior Court at Harrisburg, in the matter of income and capital tax, a balance of last year's tax, with the cost of the suits, amounting to $2,010.20, was charged to profit and loss this year. The amount paid in taxes for the year 1868 was—

| | |
|---|---|
| Income Tax | $10,316.01 |
| Capital " | 14,950.00 |
| Balance from last year | 2,010.21 |
| State, County, School, and Poor tax on Story Farm | 1,125.00 |
| | $28,401.22 |

Also, the increased number of wells put down by the Company, eleven in number, ranging in depth from 542 to 797 feet, at a cost of $23,628.40.

The amount expended on movable property and permanent improvements was as follows:

For machinery, tools, pipe, materials, &c., $27,827.52.

The amount of unavailable assets carried to profit and loss, was—

| | |
|---|---|
| Movables | $2,392.16 |
| Farm Fixtures | 180.64 |
| Tanks | 9,773.40 |
| | $12,349.20 |

The miscellanies, which included attorney's fees, advertising, rent, stationery, printing, stamps, meetings, travelling expenses, &c., were less than for the three preceding years. They were—

| | |
|---|---|
| For 1864 | $3,102.07 |
| " 1865 | 3,699.93 |
| " 1866 | 4,692.27 |
| " 1867 | 3,595.29 |
| " 1868 | 3,307.08 |

The amount paid in salaries to Secretary, Superintendent, Manager, Bookkeeper, and Messenger, was—

| | |
|---|---|
| For 1864 | $ 8,174.96 |
| " 1865 | 12,523.14 |

| | |
|---|---|
| For 1866 | $10,642.29 |
| " 1867 | 8,841.98 |
| " 1868 | 8,425.32 |

The special expenses incurred in producing oil were $84,364.30, or 61½ cents per barrel, as follows:

| | |
|---|---|
| Teams, feed, &c | $ 2,222.72 |
| Fuel | 9,978,86 |
| Regular labor | 60,013.52 |
| Materials, Seed Bags, Repairs, &c | 7,249.69 |
| Hauling, Freight, &c | 2,295.01 |
| Extra labor | 2,604.59 |
| | $84,364.30 |

The general expenses and losses amounted to $76,-111.22, or 55½ cents per barrel, as follows:

| | |
|---|---|
| Miscellanies | $ 3,307.08 |
| Taxes | 28,401.22 |
| Salaries | 8,425.32 |
| Well Account | 23,628.40 |
| Depreciation, carried to profit and loss | 12,349.20 |
| | $76,111.22 |

Making the whole cost $1.17 per barrel.

| | | |
|---|---|---|
| The gross profit for the year 1868, as seen by the Treasurer's Report, amounts to | | $381,103.83 |
| Surplus from 1867 | | 61,288.53 |
| | | $442,392.38 |
| Dividends declared in 1868: | | |
| No. 29, | 75,000 | |
| " 30, | 150,000 | |
| " 31, | 100,000 | |
| | | 325,000.00 |
| Balance of gross profit | | $117,392.38 |

| | |
|---|---|
| The cash in hands of the Treasurer and Superintendent on Dec. 31, was.. | $83,884.03 |
| Oil sold and in process of delivery.... | 26,018.50 |
| Cash accounts on the farm........... | 6,650.32 |
| Estimated available assets,........... | $116,552.85 |
| Against this we have liabilities amounting to.......................... | 30,565,85 |
| Leaving a balance of................ | $85,987.00 |

The business of the Farm has been thoroughly systematized, and is in the hands of experienced and faithful men, and we feel, all things considered, that we are entering upon a year of great promise.

On behalf of the Board,
D. A. STEWART, *President.*

---

SUPERINTENDENT'S REPORT.

STORY FARM, January 4, 1869.

*To the President and Board of Directors of the Columbia Oil Company:*

GENTLEMEN:—An account of the operations on the Farm during 1868, is herewith respectfully submitted.

In the following statement will be found the average daily yield of oil for the Company, during the past five years. The production of one Lease, in which the Company has three quarters of the working interest, is included in the amount of oil received from the Lessees:

| | Received Daily From Lessees. | Received Daily From Co.'s Wells. | Total Daily Average for the Co. |
|---|---|---|---|
| In 1864.. | 334 bbls....... | 53 bbls....... | 387 bbls. |
| 1865.. | 223 " ...... | 160 " ...... | 383 " |
| 1866.. | 155 " ...... | 190 " ...... | 345 " |
| 1867.. | 90 " ...... | 213 " ...... | 303 " |
| 1868.. | 51 " ...... | 323 " ...... | 374 " |

The Company's proportion of oil for 1868, was 137,099 barrels; 18,735 of which were received from the Lessees. The production from wells owned exclusively by the Company, was 118,364 barrels, against 77,775 the previous year; an increase of 40,589 barrels, or 52 per cent.

This and former reports give the production in barrels without reference to gallons, and in this way it appears that the Company's proportion of oil for 1868 was less, by 4,509 barrels, than in 1864, while in fact it was 230,937 gallons greater than during any former year in the history of the Company. In 1864, the standard measure of oil on Oil Creek was 40 gallons; during the past two years it has been 43 gallons.

| | |
|---|---|
| In 1868, the Company's proportion of oil was 137,099 bbls., of 43 galls........................ | =5,895,257 galls. |
| In 1864, the Company's production of oil was 141,608 bbls., of 40 galls........................ | =5,664,320 " |
| Making a difference in favor of 1868 of...................... | 230,937 galls. |
| or 5,773 barrels of 40 gallons each. | |

The production from the Company's wells in 1868 was made up as follows:

| | |
|---|---|
| From wells completed in 1865......... | 9,071 bbls. |
| " " " " 1866......... | 42,937 " |
| " " " " 1867......... | 37,212 " |
| " " " " 1868......... | 29,144 " |
| | 118,364 bbls. |

The wells drilled in 1865 are, with two exceptions, on the flat, where the territory was not properly preserved. During 1866, and since then, our developments have been on the hill, where great care is taken to protect the wells by locating them at least 300 feet apart. The beneficial results arising from this system are clearly seen in the above statement by the amount of oil produced last year from wells completed in 1866.

All the Leases on the Farm have been forfeited to the Company, except Nos. 4, 8, 9, 10, 11, 12, 13, 17, 21, and 33. Nos. 4, 10, 13, and 17 have been idle most of the year; the other Leases mentioned have been worked steadily.

One well was drilled on each of the lots 9, 12, 13, and 21. The well on lot 9 was completed last November, and is producing 12 barrels of oil per day. The other three producing no oil.

The developments of the Company will compare favorably with former years. Eleven wells, averaging in depth 704 feet, were completed; three of which were commenced the year previous. Five of these are producing oil, one is abandoned, four are being tested, and the remaining one has produced as high as 60 barrels of oil per day, but owing to temporary causes is not now in proper condition for pumping.

Five of the wells were permanently cased, and the water shut off before drilling into the oil bearing rock. The expense of drilling a well on this plan is from $150 to $200 greater than the ordinary way, not including the price of casing, but the advantages are incalculable.

Owing to the increased amount of labor and material required to carry on the business of the Farm, the expenditures were greater than during the previous year. The wells of the Company are steadily increasing in number. At the close of the year we were pumping 23, and for several months past have been pumping from 19 to 20. The average cost, per foot, of the eleven wells completed last year was 74 cents greater than of the eight wells drilled in 1867. This increase resulted from various causes. The timber used in our rigs was more expensive, gas was scarce part of the time, and wood and coal were required for fuel. Five of the wells were drilled to the depth necessary to shut off the water, 2¼ inches larger in diameter than any we put down during former years, and the average depth of the wells was 45 feet greater than the average depth of those drilled in 1867.

Three fires occurred on the Farm during the year, by which two engine-houses, and one small tank, containing ten barrels of oil, were burned. Loss about $500.

For ten months of the year we received from the wells a large amount of gas, which was used for fuel, but within the past two months has partially failed, requiring us to purchase much more than the usual amount of wood and coal.

The substantial manner in which the Company's rigs are built, and the excellent condition of the engines used, enable us to pump the wells very regularly. For the past ten months, a number of stoppages occurred, on account of the frequent parting of tubing. The difficulty has been overcome by resting the tubing on the bottoms of the wells.

The Company have on hand 29 engines; 24 are at pumping wells, 2 at rigs where we are ready to commence driving-pipes; 1 is used in the machine shop, and the remaining two are being repaired.

We are using 550 feet of 3½-inch, and 8,260 feet of 2-inch pipe for leading oil; 13,111 feet of various sizes for water; and 10,885 feet for gas; all of which is buried sufficiently deep to prevent the frost from reaching it.

The principal property of the Company on the Farm is embraced in the following list:

10 Dwelling Houses, Office, and Furniture.
29 Steam Engines, 2 Steam Pumps, and 2 Rotary Pumps.
Machinery and Tools for Machine and Blacksmith Shops.
5 Sets of Drilling Tools, 5 Horses, 3 Wagons, and 2 Sleds.
45 Wooden Tanks—capacity, 19,610 barrels.
4 Iron Tanks—capacity, 26,000 barrels.
15,471 feet of Casing for wells.
16,631 feet of Tubing for wells.
34,541 feet of Pipe of various sizes.

The Company's developments for the past three years have been connfined almost exclusively to that portion of the land lying between the run and the Refining ravine. Recently we erected two rigs north of the run leading toward Wood and Stevenson farms, in which direction lies about 130 acres of the undeveloped part of the Farm. This lot of ground, from its surroundings, cannot well prove otherwise than valuable for oil purposes.

Very respectfully, yours,
GEORGE BOULTON, *Superintendent.*

# I. TABLE OF PRICES,

SHOWING THE AVERAGE PRICE PER BARREL OF OIL DELIVERED, AND THE RANGE OF PRICES AT WHICH IT WAS SOLD, ON STORY FARM, DURING EACH MONTH OF THE FIVE PRECEDING YEARS.

| MONTH. | 1864 | | 1865. | | 1866. | | 1867. | | 1868. | |
|---|---|---|---|---|---|---|---|---|---|---|
| | Oil Sold. | Delivered. | Oil Sold. | Delivered. | Oil Sold. | Delivered. | Oil Sold. | Delivered. | Oil Sold. | Delivered. |
| January | $4.00 to $4.00 | $3.39 | $6.50 to $10.00 | $9.37 | $4.00 to $5.00 | $5.07 | $1.65 to $2.10 | $1.83 | $1.80 to $2.10 | $1.90 |
| February | 3.75 to 5.00 | 3.52 | 7.00 to 8.00 | 7.87 | 4.00 to 4.80 | 4.79 | 1.70 to 2.00 | 1.90 | 2.00 to 2.50 | 2.17 |
| March | 5.00 to 6.00 | 5.18 | 5.00 to 7.00 | 6.80 | 3.00 to 4.50 | 4.03 | 1.50 to 2.00 | 1.89 | 2.25 to 2.8[illegible] | 2.30 |
| April | 6.00 to 7.12 | 5.80 | 5.00 to 7.00 | 5.99 | 3.50 to 4.40 | 3.75 | 1.95 to 2.20 | 1.95 | 2.65 to 3.00 | 2.74 |
| May | 6.50 to 7.75 | | 6.50 to 8.25 | 6.54 | 4.00 to 5.00 | 4.05 | 2.20 to 2.50 | 2.22 | 3.50 to 4.00 | 3.69 |
| June | 7.50 to 11.50 | 7.84 | 5.00 to 6.25 | 5.34 | 3.50 to 4.25 | 4.00 | 1.50 to 2.30 | 2.05 | 4.00 to 5.00 | 4.24 |
| July | 11.25 to 13.00 | | 5.00 to 5.25 | 5.08 | 2.50 to 3.50 | 3.34 | 1.75 to 3.50 | 2.31 | 4.75 to 5.50 | 4.86 |
| August | 9.00 to 11.25 | 10.00 | 4.00 to 5.25 | 5.03 | 2.50 to 5.00 | 3.25 | 2.80 to 3.50 | 3.14 | 4.00 to 5.15 | 4.87 |
| September | 6.50 to 11.25 | 9.66 | 4.25 to 9.25 | 5.04 | 4.00 to 5.00 | 4.27 | 2.80 to 4.00 | 3.54 | 3.75 to 4.25 | 4.17 |
| October | 6.00 to 9.50 | 7.04 | 6.75 to 0.50 | 7.00 | 2.50 to 4.25 | 4.02 | 3.10 to 4.00 | 3.55 | 4.00 to 4.25 | 4.16 |
| November | 9.00 to 11.00 | 9.16 | 6.50 to 8.00 | 6.92 | 2.60 to 3.60 | 2.90 | 2.00 to 3.00 | 2.66 | 3.00 to 4.50 | 4.08 |
| December | 10.50 to 11.50 | 9.48 | 5.00 to 8.00 | 7.33 | 1.65 to 2.60 | 2.35 | 1.75 to 2.00 | 1.93 | 4.10 to 4.60 | 4.42 |
| Average for year | | 7.62 | | 6.18 | | 3.78 | | 2.54 | | 3 95 |

## II. TABLE OF DIVIDENDS.

| No. | When Declared. | Number of Shares. | Amount per Share. | Capital. | Percentage. | Amount of Each Divid'd. | Amount Declared Each Year. |
|---|---|---|---|---|---|---|---|
| 1 | July 8, 1863.... | 9,000 | $ 6 00 | $ 192,000 | 30 | $ 57,600 | |
| 2 | August 12, " .... | " | 5 00 | " | 25 | 48,000 | |
| 3 | September 9, " .... | " | 5 00 | " | 25 | 48,000 | |
| 4 | October 14, " .... | " | 10 00 | " | 50 | 96,000— | $ 249,600 |
| 5 | January 13, 1864.... | " | 5 00 | " | 25 | 48,000 | |
| 6 | April 13, " .... | 10,000 | 7 00 | 200,000 | 35 | 70,000 | |
| 7 | May 11, " .... | " | 10 00 | " | 50 | 100,000 | |
| 8 | June 8, " .... | " | 10 00 | " | 50 | 100,000 | |
| 9 | July 13, " .... | 50,000 | 2 50 | 2,500,000 | 5 | 125,000 | |
| 10 | August 10, " .... | " | 2 00 | " | 4 | 100,000 | |
| 11 | September, 14, " .... | " | 2 50 | " | 5 | 125,000 | |
| 12 | October 12, " .... | " | 1 50 | " | 3 | 75,000 | |
| 13 | November 9, " .... | " | 2 00 | " | 4 | 100,000 | |
| 14 | December 14, " .... | " | 2 00 | " | 4 | 100,000— | — 943,000 |
| 15 | January 11, 1865.... | " | 1 00 | " | 2 | 50,000 | |
| 16 | March 6, " .... | " | 1 50 | " | 3 | 75,000 | |
| 17 | April 12, " .... | " | 1 00 | " | 2 | 50,000 | |
| 18 | May 10, " .... | " | 1 00 | " | 2 | 50,000 | |
| 19 | June 14, " .... | " | 1 00 | " | 2 | 50,000 | |
| 20 | August 9, " .... | " | 1 00 | " | 2 | 50,000 | |
| 21 | October 11, " .... | " | 1 50 | " | 3 | 75,000 | |
| 22 | November 8, " .... | " | 1 00 | " | 2 | 50,000 | |
| 23 | December 13, " .... | " | 1 00 | " | 2 | 50,000— | — 500,000 |
| 24 | January 10, 1866.... | " | 1 00 | " | 2 | 50,000 | |
| 25 | July 11, " .... | " | 2 00 | " | 4 | 100,000 | |
| 26 | October 10, " .... | " | 2 00 | " | 4 | 100,000— | — 250,000 |
| 27 | July 10, 1867.... | " | 75 | " | 1½ | 37,500 | |
| 28 | October 9, " .... | " | 1 50 | " | 3 | 75,000— | — 112,500 |
| 29 | January 8, 1868.... | " | 1 50 | " | 3 | 75,000 | |
| 30 | July 8, " .... | " | 3 00 | " | 6 | 150,000 | |
| 31 | October 14, " .... | " | 2 00 | " | 4 | 100,000— | — 325,000 |
| 32 | January 13, 1869.... | " | 2 00 | " | 4 | | 100,000 |
| | | | | | | | $2,480,100 |

## III. TABLE OF EXPENSES.

*Showing the total expenses and losses of the Company from 1863 to 1868, inclusive, with the amount of taxes, the cost of wells, and the outlay in producing oil, for each year specified.*

| Year. | Taxes. | Wells. | Oil. | General. | Yearly Total. |
|---|---|---|---|---|---|
| 1863. | $ 3,479.99 | $ 19,046.60 | $ 4,890.14 | $ 17,690.21 | $ 45,106.94 |
| 1864. | 44,790.21 | 8,224.14 | 34,840.65 | 32,680.28 | 120,535.28 |
| 1865. | 158,691.73 | 49,794.45 | 74,935.12 | 37,641.79 | 321,063.09 |
| 1866. | 68,409.71 | 22,093.10 | 96,853.11 | 41,377.47 | 228,733.39 |
| 1867. | 13,635.59 | 12,132.50 | 64,302.54 | 19,805.37 | 109,876.00 |
| 1868. | 28,401.22 | 23,628.40 | 84,364.30 | 24,081.60 | 160,475.52 |
| | $317,408.45 | $134,919.19 | $360,185.86 | $173,276.72 | $985,790.52 |

## IV. TABLE OF DAILY PRODUCTION.

*Showing the average daily production, belonging to the Company, from all the wells, during each month of the five preceding years.*

| MONTH. | 1864. | 1865. | 1866. | 1867. | 1868. |
|---|---|---|---|---|---|
| January...... | 146 bbls. | 345 bbls. | 290 bbls. | 200 bbls. | 310 bbls. |
| February..... | 316 " | 272 " | 329 " | 207 " | 330 " |
| March........ | 311 " | 329 " | 276 " | 212 " | 353 " |
| April......... | 521 " | 230 " | 263 " | 276 " | 388 " |
| May.......... | 545 " | 348 " | 347 " | 252 " | 332 " |
| June.......... | 509 " | 392 " | 400 " | 280 " | 443 " |
| July........ | 463 " | 462 " | 409 " | 289 " | 458 " |
| August....... | 390 " | 463 " | 351 " | 286 " | 468 " |
| September.... | 365 " | 444 " | 402 " | 393 " | 379 " |
| October....... | 330 " | 493 " | 337 " | 351 " | 353 " |
| November .... | 342 " | 473 " | 450 " | 440 " | 365 " |
| December..... | 426 " | 454 " | 284 " | 441 " | 315 " |

## V. TABLE OF ANNUAL AND SEMI-ANNUAL PRODUCTION.

*Showing the amount of oil produced for the Company, from our own and the Lessees' wells, for each annual and semi-annual period during the five preceding years.*

| Annual and Semi-Annual Periods. | | Lessees' Wells. | | Company Wells. | | Total. | |
|---|---|---|---|---|---|---|---|
| | | Half Y'rly. | Yearly. | Half Y'rly. | Yearly. | Half Y'rly. | Yearly. |
| 1864 | First Half.......... | 46,475 | 98,177 | 24,126 | 43,431 | 70,583 | 141,608 |
| | Last " .......... | 51,720 | | 19,305 | | 71,025 | |
| 1865 | First " .......... | 34,471 | 69,532 | 23,875 | 70,259 | 58,346 | 139,791 |
| | Last " .......... | 35,061 | | 46,384 | | 81,445 | |
| 1865 | First " .......... | 27,731 | 50,957 | 29,668 | 74,831 | 57,399 | 125,785 |
| | Last " .......... | 23,226 | | 45,163 | | 68,389 | |
| 1867 | First " .......... | 16,743 | 30,243 | 26,526 | 80,412 | 43,269 | 110,655 |
| | Last " .......... | 13,500 | | 53,886 | | 67,386 | |
| 1868 | First " .......... | 12,084 | 18,735 | 53,250 | 118,364 | 65,334 | 137,099 |
| | Last " .......... | 6,651 | | 65,114 | | 71,765 | |
| Total............... | | 267,644 | 267,644 | 387,297 | 387,297 | 654,941 | 654,941 |

# VI. TABLE OF WELLS.

SHOWING THE NUMBER, DEPTH, PRODUCTION, &C., OF ALL THE WELLS ON STORY FARM, DECEMBER 31, 1868.

| Company's Per Centage. | | Depth in Feet. | Number of Completed Wells. | NUMBER OF BARRELS OF OIL PRODUCED. | | | | | | | | | | | | | | | | | | Total Production of the Farm. |
|---|---|---|---|---|---|---|---|---|---|---|---|---|---|---|---|---|---|---|---|---|---|---|
| | | | | FOR LESSEES. | | | | | | | | | FOR THE COMPANY. | | | | | | | | | |
| Of Working Interest | Of Oil Produced. | | | 1861. | 1862. | 1863. | 1864. | 1865. | 1866. | 1867. | 1868. | Total. | 1861. | 1862. | 1863. | 1864. | 1865. | 1866. | 1867. | 1868. | Total. | |
| 00 | 50–52 | 488–537 | 115 | 20,800 | 89,504 | 70,990 | 96,936 | 69,524 | 50,990 | 30,146 | 16,817 | 445,707 | 20,800 | 89,602 | 71,066 | 93,177 | 69,532 | 50,957 | 30,243 | 17,817 | 447,194 | 892,901 |
| 75 | 87½ | 193–925 | 4 | | | 168 | 3,432 | 1,681 | 789 | 380 | 274 | 6,724 | | | 1,138 | 24,057 | 11,775 | 5,524 | 2,657 | 917 | 46,068 | 52,792 |
| 100 | 100 | 480–815 | 55 | | | | | | | | | | | | 16479 | 19,374 | 58,484 | 69,307 | 77,755 | 118,365 | 360,044 | 360,782 |
| | | | 174 | 20,800 | 89,504 | 71,158 | 100,368 | 71,205 | 51,779 | 30,526 | 17,091 | 452,431 | 20,800 | 89,602 | 88,701 | 141,608 | 139,791 | 125,788 | 110,655 | 137,099 | 854,044 | 1,306,475 |

PITTSBURGH, December 31, 1868.

*To the President and Directors of the Columbia Oil Company:*

The accounts herewith submitted for your consideration, exhibit briefly the present resources and liabilities, together with the losses, expenses, and income of the Company for the past year.

DAVID RICHEY, *Treasurer.*

## RESOURCES. December 31,

| | | | |
|---|---|---|---|
| **AVAILABLE.** | | | |
| Cash in hands of Treasurer and Superintendent | | $83,884.03 | |
| Accounts Receivable | | 6,650.32 | |
| Oil on hand and partly delivered | | 25,018.50 | $116,552.85 |
| **UNAVAILABLE.** | | | |
| STORY FARM— | | | |
| Nominal value of land | | $2,428,000.00 | |
| Valuation of Buildings, January 1, 1868 | $9,325.00 | | |
| Added during the year | 100.00 | $9,425.00 | |
| MOVABLES— | | | |
| Machinery, Tubing, &c., January 1, 1868 | 49,500.00 | | |
| Added during the year | 27,827.52 | | |
| | 77,327.52 | | |
| Carried to Profit and Loss | 2,395.16 | $74,932.36 | |
| TANKS— | 25,421.40 | | |
| Carried to Profit and Loss | 9,773.40 | $15,648.00 | |
| Fuel on hand | | 600.00 | |
| Lumber on hand | | 1,600.00 | |
| M. Graver | | 1,200.00 | $2,531,405.30 |
| | | | $2,647,958.25 |

1868. LIABILITIES.

| | | | |
|---|---|---|---|
| REAL. | | | |
| Dividends uncalled for.................... | | $ 871.50 | |
| Accounts Payable......................... | | 29,694.35 | $30,565,85 |
| UNREAL. | | | |
| Capital Stock—50,000 Shares at $50........ | | 2,500,000.00 | |
| Surplus profit from December 31, 1867..... | $61,288.53 | | |
| Net profit for 1868......................... | 381,103.83 | | |
| | $442,392.36 | | |
| Dividends No. 29, 30, and 31............... | 325,000.00 | $117,392.36 | $2,617,392.36 |
| | | | $2,647,958.21 |

## EXPENSES AND LOSSES. December 31,

| | | | |
|---|---|---|---|
| MISCELLANIES— | | | |
| Attorney's fees.........................$ | 124.00 | | |
| Office Expenses, Rent, &c............. | 374.42 | | |
| Advertising, Printing, Stationery, and Papers............................. | 424.66 | | |
| Telegrams, Stamps, and Postage...... | 531.54 | | |
| Travelling Expenses, Meetings, &c.... | 1,852.46 | $ 3,307.08 | |
| TAXES— | | | |
| County, State, School, and Poor tax on Story Farm..................... | 1,125.00 | | |
| Balance due State Treasurer from 1867............................. | 2,010.21 | | |
| State Treasurer...................... | 25,266.01 | 28,401.22 | |
| SALARIES— | | | |
| Paid Secretary, Superintendent, Manager, Bookkeeper and Messenger............................ | | 8,425.32 | |
| WELL ACCOUNT— | | | |
| Beam Well.......750 feet deep........ | 2,070.00 | | |
| Morrison.........772 " ........ | 1,582.00 | | |
| Greeley..........758 " ........ | 1,627.50 | | |
| Blocker..........570 " ........ | 1,737.50 | | |
| Hoffman.........665 " ........ | 1,503.00 | | |
| Boulton..........797 " ........ | 2,480.00 | | |
| Gratz............780 " ........ | 2,278.00 | | |
| Logan...........770 " ........ | 3,240,00 | | |
| Say.............765 " ........ | 2,972.00 | | |
| Shriver..........580 " ........ | 2,322.40 | | |
| No. 55..........542 " ........ | 1,916.00 | 23,628.40 | |
| OIL— | | | |
| Expense of Teams, Feed, &c.......... | 2,222.72 | | |
| Fuel.................................. | 9,978.86 | | |
| Regular Labor........................ | 60,013.52 | | |
| Materials, Seed Bags, Repairs, &c..... | 7,249.69 | | |
| Hauling, Freight, &c................. | 2,295.01 | | |
| Miscellanies.......................... | 2,604.50 | 84,364.30 | |
| Movables............................ | | 2,395.16 | |
| Farm Fixtures........................ | | 180.64 | |
| Tanks............................... | | 9,773.40 | |
| NET PROFIT— | | | |
| Surplus from 1867.................... | 61,288.53 | | |
| Net profit for 1868.................... | 381,103.83 | 442,392.36 | |
| | | | $602.867.83 |

## 1868. GROSS INCOME OR PROFIT.

| | | |
|---|---|---|
| Surplus from 1867 | | 61,288.53 |
| Interest Account | 1,467.54 | |
| Rents on Story Farm | 1,127.41 | |
| Oil—137,099 barrels | 538,984.40 | |
| Total income on 137,099 barrels, at $3.95 | | 541,579.35 |
| | | $602,867.88 |

The foregoing report gives in detail the best general information upon the topics embraced, that we could furnish the reader.

Our work being issued at a later date than anticipated, enables us to give a brief outline of the oil operations of 1869. Prices averaged $5.50 per barrel on the Creek, to date of September 1, being an increase over the same of 1868. The principal new development has been on Upper Cherry Tree Run, and Parker's Landing, on the Lower Allegheny; those being the only localities where any considerable number of operators have congregated. At Parker's Landing, considerable excitement prevails, almost equalling Pleasantville in 1868. The new operations, as a whole, have been scattering, embracing localities on the Upper Allegheny, at Eagle Rock, and Tionesta, Cranberry Township, near Oil City, and other points. The general features of the sandrocks in the new localities vary but slightly from those before described.

Rapid progress has been made in the methods of drilling oil wells. The diameter has been increased with good results. The size being increased to the first sandrock, enables the operator to more successfully shut off the surface water. So far the year has been a highly prosperous one to all classes, and no unhealthy excitements, like those of previous years, have occurred.

From the sound basis attained, the prosperous future of the business is assured. Many of the old wells drilled in former years have been cleaned out and pumped this summer and approximated in yield to their former standard, giving ample proof that the supply of no locality becomes entirely exhausted. The Clapp farm, adjoining Oil City, and the H. M'Clintock farm, next above, are the most notable localities. Arrangements are being made to work old wells at other points, with good prospects of success. The new oil fields have been only moderate in production, as compared with those of previous years, while the general production shows a full average. The latter can be increased whenever the provocation of higher prices is furnished.

Where so much depends upon the management, it is impossible to give the average cost of sinking and working the wells over so large an extent of territory. Approximation is the best that can be done. The addition of from 100 to 300 per centum to the detailed cost of production, as given in the foregoing report, is as near as is necessary for all practical purposes, in a general average.

The area of the oil producing, or what was considered oil territory in 1862, consisted of from 3,000 to 4,000 acres. Then operations were chiefly confined to the flat lands of the valley of Oil Creek. The successful development of the hills and elevated table lands during 1867 and 1868, has increased the amount of area susceptible of Petroleum development to nearly as many square miles, embracing large portions of Venango, Warren, Clarion, Armstrong, Butler, Mercer, and Crawford Counties. As yet the bulk of oil operating has been confined to Venango County, but in all the counties named evidences of Petroleum have been found, possessing the general characteristics of sand-rocks and quality of oil as found in Venango. The extent of territory that will prove productive, can only be determined by the test of the miner's drill. To go out of our own State, it is found in nearly all the other States, from the Alleghenies to the Gulf stream. The territory embraced in the several counties named is sufficient to supply all the needed wants for development for many years to come, and the field may be extended to an almost indefinite extent. The supply yielded in 1868 has been sufficient for the wants of consumption, leaving no surplus, while the foreign demand for export has increased nearly forty per cent. In order to meet the demand it was found necessary to drill about 1,000 wells in 1868, at a cost of quite $5,000,000. The necessity of a still larger production for 1869 to meet the foreign and home wants is clearly indicated. The old field will supply a large proportion of all that will be needed, while preparations have been made to develop new fields, far greater in area than any of former years, and giving as rich promise of abundant production.

The experience of former years has been improved upon. The skill devoted to development has been more productive of successful results in 1868 than during any previous year. The proportion of "dry holes" during the same period, has not exceeded, we should judge, scarcely twenty per cent. In preceding years this per centum of successes was scarcely attained. It is fair to assume that fifty per cent. of the wells drilled and properly tested in any of the known producing localities, paid for cost of drilling and a handsome profit besides. A further improvement in the mode of oil operating in 1869 will not surely lessen these chances. Never before in any branch of mining has greater progress been made. Certain contingencies may, of course, greatly enhance or depress the chances for obtaining paying wells, but we can only give the experience of the past year as a criterion for the next one. The cost of the deep wells of the present day, owing to improved machinery, amount to no more comparatively, than the former shallow ones.

We have no desire to present inducements to any

parties not thoroughly conversant with the subject, to make any investments in Petrolia. Taking the average through for the different years, the chances for attaining wealth or competence are no greater or less than is presented in any other general business. But the Petroleum possesses these manifest advantages over any other branch of business. No long years of business training are necessarily essential to success, nor is the same chance for realization offered in any branch of trade on so small an amount of capital invested, in so short a time. A fortune, a competence, and even the first million has often been obtained in a few months from date of first investment. What has been done is not impossible in the future, and the chances are equally as good now as at any previous time for the rapid realization of wealth. Something else is required beyond the cash capital to ensure success, for instance, untiring industry and perseverance. With the two latter qualifications, and a moderate cash capital, the man who fails to achieve wealth or comfortable competence here, would in a majority of cases fail to do so any where else, or in any other business. The number of wealthy oil operators would be greatly increased were there any limits to human cupidity or greed. When a good well is obtained, the operator, in a large majority of cases, commences the drilling of new ones, in the immediate vicinity if possible, before realizing the full amount of investment, using the proceeds of the producing well to sink the new ones. It not unfrequently happens that the producing well fails before the new ones are completed. In this case financial ruin becomes imminent, and the blame is laid upon any thing but the right source. A portion of the same expense devoted to one producing well, would, we have no doubt, in a great number of cases yield a handsome return on the investment, and in any case would prevent the entire loss of the same. The fear of losing their oil when other parties commence the drilling of wells in close proximity, frequently causes even more prudent operators to drill one or more wells in order to prevent the others from diverting the current of oil from their own. We have no evidence that the life of any well has been prolonged by such means of precaution. In our own case we should prefer to take the chances on the first investment.

The greatest drawback, in our opinion, to the entire business, is the want of a more permanent element among those engaged in the same. Although great improvement in this respect is manifest from year to year, still it is difficult to find many who are willing to admit that they intend to make any permanent stay in the Oil Region. From this reason, the same measures are not taken to build up comfortable homes, and people endure many abuses of public morality that would not be tolerated in older settlements.

The most serious result occasioned by the lack of this element, so essential to lasting prosperity, is the want of concert of action upon matters vitally affecting the general interests of the trade. Designing men find but little difficulty in procuring by legislative enactments the most outrageous monopolies upon the industry of the operators. So flagrant has this become, that the oil producers are now engaged in the formation of an association for self-protection against this swindling class. Several efforts have been made by the Oil Pipe Companies, aided in some instances by the rival railroads of the Oil Region, to obtain a monopoly of the carrying trade, but so far the success has only been partial. The new association will doubtless give all these matters their attention in future.

After repeated experiment it became evident to the General Government that any tax laid upon the crude material could only result in ruin to the business, and the excise duty upon crude Petroleum was very properly repealed. Drilling for oil, so far as prosecuted, can be strictly termed *experimental mining.* There is perhaps an absolute certainty of obtaining a given percentage of paying wells, but the amount is liable to vary with both the season and locality. Again, the risk is—$5,000, on a decided uncertainty. The amount of supply is liable to vary so much as to make prices for the oil irregular at any and all seasons of the year. The prudent sagacious business man may doubtless wonder that people will embark their means in such an uncertain enterprise, and one subject to so many contingencies and fluctuations. Nothing strange in the matter. The fact of the business being free from taxation is the chief inducement for investment in it. The aggregate of failures are not deemed to involve so great a loss as a burdensome tax. In order to procure the necessary capital for development, the business must be kept free, so far as the crude article is concerned, from any excise duty. The oil producing business, or in fact any other, cannot be made profitable, unless the consumer pays the tax. From the reasons above given it will be seen on reflection that this cannot be done at present as concerns Petroleum, while the producer cannot afford to be thus burdened unless he has a large flowing well, and of this class there are at present none in the Oil Region.

The "royalty," as it is called, or share of the oil, demanded by the landowner or lessee for the privilege of drilling a well, varies according to the proved productiveness of the territory. The average at the present is from one fourth to three-eighths, and in some localities one half of all the oil obtained, the lessee being at the expense of drilling the well, pumping the same, and delivery at the well of the royalty or landowner's share, free of expense. The amount of royalty could be lessened considerably, we think, with manifest benefit to all parties concerned.

Associations of individuals or oil stock companies, have not been successful, save in a few exceptional cases. The success of Petroleum development will have to depend in the future as in the past upon individual enterprise. The amount of working capital required to develope thoroughly even a few acres, is more than any joint stock company can afford, whose stockholders expect annual dividends. The aggregation of skilled labor and energy required for successful development, cannot be congregated in any single locality by inducements that a company could offer.

The periodical excitements consequent upon the discovery of the different new oil fields, have been productive of much injury to the general business, causing operators to forsake the older localities to invest their means in the new, concentrating the general working capital at one locality to the detriment and neglect of all the rest, and enhancing the cost of operating to an almost ruinous figure. We had hoped that the experience of Pithole in 1865, would be sufficient for all succeeding time, but each subsequent year has had its sensation. But for these feverish, reckless excitements, the cost of development could be reduced one half. Pleasantville furnished the sensation for 1868. The favored locality for that of 1869 has not as yet been clearly indicated, but will doubtless be painfully apparent by the month of August next to many an unlucky operator.

The most reckless waste and extravagance has characterized the development, and, in fact, the general business, from its beginning. From many causes this waste was unavoidable. We should estimate the amount wasted, on a general average, fully one seventh of the entire production. With the present facilities of transportation and approved vessels for holding it secure from leakage and evaporation, the waste can be greatly lessened. The amount of loss now by transportation is but a trifle. The oil-pipe companies have a draw-back of only two per cent. on the oil transported by them from the wells to the shipping-points. The cost of fuel is becoming of chief importance to the oil operator. The gas from the wells is used for fuel whenever it can be obtained in sufficient quantity. But the supply is not permanent nor reliable. Oil can be used to advantage only when extremely low prices prevail. The entire average cost of running a well with coal for fuel will not exceed fifteen dollars per day. By the employment of the best apparatus for burning oil yet devised, it requires three to four barrels of oil for fuel each per *diem*. Many devices have been conceived for supplying the necessary motive power for pumping and drilling at a less cost of fuel. As yet nothing practical has been arrived at; but it will be.

We have thus far accompanied the pioneer oil operator in his chequered journey. Occasionally he would be absent from us for a short interval, having made his "pile," and concluded to go out "into America," and take his ease for the balance of his earthly sojourn. But ease and indolence for him had lost their charm, and the ordinary channels of trade ran with too slow a current. One short year of such life has generally sufficed, and he came back and renewed operations with unabated energy. He commenced drilling his first well on Oil Creek about the time of

Drake's successful venture, "kicking it down" with a spring-pole, then selling interests to obtain a horse-power, and finally striking oil and wealth by means of an old rattle-trap of an engine. He was among the first to set up a derrick at Pithole, and afterward played his part well at Triumph Hill and Shamburg. Though deserted oil wells and lonely derricks beset him at every step, he is still as energetic and full of hope as ever. Whether a millionaire or without the means to procure a meal, he floats upon the seething bubbles of glorious successes and uncertainties like a cork, and never loses his general bearings amid all the various changes of fortune to which he is subject.

Last summer he was the possessor of some hundred barrel wells at Pleasantville, and at Shamburg. These having "played out," he is now prospecting for other fields to develop, and will be heard from in some new locality, "eight hundred feet down with a good show," ere the early roses bloom.

With a rapidity of growth and extent of development never equalled, and possessed of resources vast beyond conception, it is difficult to imagine any but a brilliant future for the entire Petroleum Region and a business, that has added so largely to the already stupendous revenues derived by Pennsylvania from her rock-ribbed hills, studded with mineral wealth.

In a few short years the development of the present day will seem like the distorted dream of a dyspeptic, the machinery now so ponderous to the sight, toy-like, and the language of the Scripture be literally fulfilled, "And the rocks poured me out rivers of oil."

# INDEX OF FARMS.

ALLEGHENY RIVER AND FRENCH CREEK.


| Name of Farms. | PAGE. |
|---|---|
| Haliday | 196 |
| Nevins | 197 |
| Bastian | 200 |
| Lee | 200 |
| Moran | 201 |
| Piaget | 203 |
| Hickory Island | 204 |
| Hayes | 205 |
| Wilson, Shaffer & Ormsby | 206 |
| Huff | 207 |
| Plowman | 208 |
| Neely | 208 |
| Milton | 209 |
| Bowser | 209 |
| Hoge Island | 210 |
| Martin & Epley Tract | 210 |
| Chambers | 211 |
| Franklin Borough | 212 |
| Booth | 214 |
| Fuller | 214 |
| Blakely | 215 |
| Hastings | 215 |
| Bissell & Stewart | 216 |
| Irvine | 216 |
| Cochran | 217 |
| Hoover | 218 |
| Harmon | 218 |
| Martin | 219 |
| Nicklin | 220 |
| Rice | 220 |
| Hoover | 221 |
| Pope | 222 |
| Smith | 223 |
| Miller | 223 |
| Upper Two Mile Run | 223 |
| M'Cormick | 224 |
| Dale | 225 |
| Hayes | 226 |
| Bowman | 226 |
| Longwell | 227 |
| Sutley | 228 |
| Dubbs | 228 |
| Newell | 230 |
| Brown | 231 |
| M'Elrath | 231 |
| Brown | 232 |
| Roberts | 233 |
| Hayes | 234 |
| Blakely & Clark | 235 |
| Neely | 235 |
| Homan | 235 |
| Shotwell | 236 |
| Russell | 237 |
| Homan | 237 |
| M'Calmont | 237 |
| Frazier | 237 |


OIL CREEK VALLEY.


| Name of Farms. | PAGE. |
|---|---|
| Graff, Hasson | 238 |
| Clapp | 241 |
| Ham M'Clintock | 243 |
| Buchanan Farms | 245 |
| J. M'Clintock | 248 |
| Widow M'Clintock & Steele | 250 |
| Rynd | 252 |
| Blood | 254 |

| NAME OF FARMS. | PAGE. |
|---|---|
| Tarr | 256 |
| Story | 258 |
| Hayes | 261 |
| Hyde & Egbert | 263 |
| G. W. M'Clintock | 264 |
| Stevenson | 266 |
| M'Cray | 267 |
| Boyd | 268 |
| Lower M'Elhenny | 268 |
| Benninghoff | 270 |
| Espy | 271 |
| Upper M'Elhenny | 272 |
| Caldwell | 273 |
| Foster | 274 |
| Faarrel | 275 |
| Beaty | 277 |
| Gregg | 278 |
| Sanney | 278 |
| Shaffer | 279 |
| Miller | 281 |
| Fleming | 283 |
| Jones | 284 |
| Henderson | 284 |
| Fleming | 284 |
| Stackpole | 284 |
| Shreve | 285 |
| Pott | 285 |
| Stackpole | 285 |
| Stackpole | 285 |
| Bissell | 286 |
| Conley | 286 |
| Griffin | 286 |
| Bissell | 287 |
| Watson Petroleum Co | 288 |
| Kingsland | 289 |
| Watson Flats | 290 |
| Parker | 291 |
| Guild | 294 |
| Brewer | 294 |
| Patterson | 295 |
| Skinner | 295 |
| Benninghoff Run | 296 |
| Warner | 298 |
| Weiss | 299 |
| Widow Stevenson | 299 |
| Wykle Run | 300 |
| Cherry Tree Run | 300 |
| Tarr | 301 |
| Ward | 301 |
| Boyd | 302 |
| Brown | 302 |
| Sutley | 303 |
| Black | 303 |
| Dempsey | 304 |
| Mallory | 304 |
| Baney | 305 |
| Pierson | 305 |
| Irwin | 306 |


CHERRY RUN.


| NAME OF FARMS. | PAGE. |
|---|---|
| John M'Clintock | 308 |
| Knapp | 309 |
| Tolls | 310 |
| Rynd | 310 |
| Part of | 312 |
| Smith | 313 |
| Cherry Run Petroleum Co. | 314 |
| Widow M'Fate | 315 |
| M'Fate | 315 |
| Hogg | 316 |
| Humboldt Petroleum Co. | 316 |
| Lamb | 317 |
| M'Calmont | 319 |
| Prather | 319 |
| Prathers' | 320 |
| Ricketts | 321 |
| Stowell | 321 |
| Huidekoper | 322 |
| Nevins | 323 |
| Shaw | 323 |

UPPER ALLEGHENY.

| NAME OF FARMS. | PAGE. |
|---|---|
| Downing | 324 |
| Graff, Hasson & Co | 324 |
| Siverly | 325 |
| Clapp | 326 |
| Crammond | 327 |
| Alcorn | 327 |
| Alcorn | 328 |
| Downing | 329 |
| Eaker | 329 |
| Evans | 330 |
| Cary | 331 |
| Lamb | 332 |
| Rennoff | 332 |
| Blakely | 333 |
| Nellis | 334 |
| Krotzer | 334 |
| Dille | 335 |
| Dotson | 335 |
| Barr | 336 |
| Krotzer | 337 |
| Tolls | 337 |
| Conner | 338 |
| Kintzler | 338 |
| Russell | 339 |
| M'Fate | 339 |
| M'Mahan | 340 |
| Bruner | 341 |
| Lower Walnut Island | 342 |
| Middle and Upper Island | 342 |
| Anderson | 343 |
| M'Calmont | 344 |
| N. Y. Phila. & Balt. Co | 345 |
| Foster | 347 |
| Heydrick | 347 |
| Howe | 347 |
| Alexander | 348 |
| Heydrick | 348 |
| Shaffer | 349 |
| M'Calmont | 350 |
| Armstrong | 350 |
| Henry | 351 |
| Culbertson | 352 |
| Culbertson | 353 |
| M'Crea | 354 |
| Clapp | 355 |
| Harper | 356 |
| President | 356 |
| Elliott | 371 |

PITHOLE CREEK.

| NAME OF FARMS. | PAGE. |
|---|---|
| Caldwell | 358 |
| Shaw Tract | 359 |
| Shaw | 359 |
| M'Kissick | 360 |
| Rickets | 361 |
| Prather | 362 |
| Lake | 362 |
| Balliett | 363 |
| Reynolds | 364 |
| Ricketts | 364 |
| Wood & Copeland | 364 |
| Blackmer | 365 |
| Rooker | 365 |
| Holmden | 367 |
| Hyner | 371 |
| Copeland | 372 |
| M'Kinney | 373 |
| Ball | 374 |
| Dawson | 375 |
| S. Blank | 376 |
| J. Blank | 377 |
| Haworth | 377 |
| M'Caslaw | 378 |
| Stewart | 379 |
| Haworth | 379 |
| Conley | 379 |
| Siggins J. | 380 |
| Pratt | 380 |
| Van Wykle | 380 |

| NAME OF FARMS. | PAGE. |
|---|---|
| Tyrrell | 381 |
| Widow Holmden | 381 |
| Lyons | 382 |
| Clark | 383 |
| Steen | 383 |
| Turner | 383 |
| Vose | 384 |
| Brumegin | 385 |
| Austin | 385 |
| Elliott | 386 |
| M'Crea | 386 |
| Griffin | 387 |
| M'Calmont | 387 |
| D. Smith | 388 |
| M'Calmont | 389 |
| J. M'Calmont | 389 |
| J. Brewer | 390 |
| Myres | 390 |
| Van Giesen | 390 |
| Sager | 391 |
| Van Giesen (heirs) | 391 |
| J. S. M'Calmont | 392 |
| Holeman, Alex | 392 |
| Holeman Island | 393 |
| Maple Islands | 393 |
| Keeler | 393 |
| Alex. Holeman, jr | 394 |
| J. M. Clapp | 394 |
| Pierson & Elder | 394 |
| D. Hunter | 395 |
| Hunter (heirs) | 395 |
| Dustin | 396 |
| Watson | 396 |
| Chas. Holeman | 396 |
| M'Clatchey | 397 |
| S. M'Cray | 397 |
| Towner | 397 |
| Hilands | 398 |
| A. Dustin | 398 |
| Hulings | 399 |
| Shreeves | 399 |
| Watson, Jas | 400 |
| Hunter Island | 400 |
| A. M'Calmont | 400 |
| M'Calmont (heirs) | 401 |
| Wm. Hunter | 401 |
| May | 402 |
| May (heirs) | 402 |
| G. S. Hunter | 402 |
| M'Calmont | 403 |
| Sowers | 403 |
| Ensign & Jamison | 403 |
| Jamison | 404 |
| Jamison G. | 404 |
| Johnson | 405 |
| Dale | 405 |
| Dawson | 405 |
| Stave Factory Tract | 406 |
| Gordon | 406 |
| Green | 407 |
| Sibbald | 408 |
| Woolcott | 408 |
| Ball | 409 |
| Bennett | 409 |
| Riddles | 409 |
| Walter | 410 |
| Richardson | 410 |
| S. Hunter | 410 |
| Prather | 411 |
| Hickory Village | 411 |
| J. Bull | 412 |
| J. Green | 412 |
| G. S. Siggins | 412 |
| J. C. Siggins | 413 |
| Fleming | 413 |
| Hawthorn | 414 |
| Jones | 414 |
| Scott | 414 |
| E. L. Jones | 414 |
| J. Jones | 415 |

| NAME OF FARMS. | PAGE. |
|---|---|
| S. Smith | 415 |
| Prather | 415 |
| Brown | 415 |
| Alcorn | 416 |
| Hunter | 416 |
| Bozer | 416 |
| Kortman | 417 |
| Kortman, P. | 417 |
| Dunn | 417 |
| Magee | 418 |
| Wm. Dale | 418 |
| M. Hunter | 418 |
| Richardson | 418 |
| Tipton | 419 |
| Cohell | 421 |
| Economy Oil Co.'s Lands | 421 |
| Henry | 423 |
| Parshall | 425 |
| State Land | 426 |
| Tideoute Borough | 426 |
| M'Guyr Run | 427 |
| Tideoute Creek | 428 |
| Gordon Run | 429 |
| East Hickory Creek | 429 |
| Manross | 430 |
| White | 431 |
| Fleming, No. 1 sub | 432 |
| Fleming, No. 2 " | 432 |
| Fleming, No. 3 " | 433 |
| Fleming, No. 4 " | 433 |
| Fleming, No. 5 " | 433 |
| Giles | 433 |
| Gorman | 434 |
| Miles | 434 |
| Gorman M. | 434 |
| Walter | 436 |
| Barnes | 436 |
| Barnes & Church | 437 |
| Hatch | 437 |
| Jamison | 437 |
| Tionesta Creek | 437 |
| G. G. Sickles | 437 |
| Noble | 438 |
| Dale Bros | 438 |
| Towner & Brett | 438 |
| Keyser | 438 |
| Dale Bros | 438 |
| Dale Bros | 438 |
| Dale & Bro | 438 |
| D. Stow | 438 |
| J. A. Dale | 438 |
| John Noble | 438 |
| Brett & Co | 438 |
| May Bros | 438 |
| May Bros | 438 |
| Stow, Blakely | 438 |
| Hyner | 439 |
| H. Pierson | 439 |
| Stephen Smith | 439 |
| Tionesta Lumb. & Mining Co. | 439 |
| Wm. White | 439 |
| A. Purdy | 439 |
| Harrington | 439 |
| Green & Gordon | 439 |
| Stow, Wheeler & Co | 439 |
| Johnson | 440 |
| Salmon Creek Pet. Co | 440 |
| May | 440 |
| W. W. Crook | 441 |
| Dale & Irwin | 441 |
| Hunter | 441 |
| Little Tionesta Creek | 442 |
| Edmonson & Sligo | 442 |
| Proper | 442 |
| Stewart's Run | 443 |
| T. M'Calmont | 443 |
| J. S. M'Calmont | 443 |
| Taylor | 443 |
| M'Kinley | 443 |
| Heckard | 443 |

NAME OF FARMS. PAGE.

Fair ..................... 444
Dawson ..................... 444
Abbott ..................... 445
M'Kinley..................... 445
Wm. M'Kinley............ 446
Range..................... 446
Dawson..................... 446
Green..................... 446
Johnson's Run............ 447
Johnson..................... 447
Leedham..................... 447
Sugar Camp Run............ 449
Vandine..................... 449
W. Willings............ 449
Hoffman..................... 449
Youngs..................... 450
Switzer..................... 450
Pine Run..................... 450
J. S. M'Calmont............ 450
Brown Tract............ 450
Thos. M'Calmont............ 450
Tideoute..................... 451
Noble Tract..................... 452
Triumph Oil Co.'s Lands.... 453
Pittsburgh & Cherry Run Pet. Co. Tract............ 459
Shamburg Pet. Co............ 461
Tallman..................... 462
Atkinson..................... 464
Pleasantville............ 466
S. Q. Brown, Bates & others. 470
Brown Bros. Tract............ 471
Holeman & Newkirk............ 471
Brown & House ............ 472
Tyrrell..................... 472
Porter..................... 473
J. H. Jack..................... 473
Hebert..................... 474
Byles..................... 474
Beebe ..................... 475

NAME OF FARMS. PAGE

Mill ..................... 476
Ensign..................... 476
Gerow ..................... 476
Marcy..................... 477
Drake..................... 477
Armstrong..................... 477
S. M. Dunham............ 477
National Petroleum Co...... 478
Lyle..................... 479
Valley of Oil Creek........ 480
Woods ..................... 485
Canusett ..................... 486
John Pearson............ 486
Hyde & Egbert............ 488
Tarr..................... 489
Blood..................... 490
John M'Clintock............ 491
A. & J. Buchanan.......... 491
H. M'Clintock............ 492
Cornplanter Tract.......... 492
Clapp ..................... 492

OIL CITY AND VICINITY.

Hickman..................... 493
Lackawanna Oil Co.'s Tract.. 494
Oil City Land Co.'s Tract.... 494

SCRUBGRASS.

M'Millan..................... 498
Belle Island..................... 499
Philadelphia & Boston Pet. Land Co.'s Lands......... 500
Tract No. 1..................... 500
Foster..................... 501
Foster's Island............ 502
Miller..................... 502
Lewis, Bonsall & Co.'s Tract. 502
Snider..................... 504
Excelsior Oil Co.'s Tract.... 504
Ogden Mining Co.'s Tract... 505